Ordinary Differential Equations

Ordinary Differential Equations

Robert H. Martin, Jr.

Professor of Mathematics
North Carolina State University, Raleigh

McGraw-Hill Book Company

New York St. Louis San Francisco Auckland Bogotá Hamburg Johannesburg London Madrid Mexico Montreal New Delhi Panama Paris São Paulo Singapore Sydney Tokyo Toronto

ORDINARY DIFFERENTIAL EQUATIONS

1234567890 HALHAL 89876543

ISBN 0-07-040687-1

This book was set in Times Math. The editors were John J. Corrigan and James S. Amar; the designer was Elliot Epstein; the production supervisor was Diane Renda. The drawings were done by J & R Services, Inc.
Halliday Lithograph Corporation was printer and binder.

Library of Congress Cataloging in Publication Data

Martin, Robert H.
Ordinary differential equations.

Bibliography: p.
Includes index.
1. Differential equations. I. Title.
QA372.M377 1983 515.3′52 82-20797
ISBN 0-07-040687-1

To my wife, Elaine,
and my children, Libby, Bobby, and Sims

Contents

Preface

This book is written for the one- or two-semester course in ordinary differential equations and was developed specifically with students of engineering and the physical sciences in mind. A knowledge of elementary calculus that is normally learned in a standard two- or three-semester university calculus sequence is assumed. The book has grown out of my experiences over the past several years in teaching various topics from the book in the ordinary differential equations course at North Carolina State University. During this time I have formed definite opinions as to the most effective instruction techniques which enable students to learn both the applications and the underlying theory of ordinary differential equations. It should be emphasized that without theory applications alone become a rote use of memorized techniques which are applied only with difficulty to nonstandard or previously unencountered problems.

This experience has led me to adopt a philosophy of instruction which employs so-called qualitative methods as adjuncts to the more computational techniques for solving equation systems. An example of such a qualitative method is the procedure for sketching the graphs of solutions, introduced in Chapter 1 to complement the techniques for solving first order linear equations. Other qualitative methods are used throughout the book. The emphasis of these qualitative methods is to show how the methods actually apply to specific problems and models; they are not intended as substitutes for the standard methods of solution computation.

My experience in teaching applied science students has also influenced both the writing style and the general organization of this book. I have endeavored to keep my writing style simple, clear, and to the point but with no sacrifice of rigor. Students are introduced to proofs, but are not overburdened with technicalities. Where a proof is beyond the mathematical knowledge of students at this level, either an intuitive proof or a simple explanation of the general direction a proof would take is substituted. In general, explanations of both theories and applications are very detailed, probably more so than in any other book at this level. For those who wish further to pursue a given topic, a short bibliography is provided at the end of the book, together with appropriate in-chapter references.

Examples are usually used to introduce a given topic. The approach to any given example is usually natural, and with a little thought would probably even occur to most students, although at this stage of their mathematical development many students would not likely be able to carry the mathematics through to a conclusion. In this respect the book is designed to be partially self-teaching, and by reducing the class time needed for basic explanations should free class time for analysis of topics especially pertinent to a given class's interests.

In keeping with my goal of making this book useful for as broad a range of students as practical, I have provided a wide variety of applications with clear, detailed explanations of the physical properties involved in the examples and problems. This variety allows the instructor to choose applications which most closely match the mathematical backgrounds and interests of the class. There is a generous number of problems. All problems are intended to be instructive, and not merely drill. Some problems are so structured as to require the student to bring together material from several different sections in order to develop a strategy for solving the problem. In effect, such problems are brief tests.

Finally, the inclusion of several not so standard but illuminating topics, such as the thorough discussions of critical points in Section 1.5 and limits of solutions in Section 2.5*b*, encourages the growth of the students' mathematical sophistication.

Chapters 1 and 2 emphasize basic methods of explicitly solving standard types of equations and also indicate some fundamental applications. Methods for first order linear and nonlinear equations are contained in Chapter 1 and methods for second order linear equations are contained in Chapter 2. A procedure for sketching the graph of solutions, the first of several qualitative methods, is introduced in these two chapters. A knowledge of the fundamental terminology and solution methods for first and second order equations in Chapters 1 and 2 is the basis for the entire book. The remaining chapters are mutually independent of one another and depend only on the ideas in the first two chapters. This allows the instructor to tailor the bulk of the course to the interests and abilities of the class.

Chapter 3 contains the basic properties of the Laplace transform and indicates the procedure for applying transform methods to solve nonhomogeneous second order equations with constant coefficients. The stress here is on the elementary properties of the Laplace transform and its application in determining solutions to second order linear equations that have discontinuous nonhomogeneous terms. The case of a jump discontinuity is considered, and some of the procedures and interpretations for nonhomogeneous terms involving the Dirac delta function are also indicated. Power series methods for second order linear equations with nonconstant coefficients are developed in Chapter 4. Equations having ordinary power series solutions and those having series solutions of the Frobenius type are both considered. Also, one section is devoted to analyzing some of the most important equations where power series methods are used (Bessel's equation, Legendre's equation, and the hypergeometric equation).

Elementary concepts and techniques for first order systems of two equations (linear and nonlinear) are developed in Chapter 5. This chapter divides naturally into two parts: the explicit computation of solutions to linear equations with constant coefficients, and the sketching of solution curves in the plane for time-independent nonlinear equations. Several interesting and important models involving planar systems are included in the chapter: mixing in interconnected tanks, epidemics, interacting populations, and nonlinear oscillations.

Chapter 6 introduces some of the basic concepts and methods associated with the numerical approximation of solutions to differential equations. These methods include Taylor series expansions, one-step (Runge-Kutta) methods, and multistep (predictor-corrector) methods. The techniques are mainly discussed relative to single first order equations; however, some consideration is given to approximation of systems of two first order equations. Linear differential equations of arbitrary order are discussed in Chapter 7. The main emphasis is on equations with constant coefficients, and it is shown that many of the properties and techniques for second order equations developed in Chapter 2 have natural extensions to higher-order equations.

Chapter 8 introduces the basic methods involving the concepts and techniques for matrix and vector algebra. The type of differential equation studied is a first order linear system with constant coefficients. A rather detailed development of the techniques needed from matrix theory is given, and the student should be able to grasp these ideas without any previous study in matrix algebra. Solution computations using eigenvalue-eigenvector techniques are stressed, but other matrix methods are also indicated.

In the one-semester course in introductory differential equations at North Carolina State, I usually cover the basic solution methods in Chapter 1 (Sections 1.2, 1.4, 1.6, and 1.7), curve sketching of solutions (Section 1.5), and some applications (e.g., radioactive decay, mixing problems, and population models). The first five sections in Chapter 2 on second order equations are then covered, with the emphasis on Sections 2.3 to 2.5. This chapter is concluded with an application (usually the analysis of vibrations in linear springs—Section 2.6) and an analysis of Euler's equation in Section 2.7.

After the first two chapters are covered, the rest of the course can be taken from any of the remaining chapters and in any order. In the previously mentioned course I usually go from Chapter 2 directly into Chapter 5, emphasizing solving linear equations (Section 5.3) and sketching solution curves for nonlinear equations (Section 5.4). In the remaining time (which varies considerably, depending on how much detail is covered in Chapter 5)

I try to cover as much as possible from either Laplace transform methods or power series methods.

Of course, once Chapters 1 and 2 are covered there is great flexibility in the choice of additional material. For example, one could cover Laplace transform methods (Chapter 3), power series methods (Chapter 4), and, depending on the remaining time, systems of two linear equations (Sections 5.2 and 5.3) or higher-order linear equations (Chapter 7). Another possibility is to go directly from Chapter 2 to higher-order linear equations (Chapter 7) and conclude with a detailed development of matrix methods (Chapter 8). Most of the topics in this text can be covered in a two-semester course.

The chapters in this text are divided into sections, and some of the sections are further divided into subsections. An asterisk preceding the section or subsection number indicates that the material in this section is perhaps more difficult to understand than what might be expected of an average student at the level of this book. Therefore, some care should be taken in covering those topics. Each section concludes with a set of problems that pertains to the topics of that section. Generally speaking, the first few problems are more basic and the latter ones more difficult. Problems requiring a good deal more than basic procedures are indicated by an asterisk. Answers to selected problems are included at the end of the book, and almost all of the computational exercises have answers in this section.

No book emerges fully formed from an author's forehead. I would like to acknowledge the inspiration and encouragement I received from my colleagues at North Carolina State and the help of my students, who class-tested early versions of the book. I am especially grateful to Professor James Selgrade who carefully read the entire manuscript and provided answers to most of the problems. Special thanks also are due to Margaret Memory and Dale Boger for checking the examples and answers for accuracy in several of the chapters.

Additionally, I would like to thank Vasilios Alexiades, University of Tennessee; Prem N. Bajaj, Witchita State University; Peter W. Bates, Texas A & M University; John Bradley, University of Tennessee; Hsim Chu, University of Maryland; Maurice Eggen, Trinity University; Donald L. Goldsmith, Western Michigan University; Herman Gollwitzer, Drexel University; Ronald B. Guenther, Oregon State University; Terry Herdman, Virginia Polytechnic Institute; Allan M. Krall, Pennsylvania State University; Kenneth R. Meyer, University of Cincinnati; James A. Morrow, University of Washington; David A. Sánchez, University of New Mexico; Thomas J. Smith, Manhattan College; Ralph E. Showalter, University of Texas at Austin; and Jacob Towber, DePaul University; who read all or parts of the draft manuscript. Their comments and occasional criticisms were always to the point and contributed to what I hope is an excellent introduction to the study of ordinary differential equations. Any remaining errors of omission or commission are solely my responsibility. I would be pleased if readers would write to me, to deliver either compliments or criticisms. I would also like to thank Sharon Jones for her help in turning rough drafts into beautifully prepared final manuscript. Thanks also to John Corrigan and James Amar of McGraw-Hill for their help throughout the project.

Finally, I would like to thank my family, without whose loving support this book would not have been written.

Robert H. Martin, Jr.

1 First Order Equations

The purpose of this chapter is to develop elementary methods for determining solutions to simple first order ordinary differential equations and to indicate some basic applications of such equations. The principal methods are integration of linear equations in Section 1.2, separation of variables in Section 1.4, change of variables in Section 1.6, and exact equations in Section 1.7. In Section 1.5 elementary techniques for sketching the graphs of solutions for autonomous equations are developed and these ideas are used in several applications and models (see, for example, the population models analyzed in Section 1.8). The applications considered in this chapter come not only from physics and engineering but also from the social sciences, chemistry, and economics.

1.1 INTRODUCTORY CONCEPTS AND EXAMPLES

A *differential equation* is an equation that involves an unknown function and its derivatives. An *ordinary differential equation* is a differential equation whose unknown is a function of a single independent variable. In this chapter we consider only ordinary differential equations that are *real* and *first order*: the unknown is a real-valued function of a single real variable, and the only derivative appearing in the equation is the first. When an ordinary differential equation arises as a model or description of a scientific phenomenon, the independent variable is often time. Therefore the independent variable in this text is usually denoted by t. Also, if y is a real-valued function of the real variable t, then y' or dy/dt denotes the first derivative of y. The second derivative of y is denoted by y'' or d^2y/dt^2 and the third derivative by y''' or d^3y/dt^3. In general, for each positive integer n, $y^{(n)}$ or d^ny/dt^n is used to denote the nth derivative of the function y. A differential equation is said to be of *order n* if the nth derivative $y^{(n)}$ appears in the equation and no derivatives of y larger than n appear in the equation. For example, the equation

$$y'' + 2y^4 - e^{-y'} \sin t = 5$$

is a differential equation of order 2 since y'' occurs in the equation and no higher derivatives of y appear. As a further example,

$$\cos ty + e^t \frac{dy}{dt} = \frac{t}{1 + y^2}$$

is a first order differential equation.

In the general case, a *real first order ordinary differential equation* has the form

$$\Phi(t, y, y') = 0$$

where the function Φ defines the relationship of the derivative $y' = dy/dt$ with the dependent variable y and the independent variable t. In this chapter, however, only those equations where the derivative y' can be explicitly written in terms of t and y are considered. Therefore, it is assumed that f is a continuous function of two variables and the differential equation

$$y' = f(t, y) \tag{1}$$

is considered. A differentiable function y on an interval I is said to be a solution to (1) on I if $y'(t) = f(t, y(t))$ for each t in I. As a simple example, a solution to the differential equation

$$y' = 3y$$

is the function $y(t) = e^{3t}$ for all t. For if $y(t) = e^{3t}$, then

$$y'(t) = 3e^{3t} = 3y(t)$$

and $y = e^{3t}$ is a solution to $y' = 3y$ by definition. In fact, the student should verify that $y(t) = ce^{3t}$ is a solution to $y' = 3y$ for any constant c. One of the fundamental problems associated with equation (1) is developing methods for special types of functions f that lead to the determination of the solutions to (1) on a given interval I. In general, there are an infinite number of solutions to (1) on any given interval. For example, if $f(t, y) \equiv 0$, then $y(t) \equiv c$ on I is a solution to (1) for any constant c and any interval I.

▶ **EXAMPLE 1.1-1**

Consider the equation $y' = y - t$ and let I be any interval. For each real constant c the function $y = ce^t + t + 1$ for t in I is a solution to this equation, since

$$y'(t) = ce^t + 1 = (ce^t + t + 1) - t = y(t) - t$$

for all t in I. At this time the reader is not expected to produce a solution to equation (1). However, one should be able to determine if a *given* function is a solution to (1) (see Problem 1.1-1).

In actually trying to explicitly determine a solution to equation (1), the simplest case is when the function f does not depend on y. Therefore, assume that g is a continuous real-valued function on an interval I and consider the equation

$$y' = g(t) \tag{2}$$

In order to solve (2) on I, one needs to determine all differentiable functions y on I such that $y'(t) = g(t)$ for all $t \in I$: that is, the solutions to (2) on I are precisely the antiderivatives of g. The solutions to the equation $y' = t^2$ are all functions y of the form $y = c + t^3/3$, where c is any constant. In general, the solutions to equation (2) on I are precisely the functions y on I having the form

$$y(t) = c + \int_{t_0}^{t} g(s)\, ds \qquad \text{for all } t \text{ in } I \tag{3}$$

where c is any constant.

Therefore, if y is a solution to (2) on I, then y has the form indicated in (3) for some constant c [and, in fact, $c = y(t_0)$]; and conversely, if y has the form in (3), then y is a solution to (2). It should be noted that if G is *any* antiderivative of g on I [that is, $G'(t) = g(t)$ for all t in I] then the family

$$y(t) = \bar{c} + G(t) \qquad \text{for } t \text{ in } I \tag{3'}$$

where $\bar{c}$ is any constant, describes precisely the same family of functions as (3). The family (3) of all solutions to equation (2) on I is called the *general solution to* (2) *on* I.

► EXAMPLE 1.1-2

Consider the equation $y' = 4 - 3t^2$ on $(-\infty, \infty)$. Since $4t - t^3$ is an antiderivative of $4 - 3t^2$ on $(-\infty, \infty)$, the family of functions

$$y(t) = c + 4t - t^3 \qquad \text{for } t \text{ in } (-\infty, \infty),\ c \text{ a constant}$$

is the general solution to this equation on $(-\infty, \infty)$.

The solution set to equation (2) indicates that there may be many different solutions to a first order ordinary differential equation. Usually one is interested in determining a function y on I that is not only a solution to (1) but also satisfies some additional property or condition. [For example, does (1) have a solution y on $(-\infty, \infty)$ that is periodic in t: $y(t + T) = y(t)$ for some $T > 0$ and all t in $(-\infty, \infty)$?] The most important case is determining a solution to (1) that assumes a specified value y_0 at some given time t_0 in I. Such a problem is called an *initial value problem* and is denoted in the following manner:

$$(4) \qquad y' = f(t, y) \qquad y(t_0) = y_0$$

where (t_0, y_0) is some given pair in the domain of f. If I is an interval and t_0 is in I, then a solution y to (4) on I is defined to be a solution to (1) on I that also has the value y_0 at time t_0: $y'(t) = f(t, y(t))$ for t in I and $y(t_0) = y_0$.

▶ **EXAMPLE 1.1-3**

Consider the initial value problem $y' = y - t$, $y(0) = 4$. According to Example 1.1-1, the function $y(t) = ce^t + t + 1$ is a solution to the corresponding differential equation for each constant c. Selecting c so that $y(0) = c + 1 = 4$ it follows that $y(t) = 3e^t + t + 1$ is a solution to the given initial value problem.

EXAMPLE 1.1-4

Suppose that the position of an object on the x axis is denoted by $x(t)$ for all $t \geq 0$, and that the instantaneous velocity of this object is $\sin t$ for all $t \geq 0$. If initially (at time $t = 0$) the object is 2 units to the right of the origin, determine the position $x(t)$ for all times $t \geq 0$. Since the instantaneous velocity is $x'(t)$, the function x should be a solution to the initial value problem

$$x' = \sin t \qquad x(0) = 2$$

By (3′) x belongs to the family of functions $c - \cos t$ on $[0, \infty)$ where c is a constant. Therefore, since $c - \cos 0 = 2$ implies that $c = 3$, the position $x(t)$ of this object is given by $x(t) = 3 - \cos t$ for all $t \geq 0$.

It is sometimes convenient in studying the behavior of solutions to look at equation (1) from a geometric point of view. At each point (t, y) in the plane the value $f(t, y)$ is the slope of a solution at this point. Therefore, in order to estimate the graphs of solutions to (1), it is helpful to select "appropriate" points in the ty plane and indicate the slope of the solutions at these points by drawing a short line with slope $f(t, y)$. The graph of these slopes is called the *direction field* for the equation (1).

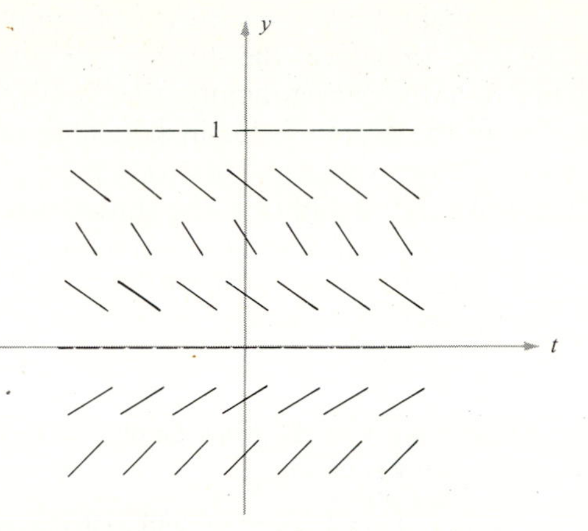

Figure 1.1 Sketch of direction field.

EXAMPLE 1.1-5

Consider the equation $y' = 5y(y-1)$ where $I = (-\infty, \infty)$. Since $f(t, y)$ in this case is independent of t, for any given y_0 the slopes of the solutions at (t, y_0) are the same for all t. Noting that the right-hand side of this equation is zero when y is 0 or 1, positive when y is in $(-\infty, 0) \cup (1, \infty)$, and negative when y is on $(0, 1)$, one can readily verify that the sketch of the slopes given in Figure 1.1 gives a reasonable indication of the direction field. For example, if $y = \frac{1}{2}$ then $y' = 5(\frac{1}{2})(-\frac{1}{2}) = -\frac{5}{4}$, so the solutions have slope $-\frac{5}{4}$ when they cross the line $y = \frac{1}{2}$. Also, if $y = \frac{1}{4}$ then $y' = -\frac{15}{16}$ and if $y = -\frac{1}{4}$ then $y' = \frac{25}{16}$ (these values are indicated in Figure 1.1).

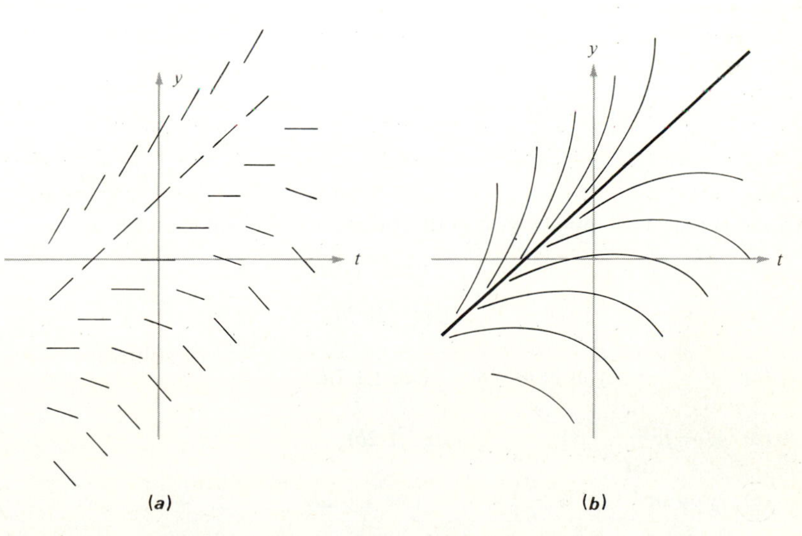

Figure 1.2 ***a*** Sketch of direction field. ***b*** Sketch of solution curves.

EXAMPLE 1.1-6

Consider the equation $y' = y - t$ where $I = (-\infty, \infty)$ (see Example 1.1-1). Notice that the slopes are all the same on the lines $y - t = k$, k any constant. As shown in Figure 1.2, indicating lightly the family of lines $y = t + k$ for various values of k on the graphs helps in sketching the slopes for the solutions to this equation. Since solutions to this equation have already been indicated in Example 1.1-1, we indicate the directional field in Figure 1.2*a* and sketch a few representative solutions in Figure 1.2*b*.

PROBLEMS

1.1-1. In parts *a* to *d* verify that each member y of the given family of functions is a solution to the corresponding differential equation.

(a) $y' = 2y - 4$ $\quad y(t) = ce^{2t} + 2$ $\quad$ for $t \in (-\infty, \infty)$ and each constant c

(b) $y' = -y^3$ $\quad y(t) = \dfrac{\pm 1}{\sqrt{2t + c}}$ $\quad$ for $t > -\dfrac{c}{2}$ and each positive constant c

(c) $y' = 2ty$ $\quad y(t) = ce^{t^2}$ $\quad$ for $t \in (-\infty, \infty)$ and each constant c

(d) $y' = \dfrac{e^t}{y}$ $\quad y(t) = \pm\sqrt{c + 2e^t}$ $\quad$ for $t \in (-\infty, \infty)$ for each constant $c \geq 0$, and $t > \ln|c| - \ln 2$ for each constant $c < 0$

1.1-2. Determine the general solution for each of the following differential equations.

(a) $y' = t^2 + 3$

(b) $y' = te^{2t}$

(c) $y' = \sin(3t)$

(d) $y' = \sin^2 t$

(e) $y' = \dfrac{t}{t^2 + 4}$

(f) $y' = \ln t$ $\quad (t > 0)$

(g) $y' = \dfrac{t}{\sqrt{t} + 1}$ $\quad (t > 0)$

1.1-3. Determine a solution to each of the following initial value problems.

(a) $y' = 2y - 4$ $\quad y(0) = 5$ $\quad$ (see 1.1-1*a*)

(b) $y' = -y^3$ $\quad y(1) = 3$ $\quad$ (see 1.1-1*b*)

(c) $y' = \dfrac{e^t}{y}$ $\quad y(\ln 2) = -8$ $\quad$ (see 1.1-1*d*)

(d) $y' = te^{2t}$ $\quad y(1) = 5$ $\quad$ (see 1.1-2*b*)

(e) $y' = \sin^2 t$ $\quad y\left(\dfrac{\pi}{6}\right) = 3$ $\quad$ (see 1.1-2*d*)

(f) $y' = 8e^{4t} + t$ $\quad y(0) = 12$

1.1-4. Sketch a representative direction field for the differential equations **(a)** $y' = y/t$ and **(b)** $y' = -t/y$. [Notice that these direction fields are perpendicular at each point (t, y) where both t and y are nonzero.]

1.1-5. Show that for each constant c the function $y(t) = 1/(1 - ce^t)$ is a solution to the equation $y' = y^2 - y$. Plot the graph of a few of these functions and compare with the direction field.

1.1-6. Sketch a representative direction field for the following differential equations.

(a) $y' = y - 1$ **(c)** $y' = y^3 - y^2$

(b) $y' = 1 - y$ **(d)** $y' = 1 - y^2$

1.2 FIRST ORDER LINEAR EQUATIONS

In this section a special and important case of the differential equation (1) in Section 1.1 is studied. The differential equation (1) in the previous section is called a *first order linear differential equation* if for each fixed t the graph of the equation $z = f(t, y)$ is a straight line in the yz plane. Therefore, in this section it is assumed that I is an interval, that a and b are continuous real-valued functions on I, and that the differential equation has the form

$$y' = a(t)y + b(t) \tag{1}$$

Equation (1) is the general form of a first order linear differential equation on I and the function b is called the *nonhomogeneous* (or *inhomogeneous*) term. If $b(t) \equiv 0$ on I, equation (1) is said to be *homogeneous*. Therefore, the homogeneous equation associated with (1) is the equation

$$y' = a(t)y \tag{2}$$

The solution to equation (1) [and hence (2)] can be explicitly computed in terms of antiderivatives, and the purpose of the section is to develop a method for solving these equations.

As an illustration of this procedure, consider first the equation

$$y' = 3y + e^t$$

By transposing the term $3y$ to the left side the equation becomes $y' - 3y = e^t$. Although the unknown function occurs only on the left side, we cannot simply antidifferentiate both sides as was done for equation (2) in the previous section (why?). However, *multiplying both sides of this equation by* e^{-3t} leads to the equation

$$e^{-3t}y' - e^{-3t}3y = e^{-3t}e^t$$

Noting that

$$\frac{d}{dt}[e^{-3t}y] = e^{-3t}y' - 3e^{-3t}y$$

by the product rule for differentiation, one sees that the original equation can be written in the form

$$\frac{d}{dt}[e^{-3t}y] = e^{-2t}$$

Now that the left-hand side of this equation is simply the derivative of the function $e^{-3t}y(t)$, we can indeed antidifferentiate each side to obtain

$$e^{-3t}y = \int e^{-2t} + c = -\tfrac{1}{2}e^{-2t} + c$$

where c is any constant. Upon multiplying both sides by e^{3t} it follows that

$$y = -\tfrac{1}{2}e^{t} + ce^{3t}$$

is a solution for any constant c. This can and should be checked by substituting into the original equation. Even in the homogeneous linear case (2) the solution cannot be directly integrated, and an appropriate "integrating factor" must be found in order to reduce the problem to one of antidifferentiation. For example, to solve the equation

$$y' = ty$$

transpose ty to the left side to obtain $y' - ty = 0$. Multiplying each side by $e^{-t^2/2}$ it follows again from the product rule that

$$0 = e^{-t^2/2}y' - te^{-t^2/2}y = \frac{d}{dt}[e^{-t^2/2}y]$$

Thus, $e^{-t^2/2}y = c$, and $y = ce^{t^2/2}$ is a solution for any constant c.

The procedure indicated by the preceding paragraph applies to first order linear equations in general. In order to obtain the form of the solution to (1), let $A(t)$ be an antiderivative of $a(t)$:

$$A(t) = \int a(t)\, dt$$

Transposing $a(t)y$ to the left side of (1) and multiplying each side of the resulting equation by $e^{-A(t)}$ shows that

$$e^{-A(t)}y'(t) - e^{-A(t)}a(t)y(t) = e^{-A(t)}b(t)$$

By the product rule for differentiation, the *left side of this equation is precisely*

$$\frac{d}{dt}[e^{-A(t)}y(t)]$$

and therefore each solution y to (1) must also satisfy

$$\frac{d}{dt}[e^{-A(t)}y(t)] = e^{-A(t)}y(t) \qquad \text{for all } t \text{ in } I$$

Now, since the left-hand side of this equation is the derivative of the function $e^{-A(t)}y(t)$, we can antidifferentiate each side to obtain

$$e^{-A(t)}y(t) = c + \int e^{-A(t)}y(t)\,dt$$

Multiplying each side of this equation by $e^{A(t)}$, we have that any solution to (1) is of the form

$$y(t) = ce^{A(t)} + e^{A(t)} \int e^{-A(t)}y(t)\,dt \tag{3}$$

where $A(t) = \int a(t)\,dt$ is any antiderivative of $a(t)$ and c is a constant. Conversely, any function y of the form (3) is a solution to (1) (see Problem 1.2-6), and so the family of all solutions to (1) is precisely the functions of the form (3).

Formula (3) can be used effectively for computing and studying the solutions to (1). It is, however, somewhat difficult to remember, and the student may find it helpful to learn the techniques used in deriving this formula. This method is outlined for convenience:

(4)

(a) Write the equation in the form of (1) and transpose $a(t)y$ to the left side of (1).

(b) Determine $A(t) = \int a(t)\,dt$ and multiply each side of the equation by $e^{-A(t)}$:

$$e^{-A(t)}y' - e^{-A(t)}a(t)y = e^{-A(t)}b(t)$$

(c) Observe that the left side of the equation is the derivative of the product $e^{-A(t)}y(t)$:

$$\frac{d}{dt}[e^{-A(t)}y(t)] = e^{-A(t)}b(t)$$

(d) Determine an antiderivative of each side of the equation:

$$e^{-A(t)}y(t) = c + \int e^{-A(t)}b(t)\,dt$$

The function $e^{-\int a(t)\,dt}$ is called an *integrating factor* for (1) and the family of functions defined by (3) is called the *general solution* of (1).

EXAMPLE 1.2-1

Consider the equation $y' = 2y - 6$. Since this has the form (1) and $-\int 2\,dt = -2t$, an integrating factor is e^{-2t}. Hence this equation can be solved as follows:

$$y' - 2y = -6$$

$$e^{-2t}y' - 2e^{-2t}y = -6e^{-2t}$$

$$\frac{d}{dt}[e^{-2t}y] = -6e^{-2t}$$

$$e^{-2t}y = c + 3e^{-2t}$$

and so the general solution is $y = ce^{2t} + 3$.

EXAMPLE 1.2-2

Consider the equation $ty' + 3y - t^2 - 4t = 0$ on $I = (0, \infty)$. Rewriting in the form

$$y' = -\frac{3}{t}y + t + 4$$

and noting that

$$\int \frac{3}{t}\,dt = 3 \ln t = \ln t^3$$

an integrating factor is $e^{\ln t^3} = t^3$. Hence this equation can be solved as follows:

$$y' + \frac{3}{t}y = t + 4$$

$$t^3y' + 3t^2y = t^4 + 4t^3$$

$$\frac{d}{dt}[t^3y] = t^4 + 4t^3$$

$$t^3y = c + \tfrac{1}{5}t^5 + t^4$$

and the general solution is

$$y = \frac{c}{t^3} + \tfrac{1}{5}t^2 + t$$

The initial value problem corresponding to equation (1) has the form

$$y' = a(t)y + b(t) \qquad y(t_0) = y_0 \tag{5}$$

where t_0 is in I. Using formula (3) one obtains (5) has exactly one solution (for each given t_0 and y_0) and this solution has the form

$$y(t) = y_0 e^{\int_{t_0}^{t} a(r)\,dr} + \int_{t_0}^{t} e^{\int_{s}^{t} a(r)\,dr} b(s)\,ds \tag{6}$$

The representation (6) [as well as (3)] is known as the *variation of constants* or *variation of parameters* formula. Although formula (6) is effective in analyzing the solution to (5), one can also determine the solution to (5) by first obtaining the general solution to (1) and then using the initial condition to compute the constant c. This technique is illustrated with two examples.

EXAMPLE 1.2-3

Consider the linear initial value problem $y' = -y + e^t$, $y(0) = 2$. An integrating factor is e^t and the solution can be computed as follows:

$$y' + y = e^t$$

$$\frac{d}{dt}[e^t y(t)] = e^{2t}$$

$$e^t y(t) = c + \tfrac{1}{2}e^{2t}$$

Thus $y(t) = ce^{-t} + \frac{1}{2}e^t$, and since $y(0) = c + \frac{1}{2} = 2$, we have that $c = \frac{3}{2}$ and that $y(t) = \frac{3}{2}e^{-t} + \frac{1}{2}e^t$ is the solution.

EXAMPLE 1.2-4

Consider the linear initial value problem

$$y' = \frac{1}{t}y + 1 + t \qquad y(2) = -4$$

Since

$$-\int \frac{1}{t}\,dt = -\ln t \qquad \text{and} \qquad e^{-\ln t} = \frac{1}{t}$$

the general solution can be computed as follows:

$$\frac{1}{t}y' - \frac{1}{t^2}y = \frac{1}{t} + 1$$

$$\frac{d}{dt}\left[\frac{1}{t}y\right] = \frac{1}{t} + 1$$

$$\frac{1}{t}y = c + \ln t + t$$

Thus $y = ct + t \ln t + t^2$ is the general solution. Since $y(2) = 2c + 2 \ln 2 + 4 = -4$, it follows that $c = -4 - \ln 2$ and so

$$y(t) = -(4 + \ln 2)t + t \ln t + t^2$$

is the solution to the given initial value problem.

The procedures indicated in this section show that the following theorem is valid:

Theorem 1.2-1

Suppose that a and b are continuous functions on I, that t_0 is in I, and that $A(t) \equiv \int a(t)\,dt$ is an antiderivative of a on I. Then each solution y to (1) has the form

$$y(t) = ce^{A(t)} + e^{A(t)} \int_{t_0}^{t} e^{-A(s)} b(s)\,ds \qquad \text{for } t \text{ in } I \tag{7}$$

where c is constant. Conversely, for each constant c the function y defined by (7) is a solution to (1).

Note that if y is given by (7) then $y(t_0) = ce^{A(t_0)}$, and hence taking $c = y_0 e^{-A(t_0)}$ in (7) gives the solution to the initial value problem (5) (see also Problem 1.2-7).

PROBLEMS

1.2-1. Determine the general solution to each of the following linear equations.

(a) $y' = (t^2 + 1)y$

(b) $y' = -y$

(c) $y' = 2y + e^{-3t}$

(d) $y' = 2y + e^{2t}$

(e) $y' = -y + t$

(f) $ty' + 2y = \sin t \qquad t > 0$

(g) $y' = (\tan t)y + (\sec t) \qquad -\frac{\pi}{2} < t < \frac{\pi}{2}$

(h) $y' = \dfrac{2t}{t^2 + 1} y + t + 1$

(i) $y' = (\tan t)y + \sec^3 t \qquad -\frac{\pi}{2} < t < \frac{\pi}{2}$

1.2-2. Determine the solution to each of the following initial value problems.

(a) $y' = y \qquad y(0) = 2$

(b) $y' = 2y \qquad y(\ln 3) = 3$

(c) $ty' = y + t^3$ $\quad y(1) = -2$

(d) $y' = (-\tan t)y + \sec t$ $\quad y(0) = 0$

(e) $y' = \dfrac{2}{t+1}y$ $\quad y(0) = 6$

(f) $ty' = -y + t^3$ $\quad y(1) = 2$

1.2-3. Determine the solution to each of the following initial value problems.

(a) $y' + 4(\tan 2t)y = \tan 2t$ $\quad y\left(\dfrac{\pi}{8}\right) = 2$

(b) $t(\ln t)y' = t \ln t - y$ $\quad y(e) = 1$

(c) $y' = \dfrac{2}{1-t^2}y + 3$ $\quad y(\tfrac{1}{2}) = 1$

(d) $y' = (-\cot t)y + 6\cos^2 t$ $\quad y\left(\dfrac{\pi}{4}\right) = 3$

***1.2-4.** Suppose that $T > 0$ and that a and b are continuous functions on $(-\infty, \infty)$ that are T-periodic: $a(t + T) \equiv a(t)$ and $b(t + T) \equiv b(t)$ for all $t \in (-\infty, \infty)$. Consider the differential equation

$$(*) \qquad y' = a(t)y + b(t)$$

(a) Show that if y is a solution to $(*)$ and $\bar{y}(t) \equiv y(t + T)$ for all $t \in (-\infty, \infty)$, then $\bar{y}$ is also a solution to $(*)$.

(b) Deduce that a solution y to $(*)$ is T-periodic if and only if $y(0) = y(T)$.

(c) Show that if $b(t) \equiv 0$ on $(-\infty, \infty)$ and $\int_0^T a(s)\, ds = 0$, then *every* solution to $(*)$ is T-periodic.

(d) Show that if $a(t) \equiv 0$ on $(-\infty, \infty)$ and $\int_0^T b(s)\, ds = 0$, then *every* solution to $(*)$ is T-periodic.

(e) Show that if $b(t) \equiv 0$ on $(-\infty, \infty)$ and $\int_0^T a(s)\, ds < 0$, then $\lim_{t \to +\infty} y(t) = 0$ for *every* solution y to $(*)$.

(f) Show that if $\int_0^T a(s)\, ds \neq 0$, then $(*)$ has exactly one T-periodic solution. {Observe that $\exp\left[\int_0^T a(s)\, ds\right] \neq 1$ and use the variation of constants formula (6) with $t_0 = 0$, $y_0 = y(0)$, and $t = T$ to show $y(0) = y(T)$ has exactly one solution—see part *b*.}

1.2-5. Determine all T-periodic solutions to the following equations.

(a) $y' = (\sin t)y$ $\quad T = 2\pi$

(b) $y' = (\sin^2 t)y$ $\quad T = \pi$

(c) $y' = -y + \cos \pi t$ $\quad T = 2$

(d) $y' = (\sin t)y + e^{-\cos t}$ $\quad T = 2\pi$

1.2-6. Show that the function defined by (3) is a solution to equation (1).

1.2-7. Show that equation (6) is valid. {*Hint:* If $A(t) \equiv \int_{t_0}^t a(s)\, ds$, then $A(t) - A(s) = \int_s^t a(r)\, dr$ and hence $e^{A(t)}e^{-A(s)} = \exp\left[\int_s^t a(r)\, dr\right]$.}

1.3 APPLICATIONS OF LINEAR EQUATIONS

1.3a Radioactive Decay

One of the simplest and most important examples involving first order linear differential equations is radioactive decay. The fundamental principle of radioactive decay is the following law: *The instantaneous rate of disintegration of a radioactive substance is proportional to the amount of the substance present.* Assume that a radioactive substance is under consideration, and for each time t let $N(t)$ denote the amount of this substance present at time t. It is *assumed* that N is a differentiable function of t so that by the above-mentioned principle, $N'(t)$ is proportional to $N(t)$. Therefore there is a positive constant k such that N is a solution to the initial value problem

(1) $$N' = -kN \qquad N(t_0) = N_0$$

where N_0 is the amount of the substance present at time t_0. The constant k is called the *decay constant* of the substance, and since this substance is disintegrating, the decay constant is necessarily positive. Also, the decay constant depends on the specific radioactive material: the more active the disintegration the larger the decay constant. Since (1) is a homogeneous linear equation, the solution N can be explicitly computed. Since $\int k\,dt = kt$, it follows that e^{kt} is an integrating factor of (1), and hence $N(t) = ce^{-kt}$ for some constant c. Since $N(t_0) = N_0$, it follows that $c = e^{kt_0}$ and

(2) $$N(t) = N_0 e^{-k(t-t_0)} \qquad \text{for all times } t$$

Moreover, if the number T is defined by

(3) $$kT = \ln 2$$

then T is the *half-life* of this substance: T is the time that it takes for one-half of the substance to disintegrate. In order to establish (3) let the time t be fixed and consider the possibility of determining a number $T > 0$ such that $N(t + T) = \frac{1}{2}N(t)$. Using (2),

$$N_0 e^{-k(t+T-t_0)} = \tfrac{1}{2}N_0 e^{-k(t-t_0)}$$

and multiplying each side of this equation by $N_0^{-1}e^{k(t-t_0)}$ one obtains that $e^{-kT} = \frac{1}{2}$, and hence that $e^{kT} = 2$. Formula (3) follows by taking the logarithm of each side of this equation. Note that since T is independent of the time t and the initial amount N_0, the term *half-life* of the substance is well-defined and depends only on the type of material (and not on the amount of the material or the time).

▶ **EXAMPLE 1.3-1**

Suppose that 4 percent of a radioactive material has disintegrated after 2 months. In order to determine the decay constant and the half-life for this material, let t denote the time in months and use equation (2) to obtain $N(t_0 + 2) = N_0 e^{-2k}$. Now use the fact that 4 percent has disintegrated in 2 months to obtain $N(t_0 + 2) = 0.96N_0$. Thus $N_0 e^{-2k} = 0.96N_0$, and it follows that $e^{-2k} = \frac{24}{25}$. Therefore,

$$k = \tfrac{1}{2}(\ln 25 - \ln 24) \quad \text{and} \quad T = \frac{2 \ln 2}{\ln 25 - \ln 24}$$

and hence $k \approx 0.0204$ and $T \approx 33.96$ months.

1.3*b* Mixing in a Continuously Stirred Liquid

A second example that illustrates the application of first order linear differential equations is a mixture problem in a continuously stirred liquid. It is assumed that a tank can hold a maximum of T_m gallons and that initially (at time t_0) there is V_0 gallons ($0 \le V_0 \le T_m$) of brine (a mixture of salt dissolved in water) that contains a total of Q_0 pounds of salt. At each time $t \ge t_0$ brine containing $m_1(t)$ pounds of salt per gallon is pumped into the tank at a rate of $r_1(t)$ gal/min (m_1 and r_1 can perhaps vary with time). The entering mixture is assumed to be "instantaneously" stirred and the well-stirred mixture leaves the tank at a rate of $r_2(t)$ gal/min (see Figure 1.3). The problem is to determine the quantity $Q(t)$ pounds of salt in the tank for all times $t \ge t_0$. This can be computed in the following manner: If $V(t)$ denotes the number of gallons of brine in the tank at time $t \ge t_0$, then $V'(t) = r_1(t) - r_2(t)$, and, by integration,

$$V(t) = V_0 + \int_{t_0}^{t} [r_1(s) - r_2(s)]\, ds \tag{4}$$

Thus, so long as $0 < V(t) \le T_m$ the quantity $Q(t)$ of salt in the tank satisfies the linear initial value problem

$$Q' = -\frac{r_2(t)}{V(t)} Q + m_1(t)r_1(t) \qquad Q(t_0) = Q_0 \tag{5}$$

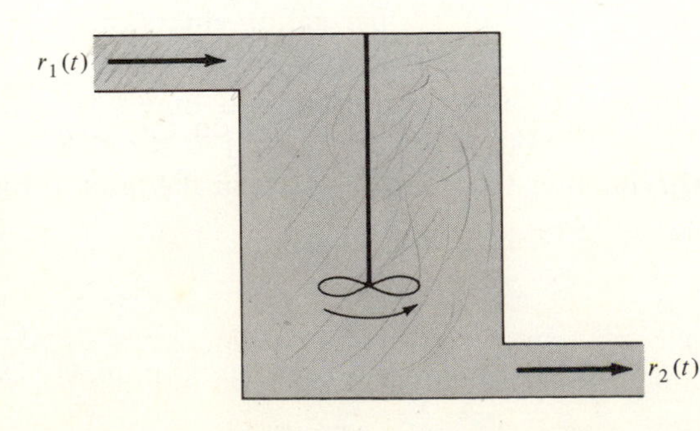

Figure 1.3 Continuously stirred mixture.

The fact that $Q(t)$ satisfies equation (5) follows from the following observation: *The instantaneous rate of change of $Q(t)$ equals the instantaneous rate of salt into the tank less the instantaneous rate of salt out of the tank.* Since $m_1(t)r_1(t)$ is the rate in and $[Q(t)V(t)^{-1}]r_2(t)$ is the rate out, equation (5) follows immediately.

EXAMPLE 1.3-2

Consider a tank with 50-gallon capacity that initially contains 50 gallons of brine with $\frac{1}{10}$ pound of salt per gallon. Assume that brine with $\frac{1}{2}$ pound of salt per gallon flows into the tank at 4 gal/min and the well-stirred mixture flows out at the same rate. From the principle indicated in the preceding paragraph, if $Q(t)$ is the amount of salt in the tank at time $t \geq 0$, then

$$Q'(t) = \text{rate in} - \text{rate out} = \frac{1}{2} \cdot 4 - \frac{Q(t)}{50} \cdot 4$$

(note that the number of gallons of brine in the tank remains 50 gallons at all times $t \geq 0$). Therefore, Q is the solution of

$$Q' = -\tfrac{2}{25}Q + 2 \qquad Q(0) = 5$$

Since $e^{2t/25}$ is the integrating factor, it follows that

$$\frac{d}{dt}[e^{2t/25}Q] = 2e^{2t/25}$$

Hence, upon antidifferentiation and multiplying each side of the resulting equation by $e^{-2t/25}$, it follows that $Q(t) = 25 + ce^{-2t/25}$. Using the initial condition $Q(0) = 5$ shows that

$$Q(t) = 25 - 20e^{-2t/25}$$

for all $t \geq 0$. Note in particular that $Q(t) \to 25$ as $t \to \infty$.

EXAMPLE 1.3-3

Suppose that a tank has a 60-gallon capacity and initially contains 30 gallons of pure water. Brine containing $\frac{1}{6}$ pound of salt per gallon enters the tank at a rate of 3 gal/min, and the well-stirred mixture flows out at a rate of 2 gal/min. If $V(t)$ denotes the gallons of water in the tank at time $t \geq 0$, then $V(0) = 30$ and $V'(t) = 3 - 2 = 1$, so $V(t) = 30 + t$ for all $0 \leq t \leq 30$ [see (4)]. By the principle (5), if $Q(t)$ denotes the pounds of salt in the tank at time t, then

$$\begin{aligned} Q' &= \text{rate in} - \text{rate out} \\ &= 3 \cdot \frac{1}{6} - \frac{Q(t)}{30 + t} \cdot 2 \end{aligned}$$

Therefore Q is a solution to the initial value problem

$$Q' = \frac{-2}{30+t}Q + \frac{1}{2} \qquad Q(0) = 0$$

Since

$$\int \frac{2\,dt}{30+t} = \ln\,[(30+t)^2]$$

it follows that $(30 + t)^2$ is an integrating factor, and hence

$$\frac{d}{dt}[(30+t)^2 Q(t)] = \tfrac{1}{2}(30+t)^2$$

$$(30+t)^2 Q(t) = \tfrac{1}{6}(30+t)^3 + c$$

$$Q(t) = \tfrac{1}{6}(30+t) + \frac{c}{(30+t)^2}$$

Since $Q(0) = 5 + c/30^2 = 0$, we have $c = -4500$ and

$$Q(t) = \tfrac{1}{6}(30+t) - \frac{4500}{(30+t)^2}$$

for all $0 \le t \le 30$.

1.3c Newton's Second Law of Motion

As the final illustration in this section we use *Newton's second law of motion: The rate of change of momentum of a body is equal to the net force acting on the body.* If m denotes the mass and v the velocity of an object, then mv is the momentum and Newton's second law has the form

$$\frac{d}{dt}[mv] = F \tag{6}$$

where F is the applied force. In particular, if m is constant then equation (6) has the form

$$m\frac{dv}{dt} = F \tag{7}$$

If the applied force F depends on the time and the velocity v of the object, then (7) is a first order ordinary differential equation with respect to the velocity v. Once the velocity has been determined one can compute the position $y(t)$ of the object by using the fact that y is an antiderivative of v.

As a specific example, consider an object of constant mass m moving vertically with velocity $v(t)$ under the influence of gravity. Letting g denote the gravitational constant (on the earth's surface g is approximately 32 ft/sec^2), the force on the object due to gravity is $-mg$ and the equation becomes $mv' = -mg$ [and hence $v(t) = -(t - t_0)g + v_0$, where v_0 is the velocity at time t_0]. In this case, however, the effects of air resistance have been neglected. A simple way to include the force F_R on the object due to air resistance is to assume that F_R is proportional to the velocity. Therefore, if it is assumed that there is a positive constant ρ such that $F_R = -\rho v$ (note that the resistance is in the *opposite* direction of velocity), the equation of motion becomes $mv' = -mg - \rho v$. Assuming that the velocity has the value v_0 at a given time t_0, it follows that v is a solution to the first order linear initial value problem

$$v' = -\frac{\rho}{m} v - g \qquad v(t_0) = v_0 \tag{8}$$

According to formula (6) in the previous section the solution to (8) is

$$v(t) = \left[\frac{mg + \rho v_0}{\rho}\right](e^{-\rho(t-t_0)/m}) - \frac{mg}{\rho} \qquad \text{for all } t \tag{9}$$

Note that $v(t) \to -mg/\rho$ as $t \to +\infty$, and this velocity $v_l = -mg/\rho$ is called the *limiting velocity*. A sketch of a few representative curves for v is given in Figure 1.4. If $y(t)$ denotes the position of the object for each time $t \geq t_0$ (the earth's surface is assumed to be at the position $y = 0$ and the positive direction is upward) and y_0 is the position at time $t = t_0$, then $y'(t) = v(t)$ for all $t \geq t_0$, and so

$$y(t) = y_0 + \int_{t_0}^{t} v(s)\, ds$$

and it follows from (9) that

$$y(t) = \left[\frac{m^2 g + \rho m v_0}{\rho^2}\right](1 - e^{-\rho(t-t_0)/m}) - \frac{mg}{\rho}(t - t_0) + y_0 \tag{10}$$

Of course, (10) and (9) are valid only so long as $y(t) \geq 0$.

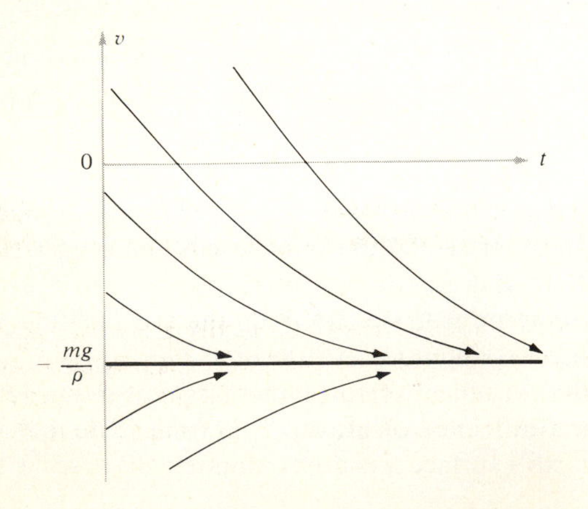

Figure 1.4 Velocity curves with resistance.

▶ **EXAMPLE 1.3-4**

A ball weighing $\frac{1}{2}$ pound is tossed upward with initial velocity 50 ft/sec and there is a force of resistance proportional to the velocity, with proportionality constant ρ. Let $v(t)$ denote the velocity and $y(t)$ the position of the ball for each time t (ground level is $y = 0$ and the positive direction is upward). With the gravitational constant $g = 32$ and the initial time $t_0 = 0$, equation (8) becomes

$$v' = -64\rho v - 32 \qquad v(0) = 50 \tag{11}$$

Therefore, the velocity v is given by

$$v(t) = \left[\frac{100\rho + 1}{2\rho}\right] e^{-64\rho t} - \frac{1}{2\rho} \tag{12}$$

and since $y(0) = 0$ the position y is given by

$$y(t) = \left[\frac{100\rho + 1}{128\rho^2}\right] (1 - e^{-64\rho t}) - \frac{t}{2\rho} \tag{13}$$

for all $t > 0$ such that $y(t) > 0$. Noting that the maximum height $y_{\max}$ is obtained at the time t_m where $v(t_m) = 0$, it follows from (12) that

$$\left[\frac{100\rho + 1}{2\rho}\right] e^{-64\rho t_m} = \frac{1}{2\rho}$$

and hence that

$$t_m = \frac{\ln (100\rho + 1)}{64\rho}$$

Therefore,

$$y_{\max} = y(t_m) = \frac{100\rho - \ln (100\rho + 1)}{128\rho^2}$$

PROBLEMS

1.3-1. It takes 3 years for 20 percent of a radioactive material to disintegrate. What is the half-life? How long does it take for 75 percent of this material to disintegrate?

1.3-2. A scientist with a Geiger counter records that a radioactive substance has a disintegration rate of 135 counts per minute now and had a disintegration rate of 150 counts per minute 6 months ago. Determine the half-life of this material. What disintegration rate should the scientist expect to record 1 year from now?

1.3-3. Suppose that the half-life of a radioactive substance is 500 years. What percentage of this substance should be expected to remain after 100, 250, and 1000 years?

1.3-4. Let $N(t)$ denote the amount of a radioactive substance present at any time t. Suppose that at distinct times T_0 and T_1, it is known that $N(T_0) = N_0$ and $N(T_1) = N_1$. Determine $N(t)$ at all times t in terms of t, T_0, T_1, N_0, and N_1.

1.3-5. Suppose that a tank initially contains 30 gallons of pure water and that brine containing $\frac{1}{3}$ pound of salt per gallon is pumped into the tank at a rate of 6 gal/min and that the well-stirred mixture flows out at the same rate. Determine the amount $Q(t)$ of salt in the tank at each time $t \geq 0$. Show further that $\lim_{t \to \infty} Q(t) = 10$ and find a time $t_1 > 0$ such that $Q(t_1) = 5$.

1.3-6. A tank initially contains 40 gallons of brine containing 1 pound of salt per gallon. Pure water is pumped into the tank at a rate of 2 gal/min and the well-stirred mixture flows out at the same rate. Determine the amount $Q(t)$ of salt in the tank at all times $t \geq 0$ and determine the time $t_1 > 0$ such that the brine has a concentration of $\frac{1}{10}$ pound per gallon.

1.3-7. Suppose that a tank initially contains 50 gallons of pure water and brine containing b pounds of salt per gallon is pumped into the tank at a rate of 5 gal/min and the well-stirred mixture leaves at the same rate. Determine b so that there are 5 pounds of salt in the tank after 1 hour. Determine b so that there are 10 pounds of salt in the tank after 1 hour.

1.3-8. A tank with a 60-gallon capacity initially contains 20 gallons of brine that contains 1 pound of salt per gallon. Pure water is pumped into this tank at the rate of 3 gal/min, and the well-stirred mixture flows out at a rate of 2 gal/min. How many pounds of salt are in the tank at the instant the tank overflows?

1.3-9. A tank initially contains 64 gallons of pure water and brine containing $\frac{1}{2}$ pound of salt per gallon is pumped into the tank at a rate of 1 gal/min, and the well-stirred mixture flows out at a rate of 2 gal/min. Determine the quantity $Q(t)$ of salt in the tank for each time t, $0 \leq t \leq 64$. What is the maximum value of $Q(t)$ and at what time is the maximum value attained?

1.3-10. An object weighing 16 pounds is dropped from rest and the air resistance is 2 times the magnitude of the velocity. Determine the limiting velocity and the time it will take for the object to reach a velocity of 75 percent of the limiting velocity (gravitational constant $g = 32$ ft/sec^2).

1.3-11. An object weighing 8 pounds is tossed vertically upward from ground level with an initial velocity of 16 ft/sec. Assuming that the air resistance is $\frac{1}{2}$ times the magnitude of the velocity and that the gravitational constant g is 32 ft/sec^2, determine the velocity $v(t)$ and the distance $y(t)$ of the ball from ground level as functions of time. Determine also the length of time the ball travels upward and the maximum height it obtains.

1.3-12. Suppose that a ball of mass m is initially at ground level and is tossed vertically upward with speed $v_0 > 0$. Suppose also that air resistance is negligible and that gravitational attraction is a constant g. Show that this ball reaches a maximum height above ground level of $H = v_0^2/2g$ and that this maximum height is reached after a time of v_0/g. How long does it take the ball to return to ground level?

1.3-13. Let $y(t)$ be defined by (13) in Example 1.3-4 and show that it takes the ball *longer* to come back to the ground than it did to reach its maximum height [it may help to show that if t_m is as in Example 1.3-4, then $y(2t_m) > 0$].

1.3-14. Carry out the details in showing that the solution to equation (8) is given by (9). Show also that the position of this object is given by formula (10).

1.3-15. For each $\rho > 0$ let $v_\rho(t)$ denote the function defined by equation (9) and let $y_\rho(t)$ be defined by equation (10). Compute $\bar{v}(t) = \lim_{\rho \to 0+} v_\rho(t)$ (it may help to apply L'Hospital's rule). Is $\bar{v}$ the solution to equation (8) with $\rho = 0$? Compute $\bar{y}(t) = \lim_{\rho \to 0+} y_\rho(t)$. Is $\bar{y}' = \bar{v}$?

1.3-16. Let the velocity $v(t)$ and the position $y(t)$ of an object be described by equation (9) and equation (10), where $t_0 = 0$, $v_0 > 0$, and $y_0 = 0$.

(a) Show that the object travels upward for a time $t_m = m\rho^{-1} \ln(1 + \rho v_0/mg)$.

(b) Show that the maximum height obtained is $y_m = mv_0/\rho - m^2 g \rho^{-2} \ln(1 + \rho v_0/mg)$.

(c) Analyze t_m and y_m as $\rho \to 0+$ and compare with Problem 1.3-12.

1.3-17. *Newton's law of cooling* asserts that the instantaneous rate of change of the temperature of a cooling body is proportional to the difference between the temperature of the body and the temperature of its surroundings. Therefore, if $T(t)$ denotes the temperature of a cooling body at time t and $S(t)$ denotes the temperature of its surroundings at time t, then there is a constant $\rho > 0$ such that

$$(*) \qquad T'(t) = -\rho[T(t) - S(t)] \qquad T(t_0) = T_0$$

where T_0 is the temperature at time $t = t_0$.

(a) If the temperature of the surroundings $S(t)$ does not change appreciably, it can be assumed that $S(t) \equiv S_0$ in $(*)$ for some constant temperature S_0. Determine the solution to $(*)$ when $S(t) \equiv S_0$ and discuss the behavior of $T(t)$ as $t \to +\infty$.

(b) An object with temperature 150° is placed in a freezer whose temperature is 30°. Assume that Newton's law of cooling applies and that the temperature of the freezer remains essentially constant. If this object is cooled to 120° after 8 minutes, what will its temperature be after 16 minutes? When will its temperature be 60°?

(c) An object with temperature 100° is put in a large vat of water with unknown temperature S_0. Assume that Newton's law of cooling applies and that the water temperature remains essentially constant at S_0. If the object's temperature is 90° after 1 hour and 86° after 2 hours, what is the temperature S_0 of the water? What is the temperature of the object after 3 hours?

1.4 SEPARATION OF VARIABLES

In this section a method to solve a special class of first order nonlinear differential equations is developed. It is assumed that the initial value problem has the form

$$y' = \alpha(t)f(y) \qquad y(t_0) = y_0 \tag{1}$$

where α is continuous on an interval I and f is continuous on an interval D. If a differential equation has the form (1), it is said that the *variables are separated.* The crucial issue is that the right side of (1) is the product of a function of t with a function of y. Whenever the variables are separated a solution can always be implicitly determined in terms of antiderivatives. If we set $y' = dy/dt$, then the differential equation in (1) becomes

$$\frac{dy}{dt} = \alpha(t)f(y) \tag{2}$$

The form of equation (2) suggests the following procedure for determining a solution: divide each side by $f(y)$ and "multiply" each side by dt to obtain

$$\frac{dy}{f(y)} = \alpha(t)\,dt$$

Since the left side involves only y and the right side involves only t, we can antidifferentiate each side to obtain the expression

$$\int \frac{dy}{f(y)} = \int \alpha(t)\,dt + c \tag{3}$$

where c is an arbitrary constant. Formula (3) gives an *implicit* representation of a solution to (1). This formal procedure of obtaining the expression (3) is called *separation of variables.*

As an illustration of this technique, consider the equation

$$y' = 4te^{-y} \qquad y(0) = 3 \tag{4}$$

Setting $y' = dy/dt$ and then separating the variables, we obtain $e^y\,dy = 4t\,dt$. Therefore,

$$\int e^y\,dy = \int 4t\,dt + c$$

and so $e^y = 2t^2 + c$. The initial condition $y(0) = 3$ implies that $e^3 = c$ and hence

$$e^y = 2t^2 + e^3 \tag{5}$$

The formula (5) defines a solution to (4) implicitly. In this case we can easily solve for y explicitly by taking the natural logarithm of each side to obtain

$$y(t) = \ln(2t^2 + e^3) \tag{5'}$$

The reader should verify that this y is indeed a solution to (4). In this instance the solution y is defined for all times t. This is not always the case, however, so one must be careful and whenever possible check for the largest interval of time that the solution y can be defined.

EXAMPLE 1.4-1

Consider the initial value problem

$$y' = y^5 \sin t \qquad y(0) = \tfrac{1}{2} \tag{6}$$

Separating variables, we obtain

$$\frac{dy}{y^5} = \sin t\, dt \qquad \text{and so} \qquad \int \frac{dy}{y^5} = \int \sin t\, dt + c$$

Therefore, $-\frac{1}{4}y^{-4} = -\cos t + c$, or

$$\frac{1}{y^4} = 4\cos t + c_1 \tag{7}$$

where c_1 is a constant. Using the initial condition $y(0) = \frac{1}{2}$, we obtain from (7) that $c_1 = 2^4 - 4 = 12$, and hence, solving for y in (7),

$$y(t) = \left(\frac{1}{4\cos t + 12}\right)^{1/4} \tag{8}$$

Again the solution y is defined *for all* t since $4\cos t + 12$ is always positive. However, instead of the initial condition $y(0) = \frac{1}{2}$ use $y(0) = 1$ to obtain the problem

$$y' = y^5 \sin t \qquad y(0) = 1 \tag{9}$$

As before, equation (7) must be satisfied, but in this case the initial condition implies that $c_1 = 1 - 4 = -3$. Substituting for c_1 into (7) and solving for y explicitly, we obtain

$$y(t) = \left(\frac{1}{4\cos t - 3}\right)^{1/4} \tag{10}$$

Formula (10) makes sense only if $4\cos t - 3 > 0$, and since y must be defined on an *interval containing 0* (recall that a solution to an initial value problem is defined on an *interval* that contains the initial time), we see that if θ is the number in $(0, \pi/2)$ such that $\cos\theta = \frac{3}{4}$ (that is, $\theta = \text{Arccos}\,\frac{3}{4}$), then the solution y in (10) is defined on $(-\theta, \theta)$. Moreover, $y(t) \to +\infty$ as $t \to \theta-$ and as $t \to -\theta+$.

▶ **EXAMPLE 1.4-2**

Consider the equation $y' = y^2$, $y(0) = -2$. Separating variables, $y^{-2}\,dy = dt$, and so

$$-\frac{1}{y} = t + c$$

Since $y(0) = -2$, we have $c = \frac{1}{2}$ and $-1/y = t + \frac{1}{2}$. Therefore,

$$y(t) = \frac{-1}{t + \frac{1}{2}} \tag{11}$$

Since (11) is defined only for $t \neq -\frac{1}{2}$ and 0 must be in the interval that y is defined, it follows that $(-\frac{1}{2}, \infty)$ is the interval of definition for y.

Although we have indicated formally that solutions to (1) can be computed by separation of variables, there is a precise way to formulate this procedure, which is outlined in the following theorem.

Theorem 1.4-1

Suppose that α, f, t_0, and y_0 are as in equation (1). If $f(y_0) = 0$, then $y(t) \equiv y_0$ on I is a solution to (1). If $f(y_0) \neq 0$, then a solution y to (1) is implicitly defined by

$$\int_{y_0}^{y} \frac{1}{f(u)}\,du = \int_{t_0}^{t} \alpha(s)\,ds \tag{12}$$

and this representation is valid for t in some open subinterval I' containing t_0.

We indicate the ideas in the proof of Theorem 1.4-1. It is clear that if $f(y_0) = 0$ and $y(t) \equiv y_0$ on I then

$$y'(t) = 0 = \alpha(t) f(y_0) = \alpha(t) f(y(t))$$

and so the constant function $y(t) \equiv y_0$ is a solution to (1) by definition. Now suppose that y is a solution to (1) on an interval $I' \subset I$ and that $f(y(t)) \neq 0$ for all t in I'. We show that this solution satisfies equation (12). Since $y'(t) = \alpha(t) f(y(t))$ for all t in I', it follows that

$$\frac{y'(t)}{f(y(t))} = \alpha(t)$$

and hence that

$$\int_{t_0}^{t} \frac{y'(s)\,ds}{f(y(s))} = \int_{t_0}^{t} \alpha(s)\,ds$$

Using the substitution $u = y(s)$, we have $du = y'(s)\,ds$, $u = y_0$ when $t = t_0$ [this follows from the initial condition $y(t_0) = y_0$], and hence

$$\int_{y_0}^{y(t)} \frac{du}{f(u)} = \int_{t_0}^{t} \alpha(s)\,ds$$

This shows that any solution to (1) must satisfy (12). Conversely, if y is any differentiable function defined by (12), then one can show by implicit differentiation that y must be a solution to (1).

It is important to remember to check to see if the initial value y_0 is a zero of f *before* proceeding with separating the variables. If $f(y_0) = 0$, then the solution to (1) is simply the constant function $y(t) \equiv y_0$. For example, the solution to

$$y' = (y^2 - 2y - 3)e^t \qquad y(0) = -1$$

is the constant function $y(t) \equiv -1$.

EXAMPLE 1.4-3

Consider the equation $y' = 2ty^2$, $y(0) = y_0$, where $t \geq 0$. If $y_0 = 0$ then $y(t) \equiv 0$ is the solution (why)? If $y_0 \neq 0$ then by separation of variables we have

$$\frac{y'}{y^2} = 2t \qquad \text{and hence} \qquad \int_{y_0}^{y} \frac{du}{u^2} = \int_{0}^{t} 2s\,ds$$

Thus

$$-\frac{1}{y} + \frac{1}{y_0} = t^2 \qquad \text{and hence} \qquad \frac{1}{y} = \frac{1}{y_0} - t^2$$

Therefore,

$$y(t) = \frac{y_0}{1 - y_0 t^2}$$

so long as $y_0 t^2 \neq 1$. If $y_0 > 0$ then $y(t) \to +\infty$ as $t \to 1/\sqrt{y_0}$, and so solutions to this equation "blow up" in finite time whenever the initial data is positive. Note, however, that if $y_0 < 0$ then y exists for all $t \geq 0$ and $y(t) \to 0$ as $t \to +\infty$.

1.4a Time-Independent Equations

There is a very important subclass of equations whose variables separate that occurs when the time-dependent factor $\alpha(t)$ is identically 1. When the right-hand side of the initial value problem (1) [or (4) in Section 1.1] does not depend explicitly on time, the equation is said to be *autonomous*. Therefore, we consider the initial value problem

$$y' = f(y) \qquad y(t_0) = y_0 \tag{13}$$

where f is a continuous function on an interval D. Since autonomous equations are special cases of equations whose variables separate, if $f(y_0) \neq 0$ a solution to (4) is implicitly defined by the equation

$$\int_{y_0}^{y} \frac{du}{f(u)} = t - t_0 \tag{14}$$

so long as $t - t_0$ is sufficiently small. The equation in Example 1.4-2 is autonomous, but the equation in Example 1.4-3 is not.

▶ **EXAMPLE 1.4-4**

Consider the autonomous equation $y' = y - y^2$, $y(0) = y_0$. If $y_0 = 0$ then $y(t) \equiv 0$, and if $y_0 = 1$ then $y(t) \equiv 1$. By partial fraction techniques

$$\frac{1}{u - u^2} = \frac{1}{u(1 - u)} = \frac{1}{u} + \frac{1}{1 - u}$$

and hence

$$\int \frac{dy}{y - y^2} = \ln |y| - \ln |1 - y| = \ln \left| \frac{y}{1 - y} \right|$$

Therefore, using formula (14), if $y_0 \neq 0, 1$ the solution y to this equation satisfies

$$\ln \left| \frac{y(t)}{1 - y(t)} \right| = \ln \left(\left| \frac{y_0}{1 - y_0} \right| \right) + t$$

and so

$$\left| \frac{y(t)}{1 - y(t)} \right| = \left| \frac{y_0}{1 - y_0} \right| e^t$$

So long as $y(t) \neq 0, 1$, it is easy to check that $y(t)/[1 - y(t)]$ and $y_0/(1 - y_0)$ must have the same sign, and so

$$\frac{y(t)}{1 - y(t)} = \frac{y_0}{1 - y_0} e^t \qquad \text{or} \qquad y(t) = \frac{y_0}{1 - y_0} e^t - \frac{y_0}{1 - y_0} e^t y(t)$$

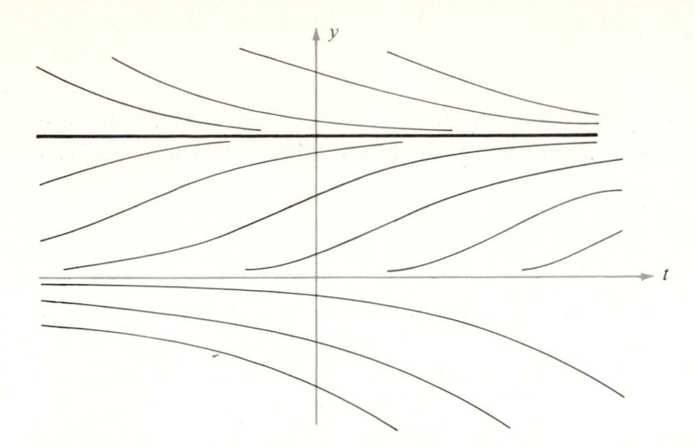

Figure 1.5 Sketch of solution curves in (15).

Solving explicitly for $y(t)$ we obtain

$$y(t) = \frac{y_0 e^t}{(1 - y_0) + y_0 e^t} \tag{15}$$

so long as $y_0 \neq 0, 1$ and $y_0 e^t \neq y_0 - 1$. If $0 \leq y_0 \leq 1$ then y_0 is defined for all t, but if $y_0 > 1$ then $y(t)$ exists only for $t > \ln [(y_0 - 1)/y_0]$, and if $y_0 < 0$ then $y(t)$ exists only for $t < \ln [(y_0 - 1)/y_0]$. Note also that if $0 < y_0 < 1$ then $\lim_{t \to +\infty} y(t) = 1$ and $\lim_{t \to -\infty} y(t) = 0$. A sketch for some of the representative curves in (15) is given in Figure 1.5.

EXAMPLE 1.4-5

Consider the equation $y' = y^2 - y^3$, $y(0) = y_0$. First $y(t) \equiv 0$ if $y_0 = 0$ and $y(t) \equiv 1$ if $y_0 = 1$. Since

$$\frac{1}{u^2 - u^3} = \frac{1}{u^2} + \frac{1}{u} + \frac{1}{1 - u}$$

by partial fractions we obtain

$$-\frac{1}{y} + \ln |y| - \ln |1 - y| = t + c$$

and hence if $y_0 \neq 0, 1$ the solution y is implicitly defined by the relation

$$\ln \left| \frac{y(t)}{1 - y(t)} \right| - \frac{1}{y(t)} = t + \ln \left| \frac{y_0}{1 - y_0} \right| - \frac{1}{y_0}$$

There seems to be no elementary way of obtaining $y(t)$ explicitly as a function of t from this relation, and so the solution y is only implicitly obtained if $y_0 \neq 0, 1$.

EXAMPLE 1.4-6

Consider the equation $y' = y^2 + 2y + 2$, $y(0) = y_0$. Since there are no real zeros of the right-hand side, there are no constant solutions, and since

$$\int \frac{dy}{y^2 + 2y + 2} = \int \frac{dy}{1 + (y + 1)^2} = \text{Arctan}\,(y + 1)$$

by separation of variables we obtain

$$\text{Arctan}\,(y + 1) = t + \text{Arctan}\,(y_0 + 1)$$

and hence so long as $-\pi/2 < t + \text{Arctan}\,(y_0 + 1) < \pi/2$,

$$y(t) = -1 + \tan\,[t + \text{Arctan}\,(y_0 + 1)]$$

In particular, if $y_0 = 0$ then $y(t) = -1 + \tan\,(t + \pi/4)$ for $-3\pi/4 < t < \pi/4$.

*1.4*b* A Fundamental Theorem on Existence

In the first part of this section it is indicated that a first order initial value problem has a solution if the variables are separated, and in Section 1.2 it is shown that first order linear equations have precisely one solution satisfying a given initial condition. The purpose here is to state a basic result on the existence and uniqueness of solutions to a reasonable class of first order initial value problems. Assume f is a function of two variables and consider the initial value problem

$$y' = f(t, y) \qquad y(t_0) = y_0 \tag{16}$$

where (t_0, y_0) is in the domain of f. We have the following fundamental result:

Theorem 1.4-2

Suppose that the domain R of f is an open region in the ty plane and that $(t_0, y_0) \in R$. Suppose further that both $f(t, y)$ and $\partial/\partial y\, f(t, y)$ are continuous in the region R. Then the initial value problem (16) has a unique solution on some open interval I_0 containing t_0.

The proof of this theorem is omitted since it is beyond the scope of this text. We do, however, discuss some important ideas related to this theorem. The uniqueness assertion in Theorem 1.4-2 means if $y_1(t)$ and $y_2(t)$ are solutions to the differential equation in (16) and both satisfy the initial condition $y_1(t_0) = y_2(t_0) = y_0$, then necessarily $y_1(t) = y_2(t)$ for t in their common interval of definition. There are examples of differential equations that have more than one solution satisfying a given initial condition. Consider the autonomous equation

$$y' = 3y^{2/3} \qquad y(0) = 0 \tag{17}$$

Separating variables, $y^{-2/3}\,dy = 3\,dt$, or $3y^{1/3} = 3t + c$. Using the condition $y(0) = 0$ implies $c = 0$, and hence $y(t) = t^3$. It is easy to check directly that $y = t^3$ is indeed a solution to (17). Of course, $y(t) \equiv 0$ is also a solution to (17), and we see that (17) has more than one solution. Notice that the function $f(t, y) = 3y^{2/3}$ does not satisfy the properties of f listed in Theorem 1.4-2 since $\partial/\partial y\,(3y^{2/3}) = 2y^{-1/3}$ does not exist at $y = 0$.

The student should also note that the existence of a solution is guaranteed only on *some* open interval I_0 containing t_0, and not necessarily on the largest possible interval in the domain of f. This problem has already occurred in the analysis of separable systems (see Examples 1.4-1, 1.4-2, and 1.4-3 for specific illustrations). Observe, however, that if the system (17) is a *linear* system and $f(t, y) = a(t)y + b(t)$ for all $(t, y) \in I \times (\infty, \infty)$, then the solution to (17) is always defined on all of I.

PROBLEMS

1.4-1. Determine an implicit solution to the following initial value problems.

(a) $y' = 2ty^2 \qquad y(0) = 4$

(b) $y' = \dfrac{(t^2 + 1)(y^2 - 1)}{ty} \qquad y(1) = 0$

(c) $y' = e^t(y + y^3) \qquad y(0) = 1$

(d) $y' = \cos t \sec y \qquad y(0) = \dfrac{\pi}{6}$

(e) $y' = \dfrac{y^2 + 4}{2y - 3} \qquad y(0) = 2$

1.4-2. Determine a solution explicitly in terms of t for the following initial value problems (indicating clearly the time interval of definition).

(a) $y' = 2ty^2 \qquad y(0) = 4$

(b) $y' = -2ty^3 \qquad y(0) = 2$

(c) $y' = \dfrac{\cos t}{y} \qquad y(0) = -4$

(d) $y' = \dfrac{\cos t}{y} \qquad y(0) = 1$

(e) $y' = 4 + y^2 \qquad y(0) = 0$

(f) $y' = \dfrac{\cos t}{\cos y} \qquad y(0) = \dfrac{\pi}{6}$

1.4-3. Determine the solution to each of the following problems, where y_0 is a given positive number and $I = (0, \infty)$.

(a) $y' = \dfrac{y}{t^{1/3}} \qquad y(1) = y_0$

(b) $y' = \dfrac{y}{t^3} \qquad y(1) = y_0$

(c) $y' = \dfrac{y}{t} \qquad y(1) = y_0$

(d) $y' = -\dfrac{y}{t^3} \qquad y(1) = y_0$

Investigate the behavior of these solutions as $t \to 0+$.

1.4-4. Determine explicitly a solution to the following autonomous equations, indicating clearly the domain of their definition.

(a) $y' = -y^3 \quad y(0) = -2$

(b) $y' = y^2 - y \quad y(0) = \frac{1}{2}$

(c) $y' = \cos^2 y \quad y(0) = 0$

(d) $y' = \cos^2 y \quad y(0) = \dfrac{\pi}{4}$

(e) $y' = y^2 - 2y - 3 \quad y(0) = 1$

(f) $y' = y^2 - 2y - 3 \quad y(0) = -2$

(g) $y' = y^2 - y - 2 \quad y(0) = 5$

(h) $y' = y^2 + 2y + 5 \quad y(0) = 1$

(i) $y' = \dfrac{y^2 + 3y + 2}{2y + 3} \quad y(0) = -3$

1.4-5. Let $I = [0, \infty)$ and consider the equation $y' = -y^{1/3}$, $y(0) = y_0$, where $y_0 > 0$. Show that there is a time T (depending on y_0) such that the solution y satisfies $y(t) \equiv 0$ for $t \geq T$. Investigate the behavior of the solutions to $y' = -y^\alpha$, $y(0) = y_0$, where $y_0 > 0$ and $0 < \alpha < 1$.

1.4-6. Suppose that a, b, c are constants with $b \neq 0$, and consider the equation $y' = F(at + by + c)$. Show that if $v = at + by + c$, then v is a solution to the autonomous equation $v' = a + bF(v)$. Use this technique to solve the following equations:

(a) $y' = (y + t + 1)^2 \quad y(0) = 0$

(b) $y' = (y - t)^2 \quad y(1) = 2$

1.4-7. The following is a mixture of linear and separable initial value problems. Also, various notations for the dependent and independent variable are used. Determine an explicit solution to each equation.

(a) $\dfrac{dx}{dt} = 7x + 14 \quad x(0) = 1$

(b) $\dfrac{dr}{ds} = r^3 \cos s \quad r(0) = -\frac{1}{2}$

(c) $\dfrac{dr}{ds} = r^3 \cos s \quad r(0) = 1$

(d) $\dfrac{dP}{du} = u \ln u \quad P(e) = 0$

(e) $\dfrac{du}{dt} = -\dfrac{u}{t} \quad u(1) = 1$

(f) $\dfrac{dQ}{dr} = e^{r + 2Q} \quad Q(0) = 0$

(g) $\dfrac{dy}{dt} = y \cot t + \sin t \quad y\left(\dfrac{\pi}{2}\right) = 0$

(h) $\dfrac{dy}{dx} = \dfrac{1 + x}{1 + y} \quad y(0) = 3$

1.5 CURVE SKETCHING FOR SOLUTIONS

A first order initial value problem is said to be autonomous if the right-hand side of this equation does not explicitly depend on t (see Section 1.4*a*). Rather than develop further procedures for explicitly or implicitly determining solutions to autonomous equations, this section develops a method that can be used to study the properties of solutions *without having to solve for the solution.* In this section the time interval I is always taken to be $[0, \infty)$, and the differential equation

$$y' = f(y) \tag{1}$$

is considered, where f is assumed to be continuously differentiable on an open interval (α, β). The basic technique is to *sketch the graphs of a representative set of solutions to* (1) by considering solutions having various initial values. Therefore, along with the differential equation (1), the corresponding initial value problem

$$y' = f(y) \qquad y(0) = y_0 \tag{2}$$

is also considered. The problem is to select the appropriate initial values y_0 in (2) and determine the crucial properties of f in order to sketch a reasonably representative set of graphs of the solutions to (1).

If z is a number and $f(z) = 0$, then $y(t) \equiv z$ on $[0, \infty)$ is a solution to (1) [note that if $y(t) \equiv z$ then $y'(t) \equiv 0 \equiv f(z) \equiv f(y(t))$]. Thus the zeros of f are the initial values for constant solutions to (1) and are called *critical points* of (1). The corresponding constant solutions are called *equilibrium solutions*. Therefore, it is crucial to determine the critical points of f, and for convenience the set of critical points of (1) is denoted CP. Note that if z is a critical point the graph of the corresponding equilibrium solution to (1) is a horizontal half-line in the right ty plane that initiates at the point $(0, z)$ on the y axis. Consider, for example, the equation

$$y' = y - y^3 \tag{3}$$

The critical points of (3) are the solutions z to $z - z^3 = 0$. Since $z - z^3 = z(1 - z)(1 + z)$, it is immediately seen that $\text{CP} = \{0, 1, -1\}$ for (3). The graphs of the corresponding equilibria are given in Figure 1.6*a*. As will soon be indicated it is important to determine the sign of the right-hand side of the equation in between the critical points. Since f is continuous, if $z_1 < z_2$ are consecutive critical points of f, then f must either satisfy $f(y) > 0$ for all $y \in (z_1, z_2)$ or $f(y) < 0$ for all $y \in (z_1, z_2)$. Therefore, f is always of

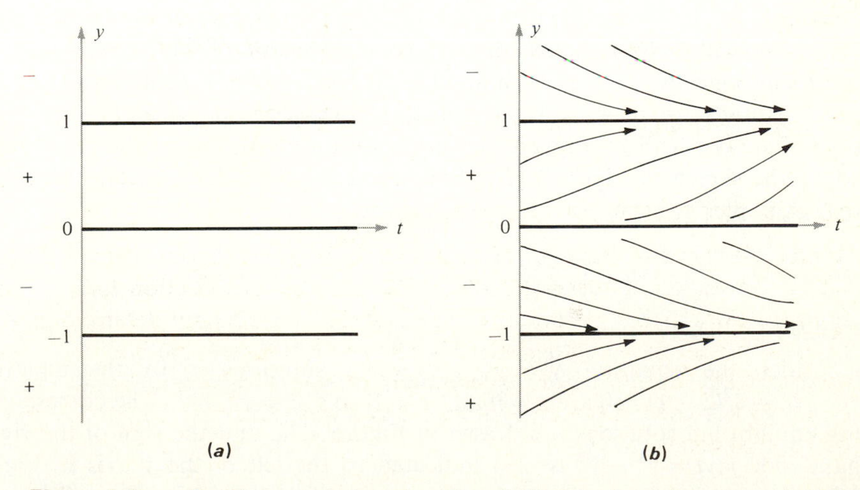

Figure 1.6 ***a*** Equilibrium solutions for (3). ***b*** Sketch of solution curves for (3).

constant sign in between consecutive critical points. In the particular example of (3), the right-hand side is negative for $y > 1$, positive for $0 < y < 1$, negative for $-1 < y < 0$, and positive for $y < -1$. This is indicated with a plus sign or a minus sign placed to the left of the y axis and in between the critical points.

Since it is assumed that f is continuously differentiable, solutions to (1) are *unique*; that is, if y_1 and y_2 are both solutions to (1) and $y_1(t_0) = y_2(t_0)$ for some $t_0 \geq 0$, then $y_1(t) \equiv y_2(t)$ for all t where they are both defined (this is asserted in Theorem 1.4-2). Note in particular that this implies that the *graphs of solutions to* (1) *do not cross.* This property of solutions to (1) is extremely helpful in sketching a representative set of graphs. Suppose that z is a critical point of (1) and y is a solution to (1). If $y(0) > z$, then necessarily $y(t) > z$ for all t such that $y(t)$ is defined. This is immediately seen from the fact that the solution y cannot ever equal the constant solution z. Similarly, if $y(0) < z$, then $y(t) < z$ for all t such that $y(t)$ is defined. Combining these properties we obtain the following:

$$\text{(4)} \quad \begin{cases} \text{Suppose that } z \text{ is a critical point of (1) and that } y \text{ is a solution to (1) with} \\ z < y(0) \text{ [or } y(0) < z\text{]. Then } z < y(t) \text{ [or } y(t) < z\text{] for all } t \geq 0. \end{cases}$$

Observe further that if $z_1 < z_2$ are consecutive critical points of (1) and $z_1 < y(0) < z_2$, then $z_1 < y(t) < z_2$ for all t; and since f does not change signs between z_1 and z_2, it follows that $f(y(t))$ has the same sign as $f(y(0))$ for all $t \geq 0$. Since $y'(t) = f(y(t))$ for all $t \geq 0$ by definition of a solution, we see that $y(t)$ is always increasing in t if f is positive between z_1 and z_2, and $y(t)$ is always decreasing in t if f is negative between z_1 and z_2. Referring back to equation (3), if y is any solution to (3) such that $0 < y(0) < 1$, then $y(t)$ is increasing in t on $[0, \infty)$ and $0 < y(t) < 1$ for all $t \geq 0$. This is indicated by the sketches of the curve in Figure 1.6*b*. In fact, not only is $y(t)$ increasing but $y(t)$ necessarily converges to the critical point $z = 1$ as $t \to \infty$ (this is also indicated in Figure 1.6*b*). As will be soon indicated, *every bounded solution to* (1) *must converge to a critical point as* $t \to \infty$. In a similar manner, if y is a solution to (3) and $-1 < y(0) < 0$, then $-1 < y(t) < 0$ and $y(t)$ decreases to -1 as $t \to \infty$. This type of analysis can be applied to every solution to (3); the results are indicated by the sketch in Figure 1.6*b*. Example 1.5-1 provides a second introductory example.

▶ **EXAMPLE 1.5-1**

Consider the equation $y' = y^3 - y^2$. The solutions z to the equation $z^3 - z^2 = z^2(z - 1) = 0$ are $z = 0$ and $z = 1$, so CP $= \{0, 1\}$. The corresponding equilibrium solutions are drawn in Figure 1.7*a*, and the sign of the right-hand side $f(y) = y^3 - y^2$ is also indicated to the left of the y axis in Figure 1.7*a*: $y^3 - y^2$ is positive if $y > 1$ and negative if $0 < y < 1$ or if $y < 0$. Since

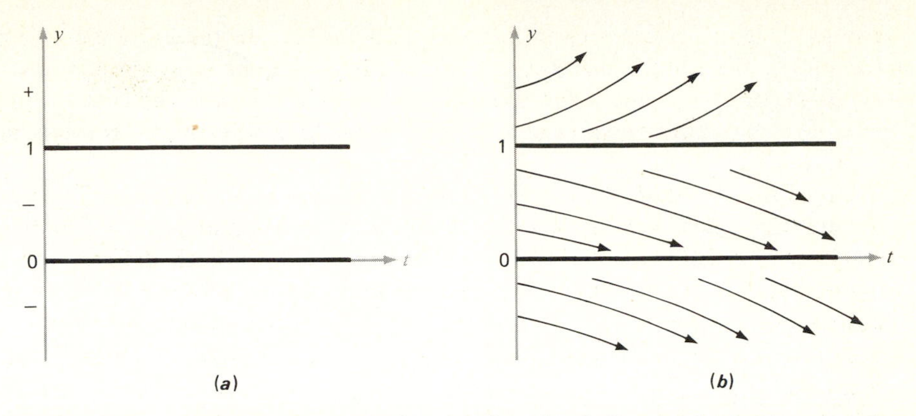

Figure 1.7 ***a*** Equilibrium solutions for $y' = y^3 - y^2$. ***b*** Sketch of solution curves for $y' = y^3 - y^2$.

this immediately indicates the monotonicity properties of the solutions relative to their initial values, the rough sketch of the solutions can be quickly made as indicated in Figure 1.7*b*. The only solutions that are trapped between critical points are those that are initially in the interval (0, 1). Since the right-hand side is negative here, these solutions must decrease to 0 as $t \to \infty$. The solutions initially larger than 1 remain larger than 1 and hence are increasing in t. Since there is no critical point larger than 1, these solutions must converge to $+\infty$. In this particular instance they "blow up" in finite time [i.e., they have a vertical asymptote at some finite time $T > 0$ that depends on the initial value $y(0)$]. Similarly, solutions that are initially negative must decrease to $-\infty$ since there is no critical point less than 0. Determining the existence of vertical asymptotes for solution can be difficult and will not be considered here.

The procedure indicated above is useful for the study of solutions to (1). If the critical points of (1) can be computed, as well as the sign of $f(x)$ for x in the intervals in between the critical points, then one can easily determine for each given initial value if the solution is increasing or decreasing and what its limiting value is as t goes to $+\infty$. This also helps in drawing a reasonable sketch of the graphs of solutions to (1) without having to compute the solution (either explicitly or implicitly). However, before considering more examples we show that the concavity of a solution to (1) can also be analyzed easily. A function y on an interval I is *concave up* if whenever $t_1, t_2 \in I$ with $t_1 < t_2$, the graph of y on $[t_1, t_2]$ lies on or below the graph of the line segment joining $(t_1, y(t_1))$ to $(t_2, y(t_2))$ (that is, the graph of y is "cupped upward"—see Figure 1.8*a*). The function y is *concave down* if whenever $t_1, t_2 \in I$ with $t_1 < t_2$, the graph of y on $[t_1, t_2]$ lies on or above the graph of the line segment joining

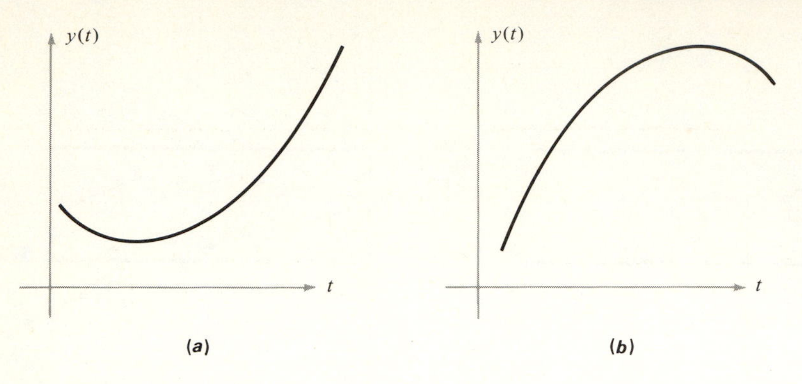

Figure 1.8 ***a*** Concave up, $y'' \geq 0$. ***b*** Concave down, $y'' \leq 0$.

$(t_1, y(t_1))$ to $(t_2, y(t_2))$—see Figure 1.8*b*. It is important to recall from calculus that if y is twice differentiable then

$$\text{(5)} \quad \begin{cases} y \text{ is concave up (or concave down) on } I \text{ if and only if } y''(t) \geq 0 \ [\text{or } y''(t) \leq 0] \\ \text{for all } t \text{ in } I. \end{cases}$$

The concavity of the solutions to (1) can be determined whenever the zeros of $f' = df/dy$ can be computed. For if y is a solution to (1), then $y'(t) = f(y(t))$, and differentiating each side of this equation and using the chain rule, it follows that

$$y''(t) = \frac{d}{dt} f(y(t)) = f'(y(t))y'(t) = f'(y(t))f(y(t))$$

Since $f(y(t))$ is always of one sign, it follows that *a solution y to (1) changes concavity only if it crosses a zero of f' and f' changes sign at this zero.* Notice that the solution $y(t)$ is concave up in each time interval where $f'(y(t))f(y(t)) > 0$ and is concave down in each time interval where $f'(y(t))f(y(t)) < 0$.

▶ **EXAMPLE 1.5-2**

Consider the equation

$$\text{(6)} \quad y' = 6 + y - y^2$$

Since $6 + y - y^2 = (3 - y)(2 + y)$, it follows that $CP = \{3, -2\}$ for equation (6). Also, the right-hand side is negative for $y > 3$, positive for $-2 < y < 3$, and negative for $y < -2$ (see Figure 1.9*a*). In this case, $f(y) = 6 + y - y^2$ and $f'(y) = 1 - 2y$. Thus $f'(\frac{1}{2}) = 0$ and $f'(y)$ changes sign at $y = \frac{1}{2}$. Therefore, the solutions to (6) change concavity when they pass through the line $y = \frac{1}{2}$ (this line is indicated with a dashed line in Figure 1.9*a*). It now follows that every

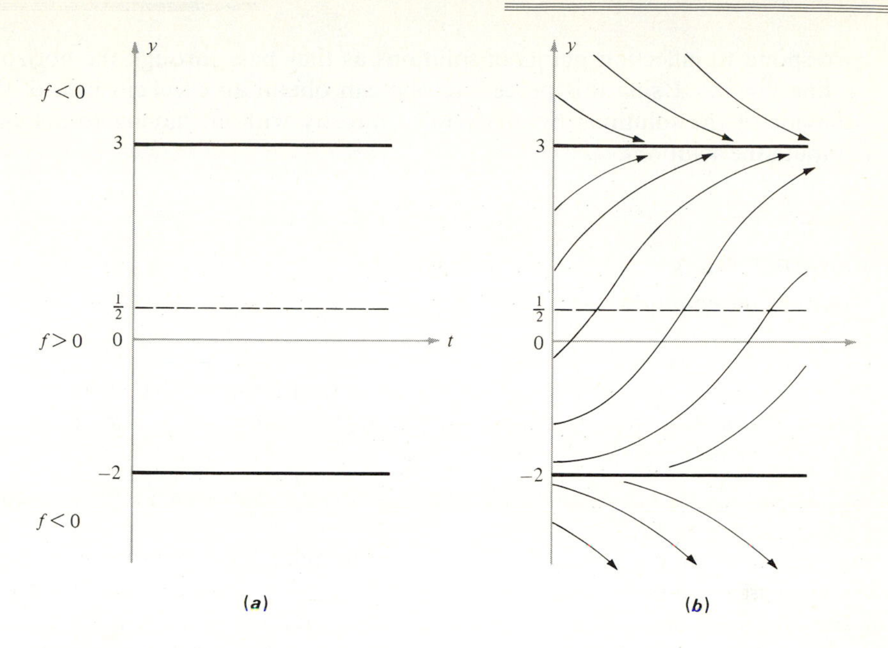

Figure 1.9 ***a*** Equilibrium solutions and inflection change for (6). ***b*** Phase-time plot for (6).

solution to (1) with $y(0) > -2$ converges to $y \equiv 3$ as $t \rightarrow \infty$, and every solution with $y(0) < -2$ must diverge to $-\infty$ (we do not have a means of determining if these solutions have a vertical asymptote or if they diverge as $t \rightarrow +\infty$). Notice that the solutions have inflection points (i.e., change concavity) if they pass through the line $y = \frac{1}{2}$. The concavity can be checked by determining the sign of $f'(y)f(y)$ and is indicated in Figure 1.9*b*.

A representative sketch of the solutions to (1) in a manner indicated by the preceding example is called a *phase-time plot* for (1). In order to sketch a phase-time plot, the following procedure is helpful:

1. Determine the critical point set CP: solve the equation $f(z) = 0$.
2. Determine the sign of f between consecutive critical points.
3. Determine the zeros of f' and the intervals where $f'(x)f(x)$ is positive and where it is negative.

The critical points correspond to horizontal lines in the phase-time plot, and the sign of f in between successive critical points determines if the solutions are increasing or decreasing and hence to which critical point the solution converges as $t \rightarrow \infty$. The points w where f' changes sign and f is not zero correspond to inflection points of solutions as they pass through the horizontal line $y = w$. Using this procedure one can obtain an excellent idea of the behavior of the solutions by analyzing f directly without having to actually compute the solutions.

EXAMPLE 1.5-3

Consider the equation

$$y' = y^2 - y^4 \tag{7}$$

Since $f(y) = y^2 - y^4 = y^2(1 - y)(1 + y)$, the critical point set CP $= \{0, 1, -1\}$ for (7). Note also that $f(y) > 0$ if $-1 < y < 0$ or if $0 < y < 1$, and $f(y) < 0$ if $y < -1$ or $y > 1$. Since

$$f'(y) = 2y - 4y^3 = 2y(1 - \sqrt{2}\,y)(1 + \sqrt{2}\,y)$$

then $f'(y) = 0$ for $y = 0,\ 1/\sqrt{2},\ -1/\sqrt{2}$. Although f' changes sign at each of these zeros, $y = 0$ is also a critical point, so $y = \pm 1/\sqrt{2}$ are the only inflection points for solutions. The phase-time plot for equation (7) now follows quite easily and is given in Figure 1.10. Although one can employ separation of variables and partial fractions to obtain an implicit representation of the solutions to (7), it is not possible to explicitly represent the solutions using elementary techniques. Thus more information about the solutions to (7) can be obtained from the phase-time plot than by trying to solve directly for solutions.

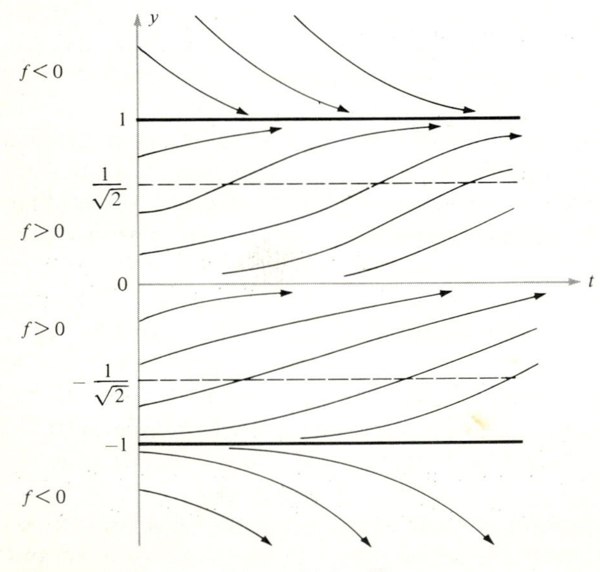

Figure 1.10 Phase-time plot for (7).

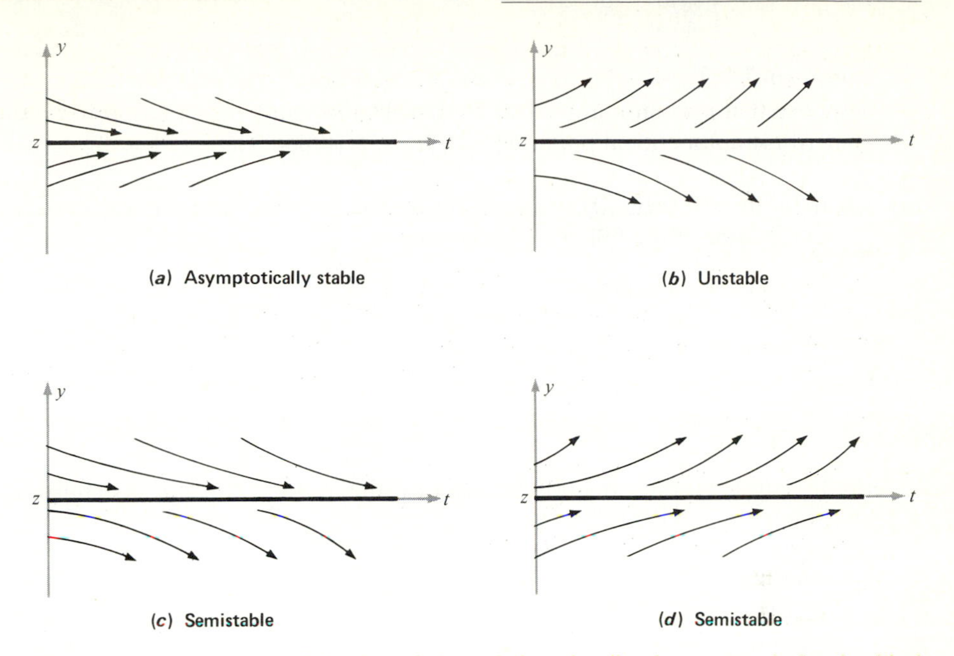

Figure 1.11 ***a, b, c, d*** Behavior of the solutions locally about z, an isolated critical point of (1).

Suppose that z is an "isolated" critical point of (1) and let $\delta > 0$ be such that z is the only critical point in the interval $(z - \delta, z + \delta)$ [that is, $f(x) \neq 0$ if $0 < |x - z| < \delta$]. Then the behavior of the solutions locally about z is determined by the sign of $f(x)$ for $z < x < z + \delta$ and for $z - \delta < x < z$. This behavior can be essentially any one of the four types depicted in Figure 1.11. The critical point z is *asymptotically stable* if whenever y is a solution to (1) with $z - \delta < y(0) < z + \delta$, it follows that $y(t) \to z$ as $t \to +\infty$ (see Figure 1.11*a*). It is said to be *unstable* if whenever y is a solution to (1) with $z - \delta < y(0) < z + \delta$ and $y(0) \neq z$, it follows that there is some time $t_1 > 0$ (depending on the solution y) such that $|y(t) - z| \geq \delta$ for all $t \geq t_1$ where $y(t)$ is defined (see Figure 1.11*b*). The semistable case is characterized by the fact that solutions beginning near and on one side of z converge to z as $t \to \infty$, and other solutions beginning near and on the other side of z move away from z as time increases (see Figure 1.11*c* and *d*). Using Figure 1.10 it follows that the critical point $z = 1$ in equation (7) is asymptotically stable, the critical point $z = 0$ is semistable, and the critical point $z = -1$ is unstable.

For completeness we formalize the procedure for determining the increasing and decreasing nature of solutions to (1) in a theorem. An indication of the proof of these assertions is also given. However, this theorem may be omitted by the reader without any loss of continuity.

Theorem 1.5-1

Suppose that f is continuously differentiable and that y is a solution to (1) such that $y(0)$ is not a critical point of (1) [that is, $f(y(0)) \neq 0$].

(i) If $f(y(0)) > 0$ then $y(t)$ is increasing on the interval where it is defined, and exactly one of the following must occur:

(a) Equation (1) has a smallest critical point z larger than $y(0)$ [that is, $f(z) = 0$ and $f(x) > 0$ for $y(0) < x < z$] and $\lim_{t\to\infty} y(t) = z$; or

(b) Equation (1) has no critical point larger than $y(0)$ and there is a $T > 0$ [which depends on $y(0)$ and may be $+\infty$] such that $\lim_{t\to T-} y(t) = +\infty$.

(ii) If $f(y(0)) < 0$ then $y(t)$ is decreasing on the interval where it is defined, and exactly one of the following must occur:

(a) Equation (1) has a largest critical point z smaller than $y(0)$ [that is, $f(z) = 0$ and $f(x) < 0$ for $z < x < y(0)$] and $\lim_{t\to\infty} y(t) = z$; or

(b) Equation (1) has no critical point smaller than $y(0)$ and there is a $T > 0$ [which depends on $y(0)$ and may be $+\infty$] such that $\lim_{t\to T-} y(t) = -\infty$.

We indicate the ideas involved only in part (i) since (ii) is established in an analogous manner. Since y is not a constant solution, it can never cross a critical point and so $f(y(t))$ can never equal zero. Since $f(y(0)) > 0$ and $t \to f(y(t))$ is continuous and never zero, it is clear that $f(y(t)) > 0$ so long as $y(t)$ is defined. Thus

$$y'(t) = f(y(t)) > 0$$

and y is increasing. Assume that y is defined on an interval $[0, T)$ where T may be $+\infty$. Suppose first that y is bounded on $[0, T)$. Since y is increasing and bounded on $[0, T)$ it must have a limit as $t \to T-$, say

$$w = \lim_{t\to T-} y(t)$$

From this it follows that T is necessarily $+\infty$ [for instance, if T were finite, the initial value problem $y' = f(y)$, $y(T) = w$ could be solved, thereby extending y to a larger interval]. Therefore, if y is bounded, it is defined on $[0, \infty)$ and $w = \lim_{t\to\infty} y(t)$ exists. Moreover, since f is continuous and y is a solution to (1),

$$\lim_{t\to\infty} y'(t) = \lim_{t\to\infty} f(y(t)) = f(w)$$

also exists. If both a function y and its derivative y' have a limit as $t \to +\infty$, then, necessarily, the limit of the derivative is 0 (see Problem 1.5-6). Therefore, $0 = y'(\infty) = f(w)$, and so w is a critical point of (1). Therefore, if $f(y(0)) > 0$ and if y remains bounded, there must be a smallest critical point w larger than y, and $y(t) \to w$ as $t \to \infty$. If there is a critical point z larger than the solution y, then $y(t) < z$ for all t, so y must remain bounded since it is increasing. Thus assertion (i), part a, holds only in case y remains bounded, and if part a does not hold then part b must hold. This establishes (i) in Theorem 1.5-1, and assertion (ii) follows in an analogous manner.

1.5*a* Motion of a Falling Object with Resistance

As an indication of the applicability of these techniques, consider an object with constant mass m moving vertically with velocity $v(t)$ and under the influence of gravity and an air resistance force (see Section 1.3). However, instead of assuming that air resistance is proportional to the velocity, suppose that it is proportional to some power σ of the velocity where $\sigma \geq 1$. Therefore, the resistance force F_R has the form $F_R = -\rho v(t)|v(t)|^{\sigma-1}$ (F_R is in the *opposite* direction of the velocity). According to Newton's second law (see Section 1.3) the equation of motion is $d/dt\,(mv) = -mg - \rho v|v|^{\sigma-1}$, and since m is constant,

$$v' = -\frac{\rho}{m} v|v|^{\sigma-1} - g \qquad v(0) = v_0 \tag{8}$$

If $\sigma = 1$ this equation is linear and was solved in Section 1.3. In general, for $\sigma > 1$ it can be difficult to determine v explicitly even though (8) is autonomous (and so the variables separate). If f is defined by the right-hand side of (8), then

$$f(v) = \begin{cases} -\dfrac{\rho}{m} v^{\sigma} - g & \text{if } v \geq 0 \\[2ex] \dfrac{\rho}{m}(-v)^{\sigma} - g & \text{if } v < 0 \end{cases}$$

and

$$f'(v) = \begin{cases} -\dfrac{\sigma\rho}{m} v^{\sigma-1} & \text{if } v \geq 0 \\[2ex] -\dfrac{\sigma\rho}{m}(-v)^{\sigma-1} & \text{if } v < 0 \end{cases}$$

Therefore f and f' are continuous, and it follows that the techniques for phase-time plots apply to equation (8). Clearly $-(mg/\rho)^{1/\sigma}$ is a critical point for (8), and since $f'(v) < 0$ for all $v \neq 0$, this is the only critical point for (8). Since $f(v) < 0$ for $v > -(mg/\rho)^{1/\sigma}$ and $f(v) > 0$ for $v < -(mg/\rho)^{1/\sigma}$, then $-(mg/\rho)^{1/\sigma}$ is an asymptotically stable critical point, and the solution to (8) converges to this critical point as $t \to +\infty$ for *every* initial value v_0.

The velocity $-(mg/\rho)^{1/\sigma}$ is called the *limiting velocity* for this object. Observe that the limiting velocity decreases as ρ increases, and that if $mg/\rho > 1$ it also decreases as σ increases.

PROBLEMS

1.5-1. Sketch phase-time plots for the following equations. Indicate critical points, where solutions increase and decrease, and where solutions change concavity.

(a) $y' = 1 - y^2$
(b) $y' = y^2 - y - 2$
(c) $y' = 1 - e^y$
(d) $y' = y - \frac{1}{4}y^3$
(e) $y' = -6y + 5y^2 - y^3$
(f) $y' = \sin y$
(g) $y' = \sin^2 y$
(h) $y' = y^3 - 2y^2 - 3y$
(i) $y' = y^4 - 2y^3 - 3y^2$
(j) $y' = 1 + y - e^y$
(k) $y' = -y^3(y - 1)^2(y + 1)$

1.5-2. Sketch representative phase-time plots and discuss stability properties of critical points for various values of the number η for the following equations:

(a) $y' = y(y - \eta)$ (*Hint:* Consider the cases $\eta > 0$, $\eta = 0$, and $\eta < 0$.)
(b) $y' = y^2(y - \eta)$
(c) $y' = y(y - 1)(y - \eta)$

1.5-3. Suppose that an object with constant mass m is moving vertically with velocity $v(t)$ and under the influence of both gravity and air resistance, where the air resistance is assumed to be proportional to the velocity squared. Assuming that v is a solution to (8), with $\sigma = 2$ and $v(0) = 0$, determine $v(t)$ explicitly as a function of time t [note that since $v(t) \leq 0$ for all $t \geq 0$, v is the solution to the equation $v' = (\rho/m)v^2 - g$, $v(0) = 0$]. Compute $\lim_{t\to\infty} v(t)$ and compare with the limiting velocity.

***1.5-4.** Suppose that f and its derivative f' are continuous, that z is a critical point of f, and that $\delta > 0$ is such that $f(x) \neq 0$ if $0 < |z - x| < \delta$.

(a) If $f(x) < 0$ for $z < x < z + \delta$ and $f(x) > 0$ for $z - \delta < x < z$, show that the critical point z is asymptotically stable.
(b) If $f'(z) < 0$, show that z is asymptotically stable. (*Hint:* Use part *a*.)
(c) If $f'(z) > 0$, show that z is unstable.

***1.5-5.** Suppose that n is a positive integer, and consider the two differential equations **(a)** $y' = y^n$ and **(b)** $y' = -y^n$. Discuss the types of stability and instability of the critical point $z = 0$ to each of these equations.

***1.5-6.** Suppose that y is a continuously differentiable function on $[0, \infty)$ and that $w = \lim_{t\to\infty} y'(t)$ exists:

(a) If $w > 0$ and $0 < \varepsilon < w$, show that there is a $T \geq 0$ such that $y(t) - y(T) \geq \varepsilon(t - T)$ for all $t \geq T$. Deduce that $y(t) \to +\infty$ as $t \to \infty$.

(b) If $w < 0$, show that $y(t) \to -\infty$ as $t \to \infty$. [*Hint:* Replace $y(t)$ by $-y(t)$ and use part *a*.]

(c) If $\lim_{t \to \infty} y(t) = z$ exists and is finite, show that $w = 0$.

1.6 CHANGE OF DEPENDENT VARIABLE: HOMOGENEOUS EQUATIONS

In this section it is shown how a change of dependent variable can be used in order to simplify certain first order equations. As before, the equation

$$y' = f(t, y) \tag{1}$$

is considered. The purpose of this section is to introduce methods involving a change in the variable y in order to reduce equation (1) to a type of equation that can be solved (e.g., a linear equation or an equation whose variables separate).

The first case studied is when equation (1) is *homogeneous.* The function f is said to be homogeneous if $f(kt, ky) \equiv f(t, y)$ for any constant $k \neq 0$ (the term *homogeneous* here is different from that used in connection with linear equations). It is easy to see, for example, that the function

$$f(t, y) \equiv \frac{y^2}{t^2 + y^2}$$

is homogeneous. For if $k > 0$, then

$$f(kt, ky) = \frac{(ky)^2}{(kt)^2 + (ky)^2} = \frac{k^2 y^2}{k^2(t^2 + y^2)} = \frac{y^2}{t^2 + y^2} = f(t, y)$$

Notice that for this particular f we can write

$$f(t, y) = \frac{y^2}{t^2 + y^2} = \frac{(y/t)^2}{1 + (y/t)^2} \qquad \text{if } t \neq 0$$

and hence this function does not depend on t and y separately, but only on the ratio y/t. This type of property is true in general for homogeneous functions: that is, *homogeneous functions $f(t, y)$ are precisely those that depend only on the ratio y/t (or the ratio t/y).* For note that if $t \neq 0$ then

$$f(t, y) = f\left(t, t\left(\frac{y}{t}\right)\right) = f\left(1, \frac{y}{t}\right)$$

and so if $g(r) = f(1, r)$ then $f(t, y) \equiv g(y/t)$. Therefore, in the homogeneous case, equation (1) has the form

$$y' = g\left(\frac{y}{t}\right) \tag{2}$$

The form of equation (2) certainly indicates that the change of variable $v = y/t$ might prove helpful, and this is indeed the case. As an illustration consider the equation

$$y' = \frac{2t - y}{t - 2y} \tag{3}$$

Note that if t is replaced by kt and y by ky in the right-hand side of (3), then

$$\frac{2(kt) - ky}{kt - 2(ky)} = \frac{2t - y}{t - 2y}$$

for all $k \neq 0$, and hence (3) is a homogeneous equation [i.e., the right-hand side of (3) is a homogeneous function]. Therefore, define the function v by the equation $v = y/t$. Then $y = tv$ and $y' = dy/dt = v + tv'$, and substituting for y and y' into (3), we obtain

$$v + tv' = \frac{2t - tv}{t - 2tv} = \frac{2 - v}{1 - 2v}$$

Thus

$$tv' = \frac{2 - v}{1 - 2v} - v = \frac{2 - v - v + 2v^2}{1 - 2v} = 2\,\frac{1 - v + v^2}{1 - 2v}$$

and it follows that v should be a solution to

$$tv' = 2\,\frac{1 - v + v^2}{1 - 2v}$$

The variables separate in this equation (see Section 1.4), and since $v' = dv/dt$, we obtain

$$\frac{(1 - 2v)\,dv}{1 - v + v^2} = 2\,\frac{dt}{t}$$

Since $d/dv\,(1 - v + v^2) = -(1 - 2v)$, we have, by antidifferentiating each side of this equation,

$$-\ln|1 - v + v^2| = 2\ln|t| + c$$

and hence

$$\ln t^2 + \ln|1 - v + v^2| = \bar{c}$$

where $\bar{c}$ is a constant. Since $1 - v + v^2 > 0$ for *all* v (why?) and since the sum of the logarithms is the logarithm of the product, we have

$$\ln(t^2 - t^2v + t^2v^2) = \bar{c}$$

Taking the exponential of each side, we obtain

$$t^2 - t^2v + t^2v^2 = A \qquad \text{where } A > 0 \text{ is constant}$$

Since $v = y/t$ the solution y to (3) is implicitly defined by the relation

$$t^2 - ty + y^2 = A \qquad \text{where } A > 0 \text{ is constant} \tag{4}$$

The student should show that these equations describe a family of ellipses in the ty plane.

The procedure just indicated with the preceding example applies equally well to homogeneous equations in general. In fact, the following result is valid: Suppose that equation (1) is homogeneous and has the form of equation (2). Then the change of variable $y = tv$ transforms equation (2) into the form

$$v' = \frac{g(v) - v}{t} \tag{5}$$

which can be solved by separation of variables.

Since $y = tv$, then $y' = tv' + v$ by the product rule for differentiation, and substituting directly for y and y' in (2) leads to the equation $tv' + v = g(v)$. Equation (5) follows immediately from this equation.

EXAMPLE 1.6-1

Consider the equation $y' = (y^2 + ty)/t^2$, $y(1) = 1$. Since this equation is homogeneous (see also Problem 1.6-6), the change of variables $y = tv$ is used. Therefore, $y' = v + tv'$ and so

$$v + tv' = \frac{t^2v^2 + t^2v}{t^2} = v^2 + v$$

Thus $v' = v^2/t$, and, by separation of variables, $-1/v = \ln t + c$. Since $y = tv = -t/(\ln t + c)$, the condition $y(1) = 1$ implies that the solution is

$$y(t) = \frac{-t}{\ln t - 1}$$

which is valid for $0 < t < e$.

In general, it may be difficult to recognize whether it is possible to make an appropriate change of variable in order to solve equation (1). Sometimes, however, the form of (1) may "suggest" a helpful variable change [as in the case when (1) is homogeneous]. A simple, yet reasonably general, example is the equation

$$y' = \frac{a(t)g(y) + b(t)}{g'(y)} \tag{6}$$

where a and b are continuous on I and g is continuously differentiable on D, with $g'(y) \neq 0$ for $y \in D$. It is clear that if $v = g(y)$, then $v' = g'(y)y'$, and so, with respect to the variable v, equation (6) is transformed into the linear equation

(7) $$v' = a(t)v + b(t)$$

This equation can be explicitly solved using the techniques in Section 1.2.

▶ **EXAMPLE 1.6-2**

Consider the equation $y' = \tan y + t \sec y$, $y(0) = 0$. Since $\tan y = (\sin y)/(\cos y)$ and $\sec y = 1/(\cos y)$, this equation has the form

$$y' = \frac{\sin y + t}{\cos y} \qquad y(0) = 0$$

This is the same form as (6), with $g(y) = \sin y$, and it follows that with the transformation $v = \sin y$,

$$v' = (\cos y)y' = v + t \qquad \text{and} \qquad v(0) = \sin 0 = 0$$

Multiplying by the integrating factor e^{-t} one obtains $d/dt\ (e^{-t}v) = te^{-t}$, and, by antidifferentiating each side of this equation, $v(t) = ce^t - t - 1$. Since $v(0) = c - 1 = 0$, it follows that $v(t) = e^t - t - 1$, and if J is any interval containing 0 such that $-1 < e^t - t - 1 < 1$ for all $t \in J$, then the function $y(t) = \arcsin\ (e^t - t - 1)$ on J is the solution to the given equation.

1.6a Bernoulli's Differential Equation

Suppose now that p and q are continuous functions on an interval I and that σ is a real constant. The equation of the form

(8) $$y' = p(t)y + q(t)y^\sigma$$

is known as *Bernoulli's differential equation.* When $\sigma = 0$ or 1, equation (8) is a linear equation and can be solved by the method in Section 1.2. For other values of σ Bernoulli's equation is in the form of (6). To see that this is so, observe that (8) can be rewritten as

$$y' = \frac{p(t)y^{1-\sigma} + q(t)}{y^{-\sigma}} = \frac{(1-\sigma)p(t)y^{1-\sigma} + (1-\sigma)q(t)}{(1-\sigma)y^{-\sigma}}$$

Therefore, equation (8) has the form of equation (6), with $g(y) = y^{1-\sigma}$. According to the discussion following equation (6), the following technique applies in solving Bernoulli's equation:

(9) If $\sigma \neq 0, 1$, the change of variable $v = y^{1-\sigma}$ (and hence $y = v^{1/(1-\sigma)}$) transforms Bernoulli's equation (8) into the linear equation

$$v' = (1 - \sigma)p(t)v + (1 - \sigma)q(t)$$

It is quite easy to check directly that the change of variable indicated in (9) reduces Bernoulli's equation to a linear equation (which can therefore be solved by the method of Section 1.2).

EXAMPLE 1.6-3

Consider the equation $y' = t^{-1}y + ty^2$, $y(1) = 2$. Since this has the form of Bernoulli's equation (8) with $\sigma = 2$, use the transformation $y = v^{-1}$ and $y' = -v^{-2}v'$ to obtain

$$-v^{-2}v' = y' = t^{-1}y + ty^2 = t^{-1}v^{-1} + tv^{-2}$$

and hence

$$v' = -t^{-1}v - t$$

This equation is linear and an integrating factor is

$$e^{\int t^{-1}\,dt} = e^{\ln t} = t$$

so that $tv' + v = -t^2$, and hence

$$\frac{d}{dt}(tv) = -t^2$$

Therefore, $tv = -t^3/3 + c$, and since $v(1) = y(1)^{-1} = \frac{1}{2}$, it follows that $c = \frac{5}{6}$ and that

$$y(t) = \frac{1}{v(t)} = \frac{6t}{5 - 2t^3}$$

for all $0 < t < (\frac{5}{2})^{1/3}$.

PROBLEMS

1.6-1. Determine a solution to each of the following homogeneous initial value problems and solve for the solution explicitly whenever possible.

(a) $y' = \dfrac{y^2 + 2ty}{t^2}$ $y(1) = 2$

(b) $y' = \dfrac{2y^2 - t^2}{2ty}$ $y(2) = 1$

(c) $y' = \dfrac{t - y}{t + y}$ $y(1) = 3$

(d) $y' = \dfrac{y^2 + 3ty}{t^2}$ $y(4) = 8$

(e) $y' = \dfrac{1}{2}\left(\dfrac{t}{y} + \dfrac{y}{t}\right)$ $y(2) = -4$

(f) $ty' = y - \sqrt{t^2 - y^2}$ $y(1) = \frac{1}{2}$

1.6-2. Determine a solution to each of the following initial value problems (it may be helpful to use the indicated change of variable).

(a) $t + yy' = t^2 + y^2 \qquad y(0) = -1 \qquad (\text{Let } v = t^2 + y^2)$

(b) $y' = ty + 2y \ln y \qquad y(0) = 1 \qquad (\text{Let } v = \ln y)$

(c) $(\cos y)y' = 2t \sin y - 2t \qquad y(0) = 0 \qquad (\text{Let } v = \sin y)$

(d) $y' = \dfrac{2y}{t} + t \tan \dfrac{y}{t^2} \qquad y(1) = 2 \qquad \left(\text{Let } v = \dfrac{y}{t^2}\right)$

1.6-3. Determine a solution to each of the following Bernoulli equations.

(a) $y' = -ty + ty^3 \qquad y(0) = 3$

(b) $y' = y + ty^2 \qquad y(0) = 1$

(c) $y' + \dfrac{1}{t} y = \dfrac{1}{t^2} y^3 \qquad y(1) = 2$

(d) $y' = y - \frac{1}{4}y^{3/2} \qquad y(0) = \frac{1}{4}$

1.6-4. If F is a continuous function on $(-\infty, \infty)$ and a_1, a_2 and b_1, b_2 are constants (a_1, $a_2 \neq 0$), show that the equation

$$(*) \qquad y' = F\left(\frac{a_1 t + b_1 y}{a_2 t + b_2 y}\right)$$

is homogeneous. Therefore, solve the equations:

(a) $y' = \dfrac{t - 2y}{2t + y}$ **(b)** $y' = \left(\dfrac{t + y}{t}\right)^2$

***1.6-5.** Suppose that a_1, a_2, b_1, b_2, and c_1, c_2 are constants, and consider the equation

$$(**) \qquad y' = F\left(\frac{a_1 t + b_1 y + c_1}{a_2 t + b_2 y + c_2}\right)$$

(a) If $a_1 b_2 - a_2 b_1 \neq 0$ and h, k are solutions to

$$a_1 h + b_1 k + c_1 = 0$$

$$a_2 h + b_2 k + c_2 = 0$$

show that the change of variables $t = \bar{t} + h$ and $y = \bar{y} + k$ reduces equation (**) to a homogeneous equation of the form (*) in Problem 1.6-4.

(b) Solve the equation $y' = \dfrac{2t + 5y + 3}{2t - y - 3}$.

***1.6-6.** Suppose that M and N are polynomials in t, y and that there is a positive integer n such that the power of each term in both M and N is n (that is, each term in both M and N has the form at^iy^j, where a is a constant and i and j are nonnegative integers with $i + j = n$). Show that the equation

$$\frac{dy}{dt} = \frac{M(t, y)}{N(t, y)}$$

is homogeneous.

1.6-7. The following initial value problems are either linear or separable, or can be transformed into such an equation by an appropriate change of dependent variable. Also, various notations for the variables are used. Determine an explicit solution to each equation.

(a) $\dfrac{du}{ds} = \dfrac{u^2}{s}$ $\quad u(1) = 1$

(b) $\dfrac{dp}{dt} = \dfrac{t - p}{t + 2p}$ $\quad p(0) = 1$

(c) $\dfrac{dx}{dr} = x + rx^2$ $\quad x(0) = \frac{1}{2}$

(d) $\dfrac{dQ}{ds} = \dfrac{se^s - Q}{s}$ $\quad Q(2) = 0$

(e) $\dfrac{du}{dt} = \dfrac{e^t + 2u^2}{2u}$ $\quad u(0) = -1$

(f) $\dfrac{du}{dv} = \dfrac{4v^2 + 2u^2}{uv}$ $\quad u(1) = \sqrt{5}$

(g) $\dfrac{dT}{ds} = \dfrac{8s^2 - 2T}{s}$ $\quad T(3) = 1$

(h) $\dfrac{dy}{dx} = \dfrac{3x + xy^2}{y + x^2y}$ $\quad y(1) = 3$

(i) $\dfrac{dy}{dt} = \dfrac{t - \rho y}{t + 1}$ $\quad y(1) = 0$ where $\rho \neq 0, 1$ is a given constant

1.7 EXACT DIFFERENTIAL EQUATIONS

In this section we study a class of nonlinear differential equations which have a more general form. Previous equations have maintained a distinction between the independent variable t and the dependent variable y. However, in this section solutions are viewed as curves in a plane, and so the concepts of independent and dependent variables play no role. Therefore, the two variables that are used in this section are considered as belonging to a plane (with the first variable denoted x and the second y). Points are denoted (x, y) and are said to be in the xy plane. Also, throughout this section D is assumed to be an open rectangle in the xy plane: $D = \{(x, y): \alpha_1 < x < \alpha_2, \beta_1 < y < \beta_2\}$, where $-\infty \leq \alpha_1 < \alpha_2 \leq \infty$ and $-\infty \leq \beta_1 < \beta_2 \leq \infty$.

Suppose that U is a real-valued function on D such that U and its first partial derivatives U_x and U_y are continuous on D. Let c be a constant and consider the expression

$$U(x, y) = c \tag{1}$$

If the constant c is such that (1) defines a curve in the xy plane, then this curve is called a *level curve of U* [for example, if $U(x, y) = x^2 + y^2$, then (1) defines a level curve only if $c > 0$]. Assume for the moment that c is such that (1) defines a level curve and let (x_0, y_0) be a point on this curve. If

$U_y(x_0, y_0) \neq 0$ and y is a differentiable function of x (locally about x_0), then by the chain rule

$$0 = \frac{d}{dx}[c] = \frac{d}{dx}[U(x, y)] = U_x(x, y) + U_y(x, y)\frac{dy}{dx}$$

and hence

$$\frac{dy}{dx} = -\frac{U_x(x, y)}{U_y(x, y)} \tag{2}$$

Similarly, if $U_x(x_0, y_0) \neq 0$ and x is a differentiable function of y (locally about y_0), then

$$\frac{dx}{dy} = -\frac{U_y(x, y)}{U_x(x, y)} \tag{3}$$

Thus the level curves of (1) define locally a solution to either (2) or (3). For example, if $U(x, y) = x^2 + y^2$ and $c > 0$, then $x^2 + y^2 = c$ defines implicitly a solution to $dy/dx = -x/y$ if $y \neq 0$ and a solution to $dx/dy = -y/x$ if $x \neq 0$.

Let both M and N be continuous real-valued functions on the rectangle D such that the first partial derivatives of M and N are also continuous on D. As opposed to considering the pair of equations

$$\frac{dy}{dx} = -\frac{M(x, y)}{N(x, y)} \qquad \frac{dx}{dy} = -\frac{N(x, y)}{M(x, y)} \tag{4}$$

we use the expression

$$M(x, y)\,dx + N(x, y)\,dy = 0 \tag{5}$$

Equation (5) is called a *total differential equation*, and the function U on D is said to be a *first integral* of (5) on D if the following holds: there is a positive function μ on D such that μ and its first partial derivatives are continuous and

$$U_x(x, y) \equiv \mu(x, y)M(x, y) \qquad \text{and} \qquad U_y(x, y) \equiv \mu(x, y)N(x, y) \qquad \text{on } D \tag{6}$$

The function μ is called an *integrating factor* for (5) on D. From (6) we see immediately that $U_x/U_y = M/N$ on D, and comparing (2) and (3) with (4) we see that the level curves of U are implicitly defined solutions of the differential equations in (4).

Consider, for example, the equation

$$(y^2 - 1)\,dx + (2xy - \sin y)\,dy = 0$$

and define $U(x, y) = xy^2 - x + \cos y$ for all x and y. Then

$$U_x(x, y) = y^2 - 1 \qquad \text{and} \qquad U_y(x, y) = 2xy - \sin y$$

and it follows that U is a first integral for this equation (the integrating factor μ is the identically constant function 1).

In order to determine a first integral of (5), an integrating factor μ must be found and then U should be constructed from (6). The concept of an *exact differential* plays an important role. The expression $M(x, y)\,dx + N(x, y)\,dy$ is said to be *exact on D* if there exists a U such that $U_x = M$ and $U_y = N$ on D: that is, $dU(x, y) = M(x, y)\,dx + N(x, y)\,dy$. Analogously, equation (5) is said to be *exact* when the differential $M\,dx + N\,dy$ is exact. Note that (5) is exact only if (5) has a first integral with $\mu \equiv 1$ as an integrating factor. Exact equations can be readily characterized, as shown in the following theorem.

Theorem 1.7-1

Suppose that M and N have continuous first partial derivatives on D. Then the equation (5) is exact if and only if $M_y(x, y) \equiv N_x(x, y)$ on D.

Note that if (5) is exact then

$$M_y = (U_x)_y = U_{xy} \qquad \text{and} \qquad N_x = (U_y)_x = U_{yx}$$

and from the continuity assumptions on the first partials of M and N, we have $U_{xy} = U_{yx}$, and so $M_y = N_x$. Conversely, suppose $M_y = N_x$ on D and let $(x_0, y_0) \in D$. Define the function U on D by

$$U(x, y) = \int_{x_0}^{x} M(r, y_0)\,dr + \int_{y_0}^{y} N(x, s)\,ds \tag{7}$$

Then, upon differentiation under the integral sign,

$$\begin{aligned} U_x(x, y) &= M(x, y_0) + \int_{y_0}^{y} N_x(x, s)\,ds \\ &= M(x, y_0) + \int_{y_0}^{y} M_y(x, s)\,ds \\ &= M(x, y_0) + M(x, y) - M(x, y_0) = M(x, y) \end{aligned}$$

By the fundamental theorem of calculus, $U_y(x, y) = N(x, y)$ and Theorem 1.7-1 is seen to be true.

As opposed to memorizing formula (7) to construct U, it is perhaps simpler to use the equations $U_x = M$ and $U_y = N$ directly. First use $U_x \equiv M$ to obtain

$$U(x, y) = \int M(x, y)\,dx + h(y)$$

and then choose h so that $U_y = N$:

$$h'(y) = N(x, y) - \frac{\partial}{\partial y}\int M(x, y)\,dx$$

To see that the right-hand side of this expression is a function only of y (i.e., is independent of x), note that

$$\frac{\partial}{\partial x}\left[N - \frac{\partial}{\partial y}\int M\,dx\right] \equiv N_x - \frac{\partial^2}{\partial y\,\partial x}\int M\,dx \equiv N_x - M_y \equiv 0$$

EXAMPLE 1.7-1

Consider the total differential equation

$$(xy^2 + y + e^{2x})\,dx + (x + x^2y - y)\,dy = 0$$

Since $M_y = 2xy + 1$ and $N_x = 1 + 2xy$, this equation is exact. From $U_x = xy^2 + y + e^{2x}$ we obtain

$$U = \tfrac{1}{2}x^2y^2 + xy + \tfrac{1}{2}e^{2x} + h(y)$$

Using this expression for U along with $U_y = x + x^2y - y$, we obtain

$$U_y = x^2y + x + h'(y) = x + x^2y - y$$

Thus $h'(y) = -y$ and $h(y) = -\tfrac{1}{2}y^2$. Therefore

$$U(x, y) = \tfrac{1}{2}x^2y^2 + xy + \tfrac{1}{2}e^{2x} - \tfrac{1}{2}y^2$$

is a first integral for this equation.

It follows directly from the definitions that a function μ on D is an integrating factor for (5) if and only if the equation $\mu M\,dx + \mu N\,dy = 0$ is exact. Since multiplying each term in (5) by an integrating factor μ makes the resulting equation exact, we see that a first integral U can be constructed exactly as above, with M replaced by μM and N by μN. Of course, the problem of finding an integrating factor can be difficult. One case where an integrating factor is apparent is when the variables separate. Note that if M is a function of x alone and N is a function of y alone, then (5) is exact since $M_y \equiv N_x \equiv 0$. If there are functions M_1, M_2, N_1, and N_2 such that $M(x, y) = M_1(x)M_2(y)$ and $N(x, y) = N_1(x)N_2(y)$, then one says that the *variables are separated* in (5) (compare with the type of equation studied in Section 1.4). In this case the following is valid:

(8) If $M(x, y) = M_1(x)M_2(y)$ and $N(x, y) = N_1(x)N_2(y)$, then $\mu(x, y) = M_2(y)^{-1}N_1(x)^{-1}$ is an integrating factor and

$$U(x, y) = \int M_1(x)N_1(x)^{-1}\,dx + \int N_2(y)M_2(y)^{-1}\,dy$$

is a first integral of (5).

EXAMPLE 1.7-2

Consider the total differential equation

$$2xe^{-y}\,dx - (1 + x^2)e^y\,dy = 0$$

Since the variables separate, $\mu(x, y) = e^y(1 + x^2)^{-1}$ is an integrating factor, and we have

$$\frac{2x}{1 + x^2}\,dx - e^{2y}\,dy = 0$$

Therefore $U(x, y) = \ln(1 + x^2) - \frac{1}{2}e^{2y}$ is a first integral.

1.7*a* Orthogonal Trajectories

Suppose that W is a function of three variables (x, y, c), where (x, y) are points in the plane and c a real parameter. Suppose further that for every value of the parameter c the equation

$$W(x, y, c) = 0 \tag{9}$$

defines a differentiable curve in the plane. A differentiable curve C in the plane is said to be an *orthogonal trajectory* of (9) if whenever C intersects a curve in (9) it intersects it at a right angle. A family of curves $U(x, y) = c$ is said to be *mutually orthogonal* to (9) if every curve defined by $U(x, y) = c$ is orthogonal to the family of curves (9). As a simple illustration, suppose that (9) is the family of all lines in the plane having slope 2: $y - 2x = c$. Since curves that are perpendicular to each member of this family must have slope $-\frac{1}{2}$ (i.e., the negative reciprocal of 2), it follows that a mutually orthogonal family is the family of lines having slope $-\frac{1}{2}$: $y + x/2 = c$. (See Figure 1.12*a*.)

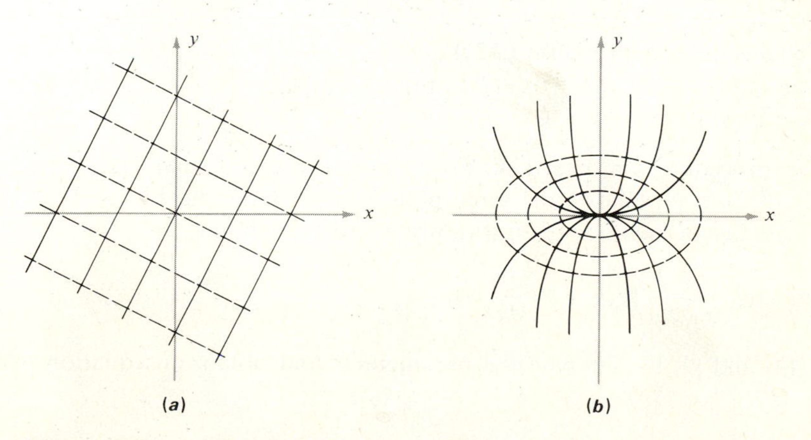

Figure 1.12 ***a*** Orthogonal families of lines. ***b*** Orthogonal family of parabolas and ellipses.

EXAMPLE 1.7-3

Consider the family of parabolic curves

$$y - cx^2 = 0 \tag{10}$$

(see the dark lines in Figure 1.12*b*). We want to determine a differential equation that has the family (10) as a solution. Differentiating each side of (10) with respect to x, we have

$$\frac{dy}{dx} - 2cx = 0 \tag{11}$$

The procedure is to use the two equations (10) and (11) to eliminate the parameter c. From (11) we have

$$c = \frac{1}{2x}\frac{dy}{dx}$$

and substituting into (10) it follows that

$$y - \frac{1}{2x}\frac{dy}{dx}x^2 = 0$$

Therefore if $y = y(x)$ is defined by (10), then $dy/dx = 2y/x$. If we want the curves that are perpendicular to those in (10), they must have negative reciprocal slopes. Thus we need to solve the equation

$$\frac{dy}{dx} = -\frac{x}{2y} \tag{12}$$

Separating variables in (12) we have $x\,dx + 2y\,dy = 0$, and hence the family mutually orthogonal to (10) is the family of ellipses

$$\tfrac{1}{2}x^2 + y^2 = c \tag{13}$$

(see the dashed lines in Figure 1.12*b*).

The procedure for determining the mutually orthogonal family for any family of form (9) follows the same pattern as in Example 1.7-3. First take the derivative of each side of (9) with respect to x to obtain

$$\frac{\partial}{\partial x}W(x, y, c) + \frac{\partial}{\partial y}W(x, y, c)\frac{dy}{dx} = 0 \tag{14}$$

Use (14) and (9) to eliminate the parameter c and obtain an equation of the form

$$\frac{dy}{dx} = g(x, y) \tag{15}$$

The family of mutually orthogonal trajectories are the solutions to the differential equations

$$\frac{dy}{dx} = -\frac{1}{g(x, y)} \tag{16}$$

For the family (10) of parabolas equation (15) has the form $dy/dx = 2y/x$, and equation (16) has the form $dy/dx = -x/(2y)$ [see (12)].

PROBLEMS

1.7-1. Check each of the following total differential equations for exactness. Determine a first integral whenever the equation is exact or the variables separate.

(a) $x^2\, dx + \cos 2y\, dy = 0$

(b) $(2xy^3 - y^2)\, dx + (3x^2y^2 - 2xy)\, dy = 0$

(c) $x(1 + y^3)\, dx + y^2(1 + x^2)\, dy = 0$

(d) $y\, dx + (1 + y)\, dy = 0$

(e) $(\sin x - ye^x)\, dx - (2y + e^x)\, dy = 0$

(f) $xye^x\, dx + (1 + y)\, dy = 0$

1.7-2. Show that each of the following total differential equations is not exact. Show, however, that each equation has an integrating factor of the form indicated, and then use this integrating factor to determine a first integral for each equation.

(a) $(y^2 + 2x^2)\, dx + xy\, dy = 0$ $\quad$ ($\mu = x^m$, where m is a constant)

(b) $(xy^3 + y^2)\, dx + (y - xy)\, dy = 0$ $\quad$ ($\mu = y^n$, where n is a constant)

(c) $(y^2 + 2x^2)\, dx - xy\, dy = 0$ $\quad$ ($\mu = x^m$, where m is a constant)

(d) $(y^2 + yx^2)\, dx + (x^3 - 3xy)\, dy = 0$ ($\mu = x^m y^n$, where m and n are constants)

***1.7-3.** Suppose that $\mu = \mu(x, y)$ satisfies the equation

$$N\mu_x - M\mu_y = (M_y - N_x)\mu$$

Show that μ is an integrating factor for (5).

(a) Show that (5) has an integrating factor $\mu = \mu(x)$ that is independent of y if $(M_y - N_x)/N = \alpha(x)$ is independent of y. Moreover, $\mu(x) = e^{\int \alpha(x)\, dx}$.

(b) Show that (5) has an integrating factor $\mu = \mu(y)$ that is independent of x if $(N_x - M_y)/M = \beta(y)$ is independent of x. Moreover, $\mu(y) = e^{\int \beta(y)\, dy}$.

***1.7-4.** Determine a first integral for each of the following total differential equations (Problem 1.7-3 may be of some help).

(a) $(y - x)\, dx + dy = 0$

(b) $(2y \sin x - \cos^3 x)\, dx + \cos x\, dy = 0$

(c) $2xy\, dx + (y^2 - x^2)\, dy = 0$

1.7-5. Sketch the graph of and determine the family of mutually orthogonal trajectories for each of the following families of curves in the plane.

(a) $x^2 + y^2 = c$ **(d)** $y = ce^x$

(b) $x^2 + 4y^2 = c$ **(e)** $y = cx^3$

(c) $-x^2 + y^2 = cx$ **(f)** $x^2 - y^2 = c$

1.7-6. The following differential equations are either linear, separable, or exact, or can be transformed into such an equation by an appropriate change of variable. Also, various notations for the variables are used. Determine an appropriate family of solutions to each equation, and if an initial condition is given, determine an explicit solution satisfying the initial condition.

(a) $\theta\, d\theta - 2r(\theta + 1)\, dr = 0 \qquad r = 0 \text{ when } \theta = -2$

(b) $\dfrac{du}{dv} = \dfrac{\tan u - 2uv}{v^2 - v \sec^2 u}$

(c) $\dfrac{du}{dt} = te^{u - t^2} \qquad u(0) = 0$

(d) $\dfrac{dr}{d\theta} = \dfrac{2r^2 + 7r\theta + 4\theta^2}{\theta^2} \qquad r = 2 \text{ when } \theta = 1$

(e) $\dfrac{dy}{dx} = \dfrac{x^3 + 2y}{3x^2 y - 6x} \qquad y = 2 \text{ when } x = 1$

(f) $\dfrac{dP}{ds} = \dfrac{-s \cos^2 P}{\tan P}$

(g) $(2y - 8x^2)\, dx + x\, dy = 0$

(h) $\dfrac{dq}{dt} = q - q^{3/2} \qquad q(0) = 1$

(i) $2q(q - 3)\, dp - 3p^3\, dq = 0$

(j) $\dfrac{dQ}{dt} = (3 \tan t)Q + 1$

(k) $\dfrac{dy}{dt} = e^{2t}u - u \ln u \qquad u(0) = e$

(l) $(s^2 + x^2)\, ds - 2sx\, dx = 0$

1.8 ELEMENTARY POPULATION MODELS

First order differential equations arise naturally in the study of population growth, and some basic ideas and techniques are indicated in this section. For each given time t let $p(t)$ denote the number of units in a population [for example, $p(t)$ could denote the number of people in North Carolina, the number of sparrows in the United States, or the pounds of fish in Lake Erie]. Instead of allowing $p(t)$ to be integer-valued (which may be the case), it is assumed that $p(t)$ is a *continuously differentiable approximation of the population at time t*. Therefore, $p'(t)$ can be considered as an approximation of the "instantaneous" rate of change of the population.

A simple growth model for a population is to assume that the rate of growth at any time is proportional to the population at that time:

(1) $$p'(t) = ap(t) \qquad p(t_0) = p_0$$

where p_0 is the population at time $t = t_0$ and $a > 0$ is the *growth coefficient* for the population. A population obeying the equation (1) is said to satisfy a *malthusian growth model*. Equation (1) is simple, and the following result regarding the solution to (1) is analogous to that for radioactive decay in Section 1.3: If the population $p(t)$ satisfies the malthusian model (1), then

(2) $$p(t) = p_0 e^{a(t-t_0)} \qquad \text{for all times } t$$

and if T is defined by the relation

(3) $$aT = \ln 2$$

then T is the *doubling time* of the malthusian model (the population doubles every T units of time).

The fact that the solution p is given by (2) is immediate. In order to determine $T > 0$ so that $p(t + T) = 2p(t)$, one needs to solve the equation $p_0 e^{a(t+T-t_0)} = 2p_0 e^{a(t-t_0)}$, and hence $e^{aT} = 2$. Equation (3) follows from this.

The growth coefficient a in (1) is one-hundredth of the instantaneous percent growth per unit time of the population. For example, a population growing at a 5 percent rate per unit time corresponds to a growth coefficient of $a = 0.05$ $(= \frac{1}{20})$. Since $20 \ln 2 \cong 13.86$, a 5 percent growth rate causes the population to double approximately every 13.86 units of time. Conversely, if a malthusian population doubles, say, every 10 units of time, then the growth coefficient is about 0.0693 $(\cong \frac{1}{10} \ln 2)$.

▶ EXAMPLE 1.8-1

Suppose that a certain population obeys a malthusian growth model, and from 1960 to 1970 this population increases from 20,000 to 30,000. Taking $t_0 = 1960$ and $t = 1970$, we obtain from (2) the equation

$$30{,}000 = p(1970) = 20{,}000e^{a(1970-1960)}$$

and hence $3 = 2e^{10a}$. Therefore, $e^{10a} = \frac{3}{2}$, $a = \frac{1}{10} \ln \frac{3}{2}$, and it follows that the population $p(t)$ at any time t satisfies

$$p(t) = 20{,}000e^{\ln(3/2)(t-1960)/10} = 20{,}000(\tfrac{3}{2})^{(t-1960)/10}$$

Note that the population in 1980 should be $p(1980) = 20{,}000(\frac{3}{2})^2 = 45{,}000$. Also, from formula (3), this population doubles every $T = (10 \ln 2)/\ln \frac{3}{2}) \cong 17.1$ years.

1.8a The Logistic Model

As long as the population has unconstrained growth (in particular, if there is sufficient food and space), the malthusian growth approximation is accurate. However, once a population becomes "crowded" or a food shortage develops, the exponential growth is no longer valid, and the growth rate of the population begins to decrease as the population becomes larger. In order to take this possibility into account, a second term is included in equation (1). This term should have only a small effect when the population is small and should decrease the growth rate as the population grows large. A simple example of this modification is the equation

$$p' = ap - \varepsilon p^2 \qquad p(t_0) = p_0 \tag{4}$$

where both a and ε are positive. Normally, ε is much smaller than a (and hence a/ε is "large") and $p_0 < a/\varepsilon$. Equation (4) is called the *logistic equation* (or the *Verhulst equation of population growth*), and its solutions (when $0 < p_0 < a/\varepsilon$) are called *logistic curves.*

Equation (4) is autonomous (i.e., the right-hand side is independent of time), so the variables separate and the techniques for sketching a graph of the solutions to (4) (i.e., a phase-time plot) developed in Section 1.5 also apply. In fact, a phase-time plot is easy to sketch. Setting the right-hand side of the differential equation in (4) to zero, we have

$$0 = ap - \varepsilon p^2 = p(a - \varepsilon p)$$

and hence $p = 0$ and $p = a/\varepsilon$ are the critical points of (4) [that is, if $p_0 = 0$ or $p_0 = a/\varepsilon$, then the solution to (4) is the constant function $p(t) \equiv p_0$]. Since the right-hand side is positive for $0 < p < a/\varepsilon$ and negative for $p > a/\varepsilon$, the population increases and approaches the critical point a/ε if $0 < p_0 < a/\varepsilon$, and the population decreases and approaches a/ε if $p_0 > a/\varepsilon$ [in order to have a meaningful interpretation, $p(t)$ must be nonnegative, and hence we are not concerned with the case $p_0 < 0$]. Therefore the critical point a/ε is asymptotically stable, and, in fact, the population approaches a/ε as time becomes infinite independent of the initial population $p_0 > 0$ (see Figure 1.13). The number a/ε is called the *limiting population* for this model. The derivative (with respect to p) of the right-hand side of (4) is $a - 2\varepsilon p$, and hence changes

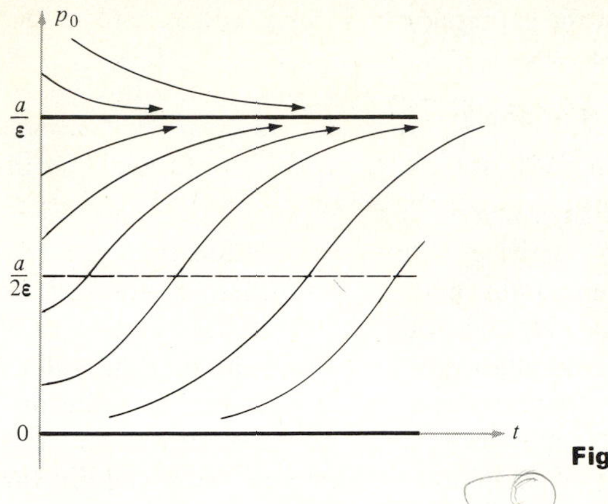

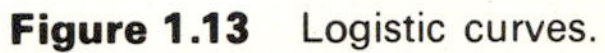

Figure 1.13 Logistic curves.

sign at $p = a/2\varepsilon$. Therefore $a/2\varepsilon$ is an inflection point for the solutions: the solution p to (4) is concave up so long as $0 < p(t) < a/2\varepsilon$ and concave down so long as $a/2\varepsilon < p(t) < a/\varepsilon$. A phase-time plot for (4) is given in Figure 1.13.

Although the phase-time plot for the logistic equation (4) gives a good description of the behavior of the solutions, it is a straightforward procedure to compute the solution to (4) explicitly. Therefore, we have the following result:

■ **Theorem 1.8-1**

Suppose that a, ε, $p_0 > 0$ and that p is the solution to the logistic equation (4). Then

$$p(t) = \frac{p_0 a}{\varepsilon p_0 + (a - \varepsilon p_0)e^{-a(t-t_0)}} \tag{5}$$

for all $t \geq t_0$.

The solution p to (4) can be computed using separation of variables; however, since (4) is also a Bernoulli equation, we use the substitution method indicated by (9) in Section 1.6*a*. Note that if $p = v^{-1}$ then v should be a solution to the linear equation $v' + av = \varepsilon$, $v(t_0) = p_0^{-1}$. Multiplying by the integrating factor $e^{a(t-t_0)}$, we obtain

$$\frac{d}{dt}\left[e^{a(t-t_0)}v(t)\right] = \varepsilon e^{a(t-t_0)}$$

and integrating each side of this equation from t_0 to t,

$$e^{a(t-t_0)}v(t) - p_0^{-1} = \frac{\varepsilon}{a}\left(e^{a(t-t_0)} - 1\right)$$

Setting $v(t) = p(t)^{-1}$ we obtain the expression

$$\frac{(a/\varepsilon)p(t)^{-1} - 1}{(a/\varepsilon)p(t_0)^{-1} - 1} = e^{-a(t-t_0)} \qquad \text{for all } t \geq t_0 \tag{6}$$

and formula (5) follows by solving for $p(t)$.

The limiting population a/ε is the maximum population that the environment of the population can maintain, and, as time goes on, the population approaches this limiting value. Note, however, that the rate of growth of the population increases until the population reaches half the limiting value, and then the growth rate begins to decrease [this is because the solution p is concave up so long as $0 < p(t) < a/2\varepsilon$ and concave down so long as $a/2\varepsilon < p(t) < a/\varepsilon$]. This type of growth behavior can be found in various populations (e.g., human populations in certain cities and countries and fish populations in lakes and streams), and therefore logistic curves can be effective in modeling (and hence predicting) population growth. However, great care must be taken in applying logistic curves to population growth, for these curves can be sensitive to any fluctuations in this pattern. In general terms, given data concerning a certain population (e.g., given the populations $p_0, p_1, \ldots, p_n$ at time $t_0, t_1, \ldots, t_n$, respectively), can one determine a logistic curve (i.e., constants a and ε) that gives a "good fit" for this data?

EXAMPLE 1.8-2

If there is plenty of space and food, a certain species of fish increases according to a malthusian growth model with growth coefficient $a = \frac{1}{10}$. Here the time t is considered in *months* and the fish population is measured in *pounds*. However, if this species is placed in a pond, the malthusian growth model is no longer valid and the species increases according to a logistic model, with the coefficient $a = \frac{1}{10}$ in (4) and the coefficient ε dependent on the size of the pond and availability of food. A farmer with a pond buys 1000 pounds of this species of fish and calculates that after 1 year in the pond there is 2000 pounds of fish in the pond. Since this species obeys a logistic model with $a = \frac{1}{10}$, $p_0 = 1000$ (the initial time is taken as $t_0 = 0$), and $p(12) = 2000$, from (6) we have

$$\frac{p(t)^{-1} - \varepsilon/a}{p(t_0)^{-1} - \varepsilon/a} = e^{-a(t-t_0)}$$

and hence, with $t = 12$ and $t_0 = 0$,

$$\frac{2^{-1}10^{-3} - 10\varepsilon}{10^{-3} - 10\varepsilon} = e^{-12/10}$$

Solving for ε we obtain

$$\varepsilon = \frac{2^{-1} - e^{-12/10}}{1 - e^{-12/10}} 10^{-4} \cong 2.84 \times 10^{-5}$$

and hence the limiting population of the fish in this pond is

$$\frac{a}{\varepsilon} = \frac{10^{-1}}{2.84 \times 10^{-5}} \cong 0.35 \times 10^4 = 3500 \text{ pounds}$$

Therefore, 3500 pounds of fish is the most the farmer could hope to sustain in this pond.

---

* 1.8*b* The Logistic Model with Constant Harvesting

The farmer with the fish pond in Example 1.8-2 would perhaps be interested in selling some fish from time to time in order to profit from the investment. Therefore, instead of analyzing the logistic equation directly, sometimes the rate of growth of the population must be altered in order to reflect the rate at which some of the population is removed. The removal of the population at a specified rate is called *harvesting*, and we are interested in modeling the simple situation when a population is harvested at a *constant rate* (e.g., the farmer in Example 1.8-2 might want to sell 100 pounds of fish per month—see Example 1.8-3 below). In general, it is supposed that a population $p(t)$ is described by a logistic growth equation and that it is subject to harvesting at a constant rate $h > 0$. This harvesting reflects the instantaneous rate of change $p'(t)$ by subtracting the constant h. Therefore, instead of obeying the logistic model (4), the population p is a solution to the initial value problem

$$p' = ap - \varepsilon p^2 - h \qquad p(t_0) = p_0 \tag{7}$$

Equation (7) is called the *logistic equation with constant harvesting*, and our concern is to investigate the behavior of the solution to this equation relative to the harvesting constant h.

Equation (7) is autonomous and the variables separate; however, it is difficult to obtain an explicit representation of the solution p. Since the right-hand side of (7) is quadratic in p, it is easy to sketch a phase-time plot. In order to determine the critical points of (7), set the right-hand side equal to zero to obtain the quadratic equation

$$\varepsilon p^2 - ap + h = 0$$

From the quadratic formula we immediately see that

(8)
$$p = \frac{a \pm \sqrt{a^2 - 4\varepsilon h}}{2\varepsilon} = \frac{a}{2\varepsilon} \pm \sqrt{\frac{a^2}{4\varepsilon^2} - \frac{h}{\varepsilon}}$$

are the critical points of (7) provided that $a^2 - 4\varepsilon h \geq 0$. There are no real critical points if $a^2 - 4\varepsilon h < 0$.

Since the derivative of the right side of (7) is $a - 2\varepsilon p$ (this is the same as in the logistic equation since the harvesting equation differs from it by a constant), we see that it changes sign at $p = a/2\varepsilon$, and hence solutions change concavity as they pass through this point. The behavior of the solution to (7) is dictated by the critical points, and from the computation in (8) it follows that the situation is divided naturally into three parts: there are two distinct critical points for (7) (that is, $a^2 - 4\varepsilon h > 0$); there is exactly one critical point for (7) (that is, $a^2 - 4\varepsilon h = 0$); and there are no critical points for (7), (that is, $a^2 - 4\varepsilon h < 0$). Therefore, the crucial value for h is when $a^2 - 4\varepsilon h = 0$, and so we define

(9)
$$h^* = \frac{a^2}{4\varepsilon}$$

Typical phase-time plots for (7) are given in Figure 1.14*a*, *b*, and *c*. These sketches are easy to obtain from the procedure indicated in Section 1.5. Note, for example, that if $0 < h < h^*$ there are positive critical points q_- and q_+ (the inflection value $a/2\varepsilon$ is exactly in the middle between these two critical points), and the right-hand side of (7) is positive for $q_- < p < q_+$ and negative for $p < q_-$ or $p > q_+$. This shows that Figure 1.14*a* is accurate. One can now easily obtain the following results:

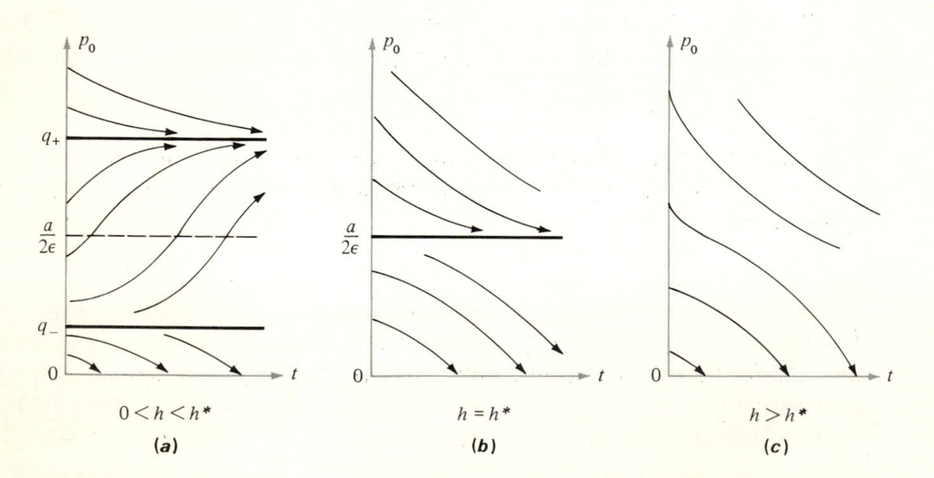

Figure 1.14 ***a, b, c*** Logistic curves with constant harvesting.

Theorem 1.8-2

Suppose that $h^* = a^2/4\varepsilon$.

(i) If $0 < h < h^*$, then (7) has two positive critical points

$$q_\pm = \frac{a}{2\varepsilon} \pm \sqrt{\frac{a^2}{4\varepsilon^2} - \frac{h}{\varepsilon}}$$

Moreover, if $p_0 > q_-$ then $p(t) \to q_+$ as $t \to \infty$, and if $p_0 < q_-$ then $p(t) = 0$ in some finite time t.

(ii) If $h = h^*$, then $p = a/2\varepsilon$ is the unique critical point to (7); if $p_0 \geq a/2\varepsilon$, then $p(t) \to a/2\varepsilon$ as $t \to \infty$, and if $p_0 < a/2\varepsilon$, then $p(t) = 0$ in some finite time t.

(iii) If $h > h^*$, then $p(t) = 0$ in some finite time t for every $p_0 > 0$.

Moreover, a phase-time plot for solutions to (7) has the appropriate form indicated in Figure 1.14*a*, *b*, and *c*.

The interpretation of the case in which $p(t) = 0$ in finite time is that the population dies out in finite time and the value $h^* = a^2/4\varepsilon$ is called the *maximum sustainable yield:* the maximum that one can harvest and not have the population die out in finite time. Observe that in all cases *any* constant harvesting (no matter how small) decreases the limiting population (if $0 < h < h^*$, the limiting population is $a/2\varepsilon + \sqrt{a^2/4\varepsilon^2 - h/\varepsilon}$, which is strictly less than a/ε). Also, if the initial population p_0 is not sufficiently large, the population may die out even if $h < h^*$.

As a simple illustration consider the logistic equation with constant harvesting h given by

$$p' = p - \tfrac{1}{4}p^2 - h \qquad p(0) = p_0 \tag{10}$$

Multiplying each side of the equation $p - \frac{1}{4}p^2 - h = 0$ by -4 gives

$$p^2 - 4p + 4h = 0$$

and hence by the quadratic formula the critical points of (10) are

$$p = \frac{4 \pm \sqrt{16 - 16h}}{2} = 2 \pm 2\sqrt{1 - h} \tag{11}$$

The equation has a double root for $h = 1$, and it follows from (11) that the maximum sustainable yield h^* is 1 for equation (10). Phase-time sketches for (10) when $h = 0, \frac{1}{4}$, and $\frac{3}{4}$ are given in Figure 1.15. Note that for $h = 0, \frac{1}{4}$, and $\frac{3}{4}$ the limiting populations are 4, $2 + \sqrt{3}$, and 3, respectively. Note further, for example, if $h = \frac{3}{4}$ and $p_0 < 1$, then the population will die out in finite time.

As another problem associated with equation (10), suppose that $p_0 = \frac{1}{2}$; determine all of the values of $h \geq 0$ such that the solution to (10) remains

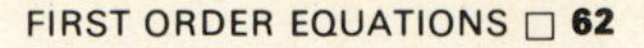

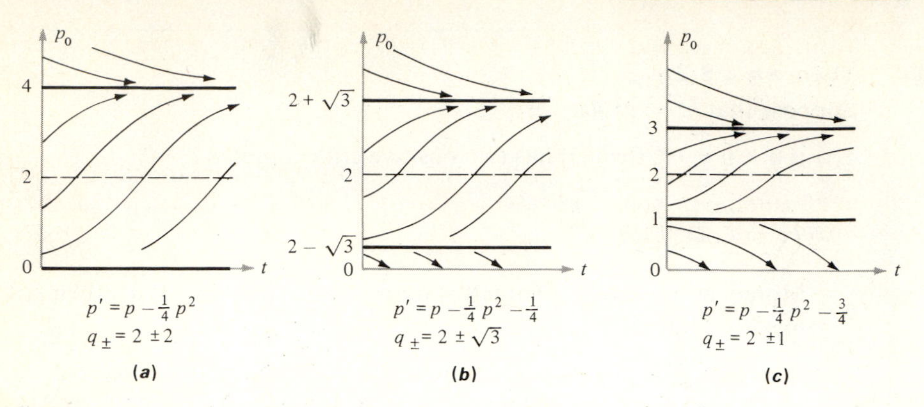

Figure 1.15 ***a, b, c*** Phase-time plots of solutions for (10) with $h = 0$, $\frac{1}{4}$, and $\frac{3}{4}$, respectively.

positive for all $t \geq 0$. It follows from the general behavior of the solutions in the phase-time plots for (10) (see Figures 1.14 and 1.15) that a solution p to (10) remains positive for all $t \geq 0$ only in case $p_0 \geq q_-$, where q_- is the smallest critical point of (10). Since $q_- = 2 - 2\sqrt{1-h}$ when $0 \leq h \leq 1$ [see (11)], we see that if $p_0 = \frac{1}{2}$ then $2 - 2\sqrt{1-h} \leq \frac{1}{2}$ only if $2\sqrt{1-h} \geq \frac{3}{2}$. Hence h must satisfy $4(1-h) \geq \frac{9}{4}$ or $h \leq \frac{7}{16}$. Therefore, if $0 \leq h \leq \frac{7}{16}$ and $p_0 = \frac{1}{2}$ (or, in fact, $p_0 \geq \frac{1}{2}$), then the solution to (10) remains positive for all $t \geq 0$.

▶ **EXAMPLE 1.8-3** (Example 1.8-2 continued)

Suppose that the farmer in Example 1.8-2 desires to sell some fish at a certain constant rate h pounds per month. Then the number $p(t)$ of pounds of fish should satisfy the equation

$$p' = 10^{-1}p - 2.84 \times 10^{-5}p^2 - h \qquad p(0) = 1000 \tag{12}$$

The farmer wants to select $h > 0$ as large as possible so that there will always be fish in the pond to sell [that is, $p(t) \geq 0$ for all $t \geq 0$]. Since the critical points of (12) are the solutions q to

$$2.84 \times 10^{-5}q^2 - 10^{-1}q + h = 0$$

or, upon multiplying by $(2.84 \times 10^{-5})^{-1} \cong 3.5 \times 10^4$,

$$q^2 - 3.5 \times 10^3 q + 3.5 \times 10^4 h = 0$$

By the quadratic formula

$$q = \frac{3.5 \times 10^3 \pm \sqrt{(3.5)^2 \times 10^6 - 4(3.5) \times 10^4 h}}{2}$$

$$= 1750 \pm 50\sqrt{1225 - 14h}$$

Therefore, since the smallest critical point should be larger than $p_0 = 1000$, h must satisfy the inequality

$$1750 - 50\sqrt{1225 - 14h} \leq 1000$$

Equivalently, $\sqrt{1225 - 14h} \geq 15$, and squaring each side, we obtain $h \leq \frac{1000}{14}$.

Thus the farmer can sell approximately 71.4 pounds of fish per month and not run out of fish.

PROBLEMS

1.8-1. Suppose that a population satisfies a malthusian growth model and that this population doubles every 34 years. How long will it take this population to triple in size? To increase fivefold?

1.8-2. Suppose that a certain population obeys a malthusian growth model and from 1975 to 1980 this population increases from 4.5×10^6 to 6×10^6. What will this population be in 1985? When will this population be 10^7?

1.8-3. Suppose that a malthusian population model has a growth coefficient $a > 0$. Suppose that at times t_0 and t_1 ($t_0 < t_1$) it is known that this population is p_0 and p_1, respectively ($p_0 < p_1$). Show that

$$a = (t_1 - t_0)^{-1} \ln \frac{p_1}{p_0}$$

and if $p(t)$ denotes this population at time $t \geq t_0$, show that

$$p(t) = p_0 \left(\frac{p_1}{p_0} \right)^{(t - t_0)/(t_1 - t_0)}$$

for all $t \geq 0$.

1.8-4. Consider a malthusian model with constant harvesting $h > 0$:

$$p' = ap - h \qquad p(t_0) = p_0$$

Investigate the behavior of the solution to this equation for values of h and p_0.

1.8-5. Show that the solution (5) to the logistic equation approaches the solution (2) to the malthusian equation as $\varepsilon \to 0+$.

1.8-6. Consider the following logistic equation with constant harvesting:

$$p' = 4 \times 10^{-2} p - 4 \times 10^{-8} p^2 - h \qquad p(0) = p_0$$

(a) Sketch a phase-time plot for this equation for $h = 0$ and $h = 9100$.

(b) Determine the maximum sustainable yield h^* and determine the solution explicitly when $h = h^*$. (*Hint:* If $h = h^*$, then the right-hand side is a perfect square.)

1.8-7. Consider the logistic equation with constant harvesting:

$$p' = p - 10^{-4}p^2 - h \qquad p(0) = p_0$$

(a) Sketch a phase-time plot for $h = 0$, $h = 900$, and $h = 2400$.

(b) Determine the largest rate $h = h_0$ so that if $p_0 \geq 2000$ the population will not die out. Also, if $h = h_0$ and $p_0 > 2000$, what is $\lim_{t \to \infty} p(t)$?

(c) Determine the maximum sustainable yield h^* and compute the solution explicitly with $h = h^*$.

1.8-8. If the farmer in Example 1.8-2 wishes to sell fish at a constant rate of 80 pounds per month and never run out of fish, how long must the farmer wait before beginning to sell the fish?

1.8-9. An equation that can also be used to model population growth has the form

$$p' = -kp \ln \frac{p}{K} \qquad p(t_0) = p_0 \qquad 0 < p_0 < K$$

where $k > 0$ is the growth coefficient and $K > 0$ is a constant. Show that the solution to this equation is

$$p(t) = K\left(\frac{p_0}{K}\right)^{e^{-k(t-t_0)}} \qquad \text{for all } t \geq t_0$$

Hence show that $0 < p(t) < K$ and $p(t) \to K$ as $t \to \infty$. Determine a constant c in $(0, 1)$ so that the solution p is concave upward if $p \leq cK$ and downward if $cK < p < K$. (This equation is known as the *Gompertz* equation.)

***1.8-10.** Suppose that $a, \varepsilon > 0$ and $\sigma > 1$. Consider the *generalized logistic equation*

$$p' = ap - \varepsilon p^{\sigma} \qquad p(t_0) = p_0 \tag{13}$$

(a) Sketch a phase-time plot for this equation.

(b) Show that (13) is a Bernoulli equation and that if the change of variables $v = p^{1-\sigma}$ is used, v is a solution to the linear equation

$$v' + (\sigma - 1)av = (\sigma - 1)\varepsilon$$

(c) Solve equation (13) explicitly for p in terms of t and p_0 if $a = 1$, $\varepsilon = \frac{1}{4}$, and $\sigma = 3$.

(d) Repeat part c for $a = 1$, $\varepsilon = \frac{1}{2}$, and $\sigma = \frac{3}{2}$.

*1.9 ELEMENTARY CHEMICAL REACTION EQUATIONS

Nonlinear first order differential equations play a fundamental role in analysis of chemical reaction kinetics, and some of the basic ideas are outlined in this section. Suppose that A and B denote two chemicals that react to form another chemical (or other chemicals) and that for each time t, $a(t)$ and $b(t)$ denote the concentration of the chemicals A and B, respectively. The units of

a and b are weight per unit of volume (e.g., pounds per gallon, grams per liter). It is supposed that each reaction involves a fixed number m of A molecules and a fixed number n of B molecules (the numbers m and n are called the *stoichiometric coefficients*), and this type of reaction is denoted

$$mA + nB \rightarrow \text{products}$$

The *law of mass action* asserts that the rate of the reaction is proportional to the product of the concentrations of the reactants, where each of the concentrations may be raised to some power. Therefore, there is a positive constant k and positive integers μ and ν such that

$$\text{Rate of the reaction} = ka^{\mu}b^{\nu} \tag{1}$$

The constant k is called the *rate constant* for the reaction, and the sum of the exponents $\mu + \nu$ is called the *order* of the reaction. The rate constant k is independent of the concentration of reactants, but it does *depend on temperature and pressure*. Therefore, in this analysis the temperature and pressure are assumed throughout to be constant. Since each reaction destroys m molecules of A and n molecules of B,

$$-\frac{1}{m}\frac{da}{dt} = -\frac{1}{n}\frac{db}{dt} = \text{rate of the reaction} = ka^{\mu}b^{\nu}$$

so if

$$y(t) \equiv \text{concentration of the products produced up to time } t \tag{2}$$

then

$$y' = -\frac{1}{m}a' = -\frac{1}{n}b'$$

and it follows that

$$y(t) = -\frac{1}{m}[a(t) - a_0] = -\frac{1}{n}[b(t) - b_0]$$

where a_0 is the initial concentration of A and b_0 the initial concentration of B (the initial time is taken to be $t_0 = 0$). Therefore, $a(t) = a_0 - my(t)$ and $b(t) = b_0 - ny(t)$ for all $t \geq 0$, and substituting this in (1) the following initial value problem for y is obtained:

$$y' = k(a_0 - my)^{\mu}(b_0 - ny)^{\nu} \qquad y(0) = 0,\ t \geq 0 \tag{3}$$

It is important to note that the interactions depend only on the specific reactions considered (and hence there are no secondary reactions affecting A and B) and also that the temperature and pressure are assumed to be constant. Equation (3) is autonomous, and its solution can be explicitly determined in many situations. However, even in the general case the following result of the behavior of y can be easily obtained from the results developed in Section 1.5 on phase-time plots.

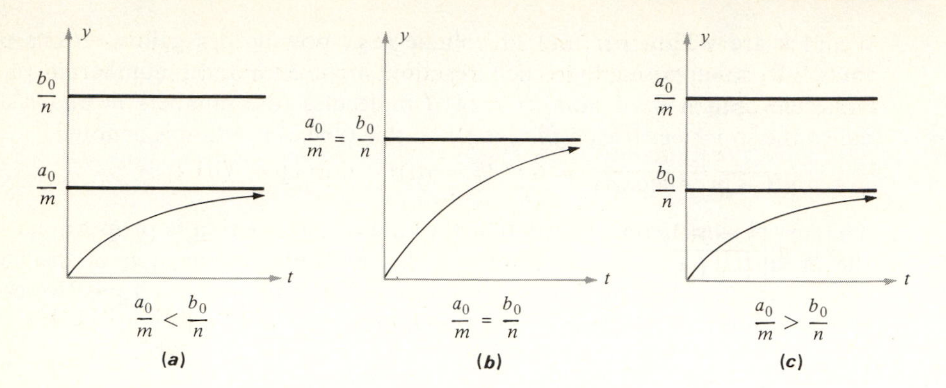

Figure 1.16 ***a, b, c*** Limiting behavior of the concentration.

Theorem 1.9-1

Suppose that y is the solution to (3). Then y exists and is bounded on $[0, \infty)$; $y(t)$ is increasing in t; and

$$\lim_{t\to\infty} y(t) = \text{minimum}\left\{\frac{a_0}{m}, \frac{b_0}{n}\right\}$$

Moreover, the rate $y'(t)$ of the reaction is decreasing in t [i.e., the graph of $y(t)$ is concave downward].

Since $y'(0) = ka_0^\mu b_0^\nu > 0$ and the right-hand side of (3) is zero only if $y = a_0/m$ or $y = b_0/n$, the first part of this theorem follows from the method of phase-time plots shown in Section 1.5. The fact that y' is decreasing also follows from the ideas in Section 1.5. The derivative with respect to y of the right-hand side of (3) is

$$-k[\mu m(b_0 - ny) + \nu n(a_0 - my)][(a_0 - my)^{\mu-1}(b_0 - ny)^{\nu-1}]$$

which remains negative since $y(t) < a_0/m$ and $y(t) < b_0/n$. Since $y'(t) > 0$ it follows that $y''(t) < 0$ for all $t \geq 0$, and hence $y'(t)$ is decreasing. A sketch of the three possible situations is given in Figure 1.16.

EXAMPLE 1.9-1

Consider a second order reaction $A + B \rightarrow$ products, where the initial concentrations are $a_0 = \frac{1}{2}$ and $b_0 = \frac{1}{3}$. Since $m = n = \mu = \nu = 1$ (the reaction is second order so $\mu + \nu = 2$, and hence $\mu = \nu = 1$), equation (3) becomes

$$y' = k(\tfrac{1}{2} - y)(\tfrac{1}{3} - y) \qquad y(0) = 0,\ t \geq 0$$

We know from Theorem 1.9-1 that $y(t) \to \frac{1}{3}$ as $t \to \infty$. This equation is simple enough to solve explicitly for y. Separating variables and using partial fractions, one obtains

$$kt = \int \frac{y'(t)\,dt}{[\frac{1}{2} - y(t)][\frac{1}{3} - y(t)]} = 6 \ln |\tfrac{1}{2} - y(t)| - 6 \ln |\tfrac{1}{3} - y(t)| + c$$
$$= 6 \ln \left| \frac{\frac{1}{2} - y(t)}{\frac{1}{3} - y(t)} \right| + c$$

Using the initial value $y(0) = 0$ and the fact that $0 < y(t) < \frac{1}{3}$, it follows that

$$\frac{3 - 6y(t)}{2 - 6y(t)} = \frac{3}{2} e^{kt/6} \qquad \text{for all } t \geq 0$$

and hence that

$$y(t) = \frac{e^{kt/6} - 1}{3e^{kt/6} - 2} \qquad \text{for all } t \geq 0$$

Given two chemicals A and B that react according to equation (3), an important problem is to determine the order and the rate constant for this reaction. These computations must be done experimentally. Suppose that the stoichiometric coefficients m and n are known. If almost all the reaction is complete in a reasonable length of time—say, after time t_1 at least 99.9 percent of the reaction is complete—then one can approximate m (or n) by setting up the initial concentrations so that b_0 (or a_0) is large enough that $a_0/m > b_0/n$ (or $b_0/n > a_0/m$). Thus, by Theorem 1.9-1, $m \cong a_0/y(t_1)$ [or $n \cong b_0/y(t_1)$]. Setting up the initial concentrations so that $a_0/m = b_0/n$ and defining

$$u(t) \equiv \frac{a_0}{m} - y(t) \qquad \text{for all } t \geq 0$$

one obtains from equation (3)

$$u'(t) = -y'(t) = -k[a_0 - my(t)]^{\mu}[b_0 - ny(t)]^{\nu}$$
$$= -km^{\mu}n^{\nu}\left[\frac{a_0}{m} - y(t)\right]^{\mu}\left[\frac{b_0}{n} - y(t)\right]^{\nu}$$
$$= -km^{\mu}n^{\nu}u(t)^{\mu+\nu}$$

for all $t \geq 0$. Since $u(0) = a_0/m$ and $\mu + \nu > 1$, we have (by separating variables)

$$u(t)^{1-\mu-\nu} = (\mu + \nu - 1)m^{\mu}n^{\nu}kt + \left(\frac{a_0}{m}\right)^{1-\mu-\nu} \tag{4}$$

Equation (4) can be used to determine the order $\mu + \nu$. Also k can be determined if either $m = n$ or $\mu + \nu = 2$. The procedure is to take readings to determine the concentration of the product $y(t)$ at certain times, say, $0 < t_1 < t_2 < \cdots < t_p$. Therefore, the quantities $u(t_1), u(t_2), \ldots, u(t_p)$ are computed. From (4) it follows that the points $(t_i, u(t_i)^{1-\mu-\nu})$, $i = 1, \ldots, p$, lie on a straight line in the tu plane. Therefore, the order of this reaction is precisely the integer σ, $\sigma \geq 2$, such that the points $(t_i, u(t_i)^{1-\sigma})$ lie on a straight line in the tu plane. Noting that the slope of this line is $(\mu + \nu - 1)km^{\mu}n^{\nu}$ and that m, n, and $\mu + \nu$ are known, one can determine k whenever $m = n$ or whenever $\mu + \nu = 2$ (and so $\mu = \nu = 1$).

One can also use the concept of half-life to help determine the order and the rate constant of a reaction. The *half-life* of the chemical reaction described by equation (3) is the time T that it takes for one-half of the product to be produced. As opposed to the case of radioactivity discussed in Section 1.3 (note that radioactive decay can be regarded as a chemical reaction of order 1), the half-life in this case *depends on the initial concentrations.* Assuming as in the previous paragraph that $a_0/m = b_0/n$ and that $T(a_0)$ is the half-life of the reaction, then $y(T(a_0)) = a_0/2m$, and since $u(t) = a_0/m - y(t)$, we have from (4) that

$$\left(\frac{a_0}{2m}\right)^{1-\mu-\nu} = (\mu + \nu - 1)m^{\mu}n^{\nu}kT(a_0) + \left(\frac{a_0}{m}\right)^{1-\mu-\nu}$$

and hence that

$$T(a_0) = \frac{2^{\mu+\nu-1} - 1}{m^{\mu}n^{\nu}(\mu + \nu + 1)k}\left(\frac{a_0}{m}\right)^{1-\mu-\nu} \tag{5}$$

Letting c denote the coefficient of $(a_0/m)^{1-\mu-\nu}$ in the right side of (5) and taking the logarithm of each side of this equation, we obtain

$$\ln T(a_0) = \ln c + (1 - \mu - \nu) \ln \frac{a_0}{m}$$

Therefore, the graph of the logarithm of the half-life against the logarithm of the initial concentration should be a straight line (recall that it is assumed that $a_0/m = b_0/n$). Since the slope of this line is $1 - \mu - \nu$, the order $\mu + \nu$ can be determined immediately from this graph (if the slope of this line is $-1, -2, \ldots$, then the order of the reaction is $2, 3, \ldots$, respectively). Also, k can be determined from (5) if either $m = n$ or $\mu + \nu = 2$.

PROBLEMS

1.9-1. Consider the second order reaction $A + B \rightarrow$ products, with initial concentrations a_0 and b_0. According to Theorem 1.9-1, if $y(t)$ is the concentration of the products produced up to time t, then

$$y' = k(a_0 - y)(b_0 - y) \qquad y(0) = 0$$

(a) Determine the solution $y(t)$ to this equation *explicitly* in terms of t, a_0, b_0, and k in the cases **(i)** $a_0 \neq b_0$, and **(ii)** $a_0 = b_0$.

(b) Show that if $y_1 = y(t_1)$ is known for some $t_1 > 0$, then the rate constant k can be computed in terms of t_1, y_1, a_0, and b_0.

1.9-2. Consider the second order reaction $A + B \rightarrow$ products, with initial concentrations $a_0 = 3$ and $b_0 = 2$.

(a) When one-half the chemical A has reacted, what percent of B has reacted?

(b) When one-half the chemical B has reacted, what percent of A has reacted?

*1.10 MISCELLANEOUS APPLICATIONS

1.10*a* Gravitational Attraction

In Section 1.3 the application of Newton's second law to the vertical motion of an object was made under the implicit assumption that the force on the object due to gravity was constant. If the object remains near the earth, this assumption presents no problems and reasonably accurate results are obtained. However, if the object travels a large distance from the earth's surface, this assumption is no longer appropriate and it is necessary to use *Newton's law of gravitation: Two objects are attracted toward each other along the line joining them with a force directly proportional to the product of their masses and inversely proportional to the square of the distance between them.* The distance between the objects is considered to be the distance between their centers of mass. This law is applied to study the vertical motion of an object either toward or away from the earth's surface.

It is assumed that the earth is a perfect sphere of radius R and mass M. Also, an object of mass m is assumed to move in a vertical line emanating from the earth's surface (see Figure 1.17). The position of this object at time t is denoted $y(t)$, where $y(t)$ is the vertical distance of the object from the earth's surface. The velocity of this object at time t is denoted by $v(t)$ [that is, $v(t) = y'(t)$]. Note that positive velocity indicates upward movement and negative velocity indicates downward movement. If F denotes the force of attraction between this object and the earth, Newton's law of gravitation asserts that

$$F = -\frac{GMm}{(R+y)^2}$$

whenever the mass m is at position $y \geq 0$. Since $F = mg$ (where g is the gravitational constant) when $y = 0$, it follows that $mg = GMm/R^2$, and hence the proportionality constant $G = R^2g/M$. Therefore,

$$F = -\frac{R^2mg}{(R+y)^2} \tag{1}$$

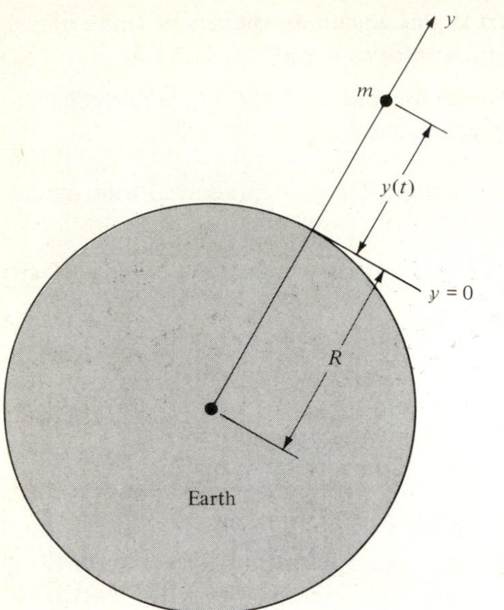

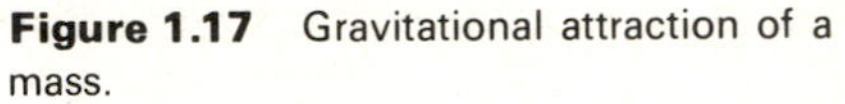

Figure 1.17 Gravitational attraction of a mass.

is the force of attraction when the object is a distance y above the earth's surface. By Newton's second law [see equation (7) in Section 1.3] we have $F = my''$, and hence, neglecting all forces except the gravitational interaction, it follows from (1) that

$$y''(t) = -\frac{R^2 g}{[R + y(t)]^2} \qquad \text{for all times } t \tag{2}$$

Note that if $y(t)$ remains much smaller than R so that $[R + y(t)]^2 \cong R^2$ is a good approximation, then equation (2) has the form $y'' = g$. Setting $v = y'$, we obtain the first order, nonhomogeneous linear equation $v' = g$. However, setting $v = y'$ in (2), the equation

$$v' = -\frac{R^2 g}{(R + y)^2} \tag{3}$$

is obtained, and this is not a first order ordinary differential equation with v considered as a function of time t. However, if the velocity v is considered as a function of the position y and the chain rule for differentiation is used,

$$v' = \frac{dv}{dt} = \frac{dv}{dy}\frac{dy}{dt} = v\frac{dv}{dy}$$

and it follows from (3) that

$$v\frac{dv}{dy} = -\frac{R^2 g}{(R + y)^2} \tag{4}$$

With y as the independent variable, equation (4) is a first order ordinary differential equation with unknown function v, and, in fact, the variables are separated (see Section 1.4). Integrating each side of (4) from the initial position y_0 to the position y, we obtain

$$\frac{1}{2}v^2 - \frac{1}{2}v_0^2 = \frac{R^2g}{R+y} - \frac{R^2g}{R+y_0}$$

where v_0 is the initial velocity. Therefore, considering the velocity and position as functions of time again,

$$v(t)^2 = v_0^2 + 2R^2g\left[\frac{1}{R+y(t)} - \frac{1}{R+y_0}\right] \tag{5}$$

for all time $t \geq 0$ such that $y(t) \geq 0$.

Formula (5) can be used directly to obtain information about the behavior of the object. Suppose, for example, that the object is initially at the earth's surface (that is, $y_0 = 0$) and is fired upward with initial velocity $v_0 > 0$. How large must v_0 be in order that the object move away from the earth forever? In particular, is it possible for v_0 to be large enough so that the object will "escape" the earth's gravitational field? An answer to this question can be obtained easily from formula (5). Since $v(0) = v_0 > 0$ and v is continuous, the object will always be moving away from the earth if v_0 is such that $v(t) \neq 0$ for all $t \geq 0$. Setting $y_0 = 0$ in (5), a number $v_0 > 0$ should be obtained so

$$v(t)^2 = v_0^2 + 2R^2g\left[\frac{1}{R+y(t)} - \frac{1}{R}\right] > 0$$

for all $t \geq 0$. This leads easily to the condition that

$$v_0^2 > 2R^2g\left(\frac{1}{R} - \frac{1}{R+y}\right)$$

for all $y \geq 0$, and hence, letting $y \to +\infty$, $v_0^2 \geq 2Rg$. Therefore, we have the following result:

$$\{\text{If } v_0 \geq \sqrt{2Rg} \text{ and } y_0 = 0, \text{ the object leaves the earth and never returns.} \tag{6}$$

The number $\sqrt{2Rg}$ is called the *escape velocity* and is approximately 7 mi/sec (note, however, that air resistance and the effects of the sun, moon, and other planets have been neglected). Using the fact that the object reaches its maximum height exactly when the velocity is zero, formula (5) can also be used effectively to determine how high the object rises before returning to earth, when $v_0 < \sqrt{2Rg}$ (see Problem 1.10-3).

Not only can formula (5) be used directly to obtain information about the behavior of the object, but, since $v(t) = y'(t)$, it is also a first order initial value problem (with dependent variable y and independent variable t) and has the form

$$y'(t) = \pm\sqrt{v_0^2 - \frac{2R^2g}{R + y_0} + \frac{2R^2g}{R + y}} \qquad y(0) = y_0 \tag{7}$$

where the plus sign is used while the object is rising and the minus sign while the object is falling. Notice that (7) is an autonomous equation (and hence the variables are separable). The antiderivatives involved in the computation of the solution seem complicated in general; however, for certain values of v_0 and y_0 these computations are routine (see Problems 1.10-2 and 1.10-4).

1.10*b* Elementary Models of Supply and Demand

The final application discussed in this section is from elementary theory of economics. Suppose that a firm is going to sell a product in a marketplace and wishes to develop a model indicating the behavior of the price of this product. This product is introduced at a time t_0 for a price $P_0 > 0$, and for each $t \geq t_0$ the price of this product is denoted by $P(t)$. The following basic assumption is made concerning the instantaneous rate of change of the price P:

(8) The rate of instantaneous change $P'(t)$ of the price $P(t)$ is directly proportional to the difference in the demand D and the supply S of this product for each time $t \geq t_0$.

Assuming further that the demand and the supply depend on the price as well as the time $t \geq t_0$—say, $D = D(t, P)$ and $S = S(t, P)$—it follows that the price P is a solution to the initial value problem

$$P'(t) = k[D(t, P) - S(t, P)] \qquad P(0) = P_0 \tag{9}$$

where k is a positive constant. Since demand decreases and supply increases when the price increases, it is also supposed that

$$\frac{\partial}{\partial P} D(t, P) \leq 0 \qquad \text{and} \qquad \frac{\partial}{\partial P} S(t, P) \geq 0 \qquad \text{for } t \geq t_0 \text{ and } P > 0 \tag{10}$$

Frequently, the demand and supply are independent of time [and hence equation (9) is autonomous]. Another case that occurs frequently is periodic behavior in time (in either the demand function, the supply function, or both).

In order to indicate the type of behavior that can occur in equation (9) consider the case where D and S are *independent of t and depend linearly on P*: there are positive numbers α, β, γ, and δ such that

$$D(P) = \alpha - \beta P \qquad \text{and} \qquad S(P) = \gamma + \delta P \qquad \text{for all } P \geq 0 \tag{11}$$

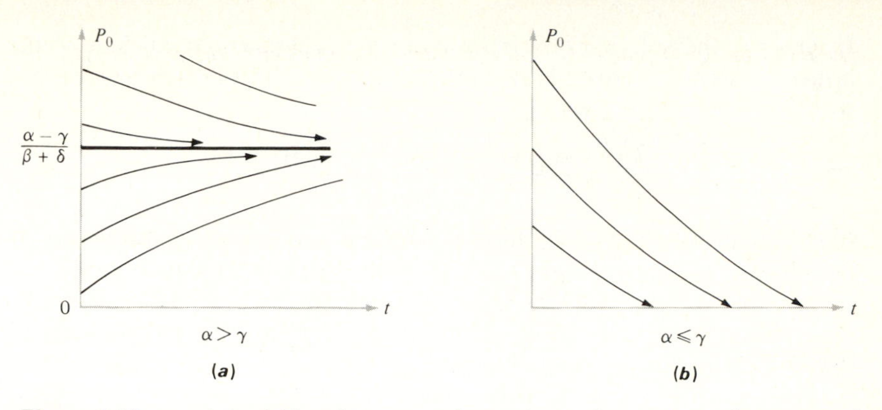

Figure 1.18 ***a*** $\alpha > \gamma$, stable price support for product. ***b*** $\alpha > \gamma$, no price support for product.

Note that $D'(P) = -\beta < 0$ and $S'(P) = \delta > 0$ so that condition (10) is fulfilled. Under supposition (11) the initial value problem (9) has the form

$$P' = -k(\beta + \delta)P + k(\alpha - \gamma) \qquad P(t_0) = P_0 \tag{12}$$

Using the method for linear equations in Section 1.2, the solution to equation (12) is easily seen to be

$$P(t) = \left(P_0 - \frac{\alpha - \gamma}{\beta + \delta}\right)e^{-k(\beta+\delta)(t-t_0)} + \frac{\alpha - \gamma}{\beta + \delta} \qquad \text{for } t \geq t_0 \tag{13}$$

Using (13) or applying the techniques for phase-time plots in Section 1.5 directly to (12) shows that the solutions to (12) separate essentially into two cases: $\alpha > \gamma$ and $\alpha \leq \gamma$ (see Figure 1.18*a* and *b*). If $\alpha > \gamma$ there is always a good demand for the product whenever the price is sufficiently low, and in this case the price tends toward the equilibrium value $(\alpha - \gamma)/(\beta + \delta)$ as $t \to +\infty$.

In the case $\alpha \leq \gamma$ there is not enough demand for the product (relative to supply) even if the price is extremely low, and so there is no price support in this case.

As an illustration of the effect of time-periodic behavior, suppose that there are positive numbers μ and ω such that

$$D(P) = \alpha - \beta P \qquad \text{and} \qquad S(t, P) = \gamma + \delta P + \mu \sin \omega t \tag{14}$$

In this case the initial value problem (9) has the form

$$P' = -k(\beta + \delta)P + k(\alpha - \gamma) - k\mu \sin \omega t \qquad P(t_0) = P_0 \tag{15}$$

and it follows that

$$P(t) = \left[P_0 - \frac{\alpha - \gamma}{\beta + \delta} - \frac{k\delta\omega}{\omega^2 + k^2(\beta + \delta)^2}\right]e^{-k(\beta+\delta)(t-t_0)} + \frac{\alpha - \gamma}{\beta + \delta} + \mu\frac{k\delta\omega \cos[\omega(t - t_0)] - k^2(\beta + \delta)\sin[\omega(t - t_0)]}{\omega^2 + k^2(\beta + \delta)^2} \tag{16}$$

In general, the solution (16) is not periodic (unless the coefficient of the exponential term is zero). However, as $t \to +\infty$, the solution approaches the so-called *steady-state term*

$$(17) \quad P^*(t) = \frac{\alpha - \gamma}{\beta + \delta} + \mu \frac{k\delta\omega \cos [\omega(t - t_0)] - k^2(\beta + \delta) \sin [\omega(t - t_0)]}{\omega^2 + k^2(\beta + \delta)^2}$$

Notice that the steady-state term is periodic and has the same period as the time periodicity in the supply (see also Problem 1.10-7).

PROBLEMS

Problems 1.10-1 to 1.10-4 refer to the vertical motion of an object of mass m above the earth's surface. Only the attraction of the earth is taken into account and the notations are the same as in equations (5) and (7).

1.10-1. For each initial height $y_0 \geq 0$ compute the escape velocity $ev(y_0)$ of the object from the height y_0: if $y(0) = y_0$ and $v(0) > ev(y_0)$, then $v(t) > 0$ for all $t \geq 0$ and $y(t) \to +\infty$ as $t \to +\infty$ [recall that $ev(0) = \sqrt{2Rg}$—see (13)]. For what values of y_0 is $ev(y_0) = \frac{1}{2}\sqrt{2Rg}$? $\frac{1}{4}\sqrt{2Rg}$? $\frac{1}{10}\sqrt{2Rg}$?

1.10-2. Suppose that $y_0 = 0$ and $v_0 = \sqrt{2Rg}$, the escape velocity. Show that the velocity v as a function of y satisfies

$$(*) \qquad v(y) = \sqrt{\frac{2R^2g}{R + y}}$$

and hence the position $y = y(t)$ is a solution to the initial value problem

$$(**) \qquad y' = \sqrt{\frac{2R^2g}{R + y}} \qquad y(0) = 0$$

(where the derivative is with respect to time t).

(a) Discuss the relationship between $v(y)$ in (*) and the escape velocity $ev(y)$ for initial height $y \geq 0$ studied in Problem 1.10-1.

(b) Determine the position y explicitly in terms of time t.

(c) How long does it take the object to travel to a height R? $2R$? $3R$?

1.10-3. Suppose that at time $t = 0$, $y(0) = 0$ and $v(0) = v_0$, where $0 < v_0 < \sqrt{2Rg}$. Show that the object rises to a maximum height $H_{max}(v_0)$ before returning to earth and that

$$H_{max}(v_0) = \frac{v_0^2 R}{2gR - v_0^2}$$

(a) Compare values of $H_{max}(v_0)$ when v_0 is $\frac{1}{2}\sqrt{2Rg}$, $\frac{1}{4}\sqrt{2Rg}$, and $\frac{1}{10}\sqrt{2Rg}$.

(b) What happens to $H_{max}(v_0)$ as $R \to +\infty$? Consider your answer relative to Problem 1.3-12.

(c) Suppose that $v_0 = \alpha\sqrt{2Rg}$, where $0 < \alpha < 1$. Using equation (7) determine $y(t)$ implicitly as a function of time t and show that the time T_{max} that it takes the object to reach its maximum height is

$$T_{max} = \sqrt{\frac{R}{2g(1-\alpha)^3}}\,(\arcsin \alpha + \alpha\sqrt{1-\alpha}\sqrt{2Rg})$$

***1.10-4.** Suppose that the object is initially at a height $y_0 > 0$ and that it is released with initial velocity $v_0 = 0$.

(a) Determine the velocity of the object as it strikes the earth's surface and discuss the behavior of this velocity as $y_0 \to +\infty$.

(b) Determine the position $y(t)$ implicitly as a function of t.

(c) What is the time required for the object to fall to the earth?

1.10-5. Instead of assuming the linear dependence of the demand and supply on the price P [see (11)], suppose that the demand D and the supply S have the form

$$D(P) = \alpha - \beta P - \varepsilon P^2 \qquad \text{and} \qquad S(P) = \gamma + \delta P$$

where α, β, γ, δ, and ε are positive numbers. Discuss the possible behavior of the solution to the initial value problem (9) in this case. (Do not try to determine P explicitly—use the method of phase-time plots discussed in Section 1.5.)

1.10-6. Discuss the behavior of the solution to the initial value problem (9) when the demand is constant and the supply grows quadratically with price: $D(P) \equiv \alpha$ and $S(P) \equiv \beta + \gamma P^2$, where α, β, and γ are positive constants.

***1.10-7.** Write the steady-state price $P^*(t)$ given in (17) in the form

$$P^*(t) = \frac{\alpha - \gamma}{\rho + \delta} + A \sin\,[\omega(t - t_0) + \theta]$$

where A and θ are constants. Determine A and θ in terms of the constants β, δ, k, μ, and ω. Set $\delta = 0$ and discuss the relationship between the times when the supply $S(t)$ is a minimum and the steady-state price $P^*(t)$ is a maximum.

Remarks The principal techniques for solving first order equations studied in this chapter are those for linear equations (Section 1.2), separable equations (Section 1.4), and exact equations (Section 1.7). More advanced texts that give detailed proofs and discussions of the validity of these methods as well as a proof of the fundamental existence theorem for solutions (Theorem 1.4-2) are Birkhoff and Rota [1], Ince [8], Kamke [9], and Petrovski [12].† Further examples of applications for such equations may be found in Wylie [15].

† The numbers in the brackets refer to listings in the bibliography.

2 Second Order Linear Equations

Ordinary differential equations where the second derivative of the unknown function appears often occur in problems arising in the physical and engineering sciences, and the main purpose of this chapter is to develop some basic methods for solving linear equations of the second order. In Section 2.2 general principles connected with these equations are discussed, and special techniques for the computation of solutions are studied in Sections 2.3, 2.4, 2.5, and 2.7. The most fundamental and important technique is that for linear equations with constant coefficients in Section 2.3. Basic procedures for solving nonhomogeneous equations are studied in Sections 2.4 and 2.5. The application of these equations to the motion of a mass-spring system is indicated in Section 2.6, and Section 2.8 includes further applications.

2.1 BASIC CONCEPTS AND EXAMPLES

This section discusses several fundamental ideas that involve *second order real ordinary differential equations:* the unknown is a real-valued function of a single real variable, and only the first and second derivatives can appear in the equation. In the general case, such equations have the form

$$\Phi(t, y, y', y'') = 0$$

where Φ is a function of four real variables. However, only those equations where the second derivative y'' can be explicitly written in terms of t, y, and y' are considered. Therefore, it is assumed that f is a continuous function of three variables, and the differential equation

$$y'' = f(t, y, y') \tag{1}$$

is considered. A twice differentiable function y on an interval I is said to be a *solution* to (1) on I if $y''(t) = f(t, y(t), y'(t))$ for all t in I [in particular, $(t, y(t), y'(t))$ must be in the domain of f for each $t \in I$]. The fundamental problem considered in this chapter is to develop techniques that lead to the determination of solutions to (1) for certain types of functions f.

Consider the simple equation $y'' = 0$. If I is any interval, the only functions on I that have a zero second derivative are those whose graphs are straight lines. Therefore, y is a solution to $y'' = 0$ on I if and only if $y(t) = c_1 t + c_2$, where c_1 and c_2 are any constants. The equation

$$y'' = e^{2t}$$

is also easy to solve. Since y' is an antiderivative of y'', we can take the antiderivative of each side of the equation to obtain

$$y' = \tfrac{1}{2}e^{2t} + c_1$$

where c_1 is any constant. Now taking antiderivatives of each side of this equation shows that the solutions are

$$y = \tfrac{1}{4}e^{2t} + c_1 t + c_2$$

where c_1 and c_2 are constants.

EXAMPLE 2.1-1

Consider the equation

$$y'' = \frac{ty' + 3y + 6}{t^2} \qquad t > 0$$

Let c_1 and c_2 be constants and define

$$y(t) = c_1 \frac{1}{t} + c_2 t^3 - 2 \qquad t > 0$$

Then

$$y'(t) = -c_1 \frac{1}{t^2} + 3c_2 t^2 \qquad \text{and} \qquad y''(t) = 2c_1 \frac{1}{t^3} + 6c_2 t$$

Substituting for y and y' in the right-hand side of the equation, we obtain

$$\frac{ty' + 3y + 6}{t^2} = \frac{t(-c_1 t^{-2} + 3c_2 t^2) + 3(c_1 t^{-1} + c_2 t^3 - 2) + 6}{t^2}$$

$$= \frac{-c_1 t^{-1} + 3c_2 t^3 + 3c_1 t^{-1} + 3c_2 t^3}{t^2}$$

$$= 2c_1 t^{-3} + 6c_2 t$$

Since the final expression is precisely $y''(t)$, this function y is a solution to the given differential equation by the definition of solution.

EXAMPLE 2.1-2

Consider the second order equation $y'' = y' + 2y - 2t$, with $I = (-\infty, \infty)$. For each real constants c_1 and c_2, the function

$$y(t) = c_1 e^{2t} + c_2 e^{-t} + t - \tfrac{1}{2}$$

is a solution to this equation since

$$\begin{aligned} y'(t) + 2y(t) - 2t &= 2c_1 e^{2t} - c_2 e^{-t} + 1 + 2(c_1 e^{2t} + c_2 e^{-t} + t - \tfrac{1}{2}) - 2t \\ &= 4c_1 e^{2t} + c_2 e^{-t} \\ &= y''(t) \end{aligned}$$

As is the case for first order equations, the simplest case for second order equations is when the right-hand side depends only on t. Therefore, assume that g is a continuous real-valued function on an interval I and consider the equation

$$y'' = g(t) \tag{2}$$

The solutions to this equation are easily described and are precisely the functions y on I having the form

$$y(t) = c_1(t - t_0) + c_2 + \int_{t_0}^{t} \left[\int_{t_0}^{s} g(r)\, dr \right] ds \qquad \text{for } t \in I \tag{3}$$

where c_1 and c_2 are any constants. To see that this representation is true, use $y'' = (y')'$ to show that y' is the solution to a first order equation of the form (2) in Section 1.1. Therefore, y' has the form

$$y'(t) = c_1 + \int_{t_0}^{t} g(r)\, dr \qquad \text{for all } t \in I$$

Applying the same method again [with $g(t)$ replaced by $c_1 + \int_{t_0}^{t} g(r)\, dr$], it follows that $y(t)$ must have the form indicated by (3). It should be noted that the family of functions described by (3) can also be written

$$y(t) = \bar{c}_1 t + \bar{c}_2 + G(t) \qquad \text{for } t \in I \tag{3'}$$

where $\bar{c}_1$ and $\bar{c}_2$ are any constants and G is *any* twice differentiable function on I such that $G''(t) = g(t)$ for $t \in I$. The family of functions described by (3) [or (3)′] is called the *general solution of* (2) *on* I.

▶ **EXAMPLE 2.1-3**

Consider the equation $y'' = 2t - \sin 2t$ on $(-\infty, \infty)$. Since a second antiderivative of $2t - \sin 2t$ is $\frac{1}{3}t^3 + \frac{1}{4}\sin 2t$, the general solution to this equation is

$$y(t) = c_1 t + c_2 + \tfrac{1}{3}t^3 + \tfrac{1}{4}\sin 2t$$

where c_1 and c_2 are constants.

The solution set (3) indicates that there may be many different solutions to a second order ordinary differential equation. As in the first order case, one is interested in determining a function y on I that is not only a solution to (1) but also satisfies some additional property or condition. Simply assigning a value for y at a given value t_0 in I is not sufficient to determine y in the simplest case [using (3), if $y(t_0)$ is required to be y_0, then $c_2 = y_0$, but c_1 is still an arbitrary constant]. There are many ways of prescribing side conditions on the function y in order to further limit the functions under consideration (see, for example, Problem 2.1-3). In this chapter, however, we are concerned almost entirely with *second order initial value problems:* given t_0 in I and numbers y_0 and v_0 such that (t_0, y_0, v_0) is in the domain of f, the equation

$$y'' = f(t, y, y') \qquad y(t_0) = y_0 \qquad y'(t_0) = v_0 \tag{4}$$

is called an initial value problem. A function y on an interval I is said to be a *solution* to the initial value problem (4) if $t_0 \in I$, if $y''(t) = f(t, y(t), y'(t))$ for $t \in I$, and if $y(t_0) = y_0$ and $y'(t_0) = v_0$. Therefore, y is a solution to (4) on I if, in addition to being a solution to (1), y also satisfies the conditions $y(t_0) = y_0$ and $y'(t_0) = v_0$.

▶ **EXAMPLE 2.1-4**

Consider the equation $y'' = t$, $y(0) = 1$, $y'(0) = 3$, with $I = (-\infty, \infty)$. The general solution to the differential equation is

$$y(t) = c_1 t + c_2 + \tfrac{1}{6}t^3$$

Since $y(0) = c_2$ and $y'(0) = c_1$, the condition $y(0) = 1$ implies $c_2 = 1$, and the condition $y'(0) = 3$ implies $c_1 = 3$. Therefore, $y(t) = 3t + 1 + \frac{1}{6}t^3$ is the solution to this problem.

▶ **EXAMPLE 2.1-5**

Consider the equation $y'' = y' + 2y - 2t$, $y(0) = -3$, $y'(0) = 5$, with $I = (-\infty, \infty)$. From Example 2.1-2 we know that

$$y(t) = c_1 e^{2t} + c_2 e^{-t} + t - \tfrac{1}{2}$$

is a solution to the differential equation for any constants c_1 and c_2. Since $y(0) = c_1 + c_2 - \frac{1}{2}$ and $y'(0) = 2c_1 - c_2 + 1$, the initial conditions are satisfied only if c_1 and c_2 satisfy the simultaneous equations

$$c_1 + c_2 = -\tfrac{5}{2}$$
$$2c_1 - c_2 = 4$$

Therefore, $c_1 = \frac{1}{2}$ and $c_2 = -3$, and a solution to this equation is

$$y(t) = \tfrac{1}{2}e^{2t} - 3e^{-t} + t - \tfrac{1}{2}$$

2.1a Special Equations Reducible to First Order

Certain types of second order equations can be reduced to first order equations by an appropriate change of variable. The most obvious type is when the function f is independent of y. Therefore, assume that g is a function of two variables and that equation (1) has the form

$$y'' = g(t, y') \tag{5}$$

Using the variable change $v = y'$ (and hence $v' = y''$), this equation takes the form

$$v' = g(t, v) \tag{6}$$

Since this is a first order equation in v, the techniques in Chapter 1 can be applied in order to compute v, and then any antiderivative of v is a solution to (2). Therefore, any second order equation of the form (5) can be transformed readily into a first order equation.

▶ **EXAMPLE 2.1-6**

Consider the equation $y'' = 2y' - 4$. Setting $v = y'$, we have $v' = 2v - 4$, which is a first order linear equation in v (see Section 1.2). Multiplying each side of this equation by the integrating factor e^{-2t} it easily follows that $d/dt\,(e^{-2t}v) = -4e^{-2t}$. Upon antidifferentiation, $v = ce^{2t} + 2$ and hence

$$y(t) = \int (ce^{2t} + 2)\,dt = \bar{c} + \frac{c}{2}e^{2t} + 2t$$

Therefore, $y(t) = c_1 + c_2e^{2t} + 2t$ is a solution to this equation for any constants c_1 and c_2.

Another type of second order equation reducible to first order is one where the function f is independent of time t [i.e., equation (1) is autonomous—see Section 1.4]. Here it is assumed that g is a function of two variables and that equation (1) has the form

$$y'' = g(y, y') \tag{7}$$

The ideas involved here are more subtle than in the preceding analysis, and we only formally indicate the procedure. We suppose that y' is a function of y and consider y as the independent variable. Setting $z = y'$ and applying the chain rule for differentiation,

$$y'' = \frac{dz}{dt} = \frac{dz}{dy} \cdot \frac{dy}{dt} = \frac{dz}{dy} \cdot z$$

and hence equation (7) becomes

$$z\frac{dz}{dy} = g(y, z) \tag{8}$$

Upon solving (if possible) equation (8) for $z(y)$, it follows that y must be a solution to the first order autonomous equation

$$y' = z(y) \tag{9}$$

which may be solved by separation of variables.

EXAMPLE 2.1-7

Suppose that $k > 0$ and consider the equation $y'' = -k^2y$. Since the variable t is missing, we set $z = dy/dt$ and obtain

$$-k^2y = \frac{dz}{dt} = \frac{dz}{dy} \cdot \frac{dy}{dt} = \frac{dz}{dy} \cdot z$$

Therefore, $z\, dz/dy = -k^2y$ and by separation of variables, $\frac{1}{2}z^2 = -\frac{1}{2}k^2y^2 + \bar{c}$, where $\bar{c} > 0$. For notational convenience, set $\bar{c} = \frac{1}{2}k^2c_1^2$ (where $c_1 > 0$ is a constant) to obtain $z^2 = k^2(c_1^2 - y^2)$. Since $z = y'$,

$$y' = \pm k\sqrt{c_1^2 - y^2}$$

and hence by separation of variables again,

$$\int \frac{dy}{\sqrt{c_1^2 - y^2}} = \pm kt + c_2$$

Therefore, if arcsine is the inverse sine, $\arcsin (y/c_1) = \pm kt + c_2$, and it follows that $y(t) = c_1 \sin (kt + c_2)$, where c_1 and c_2 are constants. (The "$\pm$" is absorbed by the constants.) It is easy to check that y is a solution to this equation (and you should verify it). Note also that if

$$y(t) = \bar{c}_1 \sin kt + \bar{c}_2 \cos kt \tag{10}$$

where $\bar{c}_1$ and $\bar{c}_2$ are constants, then y is also a solution [recall the addition formula $\sin(kt + c_2) = \cos c_2 \sin kt + \sin c_2 \cos kt$].

EXAMPLE 2.1-8

Suppose that $k > 0$ and consider the equation $y'' = k^2 y$. As in the preceding example, the variable t is missing, and if $z = dy/dt$ then $z\,dz/dy = k^2 y$; solving, we obtain $z^2 = k^2(c_1^2 + y^2)$, where $c_1 > 0$ is a constant. Therefore, for this equation,

$$\int \frac{dy}{\sqrt{c_1^2 + y^2}} = \pm kt + c_2$$

If arcsinh is the inverse hyperbolic sine function, $\operatorname{arcsinh}(y/c_1) = \pm kt + c_2$, and it follows that $y(t) = c_1 \sinh(kt + c_2)$, where c_1 and c_2 are constants. (The " $\pm$ " is absorbed by the constants in this case as in the preceding case.) From the definition of hyperbolic sine it follows that

$$y(t) = \frac{c_1 e^{kt + c_2} - c_1 e^{-kt - c_2}}{2}$$

and from this it follows that

$$y(t) = \bar{c}_1 e^{kt} + \bar{c}_2 e^{-kt} \tag{11}$$

where $\bar{c}_1$ and $\bar{c}_2$ are constants is also a solution to this equation. The reader should verify that the functions y in (11) are indeed solutions.

EXAMPLE 2.1-9

Consider the equation $y'' = (y')^3/y^2$. Setting $z = dy/dt$, we obtain from equation (8) that $z\,dz/dy = z^3/y^2$, and hence

$$\frac{dz}{dy} = \frac{z^2}{y^2}$$

The variables separate in this equation, and it follows that

$$-\frac{1}{z} = -\frac{1}{y} + c_1 \qquad \text{or} \qquad \frac{dy}{dt} = z = \frac{-1}{c_1 - y^{-1}}$$

The variables y and t also separate and we have

$$(c_1 - y^{-1})\,dy = -dt \qquad \text{or} \qquad c_1 y - \ln|y| = -t + c_2$$

Therefore, the solution y is defined implicitly by $c_1 y - \ln|y| + t = c_2$, where c_1 and c_2 are constants to be determined by initial conditions.

PROBLEMS

2.1-1. For each of the following second order equations determine the general solution, and if initial values are given compute the solution of the corresponding initial value problem as well.

(a) $y'' = 6$

(b) $y'' = te^{2t}$

(c) $y'' = \cos 3t \qquad y(\pi) = 3,\ y'(\pi) = -2$

(d) $y'' = \dfrac{1}{4 + t^2}$

(e) $y'' = e^{-t} + 3t \qquad y(0) = 4,\ y'(0) = -1$

2.1-2. For the following second order equations, verify that each member of the given family of functions is a solution, and then determine the constants c_1 and c_2 so that the given initial conditions are satisfied.

(a) $y'' = y - t \qquad y(0) = 3,\ y'(0) = 2 \qquad [y(t) = c_1 e^t + c_2 e^{-t} + t]$

(b) $y'' = \dfrac{2}{t^2} y \qquad y(2) = 8,\ y'(2) = 0 \qquad \left[y(t) = c_1 t^2 + c_2 \dfrac{1}{t}\right]$

(c) $y'' = -4y \qquad y\left(\dfrac{\pi}{4}\right) = -1,\ y'\left(\dfrac{\pi}{4}\right) = 2 \qquad [y(t) = c_1 \cos 2t + c_2 \sin 2t]$

(d) $y'' = 3y - 2y' \qquad y(0) = 2,\ y'(0) = -1 \qquad [y(t) = c_1 e^{-3t} + c_2 e^t]$

2.1-3. For the following problems determine all solutions to the differential equation that also satisfy the given side condition.

(a) $y'' = t \qquad y(0) = 0,\ y(1) = 0 \qquad (I = [0, 1])$

(b) $y'' = \sin t \qquad y(0) = 0,\ y'(\pi) = 0 \qquad (I = [0, \pi])$

(c) $y'' = \sin t \qquad y'(0) = 0,\ y'(\pi) = 0 \qquad (I = [0, \pi])$

(d) $y'' = \cos t \qquad y'(0) = 0,\ y'(\pi) = 0 \qquad (I = [0, \pi])$

2.1-4. By reducing to a first order equation, determine a solution to each of the following initial value problems.

(a) $y'' = -(y')^2 \qquad y(0) = 1,\ y'(0) = 2$

(b) $y'' = y' + 2e^t \qquad y(0) = -1,\ y'(0) = 1$

(c) $y'' = 2yy' \qquad y(0) = 2,\ y'(0) = 4$

(d) $y'' = 2yy' \qquad y(0) = -1,\ y'(0) = 0$

(e) $y'' = y^2 \qquad y(0) = \sqrt[3]{3},\ y'(0) = \sqrt{2}$

(f) $y'' = \dfrac{(y')^3}{y} \qquad y(0) = 1,\ y'(0) = 2$

2.1-5. Suppose that g is a continuous function of two variables and consider the equation

$$(*)\qquad y'' = yg\left(t, \frac{y'}{y}\right)$$

Show that the change of variable $w = y'/y$ (or $y = e^{\int w}$) transforms $(*)$ into the first order equation $w' = g(t, w) - w^2$.

(a) Use this technique to solve the equations $y'' = -k^2 y$ and $y'' = k^2 y$, where $k > 0$ is a constant (see Examples 2.1-7 and 2.1-8).

(b) Determine solutions to the equation $y'' = 2y - y'$.

2.2 SECOND ORDER LINEAR EQUATIONS

The remainder of this chapter deals almost exclusively with *second order linear differential equations*; this section discusses some of the basic concepts. Throughout this section it is assumed that I is an interval and that a, b, and c are continuous real-valued functions on I with $a(t) \neq 0$ for all t in I. Consider the second order equation

$$(1)\qquad a(t)y'' + b(t)y' + c(t)y = 0$$

Since $a(t) \neq 0$ on I, this equation is of the form of equation (1) in the preceding section [with $f(t, y, y') = -\{b(t)y' + c(t)y\}/a(t)$ for $t \in I$ and $y, y' \in \mathbb{R}$]. Equation (1) is called a *homogeneous second order linear differential equation.* The term *homogeneous* is used to indicate that the right-hand side of equation (1) is zero (and, in particular, is unrelated to the term homogeneous as applied to a certain class of nonlinear first order equations—see Section 1.6). In addition to the homogeneous equation (1), the *nonhomogeneous* (*or inhomogeneous*) *second order linear differential equation*

$$(2)\qquad a(t)y'' + b(t)y' + c(t)y = f(t)$$

is considered, where f is a continuous real-valued function on I. Observe that the homogeneous equation is a special case of the nonhomogeneous equation with $f(t) \equiv 0$ on I. Note also that the terms *homogeneous* and *nonhomogeneous* correspond to their use for first order linear equations in Section 1.2.

The purpose of this section is to indicate some basic properties of solutions to equations (1) and (2). In order to facilitate this discussion, we introduce an important notation (concept): For each twice differentiable function y on the interval I we define the function $L[y]$ on I by

$$(3)\qquad L[y](t) = a(t)y''(t) + b(t)y'(t) + c(t)y(t) \qquad \text{for all } t \in I$$

The "function" L is called a *second order linear differential operator*, and the notation

$$L[y] = a(t)y'' + b(t)y' + c(t)y$$

is often used. Note in particular that the domain of L is the twice differentiable functions on I (the terminology *L of the function y* is used to describe $L[y]$), and the range of L is the set of functions on I (and hence $L[y]$ is itself a function on I). As a specific illustration, if $I = (0, 1)$ and

$$L[y] = e^t y'' + ty' + 5y$$

then

$$L[\sin t](t) = -e^t \sin t + t \cos t + 5 \sin t$$

$$L[e^{3t}](t) = 9e^{4t} + 3te^{3t} + 5e^{3t}$$

and

$$L\left[\frac{1}{t}\right](t) = e^t \frac{2}{t^3} - \frac{1}{t} + 5\frac{1}{t}$$

The operator L defined by (3) has the following basic property: if y_1 and y_2 are twice differentiable functions on I and c_1 and c_2 are constants, then

$$L[c_1 y_1 + c_2 y_2] = c_1 L[y_1] + c_2 L[y_2] \tag{4}$$

Since a corresponding formula for differentiation is valid,

$$(c_1 y_1 + c_2 y_2)' = c_1 y_1' + c_2 y_2' \qquad \text{and} \qquad (c_1 y_1 + c_2 y_2)'' = c_1 y_1'' + c_2 y_2''$$

equation (4) follows routinely (as the student should verify).

An operator L satisfying (4) is called a *linear operator*. Since the homogeneous equation (1) can be written with the notation $L[y] = 0$, there is the following important implication for the solutions of (1):

□ **Lemma 1**

Suppose that $y = y_1$ and $y = y_2$ are solutions to the homogeneous equation (1) and that c_1 and c_2 are constants. Then $y = c_1 y_1 + c_2 y_2$ is also a solution to (1).

This follows immediately from (4). Since y_1 and y_2 are solutions to (1) we have $L[y_1] = 0$ and $L[y_2] = 0$. Therefore, by (4), $L[c_1 y_1 + c_2 y_2] = c_1 L[y_1] + c_2 L[y_2] = 0$, and so $y = c_1 y_1 + c_2 y_2$ is a solution to (1). The fact that the combination $c_1 y_1 + c_2 y_2$ is a solution to (1) whenever y_1 and y_2 are solutions is known as the *principle of superposition*.

As a simple illustration of this property, the equation $y'' = 0$ is a homogeneous linear equation and the functions $y_1(t) \equiv 1$ and $y_2(t) \equiv t$ are each solutions. According to Lemma 1, the function $y(t) \equiv c_1 1 + c_2 t$ is a solution for every constant c_1 and c_2. Consider further the equation

$$2t^2 y'' - ty' + y = 0 \qquad [I = (0, \infty)]$$

Note that this equation is of the form of (1) on any interval I that does not contain the point 0 [since the coefficient function $a(t) = 2t^2$ for y'' equals 0 when $t = 0$]. If $y_1(t) \equiv t$ and $y_2(t) \equiv \sqrt{t}$ for $t > 0$, then $y_1'(t) \equiv 1$, $y_1''(t) \equiv 0$, $y_2'(t) \equiv t^{-1/2}/2$, $y_2''(t) = -t^{-3/2}/4$, and it is easy to check that both y_1 and y_2 are solutions. Therefore, by Lemma 1, $y = c_1 t + c_2\sqrt{t}$ for $t > 0$ is a solution for any constants c_1 and c_2.

Whenever y_1 and y_2 are functions on the interval I and c_1 and c_2 are constants, the function $y = c_1 y_1 + c_2 y_2$ is called a *linear combination of* y_1 *and* y_2. Lemma 1 asserts that if y_1 and y_2 are solutions to the homogeneous equation (1), then any linear combination of y_1 and y_2 is also a solution. In order to effectively deal with linear equations we want to use a result pertaining to the existence and uniqueness of solutions. This is provided by the following theorem, which is stated without proof.

Theorem 2.2-1

Suppose that $t_0 \in I$ and y_0 and v_0 are given numbers. Then there is precisely one solution y to equation (2) such that $y(t_0) = y_0$ and $y'(t_0) = v_0$.

Since the homogeneous equation (1) is a special case of (2), Theorem 2.2-1 also guarantees the existence and uniqueness for solutions to (1) having prescribed initial values.

Suppose that y_1 and y_2 are each solutions to the homogeneous equation (1). From Lemma 1 it is known that $y = c_1 y_1 + c_2 y_2$ is a solution to (1) for all constants c_1 and c_2. The problem now is to determine when (if ever) all solutions to (1) are linear combinations of y_1 and y_2. For example, the equation $y'' = 0$ has $y_1(t) \equiv 1$ and $y_2(t) \equiv t$ as solutions, and we know from the preceding section that every solution y to $y'' = 0$ has the form $y = c_1 + c_2 t$, which is a linear combination of y_1 and y_2. However, $y_1 = 1 + t$ and $y_2 = 2 + 2t$ are also distinct solutions to $y'' = 0$, but the solution $y(t) \equiv 1 - t$ is not a linear combination of $1 + t$ and $2 + 2t$. For note that if

$$1 - t = c_1(1 + t) + c_2(2 + 2t) = (c_1 + 2c_2) + (c_1 + 2c_2)t$$

then, equating like coefficients, $c_1 + 2c_2 = 1$ *and* $c_1 + 2c_2 = -1$. This is obviously impossible.

2.2a Wronskians and Linear Independence

In order to describe exactly when every solution to (1) is a linear combination of two given solutions y_1 and y_2, we will use Theorem 2.2-1. Note that if $y = c_1 y_1 + c_2 y_2$ then $y' = c_1 y_1' + c_2 y_2'$, and hence if $t_0 \in I$ then

$$\begin{aligned} c_1 y_1(t_0) + c_2 y_2(t_0) &= y(t_0) \\ c_1 y_1'(t_0) + c_2 y_2'(t_0) &= y'(t_0) \end{aligned} \tag{5}$$

Thus if a solution y to (1) is a linear combination of y_1 and y_2, then (5) has a solution c_1, c_2 for each $t_0 \in I$. Suppose now that y is a solution to (1) and that (5) has a solution c_1, c_2 for some $t_0 \in I$. Setting $\bar{y}(t) = c_1 y_1(t) + c_2 y_2(t)$ for all $t \in I$, we have from Lemma 1 that $\bar{y}$ is a solution to (1) and from (5) that $\bar{y}(t_0) = y(t_0)$ and $\bar{y}'(t_0) = y'(t_0)$. By the uniqueness assertion in Theorem 2.2-1 we have that $\bar{y}(t) = y(t)$ [and hence that (5) has a solution for every $t_0 \in I$].

Therefore, if y_1 and y_2 are such that (5) has a solution for some $t_0 \in I$ whenever $y(t_0)$ and $y'(t_0)$ are any given numbers, then *every* solution to (1) is a linear combination of y_1 and y_2. Conversely, if every solution to (1) is a linear combination of y_1 and y_2, then (5) has a solution for some (and in fact *all*) $t_0 \in I$ and all numbers $y(t_0)$, $y'(t_0)$. It is a basic algebraic fact that (5) has a unique solution for any number $y(t_0)$, $y'(t_0)$ only in case the determinant of the coefficients is nonzero:

$$\det \begin{pmatrix} y_1(t) & y_2(t) \\ y_1'(t) & y_2'(t) \end{pmatrix} = y_1(t)y_2'(t) - y_1'(t)y_2(t) \neq 0$$

This determinant is called the *wronskian* of y_1 and y_2. Specifically, if y_1 and y_2 are *any* continuously differentiable functions on I, then the wronskian $W[y_1, y_2; \cdot]$ of y_1 and y_2 on I is defined by

$$W[y_1, y_2; t] \equiv y_1(t)y_2'(t) - y_1'(t)y_2(t) \qquad \text{for } t \in I \tag{6}$$

Thus if y_1 and y_2 are solutions to the homogeneous equation (1), then every solution to (1) is a linear combination of y_1 and y_2 if and only if the wronskian $W[y_1, y_2; t_0] \neq 0$ for some $t_0 \in I$ (and hence if $W[y_1, y_2; t_0] \neq 0$ for some $t_0 \in I$, then $W[y_1, y_2; t] \neq 0$ for *all* $t \in I$). See also Problem 2.2-3.

Consider the function $y_1(t) \equiv t$ and $y_2(t) \equiv e^{3t}$ on any interval I. Then $y_1'(t) \equiv 1$ and $y_2'(t) \equiv 3e^{3t}$, so

$$W[y_1, y_2; t] = \det \begin{pmatrix} t & e^{3t} \\ 1 & 3e^{3t} \end{pmatrix} = (3t - 1)e^{3t}$$

Similarly, if $y_1(t) \equiv \sin 2t$ and $y_2(t) \equiv \cos 2t$, then

$$\begin{aligned} W[y_1, y_2; t] &= \det \begin{pmatrix} \sin 2t & \cos 2t \\ 2 \cos 2t & -2 \sin 2t \end{pmatrix} \\ &= -2 \sin^2 2t - 2 \cos^2 2t \\ &= -2(\sin^2 2t + \cos^2 2t) = -2 \end{aligned}$$

on any interval I.

There is a second concept that is also important in these considerations. Two functions y_1 and y_2 on I are said to be *linearly dependent on I* if there are constants c_1 and c_2 such that $c_1 y_1(t) + c_2 t_2(t) \equiv 0$ on I and *at least one of the constants is nonzero.* If y_1 and y_2 are not linearly dependent on I, they are said to be *linearly independent on I*. Therefore, y_1 and y_2 are linearly independent on I only in case the only constants c_1 and c_2 such that $c_1 y_1(t) + c_2 y_2(t) \equiv 0$ on I are the constants $c_1 = c_2 = 0$. As a few specific examples, if $y_1(t) \equiv 0$ on I and $y_2(t)$ is *any* function on I, then y_1 and y_2 are linearly dependent [since $1 \cdot y_1(t) + 0 \cdot y_2(t) \equiv 0$ on I]. If $y_1(t) \equiv 1$ and $y_2(t) \equiv t$, then y_1 and y_2 are linearly independent on any interval I (since $c_1 + c_2 t \equiv 0$ on any interval only if $c_1 = c_2 = 0$). If $y_1(t) = t^2$ and $y_2(t) = t|t|$, then y_1 and y_2 are linearly dependent on any open interval I such that 0 is not in I and are linearly independent on any open interval I that contains 0. For suppose that $c_1 t^2 + c_2 t|t| = 0$. If $t > 0$ then $t|t| = t^2$ and so $c_1 = -c_2$, and if $t < 0$ then $t|t| = -t^2$ and so $c_1 = c_2$. Therefore if I contains both positive and negative numbers, it must be true that $c_1 = c_2 = 0$. The concepts of linear independence and dependence for *two* functions y_1 and y_2 has a simple interpretation: two functions y_1 and y_2 are linearly dependent on I if and only if one of the functions is a constant multiple of the other (i.e., there is a constant c such that $y_1 = cy_2$ or $y_2 = cy_1$ on I). The reader should verify this.

EXAMPLE 2.2-1

The functions $y_1(t) = 3 \sin t$ and $y_2(t) = \sin t$ are linearly dependent on any interval I since

$$\tfrac{1}{3}y_1(t) + (-1)y_2(t) \equiv \sin t - \sin t \equiv 0$$

on I. The functions $y_1(t) = e^{2t} - 3e^t$ and $y_2(t) = e^{2t} + e^t$ are linearly independent on $(-\infty, \infty)$. For suppose that c_1 and c_2 are constants such that

$$c_1(e^{2t} - 3e^t) + c_2(e^{2t} + e^t) \equiv 0$$

Differentiating each side with respect to t, we have

$$c_1(2e^{2t} - 3e^t) + c_2(2e^{2t} + e^t) \equiv 0$$

and hence c_1 and c_2 must be solutions to the simultaneous equations

$$c_1(e^{2t} - 3e^t) + c_2(e^{2t} + e^t) = 0$$

$$c_1(2e^{2t} - 3e^t) + c_2(2e^{2t} + e^t) = 0$$

for all t in $(-\infty, \infty)$. Setting $t = 0$, for example,

$$-2c_1 + 2c_2 = 0$$

$$-c_1 + 3c_2 = 0$$

Multiplying each side of the first equation by $-\frac{1}{2}$ and adding to the second, we obtain $2c_2 = 0$. Hence $c_2 = 0$, and substituting $c_2 = 0$ into the first equation implies $c_1 = 0$. Therefore both c_1 and c_2 must equal 0, and $e^{2t} - 3e^t$ and $e^{2t} + e^t$ are linearly independent by definition.

Now suppose that y_1 and y_2 are solutions on I to the homogeneous equation (1). If y_1 and y_2 are linearly dependent on I, then there are constants c_1 and c_2, not both zero, such that $c_1 y_1(t) + c_2 y_2(t) \equiv 0$ [and hence $c_1 y_1'(t) + c_2 y_2'(t) \equiv 0$] on I. Thus (5) does not have only the trivial solution $c_1 = c_2 = 0$ when $y(t_0) = y'(t_0) = 0$, and since the wronskian must then be zero at t_0, we conclude that some solution to (1) is not a linear combination of y_1 and y_2. Conversely, suppose that the solutions y_1 and y_2 are linearly independent on I, and let c_1 and c_2 be constants such that $c_1 y_1(t_0) + c_2 y_2(t_0) = 0$ and $c_1 y_1'(t_0) + c_2 y_2'(t_0) = 0$. Then, if $y(t) \equiv c_1 y_1(t) + c_2 y_2(t)$ on I, y is a solution to (1) by Lemma 1, and since $y(t_0) = y'(t_0) = 0$ we have, from the uniqueness assertion of Theorem 2.2-1, that $y(t) \equiv 0$ on I. Thus $c_1 y_1(t) + c_2 y_2(t) \equiv 0$ on I, and since y_1 and y_2 are linearly independent, it follows that $c_1 = c_2 = 0$. From this we deduce that when $y(t_0) = y'(t_0) = 0$, equation (5) has only the trivial solution, and so $W[y_1, y_2; t_0] \neq 0$ and *all* solutions to (1) are linear combinations of y_1 and y_2. Combining the results of this discussion we have the following important theorem:

Theorem 2.2-2

Suppose that y_1 and y_2 are solutions to the homogeneous equation (1). Then any one of the following four statements implies the other three.

(i) Every solution y to (1) is a linear combination of y_1 and y_2 on I (that is, $y = c_1 y_1 + c_2 y_2$).

(ii) y_1 and y_2 are linearly independent on I.

(iii) The wronskian $y_1(t_0)y_2'(t_0) - y_1'(t_0)y_2(t_0) \neq 0$ for some $t_0 \in I$.

(iv) The wronskian $y_1(t)y_2'(t) - y_1'(t)y_2(t) \neq 0$ for all $t \in I$.

It follows from Theorem 2.2-2 that in order to determine *all* solutions to the homogeneous equation (1), it is sufficient to determine *two* solutions that are linearly independent. When y_1 and y_2 are linearly independent solutions to (1), the family $y = c_1 y_1 + c_2 y_2$, where c_1 and c_2 are constants, is called the *general solution to* (1), and y_1 and y_2 are said *to generate* the general solution.

EXAMPLE 2.2-2

Suppose that $k > 0$ and consider the equation $y'' + k^2 y = 0$. It is easily verified that $y_1(t) = \cos kt$ and $y_2(t) = \sin kt$ are solutions to this equation, and since

$$\begin{aligned} W[y_1, y_2; t] &= \cos kt \; k \cos kt - (-k \sin kt)(\sin kt) \\ &= k(\cos^2 kt + \sin^2 kt) = k \end{aligned}$$

y_1 and y_2 are linearly independent. Therefore

$$y = c_1 \cos kt + c_2 \sin kt$$

is the general solution (see Example 2.1-7).

These results and techniques also apply to the nonhomogeneous equation. For let L be the differential operator defined by equation (3), and suppose that y_H is a solution to the homogeneous equation (1) and that y_p is a solution to the nonhomogeneous equation (2). Then, by (4),

$$L[y_H + y_p] = L[y_H] + L[y_p] = 0 + f = f$$

and so $y_H + y_p$ is a solution to (2). Also, if both y_1 and y_2 are solutions to (2), then $y_1 - y_2$ is a solution to (1) since

$$L[y_1 - y_2] = L[y_1] - L[y_2] = f - f = 0$$

Therefore, given any "particular" solution y_p of (2), *every* solution y to (2) is of the form $y = y_H + y_p$, where y_H is a solution to (1). For if y_p is given and y is any solution to (2), then $y - y_p = y_H$, where y_H is a solution to (1). Thus we have the following theorem:

■ **Theorem 2.2-3**

Suppose that y_p is a particular solution to the nonhomogeneous equation (2) and that y_1 and y_2 are linearly independent solutions to the homogeneous equation (1). Then every solution y of (2) has the form

$$y = c_1 y_1 + c_2 y_2 + y_p$$

where c_1 and c_2 are constants.

As in the homogeneous case, the family of functions $y = c_1 y_1 + c_2 y_2 + y_p$, where c_1 and c_2 are constants, is called the *general solution to (2)*.

EXAMPLE 2.2-3

Consider the equation $y'' + 9y = 3t + 4e^{2t}$. It should be verified that $y_1(t) = \cos 3t$ and $y_2(t) = \sin 3t$ are linearly independent solutions to the corresponding homogeneous equation $y'' + 9y = 0$ (see Example 2.2-2). If $y_p(t) = t/3 + 4e^{2t}/13$, then $y_p'(t) = \frac{1}{3} + 8e^{2t}/13$, $y_p'' = 16e^{2t}/13$, and

$$y_p'' + 9y_p = \frac{16e^{2t}}{13} + \frac{9t}{3} + \frac{36e^{2t}}{13} = 3t + \frac{52e^{2t}}{13}$$

Hence y_p is a particular solution, and it follows from Theorem 2.2-3 that the family

$$y = c_1 \cos 3t + c_2 \sin 3t + \frac{t}{3} + \frac{4e^{2t}}{13}$$

where c_1 and c_2 are constants, is the general solution.

There is a corresponding initial value problem associated with equations (1) and (2). Let t_0 be in I and let y_0 and v_0 be given numbers. We consider the nonhomogeneous initial value problem

$$a(t)y'' + b(t)y' + c(t)y = f(t) \qquad y(t_0) = y_0 \qquad y'(t_0) = v_0 \tag{7}$$

The analogous homogeneous problem is considered as a special case of (7) with $f(t) \equiv 0$ on I. Combining Theorems 2.2-1 and 2.2-3 leads to the following result.

Theorem 2.2-4

Suppose that y_1 and y_2 are linearly independent solutions to the homogeneous equation (1) and that y_p is a particular solution to the nonhomogeneous equation (2). Thus there are unique constants c_1 and c_2 such that $y = c_1 y_1 + c_2 y_2 + y_p$ is the solution to (7) [that is, $y(t_0) = y_0$ and $y'(t_0) = v_0$].

Since (6) has a solution by Theorem 2.2-1 and since every solution to (2) has the form $y = c_1 y_1 + c_2 y_2 + y_p$ by Theorem 2.2-3, there certainly exist constants c_1 and c_2 so that y is a solution to (7). In fact, c_1 and c_2 must be the solution to the system of equations

$$c_1 y_1(t_0) + c_2 y_2(t_0) = y_0 - y_p(t_0)$$

$$c_1 y_1'(t_0) + c_2 y_2'(t_0) = v_0 - y_p'(t_0)$$

Since the determinant of the coefficients of this system is nonzero by Theorem 2.2-2(iii), this system has a unique solution, and Theorem 2.2-4 is established.

EXAMPLE 2.2-4

Consider the equation $y'' - y = t^2$, $y(0) = 3$, $y'(0) = 3$. Since $y_1(t) \equiv e^t$ and $y_2(t) \equiv e^{-t}$ are solutions to the corresponding homogeneous equation $y'' - y = 0$ (verify this), and since the wronskian

$$W[e^t, e^{-t}; t] = e^t(-e^{-t}) - (e^t)(e^{-t}) = -2$$

we have that e^t and e^{-t} are linearly independent. One easily can and should check that $y_p(t) \equiv -2 - t^2$ is a solution to the nonhomogeneous equation, and hence

$$y(t) = c_1 e^t + c_2 e^{-t} - 2 - t^2$$

is the general solution. Solving the system

$$y(0) = c_1 + c_2 - 2 = 3$$

$$y'(0) = c_1 - c_2 = 3$$

for c_1 and c_2 shows that the solution to this initial value problem is $y = 4e^t + e^{-t} - 2 - t^2$.

If y_1 and y_2 are solutions to (1) on I, then $W[y_1, y_2; t] \equiv 0$ on I implies that y_1 and y_2 must be linearly dependent. However, if y_1 and y_2 are just twice continuously differentiable functions on I [and not a solution to an equation of the form (1)], then it is possible for y_1 and y_2 to be linearly independent on I and also that $W[y_1, y_2; t] \equiv 0$ on I (see Problem 2.2-6).

PROBLEMS

2.2-1. Compute the wronskian for each given pair of functions y_1 and y_2. Determine also if the functions are linearly independent or dependent on the real line.

(a) $y_1(t) = e^t \qquad y_2(t) = e^{2t}$

(b) $y_1(t) = t + 1 \qquad y_2(t) = t - 1$

(c) $y_1(t) = t^2 \qquad y_2(t) = (2t)^2$

(d) $y_1(t) = t^3 \qquad y_2(t) = |t^3|$

(e) $y_1(t) = e^t + e^{2t} \qquad y_2(t) = e^{2t}$

(f) $y_1(t) = \sin 2t \qquad y_2(t) = \sin 3t$

(g) $y_1(t) = e^t - e^{-t} \qquad y_2(t) = e^t + e^{-t}$

(h) $y_1(t) = \dfrac{1}{t} \qquad y_2(t) = \dfrac{1}{t+1} \qquad t > 0$

(i) $y_1(t) = \sin 2t \qquad y_2(t) = \sin t \cos t$

2.2-2. For each of the following equations show that the given family of functions is the general solution and then determine the constants c_1 and c_2 so that the given initial conditions are satisfied.

(a) $y'' - 4y = 0 \qquad y(0) = 1,\ y'(0) = 0 \qquad \{y = c_1 e^{2t} + c_2 e^{-2t}\}$

(b) $y'' - 4y = e^t \qquad y(0) = 1,\ y'(0) = 1 \qquad \{y = c_1 e^{2t} + c_2 e^{-2t} - \frac{1}{3}e^t\}$

(c) $t^2 y'' - 2y = 0 \qquad y(2) = 3,\ y'(2) = 0 \qquad I = (0, \infty) \qquad \left\{y = \dfrac{c_1}{t} + c_2 t^2\right\}$

(d) $(\cos t)y'' + (\sin t)y' + (\cos^3 t)y = 0 \qquad y(0) = 1,\ y'(0) = 2$

$I = \left(-\dfrac{\pi}{2}, \dfrac{\pi}{2}\right) \qquad \{y = c_1 \cos(\sin t) + c_2 \sin(\sin t)\}$

(e) $t^2y'' - 2y = 2t \qquad y(1) = 0,\ y'(1) = 0 \qquad I = (0, \infty)$

$\left\{y = \frac{c_1}{t} + c_2 t^2 - t + t^2\right\}$

(f) $y'' - y = 0 \qquad y\ (\ln 2) = 1,\ y'\ (\ln 2) = 3$

$\{y = c_1(e^t - 2e^{-t}) + c_2(3e^t + e^{-t})\}$

(g) $y'' + y = 0 \qquad y\left(\frac{\pi}{6}\right) = 1,\ y'\left(\frac{\pi}{6}\right) = -2$

$\left\{y = c_1 \cos t + c_2 \cos\left(t + \frac{\pi}{6}\right)\right\}$

2.2-3. Suppose that I is an interval and a, b, c are continuous functions on I with $a(t) \neq 0$ for $t \in I$. Let y_1 and y_2 be solutions on I to the differential equations

$$a(t)y'' + b(t)y' + c(t)y = 0$$

Let $W(t) = y_1(t)y_2'(t) - y_1'(t)y_2(t)$ be the wronskian of y_1 and y_2. Show that

$$a(t)W'(t) + b(t)W(t) = 0 \qquad \text{for all } t \in I$$

Deduce that if $t_0 \in I$ then

$$W(t) = W(t_0)\exp\left[-\int_{t_0}^{t} b(s)a(s)^{-1}\,ds\right]$$

for all $t \in I$, and hence $W(t) \equiv 0$ on I or $W(t) \neq 0$ for *all* $t \in I$. [This formula for $W(t)$ is known as *Abel's identity*.]

2.2-4. Show that $y_1 = t$ and $y_2 = t^2$ satisfy the differential equation $t^2y''(t) - 2ty'(t) + 2y(t) = 0$ for all $t \in (-\infty, \infty)$, and that y_1 and y_2 are linearly independent. Show, however, that the wronskian $W[y_1, y_2; t]$ of y_1 and y_2 satisfies $W[y_1, y_2; 0] = 0$ and $W[y_1, y_2; t] \neq 0$ if $t \neq 0$. Explain this relative to Problem 2.2-3.

2.2-5. Suppose that y_1 and y_2 are solutions to the homogeneous equation (1) at the beginning of this section and that $t_0 \in I$.

(a) If $y_1(t_0) = y_2(t_0) = 0$, are y_1 and y_2 necessarily linearly dependent? What about if $y_1'(t_0) = y_2'(t_0) = 0$?

(b) If $y_1(t_0) = 0$ and $y_2(t_0) \neq 0$, are y_1 and y_2 necessarily linearly independent? What about if $y_1'(t_0) = 0$ and $y_2'(t_0) \neq 0$?

(c) Suppose that $x_1 = y_1 + y_2$ and $x_2 = y_1 - y_2$. If y_1 and y_2 are linearly independent on I, are x_1 and x_2 linearly independent on I? If x_1 and x_2 are linearly independent on I, are y_1 and y_2 linearly independent on I?

2.2-6. Suppose that I is any open interval with $0 \in I$ and that $y_1(t) \equiv t^3$ and $y_2(t) \equiv t^2|t|$ for all $t \in I$. Show that y_1 and y_2 are twice continuously differentiable and linearly independent on I but that $W[y_1, y_2; t] \equiv 0$ on I.

2.3 HOMOGENEOUS EQUATIONS WITH CONSTANT COEFFICIENTS

In general, determining explicit representations of solutions to second order linear differential equations is impossible. This is in contrast to first order linear equations whose solutions can always be explicitly expressed in terms of integrals involving the coefficient functions (see Section 1.2). However, when the coefficients of y, y', and y'' in a homogeneous equation are *constants* (i.e., do not depend on t), then the general solution can be easily obtained as indicated below.

Assume that a, b, and c are constants, with $a \neq 0$, and consider the homogeneous second order equation

(1) $$ay'' + by' + cy = 0$$

The problem is to determine the general solution to this equation in terms of the constants a, b, and c.

In order to indicate an appropriate form for solutions to (1), we consider the first order case $\alpha y' + \beta y = 0$, where α and β are constants, $\alpha \neq 0$. Since $e^{\beta t/\alpha}$ is an integrating factor, $y = c_1 e^{-\beta t/\alpha}$ is seen to be the general solution (see Section 1.2). Therefore, for the second order case, it is *assumed that the solutions to (1) are exponential functions.* So suppose λ is a number and that

(2) $$y(t) = e^{\lambda t} \qquad \text{is a solution to (1)}$$

That is, for what numbers λ is $e^{\lambda t}$ a solution to (1)? Since $y'(t) = \lambda e^{\lambda t}$ and $y''(t) = \lambda^2 e^{\lambda t}$, we see by substituting into (1) that the equation

$$a\lambda^2 e^{\lambda t} + b\lambda e^{\lambda t} + ce^{\lambda t} = (a\lambda^2 + b\lambda + c)e^{\lambda t} = 0$$

must be satisfied in order for $e^{\lambda t}$ to be a solution. Thus if λ is any number satisfying

(3) $$a\lambda^2 + b\lambda + c = 0$$

then $y(t) = e^{\lambda t}$ is a solution to (1). The algebraic equation (3) is called the *auxiliary equation* or the *characteristic equation* for the homogeneous differential equation (1). The roots of this equation are called the *auxiliary roots* or the *characteristic roots.* Moreover, the second-degree polynomial p defined by the left-hand side of (3),

$$p(\lambda) = a\lambda^2 + b\lambda + c$$

is called the *auxiliary polynomial* or the *characteristic polynomial* of (1). The results of this section show that the general solution to (1) can be determined *directly* from the auxiliary roots. Of course, the auxiliary roots can be obtained easily using the quadratic formula:

(4) $$\lambda = \frac{-b \pm \sqrt{b^2 - 4ac}}{2a} \qquad \text{are the roots of (3)}$$

Therefore we indicate the procedure for determining two linearly independent solutions to (1) from the auxiliary roots (4). This procedure divides naturally into three cases: The auxiliary roots are real and distinct; the auxiliary root is a double root; and the auxiliary roots are complex conjugates.

2.3a The Auxiliary Roots Are Real and Distinct

This case is the most immediate and simplest. Note that the auxiliary roots (4) are real and distinct if and only if $b^2 - 4ac > 0$. Therefore, assume that this is the case, and let

$$\lambda_1 = \frac{-b + \sqrt{b^2 - 4ac}}{2a} \qquad \text{and} \qquad \lambda_2 = \frac{-b - \sqrt{b^2 - 4ac}}{2a}$$

be the auxiliary roots. As has already been indicated, $y_1(t) = e^{\lambda_1 t}$ and $y_2(t) = e^{\lambda_2 t}$ are each solutions to (1) [see (2)]. Moreover, if $W[e^{\lambda_1 t}, e^{\lambda_2 t}; t]$ is the wronskian, then

$$W[e^{\lambda_1 t}, e^{\lambda_2 t}; t] = \det\begin{pmatrix} e^{\lambda_1 t} & e^{\lambda_2 t} \\ \lambda_1 e^{\lambda_1 t} & \lambda_2 e^{\lambda_2 t} \end{pmatrix} = (\lambda_2 - \lambda_1)e^{(\lambda_1 + \lambda_2)t}$$

Since $\lambda_2 \neq \lambda_1$ the wronskian is never zero, and so y_1 and y_2 are linearly independent. Therefore,

(5) If λ_1 and λ_2 are real and distinct auxiliary roots of (1), then $y = c_1 e^{\lambda_1 t} + c_2 e^{\lambda_2 t}$ is the general solution to (1).

As an illustration consider the equation

$$y'' - y' - 2y = 0$$

The auxiliary equation is then

$$\lambda^2 - \lambda - 2 = 0$$

and since $\lambda^2 - \lambda - 2 = (\lambda - 2)(\lambda + 1)$, it follows that $\lambda = 2, -1$ are the auxiliary roots. Since these roots are real and distinct, we have from (5) that

$$y = c_1 e^{2t} + c_2 e^{-t}$$

where c_1 and c_2 are constants, is the general solution to the given equation. The student should verify directly that e^{2t} and e^{-t} are solutions.

▶ **Example 2.3-1**

Consider the initial value problem

$$y'' + 2y' - 3y = 0 \qquad y(0) = 3 \qquad y'(0) = 1 \tag{6}$$

The differential equation in (6) is second order and homogeneous, with constant coefficients. Moreover, the auxiliary equation is

$$\lambda^2 + 2\lambda - 3 = (\lambda + 3)(\lambda - 1) = 0$$

Therefore, $\lambda = -3, 1$ are the auxiliary roots, and it follows from (5) that $y = c_1 e^{-3t} + c_2 e^t$ is the general solution to the differential equation in (6). Since $y' = -3c_1 e^{-3t} + c_2 e^t$, the initial conditions in (6) are satisfied if

$$y(0) = c_1 + c_2 = 3$$

$$y'(0) = -3c_1 + c_2 = 1$$

Subtracting the two equations, we obtain $4c_1 = 2$, and so $c_1 = \frac{1}{2}$. Substituting for c_1 in the first equation shows that $c_2 = \frac{5}{2}$ and hence

$$y(t) = \frac{e^{-3t}}{2} + \frac{5e^t}{2}$$

is the solution to the initial value problem (6).

2.3b The Auxiliary Root Is a Double Root

If the auxiliary equation (3) has a double root λ_0, then λ_0 is real, and from the quadratic formula (4), $b^2 - 4ac = 0$ and $\lambda_0 = -b/2a$. It follows directly from (2) that $y_1(t) \equiv e^{\lambda_0 t}$ is a solution to (1). The problem is that we need a second linearly independent solution from $e^{\lambda_0 t}$ and we don't have an auxiliary root distinct from λ_0 to use. The simple equation

$$y'' = 0 \tag{7}$$

is of this type, and it provides insight into the general method of solution. The auxiliary equation for (7) is $\lambda^2 = 0$, and hence $\lambda_0 = 0$ is a double root. As has been stated, $y_1(t) \equiv 1$ ($\equiv e^{0t}$) is one solution to (7). However, we already have established that a second linearly independent solution to (7) is $y_2(t) \equiv t$ ($\equiv te^{0t}$). This type of second linearly independent solution also carries over to the general case of a double auxiliary root λ_0.

Before establishing this assertion, consider the equation

$$y'' + 2y' + y = 0 \tag{8}$$

The auxiliary equation is $\lambda^2 + 2\lambda + 1 = (\lambda + 1)^2 = 0$, and hence $\lambda = -1$ is a double auxiliary root. *Assume* that $y = ve^{-t}$ is a solution to (8), where v is some function of t. By the product rule, $y' = v'e^{-t} - ve^{-t}$, $y'' = v''e^{-t} - 2v'e^{-t} + ve^{-t}$, and substituting into (8) and simplifying shows that

$$y'' + 2y' + y = v''e^{-t}$$

Therefore, $y = ve^{-t}$ is a solution whenever $v'' = 0$—that is, $v = c_1 + c_2 t$. Taking $c_1 = 1$, $c_2 = 0$ and then $c_1 = 0$, $c_2 = 1$ shows that $y_1 \equiv e^{-t}$ and $y_2 \equiv te^{-t}$ are solutions to (8). Since they are linearly independent,

$$y(t) = c_1 e^{-t} + c_2 te^{-t}$$

is the general solution to (8).

For the case of a double auxiliary root the following result is valid:

(9) $\begin{cases} \text{If } \lambda = \lambda_0 \text{ is a double auxiliary root of (1), then } y = c_1 e^{\lambda_0 t} + c_2 te^{\lambda_0 t} \text{ is the} \\ \text{general solution to (1).} \end{cases}$

In order to establish (9) we proceed as in the preceding example and assume $y = ve^{\lambda_0 t}$ is a solution to (1), where λ_0 is the double auxiliary root and v is time-dependent. Then

$$y' = v'e^{\lambda_0 t} + \lambda_0 ve^{\lambda_0 t} \quad \text{and} \quad y'' = v''e^{\lambda_0 t} + 2\lambda_0 v'e^{\lambda_0 t} + \lambda_0^2 ve^{\lambda_0 t}$$

and it follows by substituting into (1) that

$$\begin{aligned} ay'' + by' + cy &= a(v'' + 2\lambda_0 v' + \lambda_0^2 v)e^{\lambda_0 t} + b(v' + \lambda_0 v)e^{\lambda_0 t} + cve^{\lambda_0 t} \\ &= [av'' + (2a\lambda_0 + b\lambda_0)v' + (a\lambda_0^2 + b\lambda_0 + c)v]e^{\lambda_0 t} \end{aligned}$$

Since λ_0 is a root of (3) the coefficient of v is zero, and since λ_0 is a double root, $\lambda_0 = -b/2a$, and the coefficient of v' is also zero. Therefore, $y = ve^{\lambda_0 t}$ is a solution to (1) if and only if $v'' = 0$, and hence $v = c_1 + c_2 t$. This establishes (9).

▶ **EXAMPLE 2.3-2**

Consider the initial value problem

(10) $4y'' - 4y' + y = 0 \qquad y(0) = 4 \qquad y'(0) = -1$

The auxiliary equation is $4\lambda^2 - 4\lambda + 1 = (2\lambda - 1)^2 = 0$, so $\lambda = \frac{1}{2}$ is a double root. According to (9),

$$y = c_1 e^{t/2} + c_2 te^{t/2}$$

is the general solution to (10). Since $y(0) = c_1$ and $y'(0) = c_1/2 + c_2$, the initial conditions imply that $c_1 = 4$, $c_2 = -3$. Thus

$$y = 4e^{t/2} - 3te^{t/2}$$

is the solution to (10).

2.3c The Auxiliary Roots Are Complex Conjugates

From the quadratic formula (4), the roots of (3) are conjugate complex only in case $b^2 - 4ac < 0$ (since the coefficients a, b, and c are real, complex roots always come in conjugate pairs). It is assumed that α and β are real numbers with $\beta \neq 0$ and that $\lambda = \alpha \pm i\beta$ where $i^2 = -1$ are the complex conjugate auxiliary roots. An illustrative example of this case is the equation

$$y'' + \beta^2 y = 0 \qquad \text{where } \beta > 0 \tag{11}$$

The auxiliary equation is $\lambda^2 + \beta^2 = 0$, and hence $\lambda = \pm i\beta$ where $i^2 = -1$ are the complex conjugate (and, in fact, pure imaginary) auxiliary roots. Thus we need an "interpretation" of $e^{\pm i\beta t}$. As is indicated in Example 2.1-7, two linearly independent solutions of (11) are

$$y_1(t) \equiv \cos \beta t \qquad \text{and} \qquad y_2(t) \equiv \sin \beta t$$

This can be verified by direct substitution. It is of interest to indicate the relationship between $e^{i\beta t}$ and a linear combination of $\cos \beta t$ and $\sin \beta t$. The appropriate formula that $e^{i\beta t}$ satisfies is

$$e^{i\beta t} = \cos \beta t + i \sin \beta t$$

so that $\cos \beta t$ is the "real part of $e^{i\beta t}$" and $\sin \beta t$ the "imaginary part of $e^{i\beta t}$." This is known as *Euler's formula* and can be established from basic techniques in complex analysis. Note further that

$$e^{-i\beta t} = \cos(-\beta t) + i \sin(-\beta t) = \cos \beta t - i \sin \beta t$$

Assuming that $e^{(\alpha + i\beta)t} = e^{\alpha t}e^{i\beta t}$ leads to the formula

$$\begin{aligned} e^{(\alpha \pm i\beta)t} &= e^{\alpha t}e^{\pm i\beta t} = e^{\alpha t}(\cos \beta t \pm i \sin \beta t) \\ &= e^{\alpha t} \cos \beta t \pm i e^{\alpha t} \sin \beta t \end{aligned}$$

and it can be shown that if $\alpha \pm \beta i$ are the auxiliary roots, then $e^{\alpha t} \cos \beta t$ and $e^{\alpha t} \sin \beta t$ are linearly independent solutions.

As a specific example, consider the equation

$$y'' + 2y' + 5y = 0 \tag{12}$$

Since $\lambda^2 + 2\lambda + 5 = 0$ is the auxiliary equation, we have from the quadratic formula that

$$\lambda = \frac{-2 \pm \sqrt{4 - 20}}{2} = -1 \pm \frac{\sqrt{-16}}{2} = -1 \pm 2i$$

are the auxiliary roots. According to the preceding paragraph, two linearly independent solutions are

$$y_1(t) = e^{-t} \cos 2t \qquad \text{and} \qquad y_2(t) = e^{-t} \sin 2t$$

Observe, for example, that

$$y_1'(t) = -e^{-t}\cos 2t - 2e^{-t}\sin 2t \qquad \text{and}$$
$$y_1''(t) = e^{-t}\cos 2t + 2e^{-t}\sin 2t + 2e^{-t}\sin 2t - 4e^{-t}\cos 2t$$
$$= -3e^{-t}\cos 2t + 4e^{-t}\sin 2t$$

and hence

$$y_1'' + 2y_1' + 5y_1 = -3e^{-t}\cos 2t + 4e^{-t}\sin 2t - 2e^{-t}\cos 2t - 4e^{-t}\sin 2t + 5e^{-t}\cos 2t$$
$$= 0$$

Therefore, y_1 is a solution to (12), and one may similarly check that y_2 is also a solution.

For the case of complex conjugate auxiliary roots the following result is valid:

(13) If $\lambda = \alpha + i\beta$ are conjugate complex auxiliary roots of equation (1), then $y = c_1 e^{\alpha t}\cos\beta t + c_2 e^{\alpha t}\sin\beta t$ is the general solution to (1).

We indicate a proof of the assertion (13). Comparing $\alpha \pm i\beta$ with the quadratic formula (4), it follows that

$$\alpha = \frac{-b}{2a} \qquad \text{and} \qquad \beta^2 = \frac{4ac - b^2}{4a^2} \tag{14}$$

Making the change of variable $y(t) = v(t)e^{-bt/2a}$, we have

$$y'(t) = \left[v'(t) - \frac{bv(t)}{2a}\right]e^{-bt/2a}$$

and

$$y''(t) = \left[v''(t) - \frac{bv'(t)}{a} + \frac{b^2v(t)}{4a^2}\right]e^{-bt/2a}$$

From these computations and (14) it follows that

$$ay'' + by' + cy = \left[v'' - \frac{b^2 - 4ac}{4a^2}v\right]ae^{-bt/2a}$$
$$= [v'' + \beta^2 v]ae^{-bt/2a}$$

Therefore, if v is a solution to $v'' + \beta^2 v = 0$, then $y = ve^{-bt/2a}$ is a solution to (1). By equation (11) we know that

$$v = c_1\cos\beta t + c_2\sin\beta t$$

and since $-b/2a = \alpha$ we see that (13) is true. For the interested reader a more complete discussion of the complex exponential is given in the first part of Section 8.4*b*.

EXAMPLE 2.3-3

Consider the initial value problem

$$y'' + 2y' + 3y = 0 \qquad y(0) = 2 \qquad y'(0) = -1 \tag{15}$$

The auxiliary equation is $\lambda^2 + 2\lambda + 3 = 0$, so by the quadratic formula,

$$\lambda = \frac{-2 \pm \sqrt{4 - 12}}{2} = -1 \pm i\sqrt{2}$$

are the auxiliary roots. From (13),

$$y = c_1 e^{-t} \cos \sqrt{2}t + c_2 e^{-t} \sin \sqrt{2}t$$

is the general solution to the differential equation in (15). Since $y(0) = c_1$ and $y'(0) = -c_1 + \sqrt{2}c_2$, we have from the initial conditions that $c_1 = 2$ and $-c_1 + \sqrt{2}c_2 = -1$. Thus $c_1 = 2$, $c_2 = 1/\sqrt{2}$, and

$$y(t) = 2e^{-t} \cos \sqrt{2}t + \frac{1}{\sqrt{2}} e^{-t} \sin \sqrt{2}t$$

is the solution to (15).

In summary, the following result is valid concerning the general solution of the homogeneous equation (1).

Theorem 2.3-1

Suppose that a, b, and c are constants with $a \neq 0$. Then the general solution to the homogeneous equation (1) has precisely one of the following forms:

(i) If $\lambda = \lambda_1$ and $\lambda = \lambda_2$ are distinct real auxiliary roots of (1) (that is, if $b^2 - 4ac > 0$), then $y = c_2 e^{\lambda_1 t} + c_2 e^{\lambda_2 t}$ is the general solution to (1);

(ii) if $\lambda = \lambda_0$ is a (real) double auxiliary root of (1) (that is, if $b^2 - 4ac = 0$), then $y = c_1 e^{\lambda_0 t} + c_2 t e^{\lambda_0 t}$ is the general solution to (1); and

(iii) if $\lambda = \alpha + i\beta$ are complex conjugate auxiliary roots of (1) (that is, if $b^2 - 4ac < 0$), then $y = c_1 e^{\alpha t} \cos \beta t + c_2 e^{\alpha t} \sin \beta t$ is the general solution to (1).

The next example indicates how one can use the representation of the solutions from the auxiliary roots when one of the coefficients in (1) depends on a parameter (see also Problem 2.3-3).

EXAMPLE 2.3-4

Let η be a real parameter and consider the equation

(16) $$y'' + 2\eta y' + y = 0$$

The auxiliary equation is $\lambda^2 + 2\eta\lambda + 1 = 0$, and by the quadratic formula the auxiliary roots are

$$\lambda = \frac{-2\eta \pm \sqrt{4\eta^2 - 4}}{2} = -\eta \pm \sqrt{\eta^2 - 1}$$

To determine the form of the solutions to (16), we need to establish when $\eta^2 - 1 = 0$, $\eta^2 - 1 > 0$, and $\eta^2 - 1 < 0$. Clearly $\eta^2 - 1 = 0$ only if $\eta = \pm 1$, and since $-\eta$ is a double root in this case,

(17) $\{y = c_1 e^{-\eta t} + c_2 t e^{-\eta t}$ is the general solution to (16) if $\eta = \pm 1$

Since $\eta^2 - 1 > 0$ only if $\eta < -1$ or $\eta > 1$, the auxiliary roots are real and distinct whenever $\eta < -1$ or $\eta > 1$, and we have

(18) $\left\{ y = c_1 e^{(-\eta + \sqrt{\eta^2 - 1})t} + c_2 e^{(-\eta - \sqrt{\eta^2 - 1})t} \right.$ is the general solution to (16) if $\eta < -1$ or if $\eta > 1$

Also $\eta^2 - 1 < 0$ only in case $-1 < \eta < 1$, and since the auxiliary roots are conjugate complex in this case, we have

(19) $\left\{ y = c_1 e^{-\eta t} \cos(\sqrt{1 - \eta^2}\, t) + c_2 e^{-\eta t} \sin(\sqrt{1 - \eta^2}\, t) \right.$ is the general solution to (16) if $-1 < \eta < 1$

Finally note that if $\eta > 0$ the solutions to (16) have limit zero as $t \to +\infty$, and if $\eta \leq 0$ then at least one solution does not have limit zero as $t \to +\infty$.

PROBLEMS

2.3-1. Determine the general solution to the following homogeneous equations with constant coefficients.

(a) $y'' + 2y' - 8y = 0$

(b) $y'' - y' - 6y = 0$

(c) $y'' + 3y' = 0$

(d) $y'' - 4y' + 4y = 0$

(e) $y'' - 3y = 0$

(f) $y'' - 2y' + 5y = 0$

(g) $2y'' + y' - 3y = 0$

(h) $y'' - 3y' + 4y = 0$

(i) $y'' + 8y = 0$

(j) $4y'' - 12y' + 9y = 0$

(k) $3y'' - 4y' - 4y = 0$

(l) $y'' + 6y' + 9y = 0$

(m) $y'' + y' + y = 0$

2.3-2. Determine the solution to the following homogeneous initial value problems.

(a) $y'' + 4y = 0 \qquad y\left(\frac{\pi}{6}\right) = 1 \qquad y'\left(\frac{\pi}{6}\right) = 0$

(b) $y'' + 3y' - 4y = 0 \qquad y(0) = 3 \qquad y'(0) = 3$

(c) $y'' - y' + y = 0 \qquad y(0) = -1 \qquad y'(0) = 2$

(d) $y'' - 2y' + y = 0 \qquad y(1) = 3 \qquad y'(1) = -2$

(e) $y'' - 4y = 0 \qquad y(\ln 2) = 1 \qquad y'(\ln 2) = 2$

(f) $y'' - 2y' + 10y = 0 \qquad y(0) = -2 \qquad y'(0) = 4$

2.3-3. For each of the following equations, determine the general solution in terms of the real parameter η. Also, indicate for what values of η all of the solutions have limit zero as $t \to +\infty$.

(a) $y'' + 2y' + \eta y = 0$ **(c)** $y'' + 2\eta y' + \eta y = 0$

(b) $y'' + 4\eta y' + 4y = 0$ **(d)** $y'' - \eta y' - \eta y = 0$

2.3-4. Show that the wronskian for the functions $y_1(t) \equiv e^{\lambda_0 t}$ and $y_2(t) \equiv te^{\lambda_0 t}$ (where λ_0 is a real constant) is never zero. Also, if α and β are real with $\beta > 0$, show that the wronskian for $y_1(t) \equiv e^{\alpha t} \cos \beta t$ and $y_2(t) \equiv e^{\alpha t} \sin \beta t$ is never zero.

2.3-5. Consider the homogeneous equation (1) where $b^2 - 4ac < 0$, and let $\alpha = -b/2a$ and $\beta = \sqrt{4ac - b^2}/2a$ so that $\lambda = \alpha \pm i\beta$ ($i = \sqrt{-1}$) are the complex conjugate auxiliary roots. Use the addition formula

$\sin(\phi_1 + \phi_2) = \sin \phi_1 \cos \phi_2 + \cos \phi_1 \sin \phi_2$

to deduce that the family $y = \bar{c}_1 e^{\alpha t} \sin(\beta t + \bar{c}_2)$, where $\bar{c}_1$ and $\bar{c}_2$ are constants, is the general solution to (1). Is the family $y = \bar{c}_1 e^{\alpha t} \cos(\beta t + \bar{c}_2)$ also the general solution to (1)?

2.3-6. Consider the homogeneous equation (1) where $b^2 - 4ac > 0$, and let $\alpha = -b/2a$ and $\beta = \sqrt{b^2 - 4ac}/2a$ so that $\lambda = \alpha \pm \beta$ are the real and distinct auxiliary roots. Recall the definitions $\sinh x \equiv (e^x - e^{-x})/2$ and $\cosh x \equiv (e^x + e^{-x})/2$.

(a) Show that $y = \bar{c}_1 e^{\alpha t} \cosh \beta t + \bar{c}_2 e^{\alpha t} \sinh \beta t$, where $\bar{c}_1$ and $\bar{c}_2$ are arbitrary constants, is the general solution to the homogeneous equation (1).

(b) Establish the addition formula

$\sinh(\phi_1 + \phi_2) = \sinh \phi_1 \cosh \phi_2 + \cosh \phi_1 \sinh \phi_2$

Is it true that the family $y = k_1 e^{\alpha t} \sinh(\beta t + k_2)$, where k_1 and k_2 are constants, is the general solution to (1)? Is the family $y = k_1 e^{\alpha t} \cosh(\beta t + k_2)$ the general solution to (1)?

2.3-7. Suppose that y is a solution to the homogeneous equation (1) on $(-\infty, \infty)$. Show that the number of zeros of y on $(-\infty, \infty)$ is either 0, 1, or ∞ [that is, show that the equation $y(t) = 0$, $t \in (-\infty, \infty)$ has either no solution, one solution, or an infinite number of solutions]. (Problems 2.3-5 and 2.3-6 may be of some help.)

2.3-8. Suppose that y is a solution to the homogeneous equation (1) where a, b, and c are all positive. Show that $y(t) \to 0$ as $t \to +\infty$.

2.4 NONHOMOGENEOUS EQUATIONS—UNDETERMINED COEFFICIENTS

In this section a method is developed for determining a particular solution to a special class of second order linear nonhomogeneous equations. Throughout it is assumed that f is a continuous function on $(-\infty, \infty)$ that has derivatives of all orders [that is, the nth derivative $f^{(n)}(t)$ of f exists on $(-\infty, \infty)$ for all positive integers n] and that a, b, and c are constants with $a \neq 0$. Consider the nonhomogeneous equation

$$ay'' + by' + cy = f(t) \tag{1}$$

as well as the corresponding homogeneous equation

$$ay'' + by' + cy = 0 \tag{2}$$

From the results of the preceding section the general solution to the homogeneous equation (2) can be computed easily. Therefore, it is assumed that y_1 and y_2 are linearly independent solutions to (2), and hence

$$y = c_1 y_1 + c_2 y_2 \qquad \text{is the general solution to (2)} \tag{3}$$

According to Theorem 2.2-3, in order to obtain the general solution to the nonhomogeneous equation (1), *it is sufficient to determine some particular solution y_p of (1).* In this case $y = c_1 y_1 + c_2 y_2 + y_p$ is the general solution to (1).

The purpose of this section is to indicate a method for determining a particular solution to (1) in the case that the function f has a special form. This special form indicates the representation of a particular solution to (1). As a simple example, consider

$$y'' - y' - 6y = 3$$

If it is *assumed* that a particular solution to this equation is a constant—say $y_p(t) \equiv d_1$—then $y_p'(t) \equiv y_p''(t) \equiv 0$, and substituting $y = y_p$ leads to the equation $-6d_1 = 3$. Thus $d_1 = -\frac{1}{2}$, and $y_p(t) \equiv -\frac{1}{2}$ is indeed a solution. As a slightly more complicated example, consider the equation

$$y'' - y' - 6y = 6t^2 - 16t + 7 \tag{4}$$

Since the right-hand side of this equation is a polynomial of degree 2, we will *assume* that a particular solution also is a polynomial of degree 2:

$$y_p(t) = d_1 t^2 + d_2 t + d_3$$

where d_1, d_2, and d_3 are constants that we hope can be chosen so that y_p is a solution to (4). Computing

$$y_p'(t) = 2d_1 t + d_2 \qquad \text{and} \qquad y_p''(t) = 2d_1$$

and substituting for y in (4), we obtain

$$(2d_1) - (2d_1 t + d_2) - 6(d_1 t^2 + d_2 t + d_3) = 6t^2 - 16t + 7$$

and hence

$$(-6d_1)t^2 - (2d_1 + 6d_2)t + (2d_1 - d_2 - 6d_3) = 6t^2 - 16t + 7$$

Equating the coefficients of like powers of t leads to the three equations

$$-6d_1 = 6 \qquad 2d_1 + 6d_2 = 16 \qquad 2d_1 - d_2 - 6d_3 = 7$$

The first equation shows that $d_1 = -1$, the second equation shows that $d_2 = 3$, and the third equation shows that $d_3 = -2$. Therefore,

$$y_p(t) = -t^2 + 3t - 2$$

and it is easy to verify that this is a solution to (4).

The type of procedure indicated by the preceding two examples applies whenever the function f has a finite *differential family:* there is a finite number of functions $\{g_1, g_2, \ldots, g_m\}$ such that f and each of its derivatives is a linear combination of $g_1, g_2, \ldots, g_m$. For convenience, the most basic functions having a finite differential family are given in the accompanying table.

$f(t)$	**Differential family**
e^{at}	$\{e^{at}\}$
$\sin bt$ or $\cos bt$	$\{\sin bt, \cos bt\}$
$a_n t^n + \cdots + a_1 t + a_0$	$\{1, t, \ldots, t^n\}$

Therefore, if the right-hand side of the nonhomogeneous equation (1) has a finite differential family, the procedure is to determine if a particular solution can be written as a linear combination of the functions in the differential family. As illustrations, if $f(t) = 6e^{-4t}$, try $y_p(t) = d_1 e^{-4t}$, and if $f(t) = 2 \cos 3t$ or if $f(x) = \sin 3t - \cos 3t$, try $y_p = d_1 \sin 3t + d_2 \cos 3t$. By substituting y_p for y in equation (1), one can determine if such a solution exists and the values of the constants d_i.

Before considering more complicated functions f, we illustrate this procedure with specific examples. Consider the equation

$$y'' + y' - 2y = e^{2t} \tag{5}$$

The differential family of e^{2t} is $\{e^{2t}\}$, and so let's assume that d_1 is a constant and that a particular solution to (5) has the form

$$y_p = d_1 e^{2t} \tag{6}$$

If y_p has the form (6), then $y_p' = 2d_1 e^{2t}$, $y_p'' = 4d_1 e^{2t}$, and substituting y_p for y in the left-hand side of (5), we obtain

$$4d_1 e^{2t} + 2d_1 e^{2t} - 2(d_1 e^{2t}) = e^{2t}$$

Collecting terms, $4d_1 e^{2t} = e^{2t}$, and hence $d_1 = \frac{1}{4}$. Therefore, assuming that y_p has the form (6) leads to $d_1 = \frac{1}{4}$, and it follows that $y_p(t) \equiv e^{2t}/4$ is a particular solution to equation (5). The auxiliary equation for (5) is $\lambda^2 + \lambda - 2 = (\lambda + 2)(\lambda - 1) = 0$, so $\lambda = -2, 1$ are the auxiliary roots, and we see that

$$y(t) \equiv c_1 e^{-2t} + c_2 e^{t} + \frac{e^{2t}}{4}$$

is the general solution to (5).

EXAMPLE 2.4-1

Consider the equation

$$y'' - 2y' + y = -5 \cos 2t \tag{7}$$

Since this equation has constant coefficients and the differential family of the right-hand side is $\{\sin 2t, \cos 2t\}$, we try to find a particular solution of the form

$$y_p = d_1 \sin 2t + d_2 \cos 2t$$

Differentiating, we obtain

$$y_p' = 2d_1 \cos 2t - 2d_2 \sin 2t \qquad y_p'' = -4d_1 \sin 2t - 4d_2 \cos 2t$$

and substituting y_p for y in (7),

$$(-4d_1 \sin 2t - 4d_2 \cos 2t) - 2(2d_1 \cos 2t - 2d_2 \sin 2t) + (d_1 \sin 2t + d_2 \cos 2t) = -5 \cos 2t$$

Collecting terms it follows that

$$(-3d_1 + 4d_2)(\sin 2t) + (-4d_1 - 3d_2)(\cos 2t) = -5 \cos 2t$$

and equating like coefficients on each side of the equation (notice that the coefficient of sin $2t$ on the right-hand side is zero), we obtain the system

$$-3d_1 + 4d_2 = 0$$

$$-4d_1 - 3d_2 = -5$$

Solving this system implies that $d_1 = \frac{4}{5}$ and $d_2 = \frac{3}{5}$, and hence

$$y_p = \frac{4 \sin 2t}{5} + \frac{3 \cos 2t}{5}$$

is a particular solution to (7) (the student should verify this directly).

Now suppose that α and β are real numbers and that the nonhomogeneous term f has the form

(8) $$f(t) = (a_n t^n + \cdots + a_1 t + a_0)(e^{\alpha t} \cos \beta t) + (b_n t^n + \cdots + b_1 t + b_0)(e^{\alpha t} \sin \beta t)$$

where $a_0, \ldots, a_n$ and $b_0, \ldots, b_n$ are constants. Assume also that either a_n or b_n (or both a_n and b_n) is nonzero. If f has the form (8), then a finite family of derivatives for f is given as follows:

(9) The function defined by equation (8) has $\{t^n e^t \cos \beta t, \ldots, te^{\alpha t} \cos \beta t, e^{\alpha t} \cos \beta t, t^n e^{\alpha t} \sin \beta t, \ldots, te^{\alpha t} \sin \beta t, e^{\alpha t} \sin \beta t\}$ as a differential family.

As specific illustrations we have the following:

$f(t) = t^2e^{2t} - e^{2t}$ has $\{t^2e^{2t}, te^{2t}, e^{2t}\}$ as a differential family [take $n = 2$, $\alpha = 2$, and $\beta = 0$ in (9)]; $f(t) = t \cos 6t$ has $\{t \cos 6t, \cos 6t, t \sin 6t, \sin 6t\}$ as a differential family [take $n = 1$, $\alpha = 0$, $\beta = 6$ in (9)]; and $f(t) = e^t(\cos t - 2 \sin t)$ has $\{e^t \cos t, e^t \sin t\}$ as a differential family [take $n = 0$, $\alpha = \beta = 1$ in (9)].

Note also that the general form of f in (8) includes as special cases the simple examples given in the table at the beginning of this section. The important fact here is the following: *If f has the form (8) and no member of the differential family in (9) is a solution to the homogeneous equation (2), then (1) has a particular solution y_p that is a linear combination of the members of the differentiable family in (9).* The case where some member of the differentiable family is a solution to (2) is treated in Section 2.4*a*. (See Theorem 2.4-1, where these procedures are precisely stated.)

The procedure in the case that f is of the form in (8) is to assume that y_p is a linear combination of the differential family in (9):

$$y_p(t) = d_n t^n e^{\alpha t} \cos \beta t + \cdots + d_1 te^{\alpha t} \cos \beta t + d_0 e^{\alpha t} \cos \beta t$$
$$+ \bar{d}_n t^b e^{\alpha t} \sin \beta t + \cdots + \bar{d}_1 te^{\alpha t} \sin \beta t + \bar{d}_0 e^{\alpha t} \sin \beta t$$

Then substitute y_p, y'_p, and y''_p into (1) and solve for the constants d_i and $\bar{d}_i$ by equating like coefficients. If none of the terms $e^{\alpha t} \cos \beta t$ or $e^{\alpha t} \sin \beta t$ is a solution to (2) [that is, if $\alpha + i\beta$ is not an auxiliary root for (2)], then we are guaranteed to be able to solve for the coefficients d_i and $\bar{d}_i$ (hence this is called the *method of undetermined coefficients*).

Consider the equation

$$y'' + y = te^{-3t} \tag{10}$$

Since $\{e^{-3t}, te^{-3t}\}$ is a differential family for (10) and -3 is not an auxiliary root for the homogeneous equation corresponding to (10), we have that

$$y_p = d_1 e^{-3t} + d_2 te^{-3t}$$

is a particular solution to (10) for some constants d_1 and d_2. Then

$$y_p' = (-3d_1 + d_2)e^{-3t} - 3d_2 te^{-3t}$$
$$y_p'' = (9d_1 - 6d_2)e^{-3t} + 9d_2 te^{-3t}$$

and substituting for y in (10), we obtain

$$[(9d_1 - 6d_2)e^{-3t} + 9d_2 te^{-3t}] + (d_1 e^{-3t} + d_2 te^{-3t}) = te^{-3t}$$

and hence that

$$(10d_1 - 6d_2)e^{-3t} + 10d_2 te^{-3t} = te^{-3t}$$

Equating like coefficients, $10d_2 = 1$ and $10d_1 - 6d_2 = 0$, so $d_2 = \frac{1}{10}$ and $d_1 = \frac{3}{50}$. Therefore,

$$y_p(t) \equiv \frac{3e^{-3t}}{50} + \frac{te^{-3t}}{10}$$

is a particular solution to (10).

This procedure can also be applied if f is the sum of two or more functions, each having the form of (8) but with different numbers α and β. For example, if

$$f(t) = te^{-2t} + 4e^{-2t} - \cos 3t$$

then one should assume that

$$y_p(t) = d_1 te^{-2t} + d_2 e^{-2t} + d_3 \cos 3t + d_4 \sin 3t$$

The first two terms for y_p come from the terms $te^{-2t} + 4e^{-2t}$ in f, and the last two terms come from the term $-\cos 3t$ in f.

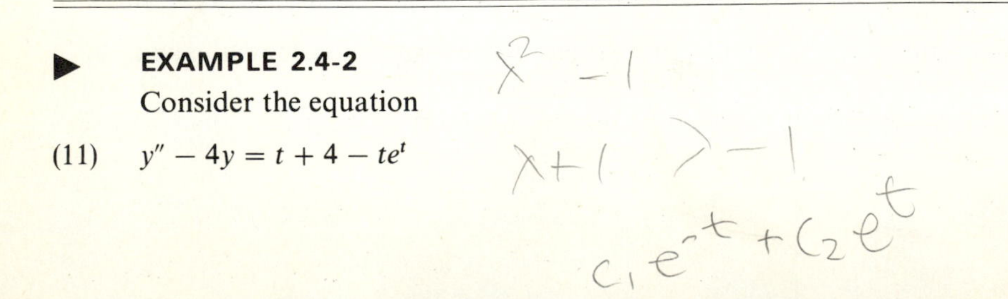

▶ **EXAMPLE 2.4-2**

Consider the equation

$$y'' - 4y = t + 4 - te^t \tag{11}$$

Since a differential family for $t + 4$ is $\{1, t\}$ and that for $-te^t$ is $\{te^t, e^t\}$ and since none of these terms is a solution to the corresponding homogeneous equation, we assume that

$$y_p = d_1 + d_2 t + d_3 te^t + d_4 e^t \tag{12}$$

Then

$$y_p'' = d_3 te^t + (2d_3 + d_4)e^t$$

and substituting for y in (11) gives

$$-3d_3 te^t + (2d_3 - 3d_4)e^t - 4d_1 - 4d_2 t = t + 4 - te^t$$

Equating coefficients of like terms shows that

$$-3d_1 = -1 \qquad 2d_3 - 3d_4 = 0 \qquad -4d_1 = 4 \qquad -4d_2 = 1$$

Therefore $d_3 = \frac{1}{3}$, $d_4 = \frac{2}{9}$, $d_1 = -1$, $d_2 = -\frac{1}{4}$, and substituting into (12) gives

$$y_p = -1 - \frac{t}{4} + \frac{te^t}{3} + \frac{2e^t}{9}$$

as a particular solution to (11).

2.4a Modifications Relative to the Homogeneous Solution

The use of the method of undetermined coefficients when some linear combination of the differential family is a solution to the corresponding homogeneous equation requires some modifications in the process. To illustrate the situation, consider the equation

$$y'' - y = e^t \tag{13}$$

The differential family of e^t is $\{e^t\}$, and if it is assumed that $y_p = d_1 e^t$, then $y_p' = d_1 e^t$, $y_p'' = d_1 e^t$, and substituting y_p into (13) leads to the equation

$$e^t = y_p'' - y_p = d_1 e^t - d_1 e^t \equiv 0$$

which is impossible. This approach failed, of course, because $d_1 e^t$ is a solution to the homogeneous equation corresponding to (13), and hence the preceding method does not apply. In this particular instance it is helpful to note that if one assumes that (13) has a solution of the form $y_p = d_1 te^t$ (that is, use the family $\{te^t\}$ instead of $\{e^t\}$), then

$$y_p' = d_1 te^t + d_1 e^t \qquad y_p'' = d_1 te^t + 2d_1 e^t$$

and, substituting y_p for y in (13),

$$(d_1 te^t + 2d_1 e^t) - d_1 te^t = e^t$$

and hence $2d_1 e^t = e^t$. Thus $d_1 = \frac{1}{2}$, and it follows that $y_p(t) \equiv te^t/2$ is a particular solution to (13) (verify this directly).

If some linear combination of members of the differential family is a solution to the corresponding homogeneous equation, then this differential family needs to be modified before applying the method of undetermined coefficients. This modification is simple if f has the special form in (8). As an elementary situation, consider the equation

$$ay'' + by' + cy = A \sin \beta t + B \cos \beta t \tag{14}$$

where $\beta > 0$ and A and B are constants. By our preceding results and the relationship of the auxiliary roots to the solution of the homogeneous equation (2), if $\pm i\beta$ are not the auxiliary roots of (2), then there is a particular solution y_p to (14) of the form $y_p = d_1 \sin \beta t + d_2 \cos \beta t$. However, if $\pm i\beta$ are the auxiliary roots of (2), instead of using the differential family $\{\sin \beta t, \cos \beta t\}$, we modify it by multiplying each member by t to obtain the family $\{t \sin \beta t, t \cos \beta t\}$. Therefore, if $\pm i\beta$ is an auxiliary root of (2), then there is a particular solution y_p to (14) of the form

$$y_p = d_1 t \sin \beta t + d_2 t \cos \beta t$$

The coefficients d_1 and d_2 may be computed by substituting y_p for y in (14) and equating like coefficients. As a specific illustration consider the following example.

▶ EXAMPLE 2.4-3

Consider the equation

$$y'' + y = \cos t \tag{15}$$

The general solution to the corresponding homogeneous equation is $y_H = c_1 \cos t + c_2 \sin t$, and the differential family of $\cos t$ is $\{\sin t, \cos t\}$. Since members of this family are solutions to the homogeneous equation, the particular solution y_p should be assumed to have the form

$$y_p = d_1 t \sin t + d_2 t \cos t$$

(that is, multiply each member of the original differential family by t). Then

$$y_p' = d_1 t \cos t + d_1 \sin t - d_2 t \sin t + d_2 \cos t$$

$$y_p'' = -d_1 t \sin t + 2d_1 \cos t - d_2 t \cos t - 2d_2 \sin t$$

and substituting y_p for y in (15), we obtain

$$y_p'' + y_p = 2d_1 \cos t - 2d_2 \sin t = \cos t$$

(In particular, all terms involving $t \cos t$ and $t \sin t$ cancel out.) Equating like coefficients, $2d_1 = 1$, $-2d_2 = 0$, and so $d_1 = \frac{1}{2}$, $d_2 = 0$, and

$$y_p = \frac{t \sin t}{2}$$

is a particular solution to (15). Therefore,

$$y = c_1 \cos t + c_2 \sin t + \frac{t \sin t}{2}$$

is the general solution to this system.

The procedure of multiplying each member of the differential family by t in order to determine the form of the solution can be used whenever the right-hand side has a particular form. Some care must be taken, however, and before stating the general result, we consider another specific example. Consider the equation

$$y'' - 2y' + y = e^t + t - 1 \tag{16}$$

Since $\lambda^2 - 2\lambda + 1 = (\lambda - 1)^2 = 0$ is the auxiliary equation, $\lambda = 1$ is a double auxiliary root and $y_H = c_1 e^t + c_2 te^t$ is the general solution to the corresponding homogeneous equation. The nonhomogeneous function divides naturally into two terms: an exponential part e^t whose differential family is $\{e^t\}$ and a polynomial term $t - 1$ whose differential family is $\{1, t\}$. Since no member of $\{1, t\}$ is a solution to the corresponding homogeneous equation, this differential family is not modified. Since a member of the family $\{e^t\}$ is a solution, this family is modified by multiplying by t: $\{te^t\}$. However, te^t is also a solution to the corresponding homogeneous equation, so the family $\{te^t\}$ must be modified further by multiplying its member by t: $\{t^2e^t\}$. Since no member of this family is a solution to the corresponding homogeneous equation, we should use the family $\{t^2e^t\}$. Therefore, there are constants d_1, d_2, and d_3 such that

$$y_p = d_1 t^2e^t + d_2 t + d_3$$

is a particular solution to (16). It is important to note that the family $\{t, 1\}$ was *not* multiplied by t or t^2. Computing y_p', y_p'' and substituting for y in (16), the equation

$$2d_1 e^t - 2d_2 + d_2 t + d_3 = e^t + t - 1$$

is obtained, after some simplification. Equating like coefficients, $2d_1 = 1$, $d_2 = 1$, and $d_3 - 2d_2 = -1$. From this it follows that $d_1 = \frac{1}{2}$, $d_2 = 1$, $d_3 = 1$, and hence

$$y_p = \frac{t^2e^t}{2} + t + 1$$

is a particular solution to (16) and

$$y = c_1 e^t + c_2 te^t + \frac{t^2e^t}{2} + t + 1$$

is the general solution to (16).

Now suppose that α and β are real numbers and that P and Q are polynomials of degree no larger than n, with at least one of them having degree exactly n. Consider the following form for f:

$$f(t) = P(t)e^{\alpha t} \cos \beta t + Q(t)e^{\alpha t} \sin \beta t \tag{17}$$

[See also (8).] It is not established here, but if f has a finite family of derivatives, then it must be the sum of a finite number of terms, each having the form indicated in (17). (See Problems 2.4-7 and 2.4-8, where it is shown that this is indeed the case if f has a differential family with one or two members.) When the nonhomogeneous term is of the form (17), then the following general result for determining the form of a particular solution is valid:

Theorem 2.4-1

Suppose that f has the form in (17):

$$f(t) = P(t)e^{\alpha t} \cos \beta t + Q(t)e^{\alpha t} \sin \beta t$$

Let k be the least nonnegative integer such that

$$y(t) = [c_0 t^k + c_1 t^{k+1} + \cdots + c_n t^{k+n}](e^{\alpha t} \cos \beta t) + [\bar{c}_0 t^k + \bar{c}_1 t^{k+1} + \cdots + \bar{c}_n t^{k+n}](e^{\alpha t} \sin \beta t)$$

is not a solution to the homogeneous equation (2) for any constants $c_0, \ldots, c_n$ and $\bar{c}_0, \ldots, \bar{c}_n$. Then $k = 0$, 1, or 2, and there are constants $d_0, \ldots, d_n$ and $\bar{d}_0, \ldots, \bar{d}_n$ such that

$$y_p(t) = [d_0 t^k + \cdots + d_n t^{k+n}](e^{\alpha t} \cos \beta t) + [\bar{d}_0 t^k + \cdots + \bar{d}_n t^{k+n}](e^{\alpha t} \sin \beta t)$$

is a particular solution to the inhomogeneous equation (1). Moreover, at least one of d_n or $\bar{d}_n$ must be nonzero.

The proof of Theorem 2.4-1 is omitted. The fact that at least one of the coefficients d_n or $\bar{d}_n$ must be nonzero is important. For it gives information on the behavior of the solutions for large t without any computation (e.g., if $\beta = 0$ then the dominating term for large t is $t^{k+n}e^{\alpha t}$). Theorem 2.4-1, along with the principle of superposition, can be applied to obtain particular solutions to (1) when f is the sum of functions, say $f_1, f_2, \ldots, f_m$, each having the form (17). For if y_{p_1} is a particular solution for $f = f_1$, y_{p_2} for $f = f_2, \ldots, y_{pm}$ for $f = f_m$, then $y_p = y_{p1} + y_{p2} + \cdots + y_{pm}$ is a particular solution to (1) for $f = f_1 + f_2 + \cdots + f_m$.

EXAMPLE 2.4-5

Consider the equation

$$y'' + 2y' = t^2 - 2 + e^{-2t} \sin 3t + te^{-2t} - 3e^{-2t}$$

The general solution to the homogeneous equation is $y = c_1 e^{-2t} + c_2$. The right-hand side may be conveniently broken up into the following parts:

$t^2 - 2$, whose differential family is $\{t^2, t, 1\}$

$e^{-2t} \sin 3t$, whose differential family is $\{e^{-2t} \sin 3t, e^{-2t} \cos 3t\}$

$te^{-2t} - 3e^{-2t}$, whose differential family is $\{te^{-2t}, e^{-2t}\}$

Comparing these families with the homogeneous solution, we see that a particular solution of the form

$$y_p = d_1 t^3 + d_2 t^2 + d_3 t + d_4 e^{-2t} \sin 3t + d_5 e^{-2t} \cos 3t + d_6 t^2 e^{-2t} + d_7 te^{-2t}$$

exists for this equation. The actual computation of the constants $d_1 - d_7$ is complicated and we omit the details.

PROBLEMS

2.4-1. Determine a particular solution to each of the following equations.

(a) $y'' + 6y = 3t - 2$

(b) $y'' + y = 3 \sin 2t$

(c) $y'' + y' + y = e^t + 4$

(d) $y'' - 4y' + 3y = t^2$

(e) $y'' + y = e^{-2t} + 6t$

(f) $y'' - y' = e^t \sin t$

2.4-2. Determine a particular solution to each of the following equations.

(a) $y'' - 4y' + 3y = 1 - e^{3t}$

(b) $y'' + 2y' + y = te^{-t}$

(c) $y'' + 2y' + 5y = e^t \cos 2t$

(d) $y'' - y = \sinh t$

(e) $y'' + y' = t^2 + 1 - e^{-t}$

(f) $y'' + 4y = \sin 2t - \cos 2t$

2.4-3. In each of the following problems use the method of undetermined coefficients to determine the form of a particular solution to the given nonhomogeneous equations. It is not necessary to evaluate the constants that appear.

(a) $y'' - y' - 2y = t^2 e^{2t} - e^{2t} \sin 2t + \cos 2t - 1$

(b) $y'' + 9y = t \cos 3t - \sin 3t + e^t \sin 3t$

(c) $y'' - 4y' + 4y = t^2 e^{2t} + te^t \cos 3t$

(d) $y'' - 2y' + 2y = t^2 e^t \sin t + t^2 e^t + t^2$

2.4-4. Determine the solution to each of the following initial value problems.

(a) $y'' - 2y' - 3y = 6 + e^{-t}$ $\quad y(0) = 0 \quad y'(0) = 0$

(b) $y'' - 4y = te^t$ $\quad y(0) = 1 \quad y'(0) = 2$

(c) $y'' + y' - 2y = -\sin t + 7 \cos t$ $\quad y(0) = 0 \quad y'(0) = 1$

2.4-5. What is a family of derivatives for $f(t) = \sin^2 \beta t$? [*Hint:* Use the half-angle formula $\sin^2 \beta t = (1 - \cos 2\beta t)/2$.] Determine a family of derivatives for the following functions:

(a) $f(t) = \sin 2t \cos 2t$ **(c)** $f(t) = \cos t \sin 2t$

(b) $f(t) = \sin^3 t$ **(d)** $f(t) = t \cos^2 t$

2.4-6. Determine the general solution for each of the following equations (see Problem 2.4-5):

(a) $y'' - 4y = \sin^2 t$

(b) $y'' + 4y = \sin^2 t$

(c) $y'' - 4y = \sin^3 t$

2.4-7. Suppose that a, b, c, A, and α are constants, with $a \neq 0$, and consider the nonhomogeneous equation

$$(*) \qquad ay'' + by' + cy = Ae^{\alpha t}$$

Let $r(\lambda)$ be the auxiliary polynomial for (*): $r(\lambda) = a\lambda^2 + b\lambda + c$. Show that a particular solution y_p to (*) has the following form:

(a) If $r(a) \neq 0$, then $y_p(t) \equiv Ae^{\alpha t}/r(\alpha)$.

(b) If $r(\alpha) = 0$ and $r'(\alpha) \neq 0$, then $y_p(t) \equiv Ate^{\alpha t}/r'(\alpha)$.

(c) If $r(\alpha) = r'(\alpha) = 0$ and $r''(\alpha) \neq 0$, then $y_p(t) \equiv At^2e^{\alpha t}/r''(\alpha)$.

2.4-8. Suppose that f has a one-member differential family $\{g_1\}$. Show that there are constants a and b such that $af' - bf = 0$, and hence $f(t) = ke^{\alpha t}$ for some constants k and α. (*Hint:* Since $f = c_1 g_1$ and $f' = c_2 g_1$, it follows that $c_1 f' - c_2 f = 0$.)

2.4-9. Suppose that f has a two-member differential family $\{g_1, g_2\}$. Show that there are constants a, b, and c such that $af'' + bf' + cf = 0$ (see Problem 2.4-7). Determine the form of the function f and compare the result with the form of f in (6).

2.4-10. Suppose that ω, β, and A are given positive numbers. Determine a particular solution to the equation

$$y'' + \omega^2 y = A \cos \beta t$$

(consider the cases $\omega = \beta$ and $\omega \neq \beta$ separately).

2.5 VARIATION OF PARAMETERS

The method of undetermined coefficients developed in the preceding section requires that the corresponding homogeneous equation have constant coefficients and that the nonhomogeneous term have a finite differential family. In this section a procedure is developed that applies to *general second order nonhomogeneous equations.* It is shown that if the general solution to the corresponding homogeneous equation is known, then the solution to the nonhomogeneous equation can be reduced to evaluating two integrals. In the second part of this section it is shown that if one nontrivial solution to the homogeneous equation is known, then the nonhomogeneous equation can be reduced to solving a first order linear equation and evaluating an antiderivative.

Suppose that I is an interval and that a, b, and c are continuous functions on I with $a(t) \neq 0$ for any $t \in I$, and that f is also a continuous function on I. Consider the nonhomogeneous equation

$$a(t)y'' + b(t)y' + c(t)y = f(t) \tag{1}$$

along with the corresponding homogeneous equation

$$a(t)y'' + b(t)y' + c(t)y = 0 \tag{2}$$

In the first part of this section it is assumed that two linearly independent solutions y_1 and y_2 to the homogeneous equation (2) are known on I, and hence

$$y = c_1 y_1 + c_2 y_2$$

where c_1 and c_2 are constants, is the general solution to (2).

The procedure for reducing the second order nonhomogeneous equation (1) to the evaluation of two antiderivatives is simple. Before indicating the general procedure, we first consider the specific example

$$y'' + y = \sec t \qquad 0 \leq t < \frac{\pi}{2} \tag{3}$$

Since $\sec t$ [$\equiv 1/(\cos t)$] does not have a finite family of derivatives, the method of undetermined coefficients does not apply. The auxiliary equation to the corresponding homogeneous equation is $\lambda^2 + 1 = 0$; hence $\lambda = \pm i$ are the auxiliary roots, and

$$y = c_1 \cos t + c_2 \sin t$$

is the general solution. The procedure is to replace the constants c_1 and c_2 by functions of t, say $c_1 = A(t)$ and $c_2 = B(t)$, and then determine how to select the functions A and B so that

$$y(t) = A(t) \cos t + B(t) \sin t$$

is a solution to (3). Therefore, we substitute this expression for y in (3). Since

$$y'(t) = -A(t) \sin t + B(t) \cos t + A'(t) \cos t + B'(t) \sin t$$

the computations become complicated, so the *simplifying assumption*

$$A'(t) \cos t + B'(t) \sin t \equiv 0$$

is made. Thus,

$$y'(t) = -A(t) \sin t + B(t) \cos t$$

and

$$y''(t) = -A(t) \cos t - B(t) \sin t - A'(t) \sin t + B'(t) \cos t$$

Substituting into equation (3) we find that

$$y'' + y = -A'(t) \sin t + B'(t) \cos t = \sec t$$

Therefore, A' and B' should satisfy the system

$$A'(t) \cos t + B'(t) \sin t = 0$$

$$-A'(t) \sin t + B'(t) \cos t = \sec t$$

Multiplying the first equation by $\sin t$, the second equation by $\cos t$, and adding, we obtain

$$B'(t)(\sin^2 t + \cos^2 t) = B'(t) = 1$$

and hence $B(t) \equiv t$ is a solution for B. Substituting $B'(t) \equiv 1$ in the first equation, it follows that

$$A'(t) = \frac{-\sin t}{\cos t}$$

and hence $A(t) = \ln(\cos t)$ ($\cos t > 0$ since $0 \leq t < \pi/2$). Therefore,

$$y_p(t) = \ln(\cos t) \cos t + t \sin t$$

is a particular solution to (3), and

$$y = c_1 \cos t + c_2 \sin t + \ln(\cos t) \cos t + t \sin t$$

is the general solution to (3).

The procedure indicated by the preceding example extends to the general case and is called the method of *variation of constants* or *variation of parameters*.

Theorem 2.5-1

Suppose that y_1 and y_2 are linearly independent solutions to the homogeneous equation (2). Then there is a particular solution y_p to the nonhomogeneous equation (1) of the form

$$y_p(t) = A(t)y_1(t) + B(t)y_2(t) \tag{4}$$

where A and B are any continuously differentiable functions on I such that their derivatives satisfy the system

$$
\begin{aligned}
&A'(t)y_1(t) + B'(t)y_2(t) \equiv 0 \\
(5) \qquad &a(t)A'(t)y_1'(t) + a(t)B'(t)y_2'(t) \equiv f(t)
\end{aligned}
$$

for all t in I.

Comparing (4) with $y = c_1 y_1 + c_2 y_2$ [the general solution to (2)], one sees that indeed if the "constants" c_1 and c_2 are allowed to "vary with t," then a particular solution to (1) may be obtained. Note also that the determinant of the left-hand side of (5) is

$$a(t)y_1(t)y_2'(t) - a(t)y_1'(t)y_2(t) \equiv a(t)W[y_1, y_2; t]$$

where $W[y_1, y_2; t]$ is the wronskian of y_1 and y_2 at t. Since y_1 and y_2 are linearly independent solutions to (2), $W[y_1, y_2; t] \neq 0$ for all $t \in I$ [see Theorem 2.2-2 (iv)]. Since $a(t) \neq 0$ for all $t \in I$ as well, the system (5) has a unique solution $A'(t)$, $B'(t)$ for all $t \in I$. In fact, this solution is easily seen to be

$$(5') \qquad A'(t) = \frac{-y_2(t)f(t)}{a(t)W[y_1, y_2; t]} \qquad \text{and} \qquad B'(t) = \frac{y_1(t)f(t)}{a(t)W[y_1, y_2; t]}$$

for all $t \in I$. Thus A and B can be computed by determining an antiderivative of each of the two expressions in (5′). In order to see that the function y_p defined by (4) is in fact a particular solution, we simply substitute this expression for y into equation (1).

Suppressing the variable t we have $y_p = Ay_1 + By_2$, and

$$y_p' = Ay_1' + By_2' + A'y_1 + B'y_2 = Ay_1' + By_2'$$

by the first equation in (5). Thus,

$$y_p'' = Ay_1'' + By_2'' + A'y_1' + B'y_2'$$

and using the second equation in (5) along with the fact that y_1 and y_2 are solutions to (2),

$$
\begin{aligned}
ay_p'' + by_p' + cy_p &= A(ay_1'' + by_1' + cy_1) + B(ay_2'' + by_2' + cy_2) + aA'y_1' + aB'y_2' \\
&= A \cdot 0 + B \cdot 0 + f
\end{aligned}
$$

and hence y_p is a solution to (1).

▶ EXAMPLE 2.5-1

Consider the equation $y'' - 2y' + y = e^t/t^2$, $t > 0$. Since 1 is a double root to the auxiliary equation, $y = c_1 e^t + c_2 te^t$ is the general solution to the corresponding homogeneous equation. Assuming that $y_p = A(t)e^t + B(t)te^t$ is a solution [see (4) and (5)], we see that A' and B' are solutions to

$$A'e^t + B'te^t = 0$$

$$A'e^t + B'(e^t + te^t) = \frac{e^t}{t^2}$$

By subtracting the first equation from the second, $B'e^t = e^t/t^2$, and hence $B' = 1/t^2$. Thus, taking $B = -1/t$ and substituting into the first equation, $A'e^t = -e^t/t$. Thus $A = -\ln t$, and it follows that

$$y_p = (-\ln t)e^t - e^t$$

is a particular solution, and

$$y = c_1 e^t + c_2 te^t - (\ln t + 1)e^t$$

is the general solution.

Even though the corresponding homogeneous equation in Example 2.5-1 has constant coefficients, the nonhomogeneous term e^t/t^2 does not have a finite differential family, and hence the method of undetermined coefficients does not apply to this equation.

EXAMPLE 2.5-2

Consider the equation $t^2y'' + ty' - y = t^2 \ln t$, $t > 0$. It is easy to verify that $y = c_1 t + c_2(1/t)$ is the general solution to the corresponding homogeneous equation. Assuming that $y_p(t) = A(t)t + B(t)(1/t)$ leads to the system

$$A't + B'\frac{1}{t} = 0$$

$$A' - B'\frac{1}{t^2} = \ln t$$

Multiplying the second equation by t and adding the resulting equation to the first equation leads to $2tA' = t \ln t$. Hence $A' = \frac{1}{2} \ln t$, and one solution for A is $A = t(\ln t - 1)/2$. Since $B' = -t^2A' = -\frac{1}{2}t^2 \ln t$, one solution for B is $B = -t^3(3 \ln t - 1)/18$ and

$$y_p = \frac{t^2(\ln t - 1)}{2} - \frac{t^2(3 \ln t - 1)}{18}$$

$$= \frac{t^2 \ln t}{3} - \frac{4t^2}{9}$$

is a particular solution. Therefore,

$$y = c_1 t + c_2 t^{-1} + \frac{t^2 \ln t}{3} - \frac{4t^2}{9}$$

is the general solution to this equation.

Although it is usually easier to use the procedure indicated in the proof of Theorem 2.5-1 to actually compute particular solutions to (1), sometimes it is convenient to have an explicit representation of a particular solution. If t_0 is any point in the interval I, we can integrate A' and B' in (5)′ from t_0 to t and substitute in (4) to obtain

$$(6) \qquad y_p(t) = \int_{t_0}^{t} \frac{y_2(t)y_1(s) - y_1(t)y_2(s)}{a(s)W[y_1, y_2; s]} f(s)\, ds$$

Sometimes formula (6) can give helpful representations of the solutions to (1) (see Problems 2.5-4 and 2.5-5). Note that if $y_p(t)$ is given by (6), then $y_p(t_0) = y_p'(t_0) = 0$.

2.5a Reduction of Order

Although the method of variation of constants applies to general second order nonhomogeneous equations, it requires that the general solution to the homogeneous equation be known. This presents no problem if the equation has constant coefficients, but in general there is no technique that can be used to obtain the homogeneous solution. In certain cases the homogeneous equation (2) may have one nontrivial solution that is simple and may be easy to determine. [For example, if $a(t) + b(t) + c(t) \equiv 0$, then $y_1(t) = e^t$ is a solution to (2); and if $b(t) + tc(t) \equiv 0$, then $y_1(t) = t$ is a solution to (2).] We show now that if one nontrivial solution to the homogeneous equation (2) is known, then a change of variables (that is completely analogous to the variation of constants procedure) transforms the nonhomogeneous equation (1) into an equivalent first order differential equation. This method is called *reduction of order* and can be used not only to determine particular solutions to nonhomogeneous equations but also to determine a second linearly independent solution to homogeneous equations.

Before stating the general result we illustrate the procedure with a specific example. Consider the equation

$$(7) \qquad 2t^2 y'' - ty' + y = 0 \qquad t > 0$$

and observe that $y(t) \equiv t$ is one solution to this equation. Now assume that $y(t) \equiv A(t)t$ is a solution and determine the possible functions $A(t)$.

Suppressing the variable t, we have $y' = A't + A$, $y'' = A''t + 2A'$, and substituting for y in (7) leads to the equation

$$(2t^3A'' + 4t^2A') - (t^2A' + tA) + tA = 0$$

Hence $2t^3A'' + 3t^2A' = 0$, and by dividing each side by $2t^3$, we obtain

$$A'' + \frac{3}{2}\frac{1}{t}A' = 0$$

This is a first order linear equation in A' and the integrating factor is

$$\exp\left(\frac{3}{2}\int\frac{1}{t}\,dt\right) = \exp\left(\frac{3}{2}\ln t\right) = t^{3/2}$$

Therefore, $t^{3/2}A'' + \frac{3}{2}t^{1/2}A' = (t^{3/2}A')' = 0$, and it follows that $A' = \bar{c}t^{-3/2}$. Antidifferentiating again, $A(t) = c_1t^{-1/2} + c_2$, and since $y(t) = A(t)t$, it follows that

$$y(t) = c_1t^{1/2} + c_2t$$

is the general solution to (7). In general, the following result is valid and can be applied whenever a nontrivial solution to the homogeneous equation can be found.

Theorem 2.5-2

Suppose that y_1 is a nontrivial solution to the homogeneous equation (2). Then a second linearly independent solution y_2 to the homogeneous equation (2) has the form

$$y_2(t) = A(t)y_1(t) \tag{8}$$

where A is any continuously differentiable function whose derivative $A' = w$ is a nontrivial solution to the first order equation

$$a(t)y_1(t)w' + [2a(t)y_1'(t) + b(t)y_1(t)]w = 0 \tag{9}$$

Also, a particular solution y_p to the nonhomogeneous equation (1) has the form

$$y_p(t) = B(t)y_1(t) \tag{10}$$

where B is any continuously differentiable function whose derivative $B' = v$ is a solution to the first order equation

$$a(t)y_1(t)v' + [2a(t)y_1'(t) + b(t)y_1(t)]v = f(t) \tag{11}$$

Moreover these representations are valid on any interval $J \subset I$, where $y_1(t) \neq 0$ for all $t \in J$.

This theorem is established in a manner similar to Theorem 2.5-1. To see that (10) is valid, set $y_p(t) = B(t)y_1(t)$, where $B(t)$ is to be determined. Suppressing the variable t, we have $y_p' = By_1' + B'y_1$ and $y_p'' = By_1'' + 2B'y_1' + B''y_1$. Therefore, using the fact that y_1 is a solution to the homogeneous equation (2),

$$a(By_1'' + 2B'y_1' + B''y_1) + b(By_1' + B'y_1) + cBy_1 = ay_1B'' + (2ay_1' + by_1)B' = f$$

whenever B' is a solution to (11). This establishes (10), and (8) follows in exactly the same way with $f = 0$ (note that y_1 and y_2 are certainly linearly independent since A is not a constant function). Of course, equations (9) and (11) can be solved by the methods in Section 1.2.

EXAMPLE 2.5-3

Consider the equation $ty'' + (1 - 2t)y' + (t - 1)y = te^t$, $t > 0$. Since $y_1 = e^t$ is a solution to the corresponding homogeneous equation, we reduce the order by assuming that $y(t) = A(t)e^t$ for all $t > 0$. Then $y' = Ae^t + A'e^t$, $y'' = Ae^t + 2A'e^t + A''e^t$, and we obtain $te^tA'' + e^tA' = te^t$, and hence $tA'' + A' = t$. Solving this equation by computing an integrating factor, we obtain $(tA')' = t$, or

$$A' = \tfrac{1}{2}t + \frac{c_2}{t}$$

Thus $A = \frac{1}{4}t^2 + c_2 \ln t + c_1$, and since $y = Ae^t$,

$$y = c_1e^t + c_2e^t \ln t + \tfrac{1}{4}t^2e^t$$

is the general solution to this equation. Note also that $y_2 = e^t \ln t$ is a solution to the corresponding homogeneous problem and that $y_1 = e^t$, $y_2 = e^t \ln t$ are linearly independent.

*2.5*b* Limits of Solutions at Infinity

As a final topic in this section we show that the methods developed for solving second order homogeneous and nonhomogeneous equations can be used effectively for the qualitative behavior of solutions as well as the actual computation of solutions. For this analysis the coefficients a, b, and c are assumed to be *constants* and the interval I is assumed to be $[0, \infty)$. Suppose that f is a continuous function on $[0, \infty)$ and that a, b, and c are constants with $a \neq 0$, and consider the nonhomogeneous equation

$$ay'' + by' + cy = f(t) \tag{12}$$

Our basic assumption is that the solutions to the corresponding homogeneous equation converge to 0 as t approaches $+\infty$. It is routine to determine that this is equivalent to assuming that the *real parts* of the auxiliary roots must be negative [see formula (14) below]. Since $a\lambda^2 + b\lambda + c = 0$ is the auxiliary equation, if α_1 and α_2 denote the real parts of the two auxiliary roots, it is easy to see that

$$(13)\quad \begin{aligned} &\alpha_1 = -\frac{b}{2a}, \alpha_2 = -\frac{b}{2a} \qquad \text{if } b^2 - 4ac < 0 \\ &\alpha_1 = \frac{-b + \sqrt{b^2 - 4ac}}{2a}, \alpha_2 = \frac{-b - \sqrt{b^2 - 4ac}}{2a} \qquad \text{if } b^2 - 4ac \geq 0 \end{aligned}$$

Taking $\beta = \sqrt{4ac - b^2}/2a$ whenever $b^2 - 4ac < 0$ and using Theorem 2.3-1, we have that if y_1 and y_2 are defined on $[0, \infty)$ by

$$(14)\quad \begin{aligned} &y_1(t) = e^{\alpha_1 t}, y_2(t) = e^{\alpha_2 t} \qquad \text{if } b^2 - 4ac > 0 \\ &y_1(t) = e^{\alpha_1 t}, y_2(t) = te^{\alpha_1 t} \qquad \text{if } b^2 - 4ac = 0 \\ &y_1(t) = e^{\alpha_1 t} \cos \beta t, y_2(t) = e^{\alpha_1 t} \sin \beta t \qquad \text{if } b^2 - 4ac < 0 \end{aligned}$$

then y_1 and y_2 are linearly independent solutions to the corresponding homogeneous equation

$$(15)\quad ay'' + by' + cy = 0$$

Therefore, the general solution to (15) is $y_H = c_1 y_1 + c_2 y_2$, where c_1 and c_2 are constants.

Using formula (13) it is easy to check that if the constants a, b, and c are all positive, then both α_1 and α_2 are negative. From formula (14) it follows that $|y_1(t)| \leq e^{\alpha_1 t}$ and $|y_2(t)| \leq e^{\alpha_2 t}$ [except for the case $b^2 - 4ac = 0$, when $|y_2(t)| = te^{\alpha_1 t}$ for all $t \geq 0$]. This implies that $|y_1(t)| + |y_2(t)| \to 0$ as $t \to +\infty$ whenever $\alpha_1, \alpha_2 < 0$. Since the general solution to (15) is the class of all linear combinations of y_1 and y_2, we have the following important fact:

Proposition 1

Suppose that the constants $a, b, c > 0$ and that y_H is a solution to the homogeneous equation (15). Then $y_H(t) \to 0$ as $t \to +\infty$.

The concern now is to determine the behavior as t approaches $+\infty$ of the solutions to (12) when the nonhomogeneous term $f(t)$ has certain properties. Suppose, for example, that there is a constant L such that $f(t) \equiv L$ for all $t \geq 0$. If $y_p(t) \equiv L/c$, then y_p is easily seen to be a particular solution to (12). Therefore, if $a, b, c > 0$ and $f(t) \equiv L$, then every solution y to (12) has the form $y(t) \equiv y_H(t) + L/c$, where y_H is a solution to (15). Since $y_H(t) \to 0$ as

$t \to +\infty$, by Proposition 1 it follows that $y(t) \to L/c$ as $t \to +\infty$. Our next result asserts that the same behavior is true if $f(t) \to L$ as $t \to +\infty$.

Theorem 2.5-3

Suppose that the constants a, b, and c in equation (12) are all positive.

(i) If there is a number $M > 0$ such that $|f(t)| \leq M$ for all $t \geq 0$, then every solution to the nonhomogeneous equation (12) is bounded on $[0, \infty)$.

(ii) If $\lim_{t \to \infty} f(t) = L$, then each solution y to the nonhomogeneous equation (12) satisfies $\lim_{t \to \infty} y(t) = L/c$ (note that $c > 0$ by hypothesis).

Theorem 2.5-3 can be readily applied to help determine the behavior of solutions to (12) without having to compute the solutions explicitly (see Problem 2.5-9). Even though the proof of Theorem 2.5-3 is difficult, it can be easily applied to important equations (see, in particular, the equation for forced vibrations in Section 2.6*a*). As a general physical interpretation, consider the function f as an external forcing function. Part (i) asserts that a bounded forcing function results in a bounded solution, and part (ii) asserts that if the external force settles down toward a constant L as time goes on, then the long-time behavior of the resulting $y(t)$ is the same as if the external force were always the constant L. Since the proof of Theorem 2.5-3 is difficult, an outline indicating the main ideas is given at the end of this section. However, the student can skip over the proof without loss since the ideas developed in the proof play no role in succeeding parts of the text.

In order to illustrate Theorem 2.5-3, consider the equation

$$y'' + 2y' + 5y = \cos\frac{1}{1+t} \qquad t \geq 0 \tag{16}$$

Since the coefficients are positive constants (notice that $\lambda = -1 \pm 2i$ are the auxiliary roots to the corresponding homogeneous equation) and since

$$\lim_{t \to \infty} \cos\frac{1}{1+t} = \cos 0 = 1$$

it follows directly from in Theorem 2.5-3 (ii) that if y is *any* solution to (16) then $y(t) \to \frac{1}{5}$ as $t \to \infty$. Similarly, every solution y to

$$y'' + 2y' + 5y = e^{-t}\sin t^2$$

satisfies

$$\lim_{t \to \infty} y(t) = \tfrac{1}{5}\lim_{t \to \infty} e^{-t}\sin t^2 = 0$$

In each case the actual solutions cannot be explicitly displayed in elementary terms.

In order to indicate a proof of (i), we use a variation of constants formula and Abel's identity for the wronskian of y_1 and y_2. Setting $k = y_1(0)y_2'(0) - y_1'(0)y_2(0) \neq 0$ and using the result in Problem 2.2-3, we obtain

$$y_1(t)y_2'(t) - y_1'(t)y_2(t) = ke^{-bt/a}$$

Using the variation of constants formula (6),

$$y_p(t) = \int_0^t \frac{y_1(s)y_2(t) - y_1(t)y_2(s)}{ake^{-bs/a}} f(s)\, ds \qquad \text{for all } t \geq 0$$

is a particular solution to (12). From the formulas in (14) along with the fact that $\alpha_2 \leq \alpha_1 < 0$ and $\alpha_1 + \alpha_2 = -b/a$ [see (13)], it follows that

$$\begin{aligned} \left| \frac{y_1(s)y_2(t) - y_1(t)y_2(s)}{ake^{-bs/a}} \right| &\leq \frac{e^{\alpha_1 s + \alpha_2 t} + e^{\alpha_1 t + \alpha_2 s}}{ake^{(\alpha_1 + \alpha_2)s}} \\ &= \frac{1}{ak} [e^{\alpha_2(t-s)} + e^{\alpha_1(t-s)}] \\ &\leq \frac{2}{ak} e^{\alpha_1(t-s)} \end{aligned}$$

whenever $t \geq s \geq 0$ and $b^2 - 4ac \neq 0$, and

$$\begin{aligned} \left| \frac{y_1(s)y_2(t) - y_1(t)y_2(s)}{ake^{-bs/a}} \right| &\leq \frac{te^{\alpha_1 s + \alpha_1 t} - se^{\alpha_1 t + \alpha_1 s}}{ake^{2\alpha_1 s}} \\ &= \frac{1}{ak} (t - s)e^{\alpha_1(t-s)} \end{aligned}$$

whenever $t \geq s \geq 0$ and $b^2 - 4ac = 0$.

Since every solution to (12) is of the form $y = c_1 y_1 + c_2 y_2 + y_p$ where $|y_1(t)|, |y_2(t)| \to 0$ as $t \to \infty$, it suffices to show that y_p is bounded on $[0, \infty)$. In the case $b^2 - 4ac \neq 0$ we have

$$\begin{aligned} |y_p(t)| &\leq \int_0^t \frac{1}{ak} e^{\alpha_1(t-s)} |f(s)|\, ds \\ &\leq \frac{e^{\alpha_1 t}}{ak} \int_0^t e^{-\alpha_1 s} M\, ds = \frac{M}{\alpha_1 ak} (1 - e^{\alpha_1 t}) \\ &\leq \frac{M}{\alpha_1 ak} \end{aligned}$$

and so y_p is bounded on $[0, \infty)$. In the case $b^2 - 4ac = 0$ we have, using integration by parts,

$$\begin{aligned} |y_p(t)| &\leq \frac{M}{ak} \int_0^t (t-s)e^{\alpha_1(1-s)}\, ds \\ &= \frac{M}{ak}\left[-\frac{(t-s)}{\alpha_1} e^{\alpha_1(t-s)} \right]_{s=0}^{t} - \frac{M}{ak}\int_0^t \frac{1}{\alpha_1} e^{\alpha_1(t-s)}\, ds \\ &= \frac{M}{ak\alpha_1} te^{\alpha_1 t} + \frac{M}{ak\alpha_1^2}[1 - e^{\alpha_1 t}] \end{aligned}$$

and since $\alpha_1 < 0$, y_p is also bounded. This establishes part (i). In order to show that (ii) is valid, it is first assumed that $\lim_{t\to\infty} |f(t)| = 0$. As above, in the case $b^2 - 4ac \neq 0$,

$$|y_p(t)| \leq \int_0^t \frac{1}{ak} e^{\alpha_1(t-s)} |f(s)|\, ds = \frac{\int_0^t e^{-\alpha_1 s}|f(s)|\, ds}{ake^{-\alpha_1 t}}$$

Since $\alpha_1 < 0$ the denominator converges to $+\infty$ as $t \to +\infty$, and since the numerator is nonnegative, increasing in t, it either is bounded on $[0, \infty)$ or converges to $+\infty$ as $t \to +\infty$. If it is bounded, then obviously $|y_p(t)| \to 0$ as $t \to +\infty$; and if it is unbounded, then L'Hospital's rule applies; and since

$$\frac{\frac{d}{dt}\int_0^t e^{-\alpha_1 s}|f(s)|\, ds}{d/dt[ake^{-\alpha_1 t}]} = \frac{e^{-\alpha_1 t}|f(t)|}{-\alpha_1\, ake^{-\alpha_1 t}} = \frac{|f(t)|}{-\alpha_1\, ak} \to 0$$

as $t \to \infty$, it follows that $|y_p(t)| \to 0$ as $t \to +\infty$ in this case as well. The case when $b^2 - 4ac = 0$ is treated in a somewhat similar manner and is left as an exercise (see Problem 2.5-7).

Part (ii) now follows easily. Suppose that $\lim_{t\to\infty} f(t) = L$, and rewrite the function f as $f = f_1 + f_2$, where $f_1(t) \equiv L$ and $f_2(t) \equiv f(t) - L$ on $[0, \infty)$. Then a particular solution y_p to (12) can be written in the form $\bar{y}_p = \bar{y}_{p1} + \bar{y}_{p2}$, where $\bar{y}_{p1}$ is a solution to (12) with $f = f_1$ and $\bar{y}_{p2}$ is a solution to (12) with $f = f_2$. Since $f_2(t) \to 0$ as $t \to \infty$, we have already shown that $\bar{y}_{p2}(t) \to 0$ as $t \to \infty$. Since f_1 is constant and a constant is not a solution to the homogeneous equation (15) (note that 0 is not an auxiliary root), we have by undetermined coefficients that $\bar{y}_{p1}(t) \equiv d_1$ is a solution for some constant d_1. Clearly $d_1 = L/c$, and hence

$$y_p(t) = \bar{y}_{p1}(t) + \bar{y}_{p2}(t) = \frac{L}{c} + \bar{y}_{p2}(t) \to \frac{L}{c}$$

as $t \to +\infty$, and part (ii) is also seen to be true.

PROBLEMS

2.5-1. Determine a particular solution to the following nonhomogeneous equations.

(a) $y'' + y = \tan t \qquad -\frac{\pi}{2} < t < \frac{\pi}{2}$

(b) $y'' + 2y' + y = e^{-t} \ln t \qquad t > 0$

(c) $y'' + 4y = \sec 2t \qquad -\frac{\pi}{4} < t < \frac{\pi}{4}$

(d) $y'' + 3y' + 2y = \dfrac{1}{1 + e^{2t}}$

(e) $y'' - 2y' + 2y = e^t \tan t \qquad -\frac{\pi}{2} < t < \frac{\pi}{2}$

(f) $y'' - y = \dfrac{1}{1 + e^t}$

(g) $y'' + y = \dfrac{1}{1 + \sin t} \qquad -\frac{3\pi}{2} < t < \frac{\pi}{2}$

(h) $y'' - y = \dfrac{e^t}{1 + e^{2t}}$

2.5-2. In each of the following problems either one nontrivial solution $\{y_1\}$ or two linearly independent solutions $\{y_1, y_2\}$ to the corresponding homogeneous equation are given. From this information determine the general solution to the given nonhomogeneous equation.

(a) $t^2y'' - ty' + y = t \qquad t > 0 \qquad \{y_1 = t\}$

(b) $ty'' + (1 - 2t)y' + (t - 1)y = e^t \qquad t > 0 \qquad \{y_1 = e^t\}$

(c) $t^2y'' - 2ty' + 2y = t \ln t \qquad t > 0 \qquad \{y_1 = t, y_2 = t^2\}$

(d) $t^2y'' - t(t + 2)y' + (t + 2)y = t^4 \qquad t > 0 \qquad \{y_1 = t\}$

2.5-3. Suppose that y_1 is a nontrivial solution to the homogeneous equation (2) and J is an interval such that $y_1(t) > 0$ for all t in J. If t_0 is in J and y_2 is a solution to

$$y_1(t)y_2' - y_1'(t)y_2 = \exp\left(\int_{t_0}^{t} b(s)a(s)^{-1}\, ds\right) \qquad y_2(t_0) = 0$$

show that y_2 is a solution to (2) on J (see Abel's identity in Problem 2.2-3). Show that y_1 and y_2 are linearly independent and determine y_2 explicitly in terms of an integral. Compare the solution y_2 obtained with this procedure with the solution y_2 defined by formula (8) in Theorem 2.5-2.

2.5-4. Suppose that $k > 0$ and consider $y'' + k^2y = f(t)$, $t \geq 0$, where f is continuous on $[0, \infty)$. Show that

$$y_p(t) \equiv k^{-1} \int_0^t \sin\,[k(t - s)]f(s)\, ds \qquad \text{for } t \geq 0$$

is a particular solution to this equation that satisfies $y_p(0) = y_p'(0) = 0$.

2.5-5. Suppose that $k > 0$ and consider $y'' - k^2y = f(t)$, $t \geq 0$, where f is continuous on $[0, \infty)$. Show that

$$y_p(t) \equiv k^{-1} \int_0^t \tfrac{1}{2}[e^{k(t-s)} - e^{-k(t-s)}]f(s)\, ds \qquad \text{for } t \geq 0$$

is a particular solution to this equation that satisfies $y_p(0) = y_p'(0) = 0$. [Recall that $\sinh(x) \equiv \frac{1}{2}(e^x - e^{-x})$ and compare this problem with the preceding one.]

2.5-6. Use Cramer's rule for solving systems of linear equations to obtain that (5′) is the solution of (5).

***2.5-7.** Suppose that a, b, and c are constants with $a \neq 0$, that $b^2 - 4ac = 0$ (so that $-b/2a$ is double auxiliary root), and that y_1 and y_2 are defined by (14). Show that if $-b/2a < 0$ and y_p is a solution of (12), where $f(t) \to 0$ as $t \to +\infty$, then $y_p(t) \to 0$ as $t \to +\infty$.

***2.5-8.** Suppose that in equation (12), one of the auxiliary roots is 0 and the other is negative (that is, $c = 0$ and $-b/a < 0$).

(a) Give an example of a bounded function f on $[0, \infty)$ so that every solution to (12) is unbounded on $[0, \infty)$.

(b) Show that if $\int_0^\infty |f(s)|\, ds < \infty$, then every solution to (12) is bounded on $[0, \infty)$.

2.5-9. Discuss the boundedness on $[0, \infty)$ and the behavior as $t \to +\infty$ of solutions to the following problems (it is assumed that $a, b, c > 0$).

(a) $ay'' + by' + cy = \dfrac{\sin t}{1 + t^2}$ for $t \geq 0$

(b) $ay'' + by' + cy = \dfrac{t^2}{1 + t^2}$ for $t \geq 0$

(c) $ay'' + by' + cy = 2 \cos e^{-t}$ for $t \geq 0$

(d) $ay'' + by' + cy = e^{\sin t}$ for $t \geq 0$

(e) $ay'' + by' + cy = (2t + 1) \sin (t + 1)^{-1}$ for $t \geq 0$

2.5-10. Using any of the previous methods of solution, determine the general solution to each of the following nonhomogeneous equations.

(a) $\dfrac{d^2s}{dt^2} - 4\dfrac{ds}{dt} + 4s = 1 - 4e^{2t}$

(b) $\dfrac{d^2r}{d\theta^2} - 2\dfrac{dr}{d\theta} + r = \dfrac{e^\theta}{(1 - \theta)^2}$

(c) $\dfrac{d^2z}{dx^2} + z = 2 \sec^3 x \tan x$

(d) $\dfrac{d^2u}{ds^2} + u = \sin s + e^s$

(e) $\dfrac{d^2\phi}{ds^2} - 3\dfrac{d\phi}{ds} + 2\phi = \dfrac{1}{1 + e^{-s}}$

2.6 VIBRATIONS IN LINEAR SPRINGS

One of the important applications of second order linear differential equations is in the description of vibrating motion or oscillations. In order to illustrate some of the basic ideas, the motion of a single mass attached to a simple spring is considered. Suppose that a vertical spring is secured at the top and has an object of mass m attached to the bottom. Figure 2.1 indicates the position of the spring at rest both with and without the mass m attached. The vertical position of the mass for each time t is denoted by $y(t)$, and the equilibrium point $y = 0$ is the place where the mass remains at rest (i.e., the force of the weight exactly balances the upward pull of the spring). The positive direction for y is upward, and positive forces are in the upward direction.

First a few basic and simplifying assumptions about the forces acting on the system are made. It is assumed that the spring is a *linear spring: the force on the mass due to the spring is proportional to the distance the mass is from equilibrium* (this is known as *Hooke's law*). Thus if $F_s(t)$ is the force due to the spring at time t, then

$$F_s(t) = -ky(t) \tag{1}$$

Since the spring force is in the opposite direction of $y(t)$ [note that $|y(t)|$ is the distance from equilibrium, with $y(t) > 0$ when above equilibrium and $y(t) < 0$ when below], the proportionality constant k is positive. The positive constant k is called the *spring constant* for the spring, and the larger the constant k the stiffer the spring. Note that in the case of Figure 2.1, $kD = mg$, where g is the gravitational constant. It is also assumed that there is a resistance force

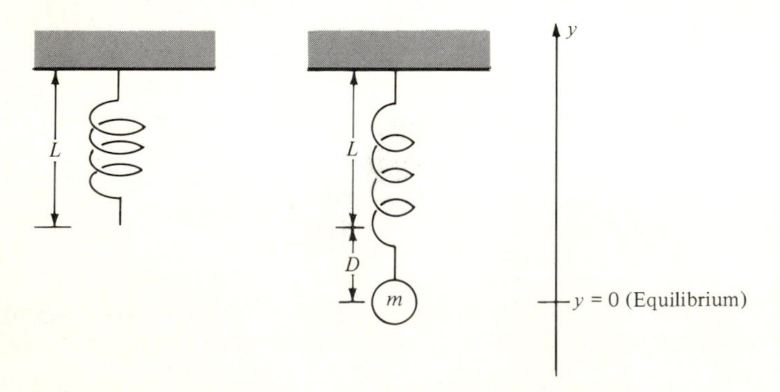

Figure 2.1 Equilibrium position for a mass-spring system.

due to friction (called a *damping* force) that is proportional to the velocity of the mass. Since $y'(t)$ is the velocity at time t, if $F_d(t)$ denotes the damping force on the mass at time t,

$$F_d(t) = -\mu y'(t) \tag{2}$$

where the *damping coefficient* μ is nonnegative, and hence the damping force is in the opposite direction of the motion. It is also assumed that the weight of the spring is small relative to the weight of the mass, and hence the weight of the spring is neglected. According to Newton's second law of motion (see Section 1.3), the rate of change of the momentum equals the total force acting on the mass, and since the mass m remains constant, the total force $F_T(t)$ equals $my''(t)$. Assuming that there are no further forces on the system, we have that $F_T(t) = F_s(t) + F_d(t)$, and hence

$$my''(t) = -ky(t) - \mu y'(t)$$

for all times t. Assuming further that the initial time is $t_0 = 0$ and the mass m has an initial displacement y_0 and an initial velocity v_0, the equation describing the motion is the second order initial value problem

$$my'' + \mu y' + ky = 0 \qquad y(0) = y_0 \qquad y'(0) = v_0 \tag{3}$$

Since this equation has constant coefficients and is homogeneous, the method in Section 2.3 applies directly and equation (3) can be explicitly solved. First we consider the situation with no damping force (that is, $\mu = 0$) and then analyze the cases of a "small" damping term and a "large" damping term.

If the damping force is negligible, the system is called *undamped*, and, since the damping coefficient μ is 0, the equation of motion is

$$my'' + ky = 0 \qquad y(0) = y_0 \qquad y'(0) = v_0 \tag{4}$$

The auxiliary equation in this case is $m\lambda^2 + k = 0$, and hence $\lambda = \pm\sqrt{k/m}\,i$. The general solution to the differential equation is

$$y = c_1 \cos \sqrt{\frac{k}{m}}\, t + c_2 \sin \sqrt{\frac{k}{m}}\, t$$

and using the initial conditions gives $c_1 = y_0$ and $c_2 = v_0\sqrt{m/k}$. Therefore the solution to (4) is

$$y(t) = y_0 \cos \sqrt{\frac{k}{m}}\, t + v_0 \sqrt{\frac{m}{k}} \sin \sqrt{\frac{k}{m}}\, t \tag{5}$$

If the motion of an object satisfies the equation (4) and hence has the form (5), then it is said to be in *simple harmonic motion*. Since y is the sum of two periodic functions, each having period $2\pi\sqrt{m/k}$, y is itself periodic. In this case, using the formula for the sine of the sum of two numbers (see Problem

2.3-5), it is easy to see that the function y described by (5) can be written in the form

$$(5') \quad \begin{cases} y(t) = \sqrt{y_0^2 + v_0^2 m/k} \sin(\sqrt{k/m}\, t + \phi), \text{ where } \phi \text{ is such that} \\ \sin\phi = \dfrac{y_0}{\sqrt{y_0^2 + v_0^2 m/k}} \quad \text{and} \quad \cos\phi = \dfrac{v_0\sqrt{m/k}}{\sqrt{y_0^2 + v_0^2 m/k}} \end{cases}$$

Thus the solution is a shift of a sine curve whose *amplitude* is $\sqrt{y_0^2 + v_0^2 m/k}$ and whose *period* is $2\pi\sqrt{m/k}$. Note that the amplitude is the maximum displacement of mass from equilibrium, and it depends on the initial conditions as well as the mass and the spring constant. The period of the motion (which is the reciprocal of the *frequency*) depends on the mass and the spring constant, but is independent of the initial conditions. The number ϕ is called the *phase shift.*

EXAMPLE 2.6-1

A suspended spring is stretched 6 inches by an attached 16-pound object. Assuming that it is a linear spring and that the damping force is negligible, describe the motion of the mass-spring system if (*a*) the mass is pulled 1 foot below equilibrium and released; and if (*b*) the mass is pulled 1 foot below equilibrium and hit upward with a velocity of 8 ft/sec. Since a weight of 16 pounds stretches the linear spring 6 inches ($\frac{1}{2}$ foot), it follows that $\frac{1}{2}k = 16$, and hence $k = 32$ lb/ft. Taking $g = 32$ ft/sec^2 and letting $y(t)$ denote the position of the mass relative to the equilibrium position (with the positive direction upward), we have that y is a solution to the differential equation

$$\tfrac{1}{2}y'' + 32y = 0$$

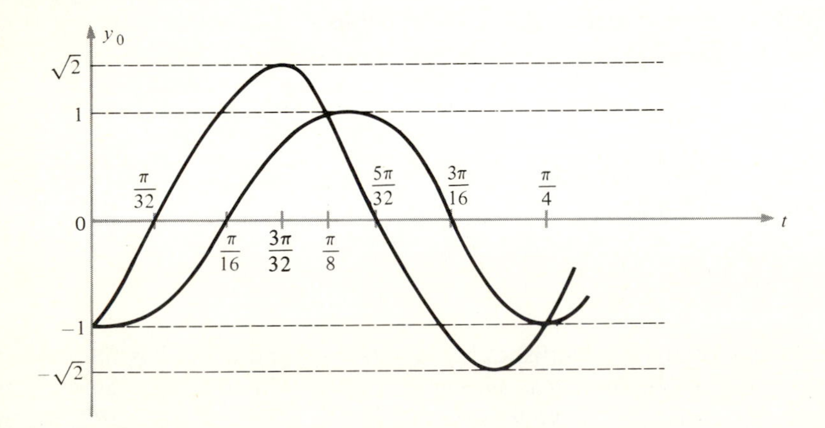

Figure 2.2 Free oscillations in mass-spring systems.

Hence $y(t) = c_1 \cos 8t + c_2 \sin 8t$ for all $t \geq 0$. In case *a* the initial conditions are $y(0) = -1$, $y'(0) = 0$, and hence $y(t) = -\cos 8t$ for all $t \geq 0$. In case *b*, the initial conditions are $y(0) = -1$, $y'(0) = 8$. Thus $y(t) = -\cos 8t + \sin 8t$ $[= \sqrt{2} \sin (8t - \pi/4)]$ for all $t \geq 0$. A sketch of the graphs of the two solutions is given in Figure 2.2.

The situation where damping forces are no longer negligible is now analyzed, so equation (3) is considered when the damping coefficient μ is positive. Since the auxiliary roots of (3) have the form

$$\lambda = \frac{-\mu \pm \sqrt{\mu^2 - 4mk}}{2m} \tag{6}$$

where μ, m, and k are positive, it is easy to see that the auxiliary roots have negative real parts. From this it is easy to conclude that the solution to (3) converges to 0 as t goes to $+\infty$ (see Proposition 1 in Section 2.5). Thus *the damping force always implies that the motion in the mass-spring system dies out after a long period of time.* A large amount of damping will affect the oscillatory properties of the mass-spring system as well. The solution to (3) can be completely described in the following manner.

Theorem 2.6-1

Suppose that the damping coefficient μ is positive. Then the solution y to (3) has exactly one of the following three forms:

(a) If $0 < \mu < 2\sqrt{mk}$, then

$$y(t) = e^{-\alpha t}\left[y_0 \cos \omega t + \left(\frac{v_0 + y_0 \alpha}{\omega}\right) \sin \omega t\right]$$

where $\alpha = \mu/2m$ and $\omega = \sqrt{4mk - \mu^2}/2m$;

(b) if $\mu = 2\sqrt{mk}$, then

$$y(t) = e^{-\alpha t}[y_0 + (v_0 + y_0 \alpha)t]$$

where $\alpha = \mu/2m$; and

(c) if $\mu > 2\sqrt{mk}$, then

$$y(t) = \left[\frac{v_0 + y_0 \alpha_2}{\alpha_2 - \alpha_1}\right] e^{-\alpha_1 t} - \left[\frac{v_0 + y_0 \alpha_1}{\alpha_2 - \alpha_1}\right] e^{-\alpha_2 t}$$

where

$$\alpha_1 = \frac{\mu - \sqrt{\mu^2 - 4mk}}{2m} \quad \text{and} \quad \alpha_2 = \frac{\mu + \sqrt{\mu^2 - 4mk}}{2m}$$

The fact that this theorem is valid follows directly from the form of the auxiliary roots in (6) and the results on homogeneous equations with constant coefficients developed in Section 2.3. As in equation (5) and (5)′, the form of y in part a of Theorem 2.6-1 can be expressed using a shift of the sine function. In particular,

(7) if $0 < \mu < 2\sqrt{mk}$, then

$$y(t) = \frac{\sqrt{\omega^2 y_0^2 + (v_0 + y_0\alpha)^2}}{\omega} e^{-\alpha t} \sin(\omega t - \phi)$$

where $\alpha = \mu/2m$, $\omega = \sqrt{4mk - \mu^2}$, and ϕ is such that $\cos\phi = (v_0 + y_0\alpha)[\omega^2 y_0^2 + (v_0 + y_0\alpha)^2]^{-1/2}$ and $\sin\phi = y_0\omega[\omega^2 y_0^2 + (v_0 + y_0\alpha)^2]^{-1/2}$

It is important to note the following observation concerning the motion of a damped mass-spring system: *If the damping coefficient satisfies* $0 < \mu < 2\sqrt{mk}$, *then the mass-spring system oscillates with constant frequency; and if* $\mu > 2\sqrt{mk}$, *then the mass-spring system passes through equilibrium at most once.* Since the crucial value for determining the oscillatory properties of the system is when $\mu = 2\sqrt{mk}$, when $\mu = 2\sqrt{mk}$ the system is said to be *critically damped.* When $\mu < 2\sqrt{mk}$ the system is said to be *underdamped*, and when $\mu > 2\sqrt{mk}$ it is said to be *overdamped.* (See Figure 2.3.)

If the mass-spring system were located in gaseous atmosphere, then the system would be underdamped, and if the mass were lowered from the equilibrium position and released, it would oscillate about the equilibrium solution and its motions would gradually die out. However, if the system were located in a thick liquid medium, then the system would probably be overdamped, and if the mass were lowered from the equilibrium position and released, it would not oscillate but would gradually return to the equilibrium position.

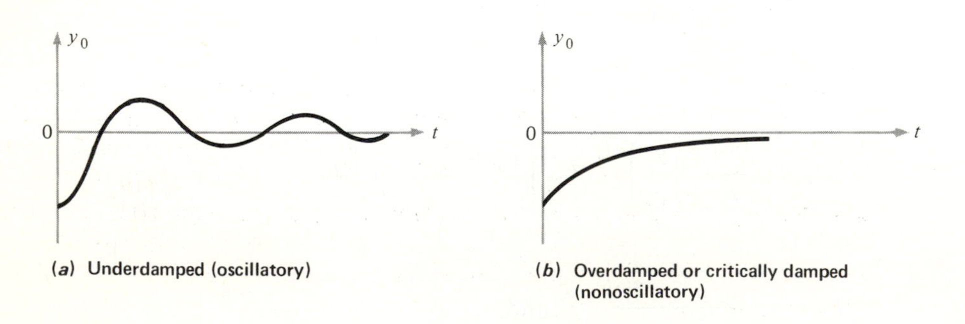

(*a*) Underdamped (oscillatory)

(*b*) Overdamped or critically damped (nonoscillatory)

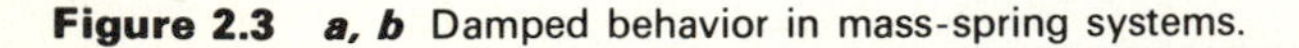

Figure 2.3 ***a, b*** Damped behavior in mass-spring systems.

EXAMPLE 2.6-2

Suppose a suspended linear spring is stretched 2 feet by an attached 16-pound object. Then $2k = 16$ and it follows that the spring constant is $k = 8$ lb/ft, and taking $g = 32$ ft/sec^2 we have that the mass $m = \frac{1}{2}$ (slug). Supposing further that there is a damping force with coefficient $\mu > 0$ and that $y(t)$ is the position of the mass relative to equilibrium, we have that y satisfies the differential equation

$$\tfrac{1}{2}y'' + \mu y' + 8y = 0 \tag{8}$$

The auxiliary roots for the equation are $\lambda = -\mu \pm \sqrt{\mu^2 - 16}$, and hence the system is critically damped when $\mu = 4$. Suppose that $\mu = 3$ (and hence the system is underdamped) and that the mass is raised 1 foot above equilibrium and hit downward with velocity 4 ft/sec. Since $\lambda = -3 \pm \sqrt{7}\,i$ are the auxiliary roots, the solution $y(t)$ to (8) has the form

$$y(t) = c_1 e^{-3t} \cos \sqrt{7}\,t + c_2 e^{-3t} \sin \sqrt{7}\,t$$

and from the initial conditions we have

$$y(0) = c_1 = 1 \qquad \text{and} \qquad y'(0) = -3c_1 + \sqrt{7}\,c_2 = -4$$

Thus $c_1 = 1$, $c_2 = -1/\sqrt{7}$, and

$$y(t) = e^{-3t} \cos \sqrt{7}\,t - (\sqrt{7})^{-1} e^{-3t} \sin \sqrt{7}\,t$$

is the solution to (8). Now suppose that $\mu = 5$ (and hence the system is overdamped), and the mass is raised 1 foot above equilibrium and hit downward with velocity 5 ft/sec. Since $\lambda = -2, -8$ are the auxiliary roots, the solution $y(t)$ to (8) has the form

$$y(t) = c_1 e^{-2t} + c_2 e^{-8t} \tag{9}$$

and from the initial conditions we have

$$y(0) = c_1 + c_2 = 1 \qquad \text{and} \qquad y'(0) = -2c_1 - 8c_2 = -5$$

Thus $c_1 = \frac{1}{2}$, $c_2 = \frac{1}{2}$, and

$$y(t) = \tfrac{1}{2}e^{-2t} + \tfrac{1}{2}e^{-8t}$$

is the solution to (8). Note that $y(t) > 0$ for all $t \geq 0$, and so the spring remains above equilibrium for all $t > 0$ and gradually moves back to the equilibrium position as $t \to +\infty$ (see Figure 2.4a). Assume again that $\mu = 5$, that the mass is raised 1 foot above equilibrium, and that, instead of being hit downward with velocity 5 ft/sec, it is hit downward with velocity 10 ft/sec. The solution still has the form (9), and from the initial conditions we have

$$y(0) = c_1 + c_2 = 1 \qquad \text{and} \qquad y'(0) = -2c_1 - 8c_2 = -10$$

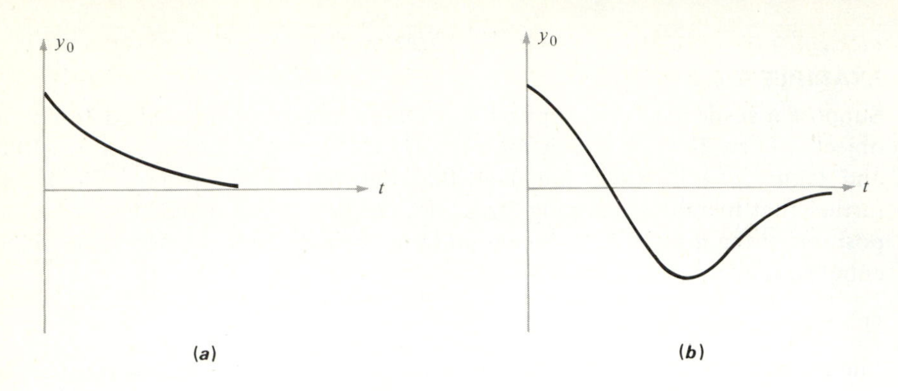

Figure 2.4 ***a, b*** Nonoscillatory behavior in overdamped systems.

Thus $c_1 = -\frac{1}{3}$ and $c_2 = \frac{4}{3}$, and we have that

$$y(t) = -\tfrac{1}{3}e^{-2t} + \tfrac{4}{3}e^{-8t}$$

is the solution to (8). This time the mass goes below equilibrium, reaches a minimum, and then moves upward, approaching equilibrium from below as $t \to +\infty$ (see Figure 2.4*b*).

2.6*a* Mechanical Vibrations with External Forcing

As a final topic it is assumed that, in addition to the force of the spring and the resistance force, an external force is acting on the system. Denoting the external force by $F(t)$ for each time t, it follows from Newton's law that

$$\text{Total force} = my''(t) = -ky(t) - \mu y'(t) + F(t)$$

and hence

$$(10) \qquad my'' + \mu y' + ky = F(t) \qquad y(0) = y_0 \qquad y'(0) = v_0$$

is the equation of motion. This equation is nonhomogeneous, and since the coefficients are constant, it can always be explicitly solved (in terms of antiderivatives) by variation of parameters (see Section 2.5). If F has a finite differential family, then undetermined coefficients may be used (see Section 2.4).

When the damping coefficient μ is positive, a considerable amount can be said about the behavior of the solution to (10) without actually solving the equation. Since the auxiliary roots to the corresponding homogeneous equation have negative real parts when $\mu > 0$, it follows from Theorem 2.5-3 that if there is a number $M > 0$ such that $|F(t)| \leq M$ for all $t \geq 0$, then the solution to (10) is bounded on $[0, \infty)$, and if $\lim_{t\to\infty} F(t) = L$, then the solution y to (10) satisfies $y(t) \to L/k$ as $t \to +\infty$. Therefore, in the damped case, the motion of the externally forced mass-spring system remains uniformly bounded on $[0, \infty)$ whenever the external force F remains bounded on $[0, \infty)$. Also, if the external force "dies out" as time becomes large [that is, $F(t) \to 0$ as $t \to \infty$], then the motion of the mass approaches equilibrium as time becomes large.

If the damping term is neglected, however, bounded external forces can lead to unbounded solutions to (10) (and hence the mass-spring system will vibrate "wildly" after a long period of time). This can be well illustrated with a particular periodic forcing function. Assume that $\mu = 0$ and that A and ω are positive numbers, and consider the equation

$$my'' + ky = A \cos \omega t \tag{11}$$

Using the method of undetermined coefficients it follows easily that if $\sqrt{k/m} \neq \omega$ the general solution of (11) is

$$y = c_1 \cos \sqrt{\frac{k}{m}}\, t + c_2 \sin \sqrt{\frac{k}{m}}\, t + \frac{A}{k - m\omega^2} \cos \omega t$$

where c_1 and c_2 are constants. If the system is initially resting at equilibrium position, then $y(0) = y'(0) = 0$, and it follows that the solution to (11) has the form

$$y(t) = \frac{A}{k - m\omega^2}\left(\cos \omega t - \cos \sqrt{\frac{k}{m}}\, t\right) \tag{12}$$

Recalling the trigonometric identities

$$\cos(\phi \pm \psi) = \cos \phi \cos \psi \mp \sin \phi \sin \psi$$

and taking $\phi = (\sqrt{k/m} + \omega)(t/2)$ and $\psi = (\sqrt{k/m} - \omega)(t/2)$ shows that formula (12) can be written in the form

$$y(t) = \frac{2A}{k - m\omega^2} \sin\left[\left(\sqrt{\frac{k}{m}} - \omega\right)\frac{t}{2}\right] \sin\left[\left(\sqrt{\frac{k}{m}} + \omega\right)\frac{t}{2}\right] \tag{12'}$$

Therefore, if $|\sqrt{k/m} - \omega|$ is small relative to $\sqrt{k/m} + \omega$, the function $y(t)$ can be considered as the product of a slowly oscillating function with a rapidly oscillating one. This type of motion illustrates a phenomenon known as a *beat* (see Figure 2.5).

The behavior of the solutions to (11) are considerably different when $\omega = \sqrt{k/m}$ (that is, when the forcing frequency is the same as the frequency of the unforced system). For if $\omega = \sqrt{k/m}$ then a member of the differential family for the right-hand side is a solution to the corresponding homogeneous equation, and this causes a factor of t to occur in the particular solution (see Theorem 2.4-2). Applying the method of undetermined coefficients, it follows easily that

$$y(t) = c_1 \cos \sqrt{\frac{k}{m}}\, t + c_2 \sin \sqrt{\frac{k}{m}}\, t + \frac{A}{2\sqrt{k/m}}\, t \sin \sqrt{\frac{k}{m}}\, t$$

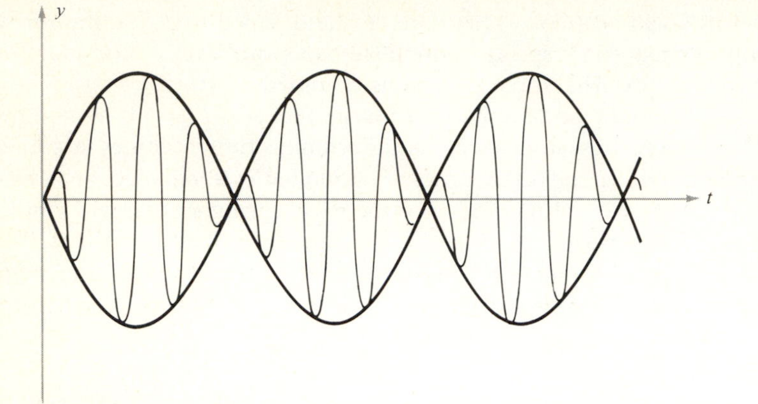

Figure 2.5 Beats.

is the general solution to (11) when $\omega = \sqrt{k/m}$. Therefore, for large t the mass-spring system vibrates wildly. This phenomenon is called *resonance*.

EXAMPLE 2.6-3

Consider equation (8) in Example 2.6-2 with external forcing:

$$\tfrac{1}{2}y'' + \mu y' + 8y = \cos \omega t \tag{13}$$

If $\mu > 0$ the method of undetermined coefficients leads to the particular solution

$$y_p = \frac{8 - \omega^2/2}{(8 - \omega^2/2)^2 + \mu^2\omega^2} \cos \omega t + \frac{\mu\omega}{(2 - \omega^2/2)^2 + \mu^2\omega^2} \sin \omega t$$

Choosing a phase shift ϕ appropriately shows that

$$y_p = \frac{1}{\sqrt{(8 - \omega^2/2)^2 + \mu^2\omega^2}} \sin (\omega t + \phi) \tag{14}$$

Since the auxiliary roots of the homogeneous equation corresponding to (13) have negative real parts for $\mu > 0$, the solution y_H must converge to 0 as $t \to \infty$. Therefore, since the solutions y to (13) are all of the form $y = y_H + y_p$, we have that each solution to (13) approaches y_p as $t \to \infty$. The solution y_p is known as the *steady-state* term and describes closely the behavior of the solutions to (13) after a long period of time. It is important to note that if the resistance term is neglected, the system is at resonance when $\omega = 4$. Also, since the amplitude A of y_p is given by

$$A = \frac{1}{\sqrt{(8 - \omega^2/2)^2 + \mu^2\omega^2}}$$

we see that A becomes large whenever ω is near 4 and μ is near 0. Therefore, resonance must be taken into account even if friction is present.

PROBLEMS

2.6-1. A 16-pound weight suspended from a linear spring stretches it 2 feet. Let $y(t)$ denote the distance of the mass from equilibrium, with the positive direction upward. Assuming that the damping force is negligible, find the differential equation that y must satisfy (assume that $g = 32$ ft/sec^2).

(a) Determine $y(t)$ explicitly if the mass is pulled 3 feet above equilibrium and released.

(b) Determine $y(t)$ explicitly if the mass is pulled 1 foot below equilibrium and hit upward with initial velocity $4\sqrt{3}$ ft/sec. Compute also the frequency and the amplitude of the motion and find the first time that the mass passes through equilibrium.

(c) Suppose that the mass is pulled 2 feet above equilibrium and hit with an initial velocity v_0. Determine all initial velocities v_0 such that the amplitude of the motion is 3 feet.

2.6-2. A 2-pound weight suspended from a linear spring stretches it 3 inches. Neglect damping and let $y(t)$ denote the distance of the mass from equilibrium, with the positive direction upward. Determine $y(t)$ explicitly if initially **(a)** the mass is raised 1 foot above equilibrium and hit upward with velocity 2 ft/sec; **(b)** the mass is raised 1 foot above equilibrium and hit downward with velocity 2 ft/sec; and **(c)** the mass is at equilibrium and hit downward with velocity 4 ft/sec.

2.6-3. A 16-pound weight suspended from a linear spring stretches it 2 feet. Let $y(t)$ denote the distance of the mass from equilibrium, with the positive direction upward, and assume that there is a damping force acting on the spring with damping constant equal to μ lb-sec/ft. Determine the differential equation that $y(t)$ must satisfy (assume that $g = 32$ ft/sec^2). **(a)** For the values $\mu = 2$, 4, and 5, compute $y(t)$ explicitly if the mass is initially at equilibrium and hit upward with velocity 2 ft/sec. **(b)** For the values $\mu = 2$, 4, and 5, compute $y(t)$ explicitly if the mass is initially pulled 2 feet below equilibrium and released.

2.6-4. Suppose that a 16-pound weight is suspended from a linear spring and that the spring has spring constants $k = 2$ lb/ft. Let $y(t)$ denote the distance of the mass from equilibrium, with the positive direction upward, and assume that there is a damping force acting on the spring with damping constant μ lb-sec/ft. Find the differential equation that y must satisfy (assume that $g = 32$ ft/sec^2).

(a) Determine $y(t)$ if $\mu = 1$ and the mass is initially pulled 1 foot below equilibrium and released. Also, find the first time $t_0 > 0$ such that the mass passes through equilibrium and the time $t_1 > 0$ such that the mass is its maximum distance above equilibrium.

(b) Suppose that $\mu = 2$ and the mass is initially pulled 1 foot below equilibrium and hit upward with initial velocity $v_0 \geq 0$. Determine $y(t)$ as a function of t

and v_0. Show that there is a $v^* > 0$ such that if $0 \leq v_0 \leq v^*$ then the mass always remains below equilibrium, and if $v_0 > v^*$ there is a time $T_{v0} > 0$ such that the mass passes through equilibrium at time T_{v0} and remains above equilibrium thereafter. Show further that if $v_0 > v^*$ there is a unique time $S_{v0} > T_{v0}$ such that the mass is its maximum distance above equilibrium at time S_{v0}.

(c) Suppose that $\mu = \frac{5}{2}$ and that the mass is initially pulled 1 foot below equilibrium and hit upward with initial velocity $v_0 \geq 0$. Determine $y(t)$ as a function of t and v_0, and discuss the various types of motion that can occur relative to the size of v_0.

2.6-5. Suppose that a mass m is suspended from a linear spring having spring constant k and that the motion is damped with damping coefficient $\mu > 0$. Consider m and k as being fixed, and for each $\mu > 0$ and $t \geq 0$ let $y_\mu(t)$ denote the distance of the mass from equilibrium, with the positive direction upward. Assume also that initial $y_\mu(0) = y_0$ and $y'_\mu(0) = v_0$, where y_0 and v_0 are independent of μ [note that $y_\mu(t)$ is described in Theorem 2.6-1].

(a) Show that $\lim_{\mu \to 0+} y_\mu(t) = y(t)$, where $y(t)$ is defined by equation (5)

(b) Show that $\lim_{\mu \to 2\sqrt{mk}-} y_\mu(t) = y_{2\sqrt{mk}}(t)$ (see parts a and b of Theorem 2.6-1 and apply L'Hospital's rule).

(c) Determine $\lim_{\mu \to 2\sqrt{mk}+} y_\mu(t)$ (see parts b and c in Theorem 2.6-1).

2.6-6. Suppose that a mass is suspended from a linear spring and that there is a resistance force with damping coefficient $\mu \geq 0$ and an external forcing function f. Assume that the mass is initially at rest at equilibrium and that the equation of motion is

$$(*) \qquad y'' + \mu y' + 4y = f(t) \qquad y(0) = 0 \qquad y'(0) = 0$$

where $y(t)$ denotes the position of the mass relative to equilibrium.

(a) Determine the solution to (*) if $\mu = 0$ and $f(t) \equiv A \sin \omega t$ where $A, \omega > 0$.

(b) Determine the solution to (*) if $0 \leq \mu < 4$ and $f(t) \equiv \sin 2t$.

(c) Determine and sketch the graph of the solution to (*) if $\mu = 0$ and $f(t) \equiv \cos \frac{9}{4}t$.

2.7 CHANGE OF INDEPENDENT VARIABLE: EULER'S EQUATION

Some linear differential equations with variable coefficients can be transformed into ones with constant coefficients, and in this section the computations involved in changing the independent variable t are indicated by solving a particular type of second order equation with variable coefficients. *Euler's equation* (or the *equidimensional equation*) is the second order linear equation

$$(1) \qquad \alpha t^2 y'' + \beta t y' + \gamma y = f(t) \qquad t > 0$$

where α, β, and γ are constants (a crucial point here is that the power of t is the number of times y is differentiated). The change of independent variable $s = \ln t$ transforms (1) into an equation with constant coefficients. In order to see this observe that if $s = \ln t$ (that is, $t = e^s$), $\dot{y} = dy/ds$, and $\ddot{y} = d^2y/ds^2$, then applications of the chain rule imply that

$$y' = \frac{ds}{dt}\frac{dy}{ds} = \frac{1}{t}\dot{y} \qquad \text{and}$$

(2)

$$y'' = \frac{d}{dt}\left(\frac{1}{t}\dot{y}\right) = -\frac{1}{t^2}\dot{y} + \frac{1}{t}\frac{ds}{dt}\cdot\frac{d\dot{y}}{ds} = -\frac{1}{t^2}\dot{y} + \frac{1}{t^2}\ddot{y}$$

Therefore,

$$\alpha t^2 y'' + \beta t y' + \gamma y = -\alpha\dot{y} + \alpha\ddot{y} + \beta\dot{y} + \gamma y$$

and equation (1) is transformed into the equation

$$\alpha\ddot{y} + (\beta - \alpha)\dot{y} + \gamma y = f(e^s) \tag{3}$$

where $\dot{y} = dy/ds$, $\ddot{y} = d^2y/ds^2$, and the coefficients α, $\beta - \alpha$, and γ are constants. In the homogeneous case (that is, when $f \equiv 0$) the solution to (3) and hence the solution to (1) can be obtained directly from the auxiliary roots, which are solutions to the auxiliary equation

$$\alpha\lambda^2 + (\beta - \alpha)\lambda + \gamma = 0 \tag{4}$$

Once the solution in terms of s has been obtained, it can be transformed to the original variable t by setting $s = \ln t$.

In order to illustrate this procedure consider the equation

$$t^2y'' + 5ty' + 4y = 0 \qquad t > 0 \tag{5}$$

According to (2) and (3), if $s = \ln t$ and "·" denotes differentiation with respect to s, then (5) becomes

$$\ddot{y} + 4\dot{y} + 4y = 0 \tag{5'}$$

Since this equation has constant coefficients and the auxiliary equation is $\lambda^2 + 4\lambda + 4 = (\lambda + 2)^2 = 0$, $\lambda = -2$ is a double root, and the general solution to (5′) (in terms of s) is

$$y = c_1e^{-2s} + c_2se^{-2s}$$

Setting $s = \ln t$ and observing that $e^{-2\ln t} = e^{\ln t^{-2}} = t^{-2}$ shows that the general solution to (5) (in terms of t) is

$$y = \frac{c_1}{t^2} + \frac{c_2 \ln t}{t^2}$$

where c_1 and c_2 are constants.

The general solution to the homogeneous equation corresponding to (3) can be written explicitly in terms of the roots to (4). Referring back to Section 2.3 gives the solution in terms of s, and the substitution $s = \ln t$ can then be used to express the general solution in terms of t. The solutions are written out here for completeness and convenience of reference:

(6) If $f \equiv 0$ and if λ_1 and λ_2 are distinct real roots to (4), then $y_1(s) = e^{\lambda_1 s}$ and $y_2(s) = e^{\lambda_2 s}$ are linearly independent solutions to (3) and $y_1(t) = t^{\lambda_1}$ and $y_2(t) = t^{\lambda_2}$ are linearly independent solutions to (1).

(7) If $f \equiv 0$ and if λ_0 is a real, double root to (4), then $y_1(s) = e^{\lambda_0 s}$ and $y_2(s) = se^{\lambda_0 s}$ are linearly independent solutions to (3) and $y_1(t) = t^{\lambda_0}$ and $y_2(t) = (\ln t)t^{\lambda_0}$ are linearly independent solutions to (1).

(8) If $f \equiv 0$ and if $\lambda = \delta \pm i\sigma$ are conjugate complex roots to (4), then $y_1(s) = e^{\delta s} \cos \sigma s$ and $y_2(s) = e^{\delta s} \sin \sigma s$ are linearly independent solutions to (3) and $y_1(t) = t^{\delta} \cos (\sigma \ln t)$ and $y_2(t) = t^{\delta} \sin (\sigma \ln t)$ are linearly independent solutions to (1).

Therefore, the homogeneous equation associated with (1) can always be solved by a change of variables [notice that example (5) fits into category (7)]. In many cases the nonhomogeneous equation can also be solved.

If the function $g(s) = f(e^s)$ has a finite family of derivatives, then the nonhomogeneous equation (3) can be solved by undetermined coefficients (see Section 2.4). However, since the general solution to the homogeneous equation associated with (1) can be computed, the method of variation of constants (see Section 2.5) can always be applied directly to the nonhomogeneous equation (1) (see, for example, Example 2.5-2). A second method for solving (1) in the homogeneous case is indicated in Problem 2.7-8.

EXAMPLE 2.7-1

Consider the equation $t^2y'' - ty' - 3y = \ln t$ for $t > 0$. Since the equation is of Euler type, the change of variable $s = \ln t$ transforms this equation into the equation $\ddot{y} - 2\dot{y} - 3y = s$ [see equations (2) and (3)]. Since $\lambda^2 - 2\lambda - 3 = (\lambda + 1)(\lambda - 3) = 0$, the auxiliary roots are $\lambda = -1, 3$, and so $y(s) = c_1 e^{-s} + c_2 e^{3s}$ is the general solution. Applying undetermined coefficients (the differential family for s is $\{s, 1\}$), it is easy to check that $y_p(s) = \frac{2}{9} - \frac{1}{3}s$ is a particular solution. Hence, since $e^{-s} = e^{-\ln t} = t^{-1}$ and $e^{3s} = e^{3 \ln t} = t^3$,

$$y(t) = \frac{c_1}{t} + c_2 t^3 + \frac{2}{9} - \frac{1}{3} \ln t$$

is the general solution to the original equation.

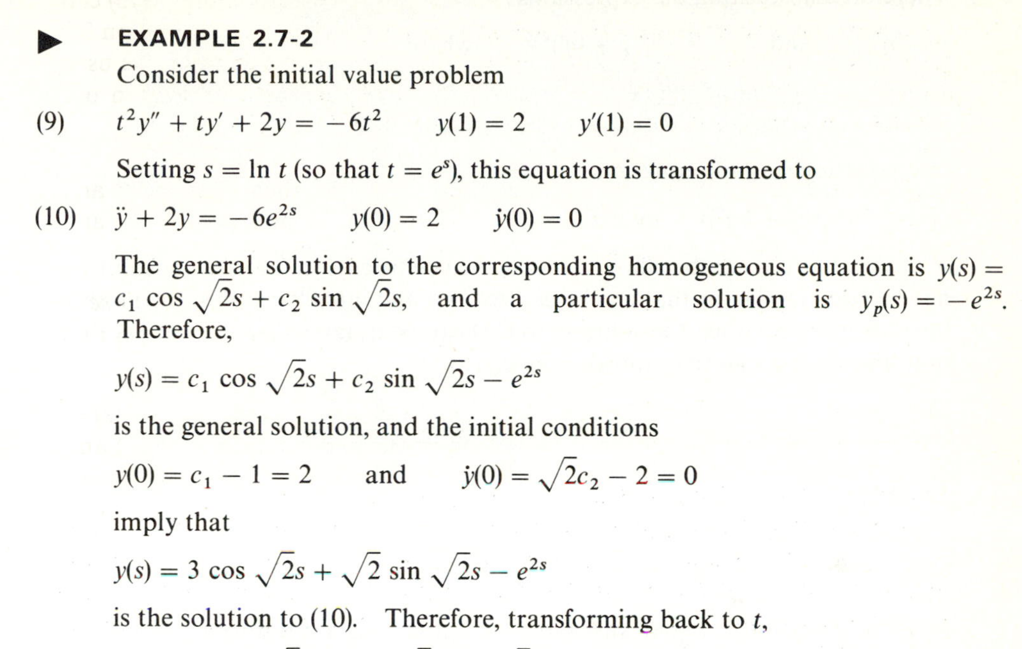

EXAMPLE 2.7-2

Consider the initial value problem

$$t^2y'' + ty' + 2y = -6t^2 \qquad y(1) = 2 \qquad y'(1) = 0 \tag{9}$$

Setting $s = \ln t$ (so that $t = e^s$), this equation is transformed to

$$\ddot{y} + 2y = -6e^{2s} \qquad y(0) = 2 \qquad \dot{y}(0) = 0 \tag{10}$$

The general solution to the corresponding homogeneous equation is $y(s) = c_1 \cos \sqrt{2}s + c_2 \sin \sqrt{2}s$, and a particular solution is $y_p(s) = -e^{2s}$. Therefore,

$$y(s) = c_1 \cos \sqrt{2}s + c_2 \sin \sqrt{2}s - e^{2s}$$

is the general solution, and the initial conditions

$$y(0) = c_1 - 1 = 2 \qquad \text{and} \qquad \dot{y}(0) = \sqrt{2}c_2 - 2 = 0$$

imply that

$$y(s) = 3 \cos \sqrt{2}s + \sqrt{2} \sin \sqrt{2}s - e^{2s}$$

is the solution to (10). Therefore, transforming back to t,

$$y(t) = 3 \cos (\sqrt{2} \ln t) + \sqrt{2} \sin (\sqrt{2} \ln t) - t^2$$

is the solution to (9).

Suppose now that I is an open interval and that a, b, c, and f are continuous functions from I into $\mathbb{R}$ with $a(t) \neq 0$ for all $t \in I$. Consider the general second order nonhomogeneous equation

$$a(t)y'' + b(t)y' + c(t)y = f(t) \qquad t \in I \tag{11}$$

Also suppose that q is a continuously differentiable function from I into $\mathbb{R}$ and that $q'(t) > 0$ for all $t \in I$. Then q maps the open interval I onto an open interval J, and we want to transform equation (1) into an equation of the independent variable $s = q(t)$ in J. Using the chain rule for differentiation,

$$y'(t) = \frac{dy}{dt} = \frac{ds}{dt} \cdot \frac{dy}{ds} = q'(t) \frac{dy}{ds}$$

and

$$\begin{aligned} y''(t) &= \frac{d}{dt} [y'(t)] = \frac{d}{dt} \left[q'(t) \frac{dy}{ds} \right] \\ &= q''(t) \frac{dy}{ds} + q'(t) \frac{d}{dt} \frac{dy}{ds} \\ &= q''(t) \frac{dy}{ds} + [q'(t)]^2 \frac{d^2y}{ds^2} \end{aligned}$$

Therefore, substituting the expressions

(12)
$$y' = q'\dot{y} \qquad \text{and} \qquad y'' = q''\dot{y} + (q')^2\ddot{y} \qquad \text{where}$$
$$\dot{y} = \frac{dy}{ds} \qquad \text{and} \qquad \ddot{y} = \frac{d^2y}{ds^2}$$

into equation (11), the equation

(13) $$a[q']^2\ddot{y} + [aq'' + bq']\dot{y} + cy = f$$

is obtained. If q can be chosen so that the coefficients in (13) are constant (or have a common factor that can be divided out and introduced into the nonhomogeneous term), then the solution to (11) can be obtained by first solving (13) in terms of $s \in J$ and then substituting $q(t)$ for s.

EXAMPLE 2.7-3

Consider the equation

(14) $$2ty'' + (1 - t^{1/2})y' - 3y = 0 \qquad t > 0$$

We show that the change of variable $s = t^{1/2}$ (or $t = s^2$) transforms (14) into an equation with constant coefficients. From equation (12)

$$y' = \tfrac{1}{2}t^{-1/2}\dot{y} \qquad y'' = -\tfrac{1}{4}t^{-3/2}\dot{y} + \tfrac{1}{4}t^{-1}\ddot{y}$$

and it follows that

$$2ty'' + (1 - t^{1/2})y' - 3y = \tfrac{1}{2}\ddot{y} - \tfrac{1}{2}\dot{y} - 3y$$

Therefore the variable change $s = t^{1/2}$ transforms (14) into the equation

$$\ddot{y} - \dot{y} - 6y = 0$$

Since the auxiliary equation for this equation is

$$\lambda^2 - \lambda - 6 = (\lambda - 3)(\lambda + 2) = 0$$

it follows that $y = c_1 e^{3s} + c_2 e^{-2s}$ is the general solution. Since $s = t^{1/2}$ we have that

$$y = c_1 e^{3\sqrt{t}} + c_2 e^{-2\sqrt{t}}$$

is the general solution to (14).

PROBLEMS

2.7-1. Use the transformation $s = \ln t$ ($t > 0$) to determine the general solution to each of the following Euler-type equations.

(a) $t^2y'' + ty' - 4y = 0$

(b) $t^2y'' - ty' + 2y = 0$

(c) $t^2y'' - 2ty' + 2y = t^2 + \frac{2}{t}$

(d) $t^2y'' + 3ty' + y = (\ln t)^2$

2.7-2. Determine a solution on $(0, \infty)$ to each of the following initial value problems.

(a) $t^2y'' + y = 0 \qquad y(1) = 2 \qquad y'(1) = -1$

(b) $t^2y'' - 2ty' + 2y = 2t^2 - 2 \qquad y(1) = 0 \qquad y'(1) = 3$

2.7-3. Suppose that α_1, β_1, γ_1, ρ, and σ are numbers with $\rho > 0$, $\alpha_1 \neq 0$, and consider the equation

$$(*) \qquad \alpha_1(\rho t + \sigma)^2 y'' + \beta_1(\rho t + \sigma)y' + \gamma_1 y = f_1(t) \qquad t > \frac{\sigma}{\rho}$$

where f is continuous for $t > \sigma/\rho$. Show that the change of variables $u = \rho t + \sigma$ transforms equation (*) into a Euler-type equation.

2.7-4. Apply Problem 2.7-3 to determine the general solution to each of the following equations.

(a) $(t + 1)^2y'' + (t + 1)y' + y = t^2$

(b) $(3t - 1)^2y'' - (3t - 1)y' - 5y = 6t + 5$

2.7-5. Determine the general solution to the equation

$$y'' - \frac{1}{t}y' + t^2y = 0 \qquad t > 0$$

(the change of variable $s = t^2$ may help).

2.7-6. Determine the general solution to the equation $y'' + (e^t - 1)y' + e^{2t}y = e^{3t}$ (use the change of variable $s = e^t$).

2.7-7. Determine the general solution to the equation $t^4y'' + 2t^3y' - 4y = 0$ (use the change of variable $s = t^{-1}$).

2.7-8. Consider the homogeneous equation

$$(*) \qquad \alpha t^2y'' + \beta ty' + \gamma y = 0 \qquad t > 0$$

and assume the solution y has the form $y = t^\lambda$ for some constant λ. Substitute $y' = \lambda t^{\lambda-1}$ and $y'' = \lambda(\lambda - 1)t^{\lambda-2}$ into (*) to show that λ must satisfy (4):

$$\alpha\lambda^2 + (\beta - \alpha)\lambda + \gamma = 0$$

Show that this immediately implies that (6) is true. When $\lambda = a \pm ib$ are conjugate complex, use the facts that

$$t^{a \pm ib} \equiv e^{(a \pm ib)(\ln t)} = e^{a \ln t}e^{\pm ib \ln t}$$

where $e^{a \ln t} = t^a$, and

$$e^{\pm ib \ln t} = \cos (b \ln t) \pm i \sin (b \ln t)$$

to show that (8) is true. When λ is a double root, use the solution $y = t^\lambda$ to reduce the order of (*) and then show that (7) is also valid.

2.7-9. The following problems are a mixture of inhomogeneous second order equations that can be solved by the methods developed in this chapter. Determine the general solution of each equation.

(a) $$\frac{d^2x}{dt^2} + x = \tan t \qquad 0 < t < \frac{\pi}{2}$$

(b) $$4 \frac{dr^2}{d\theta^2} + r = 6 \cos \theta$$

(c) $$x^2 \frac{d^2y}{dx^2} + x \frac{dy}{dx} - y = \frac{1}{1 + x^2} \qquad x > 0$$

(d) $$\frac{d^2\phi}{dz^2} - 4 \frac{d\phi}{dz} + 4\phi = e^{2z} - 8$$

(e) $$\theta^2 \frac{d^2r}{d\theta^2} + 5\theta \frac{dr}{d\theta} + 5r = 1 \qquad \theta > 0$$

(f) $$\frac{d^2\phi}{dt^2} + 2 \frac{d\phi}{dt} + \phi = \frac{1}{(e^t - 1)^2} \qquad t > 0$$

(g) $$t^2 \frac{d^2\psi}{dt^2} - 2\psi = 4 \ln t \qquad t > 0$$

(h) $$\frac{d^2\phi}{dx^2} + \frac{2}{x} \frac{d\phi}{dx} - \frac{12}{x^2} \phi = \frac{6}{x^4}$$

2.8 MISCELLANEOUS APPLICATIONS

2.8a Electric Circuits

The elementary analysis of electric circuits gives an important application of second order linear differential equations, and we indicate here some of the basic ideas. The principal aim is to give a brief derivation of a differential equation that describes the flow of current in a simple series circuit. A schematic diagram of such a circuit is shown in Figure 2.6. The *electromotive force* (whose units are *volts*) is represented by E and could typically be a battery or a generator, and the *resistance* (whose units are *ohms*) is represented by R and could typically be some household appliance such as an oven or a toaster. If only the electromotive force (emf) and the resistance are present in the circuit, then *Ohm's law* asserts that the instantaneous current I (whose

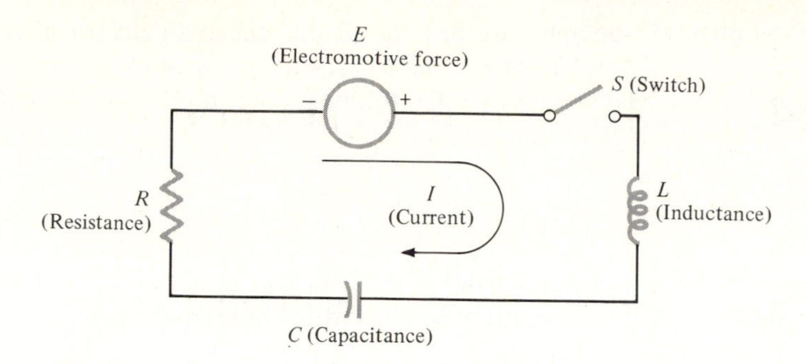

Figure 2.6 Elementary electric circuit diagram.

units are *amperes* or, simply, *amps*) is directly proportional to the emf E. Moreover, if R is the constant such that

$$E = RI \tag{1}$$

then R is the resistance.

Electric circuits often have an inductance (whose units are *henrys*) and a capacitance (whose units are *farads*). An inductor, which is represented by L in Figure 2.6, is normally a coiled wire that opposes a change in current and a capacitor (or condenser), which is represented by C in Figure 2.6, is normally two parallel and separated metal plates that become charged. In fact, if $Q(t)$ is the *charge* on the capacitor at time t, then $Q(t)$ is measured in units of *coulombs*, and the current $I(t)$ at time t is the instantaneous rate of change of the charge $Q(t)$:

$$I(t) = Q'(t) \tag{2}$$

Note in particular that if the current $I(t)$ is known for all t, then

$$Q(t) = Q_0 + \int_{t_0}^{t} I(s)\, ds$$

where Q_0 is the charge at time t_0.

The problem under consideration is to determine the charge $Q(t)$ and the current $I(t)$ in the circuit given the emf $E(t)$, the resistance R, the inductance L, and the capacitance C (the resistance, inductance, and capacitance are assumed constant throughout this discussion). Moreover, it is assumed that the charge Q_0 and current I_0 are known at some initial time t_0. Each of the elements in the circuit determine a *voltage drop* (or a *potential drop*). This drop in voltage (which can be measured with the use of a voltmeter) behaves according to the following physical laws:

(3)
- **(a)** The voltage drop across a resistance of R ohms equals RI.
- **(b)** The voltage drop across an inductance of L henrys equals LI'.
- **(c)** The voltage drop across a capacitance of C farads equals Q/C.

Finally, the fundamental principle determining the current in an electric circuit is known as *Kirchhoff's law: The sum of the voltage drops in an electric circuit equals the supplied electromotive force.* Combining Kirchhoff's law with the principles in (3) leads directly to the equation

$$RI + LI' + \frac{Q}{C} = E(t)$$

Since $I = Q'$ and hence $I' = Q''$, it follows that the charge Q on the capacitor C in the circuit in Figure 2.6 is the solution to the initial value problem

$$LQ'' + RQ' + \frac{1}{C}Q = E(t) \qquad Q(t_0) = Q_0 \qquad Q'(t_0) = I_0 \tag{4}$$

Equation (4) is a nonhomogeneous second order linear differential equation in Q with constant coefficients, and therefore the charge $Q(t)$ can be computed if the emf $E(t)$ is simple (for example, if E is constant). Once Q is computed from (4) the current I can be found simply by differentiating Q: $I(t) = Q'(t)$.

It is of interest to note that equation (4) is completely analogous to the equation of motion of a mass-spring system with the inductance L replaced by the mass m, the resistance R by the friction coefficient μ, the reciprocal of the capacitance $1/C$ by the spring constant k, and the charge Q by the displacement x [see equation (10) in Section 2.6]. Note further that the current I corresponds to the velocity v of the mass and the emf E corresponds to an external force F on the mass. Therefore, the types of behavior in simple electric circuits parallels that in vibrating mechanical systems (see the discussion at the end of Section 2.6). In particular, if the emf term $E(t)$ is periodic (and the resistance is small), the phenomenon of resonance can occur in the system. This phenomenon plays a part, for example, in the tuning of radios.

The electric circuit in Figure 2.6 is referred to as an *RLC circuit.* If the inductance L were removed, it would be called an RC circuit. As a particular illustration, consider the RLC circuit with a resistance of 4 ohms, an inductance of $\frac{1}{2}$ henry, and a capacitance of $\frac{1}{26}$ farad (see Figure 2.7). Assume first that the emf has a constant value of $E = 13$ volts and that there is no charge or current when the switch is connected. The voltage drop across the induc-

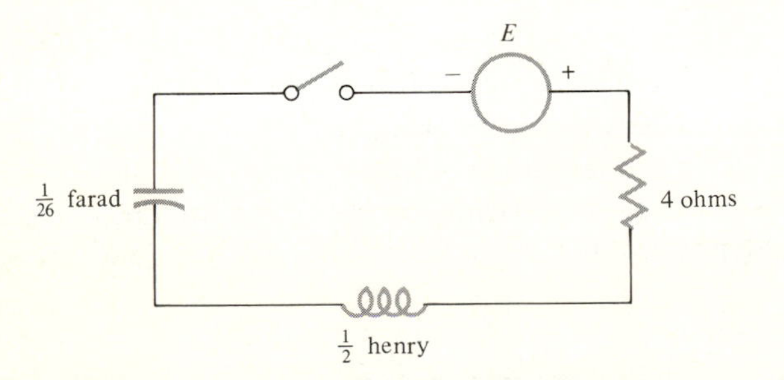

Figure 2.7 Electric circuit corresponding to (5).

tor is $I'/2$, across the resistor is $4I$, and across the capacitor is $26Q$ [see (3)]. Therefore, by Kirchhoff's law,

$$\frac{I'}{2} + 4I + 26Q = 13$$

and since $I = Q'$, it follows that the charge Q is a solution to

$$\frac{Q''}{2} + 4Q' + 26Q = 13$$

Since there is no charge or current initially, the initial value problem

$$Q'' + 8Q' + 52Q = 26 \qquad Q(0) = 0 \qquad Q'(0) = 0 \tag{5}$$

is obtained. The corresponding homogeneous equation has $\lambda^2 + 8\lambda + 52 = 0$ as its auxiliary equation; hence

$$\lambda = \frac{-8 \pm \sqrt{64 - 208}}{2} = -4 \pm 6i$$

are the auxiliary roots. Using undetermined coefficients and assuming a particular solution is of the form $Q_p = d_1$, it follows that $Q_p = \frac{1}{2}$ is a solution. Therefore,

$$Q = c_1 e^{-4t} \cos 6t + c_2 e^{-4t} \sin 6t + \tfrac{1}{2}$$

is the general solution to the differential equation in (5). Using the initial condition $Q(0) = Q'(0) = 0$ to solve for the constants c_1 and c_2, shows that

$$Q(t) = -e^{-4t} \frac{\cos 6t}{2} - e^{-4t} \frac{\sin 6t}{3} + \frac{1}{2} \tag{6}$$

is the solution to (5). The first two terms in (6) have e^{-4t} as a factor and, as in the mechanical vibration case, are called the *transient solution* (these two terms are negligible after a long period of time). The remaining term $\frac{1}{2}$ is called the *steady-state solution* and is the dominant term after a long period of time.

Suppose now that instead of a constant emf $E = 13$, the generator E has an alternating voltage given by $E = 16 \cos 2t$. The equation governing the charge Q on the capacitor in Figure 2.7 in this case has the form

$$Q'' + 8Q' + 52Q = 32 \cos 2t \qquad Q(0) = 0 \qquad Q'(0) = 0 \tag{7}$$

Using undetermined coefficients, one assumes a particular solution has the form $Q_p = d_1 \cos 2t + d_2 \sin 2t$, and it follows that $d_1 = \frac{3}{5}$ and $d_2 = \frac{1}{5}$. Therefore

$$Q = c_1 e^{-4t} \cos 6t + c_2 e^{-4t} \sin 6t + \frac{3 \cos 2t}{5} + \frac{\sin 2t}{5}$$

is the general solution to the differential equation in (7). The initial condition $Q(0) = Q'(0) = 0$ implies that $c_1 = -\frac{3}{5}$ and $c_2 = -\frac{7}{15}$, and hence

$$Q = -3e^{-4t}\frac{\cos 6t}{5} - 7e^{-4t}\frac{\sin 6t}{15} + \frac{3\cos 2t}{5} + \frac{\sin 2t}{5} \tag{8}$$

is the solution to (7). In this case the steady-state solution Q_s can be written

$$Q_s(t) = \frac{3\cos 2t + \sin 2t}{5} = \sqrt{10}\,\frac{\sin(2t + \phi)}{5}$$

where ϕ is such that $\sin\phi = 3/\sqrt{10}$ and $\cos\phi = 1/\sqrt{10}$ (that is, $\phi = \arctan 3 \cong 4\pi/10$). As in the case with constant emf the long-term behavior of the charge on the capacitor is similar to the input voltage of the generator.

2.8*b* Archimedes' Principle

As a further illustration of the applicability of second order linear equations, we consider the oscillations of an object partially submerged in water. The fundamental physical law describing this type of motion is known as *Archimedes' principle: An object that is submerged (either partially or totally) in a liquid is acted on by an upward force equal to the weight of the liquid displaced.* As a specific example to illustrate this principle, consider a cylinder of radius R, height H, and weight W and assume that the cylinder is floating with its axis vertical in a liquid having density ρ (see Figure 2.8). Assume also that $W < \pi R^2 H\rho$ (what happens if $W > \pi R^2 H\rho$?) and let $h = W/\pi R^2\rho$ (that is,

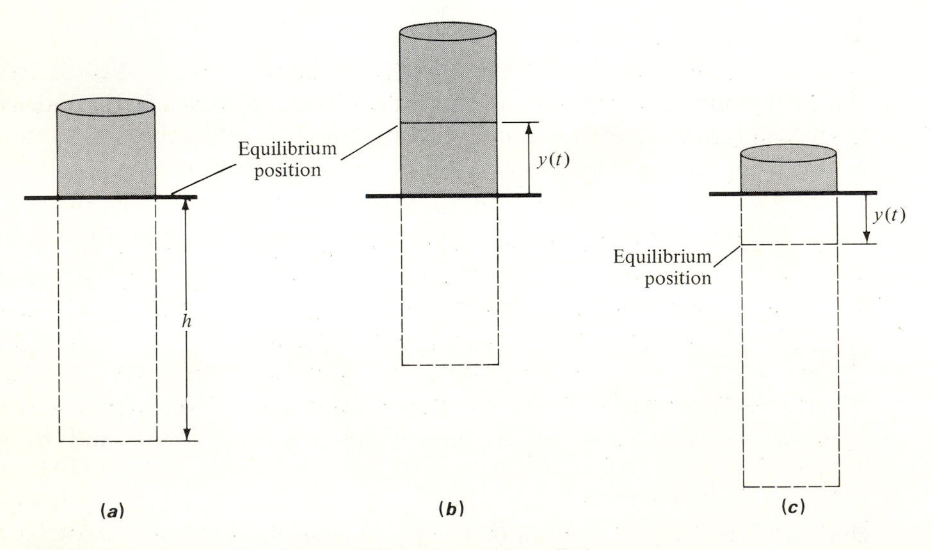

Figure 2.8 Oscillations in a partially submerged cyclinder.

$W = \pi R^2 h\rho$). Then the buoyant force upward is precisely the weight of the cylinder, and the equilibrium position for the cylinder is when h vertical units lie under the liquid (see Figure 2.8*a*). For each time $t \geq 0$ let $y(t)$ be the distance of the equilibrium from the liquid surface, with the convention that $y(t) > 0$ if the equilibrium is above the surface (Figure 2.8*b*) and $y(t) < 0$ if the equilibrium is below the surface (Figure 2.8*c*). According to Newton's second law the total force F_T acting on the cylinder can be written

$$F_T(t) = \frac{W}{g}\, y''(t) \tag{9}$$

where g is the gravitational constant. The force $F_T(t)$ can also be described as

$$F_T(t) = -W + (\text{buoyant force})$$

Since the volume of the cylinder under the liquid can be written $\pi R^2[h - y(t)]$, we have by Archimedes' principle that

$$\begin{aligned} F_T(t) &= -W + \pi R^2[h - y(t)]\rho \\ &= -W + \pi R^2 h\rho - \pi R^2 y(t)\rho \end{aligned}$$

Since h was chosen so that $W = \pi R^2 h\rho$, it follows that $F_T(t) = -\pi R^2 y(t)\rho$, and hence

$$\frac{W}{g}\, y''(t) = -\pi R^2 y(t)\rho$$

by (9). Therefore, the initial value problem describing the motion of the cylinder is

$$y'' + \frac{\pi R^2 \rho g}{W}\, y = 0 \qquad y(0) = y_0 \qquad y'(0) = v_0 \tag{10}$$

where y_0 is the initial position and v_0 the initial velocity of the cylinder. Equation (10) is second order, linear with constant coefficients, and the auxiliary roots are $\lambda = \pm iR\sqrt{\pi\rho g}/\sqrt{W}$. Hence the general solution to the differential equation in (10) is

$$y(t) = c_1 \cos\left(R\sqrt{\frac{\pi\rho g}{W}}\, t\right) + c_2 \sin\left(R\sqrt{\frac{\pi\rho g}{W}}\, t\right)$$

The initial conditions imply $c_1 = y_0$ and $c_2 = \sqrt{W}v_0/R\sqrt{\pi\rho g}$ so that

$$y(t) = y_0 \cos\left(R\sqrt{\frac{\pi\rho g}{W}}\, t\right) + \frac{\sqrt{W}v_0}{R\sqrt{\pi\rho g}} \sin\left(R\sqrt{\frac{\pi\rho g}{W}}\, t\right) \tag{11}$$

is the solution to the initial value problem (10). It follows from (11) that

$$y(t) = \sqrt{y_0^2 + \frac{Wv_0^2}{R^2\pi\rho g}} \sin\left(R\sqrt{\frac{\pi\rho g}{W}}\, t + \phi\right) \qquad \text{where } \phi = \arctan\left(\frac{y_0 R\sqrt{\pi\rho g}}{v_0\sqrt{W}}\right) \tag{11'}$$

Therefore, the motion of the cylinder is periodic, and the period T and the amplitude A are given by

$$(12) \quad T = \frac{2\sqrt{\pi W}}{R\sqrt{\rho g}} \quad \text{and} \quad A = \sqrt{y_0^2 + \frac{W v_0^2}{R^2 \pi \rho g}}$$

The amplitude depends on both the initial position and velocity, but the period is independent of the initial conditions. Notice also that the motion is independent of the height H (so long as the amplitude of the motion is smaller than $H/2$).

EXAMPLE 2.8-1

Suppose that a cylindrical tin can is floating in a pond of water (with density $\rho = 62.5$ lb/ft^3), with its axis vertical, and is oscillating up and down with a period of $\frac{1}{2}$ second per complete oscillation. If the radius of this tin can is 2 inches, determine its weight. Taking $g = 32$ ft/sec^2 we have from the first formula in (12) that

$$\frac{1}{2} = \frac{2\sqrt{\pi W}}{\left(\frac{1}{6}\right)\sqrt{62.5 \times 32}} = \frac{12\sqrt{\pi W}}{\sqrt{2000}}$$

Therefore, $\pi W = 2000/24^2 \cong 3.4722$, and it follows that $W \cong 1.105$ pounds is the approximate weight of the can. Note also that the height H of the can must be larger than 0.202 feet $\cong$ 2.43 inches (why?).

PROBLEMS

2.8-1. Consider the RLC electric circuit given in Figure 2.9a, where there is an initial charge of 2 coulombs on the capacitor and no initial current.

(a) Set up the initial value problem describing the charge $Q(t)$ on the capacitor for any given emf $E(t)$.

(b) Determine the charge $Q(t)$ explicitly if there is no emf [that is, $E(t) \equiv 0$]. What is the current $I(t)$?

(c) Determine the charge $Q(t)$ explicitly if $E(t) \equiv 20$ volts. What is the steady-state term?

(d) Determine the charge $Q(t)$ explicitly if there is an alternating emf of the form $E(t) \equiv 10 \cos 20t$. What is the steady-state term?

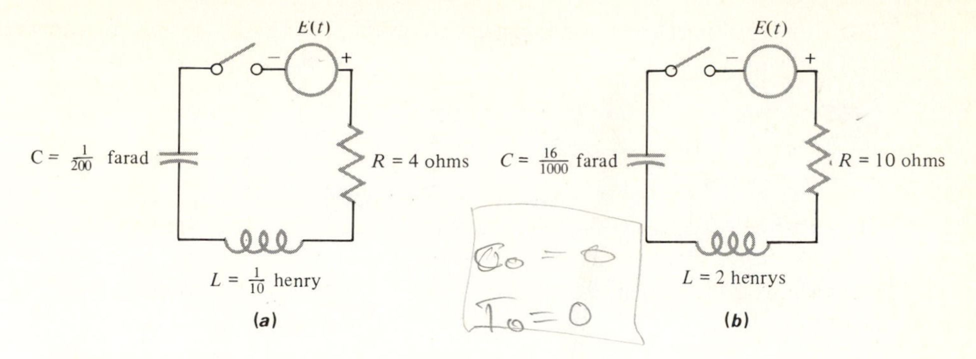

Figure 2.9 ***a*** Electric circuit for Exercise 1. ***b*** Electric circuit for Exercise 2.

2.8-2. Consider the *RLC* circuit given in Figure 2.9*b* and let $Q(t)$ be the charge on the capacitor and $I(t)$ the current in the circuit for all $t \geq 0$. Assume also that initially there is no charge or current.

(a) Determine $Q(t)$ and $I(t)$ if the emf has the form $E(t) = 200e^{-t}$ for all $t \geq 0$.

(b) Determine $Q(t)$ and $I(t)$ if there is an alternating emf of $E(t) = 200 \cos 5t$ for all $t \geq 0$. What is the steady-state form of $Q(t)$ and $I(t)$ in this case?

2.8-3. Consider the homogeneous equation that describes the charge $Q(t)$ on the capacitor C in Figure 2.6 when the emf is zero:

$$LQ'' + RQ' + \frac{1}{c}Q = 0 \qquad Q(0) = Q_0 \qquad Q'(0) = I_0$$

[see equation (4)]. Using the mechanical vibrations as an analogy, discuss the relationship among L, R, and C when this circuit would be called underdamped, overdamped, and critically damped.

2.8-4. Consider an *RLC* circuit with $L = \frac{1}{7}$ henry, $R = 4$ ohms, and $C = \frac{1}{35}$ farad, so that the differential equation describing the charge Q is

$$\tfrac{1}{7}Q'' + 4Q' + 35Q = E(t)$$

where $E(t)$ is the applied emf. Determine the general solution and the steady-state solution for each of the following emfs $E(t)$:

(a) $E(t) \equiv 7$ **(b)** $E(t) \equiv 7 \cos 7t$ **(c)** $E(t) \equiv 7 \sin 7t$

2.8-5. A cylinder with radius 3 inches and weight 5π ($\cong 15.71$) pounds is floating with its axis vertical in a pool of water (with density $\rho = 62.5$ lb/ft^3). Assuming the cylinder is high enough, determine the period of its oscillations and also the description of its position $y(t)$ relative to equilibrium if it is raised 1 inch above equilibrium and pushed downward with an initial velocity of 4 in/sec.

2.8-6. Answer the same questions posed in Problem 2.8-5 if instead of water the cylinder is floating in a liquid with density $\rho = 125$ ($= 2 \times 62.5$) lb/ft^3.

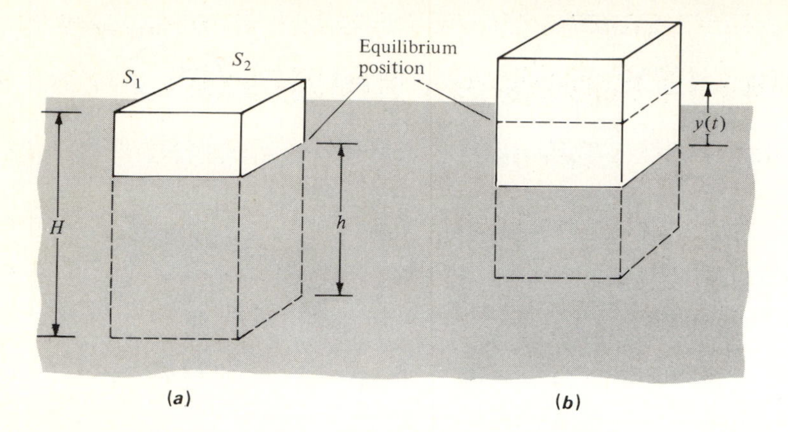

Figure 2.10 ***a, b*** Oscillations in a partially submerged box.

2.8-7. Suppose that a cylinder oscillates with its axis vertical in a certain liquid having density ρ_1. What is ρ_1 if the period of the oscillation is twice the period of oscillation in water? What is ρ_1 if the period is three times the period of oscillation in water? (The density of water is 62.5 lb/ft^3.)

2.8-8. Suppose that a rectangular box with width S_1, length S_2, and height H is floating in a liquid with density ρ as indicated in Figure 2.10*a* and *b*. As in the cylinder case let $y(t)$ denote the position of the box relative to equilibrium and suppose that W is the weight of the box.

(a) How large should H be so that the box will oscillate?

(b) Show that the position $y(t)$ is a solution to the equation

$$y'' + \frac{S_1 S_2 \rho g}{W} y = 0 \qquad y(0) = y_0 \qquad y'(0) = v_0$$

where y_0 is the initial position and v_0 the initial velocity.

(c) Determine the period and amplitude of the oscillations.

(d) Discuss the changes in the period of the oscillations if S_1 is doubled. What about if both S_1 and S_2 are doubled? Tripled?

2.8-9. The oscillations of the cylinder and the rectangular cube discussed above are distinguished by the fact that the horizontal cross section of the cylinder is circular and that of the cube is rectangular. Use Archimedes' principle to derive the equation of motion and the period of the oscillation if the object has as its cross-sectional form an equilateral triangle with sides of length L.

Remarks A more detailed discussion of solutions to second order linear equations can be found in Birkhoff and Rota [1], Coddington [3], and Ince [8]. Further basic applications of these equations are in Wylie [15].

3 Laplace Transform Methods

The Laplace transform provides a method for solving linear differential equations that is frequently used by engineers and scientists, and the purpose of this chapter is to introduce the student to some of the basic ideas and applications involved. The first two sections discuss elementary techniques and concepts connected with various properties of the transform and its inverse. The procedure for using the Laplace transform to solve nonhomogeneous linear equations with constant coefficients is indicated in Section 3.3, and applications to the forced vibrations in linear springs are studied in Section 3.4. In each of these sections discontinuous nonhomogeneous terms are considered: those with jump discontinuities in Section 3.3 and terms involving the Dirac delta function in Section 3.4.

3.1 INTRODUCTORY EXAMPLES AND CONCEPTS

The method of variation of parameters (Section 2.5) and the method of undetermined coefficients (Section 2.4) have already been developed for solving nonhomogeneous linear differential equations with constant coefficients. In this chapter a completely different method, based on the concept of an integral transformation, is investigated. In general terms a linear differential equation is "transformed" into an algebraic equation by means of an integral operation. The solution to this algebraic equation is then "transformed back" in order to obtain the solution to the original differential equation. Although this method applies to linear equations of arbitrary order (see Chapter 6), the discussion here is limited to equations of the second order.

Before giving the definition of the Laplace transform of a function, a simple example is given in order to illustrate these ideas. For simplicity a first order equation is considered and each step only formally indicated (in particular, the validity of each step is not established). Consider the initial value problem

(1) $y' + y = 3 \qquad y(0) = y_0$

Let s be a positive number, multiply each side of the differential equation in (1) by e^{-st}, and then integrate each side of the resulting equation from $t = 0$ to $t = \infty$ to obtain

$$\int_0^\infty e^{-st}y'(t)\,dt + \int_0^\infty e^{-st}y(t)\,dt = \int_0^\infty e^{-st}3\,dt$$

Note that

$$\int_0^\infty e^{-st}3\,dt = \lim_{T\to\infty}\left[-\frac{3}{s}e^{-st}\right]_{t=0}^{t=T}$$

$$= \lim_{T\to\infty}\left[\frac{3}{s} - \frac{3}{s}e^{-sT}\right]$$

$$= \frac{3}{s}$$

and, using integration by parts, that

$$\int_0^\infty e^{-st}y'(t)\,dt = \lim_{T\to\infty}\left\{[e^{-st}y(t)]_{t=0}^{T} + \int_0^T se^{-st}y(t)\,dt\right\}$$

$$= \lim_{T\to\infty}[e^{-sT}y(T) - y(0)] + s\int_0^\infty e^{-st}y(t)\,dt$$

Using $y(0) = y_0$ and *assuming that* $e^{-sT}y(T) \to 0$ as $T \to \infty$, the equation

$$s\int_0^\infty e^{-st}y(t)\,dt - y_0 + \int_0^\infty e^{-st}y(t)\,dt = \frac{3}{s}$$

is obtained. Therefore,

$$(s+1)\int_0^\infty e^{-st}y(t)\,dt = \frac{3}{s} + y_0$$

and it follows that if y is the solution to (1), then

(2) $$\int_0^\infty e^{-st}y(t)\,dt = \frac{3 + y_0 s}{s(s+1)} \qquad \text{for all } s > 0$$

A crucial issue is that the integral in (2) involves only y and not y'. In fact, y' was eliminated from the equation by using integration by parts. Of course, the question now is can one determine the solution y from equation (2)? Therefore, the integral transformation defined by the left side of equation (2) needs to be investigated, and this is the basic purpose of this chapter.

Suppose that f is a function defined on $[0, \infty)$. The *Laplace transform of f* is the function $\mathscr{L}\{f\}$ defined by

$$\mathscr{L}\{f\}(s) \equiv \int_0^\infty e^{-st} f(t)\, dt \tag{3}$$

for all numbers s such that this integral exists. Recall from the definition of the improper integral that

$$\mathscr{L}\{f\}(s) \equiv \lim_{T \to \infty} \int_0^T e^{-st} f(t)\, dt$$

Usually t is used to denote the variable of the function f. For example, $\mathscr{L}\{e^t\}$ denotes the Laplace transform of the function $f(t) = e^t$ for all $t \geq 0$ and $\mathscr{L}\{2t^3 + 3 \sin 6t\}$ denotes the Laplace transform of the function $f(t) = 2t^3 + 3 \sin 6t$ for all $t \geq 0$.

The first basic property that we establish for Laplace transforms is that of linearity. Suppose that f and g are functions on $[0, \infty)$ such that the Laplace transforms $\mathscr{L}\{f\}(s)$ and $\mathscr{L}\{g\}(s)$ exist for $s > \sigma$. If c_1 and c_2 are constants, then

$$\int_0^T e^{-st}[c_1 f(t) + c_2 g(t)]\, dt = c_1 \int_0^T e^{-st} f(t)\, dt + c_2 \int_0^T e^{-st} g(t)\, dt$$

for all $T > 0$. Letting $T \to \infty$ shows that the formula

$$\mathscr{L}\{c_1 f + c_2 g\}(s) = c_1 \mathscr{L}\{f\}(s) + c_2 \mathscr{L}\{g\}(s) \tag{4}$$

is valid. This property is known as the *linear property for the Laplace transform.*

A simple but important illustration of the Laplace transform is given by the following example.

EXAMPLE 3.1-1

Suppose that a is a number and that $s > a$. Then $\mathscr{L}\{e^{at}\}(s)$ is defined for all $s > a$, and

$$\begin{aligned}
\mathscr{L}\{e^{at}\}(s) &= \int_0^\infty e^{-st} e^{at}\, dt = \lim_{T \to \infty} \int_0^T e^{(a-s)t}\, dt \\
&= \lim_{T \to \infty} \left[\frac{1}{a-s} e^{(a-s)t} \right]_{t=0}^T \\
&= \lim_{T \to \infty} \frac{e^{(a-s)T} - 1}{a-s} = \frac{1}{s-a}
\end{aligned}$$

Note, in particular, we have $\mathscr{L}\{1\}(s) = 1/s$ by setting $a = 0$.

Using the linear property one sees, for example, that

$$\mathscr{L}\{3e^{2t} - 5e^{-t}\}(s) = 3\mathscr{L}\{e^{2t}\}(s) - 5\mathscr{L}\{e^{-t}\}(s) = \frac{3}{s-2} - \frac{5}{s+1}$$

for all $s > 2$. As a further example, use partial fractions and the definition of the Laplace transform to rewrite equation (2) as

$$\mathscr{L}\{y(t)\} = \frac{3 + y_0 s}{s(s+1)} = \frac{3}{s} + \frac{y_0 - 3}{s+1}$$

Now using Example 3.1-1 and the linear property (4), note that

$$\mathscr{L}\{3 + (y_0 - 3)e^{-t}\}(s) = 3\mathscr{L}\{1\}(s) + (y_0 - 3)\mathscr{L}\{e^{-t}\}(s)$$
$$= \frac{3}{s} + \frac{y_0 - 3}{s+1}$$

Comparing, we see that the function $y(t) = 3 + (y_0 - 3)e^{-t}$ satisfies equation (2). It is also easy to see that this is indeed the solution to the differential equation (1).

Our problem now is to describe a reasonable class of functions f such that $\mathscr{L}\{f\}(s)$ exists for s in some right half-line, say $s > s_0$. Note, for example, if $f(t) \equiv e^{t^2}$ for all $t \geq 0$, then $\mathscr{L}\{f\}(s)$ does not exist for *any* number s (see Problem 3.1-3). There are two important concepts that are needed. The function f is said to be *piecewise continuous* on an interval $[a, b]$ if there is a finite set of points $a = t_0 < t_1 < \cdots < t_n = b$ such that f is continuous on each of the intervals (t_{i-1}, t_i) for $i = 1, \ldots, n$, and the one-sided limits

$$f(t_i+) \equiv \lim_{t \to t_i+} f(t) \qquad \text{and} \qquad f(t_{i+1}-) \equiv \lim_{t \to t_{i+1}-} f(t)$$

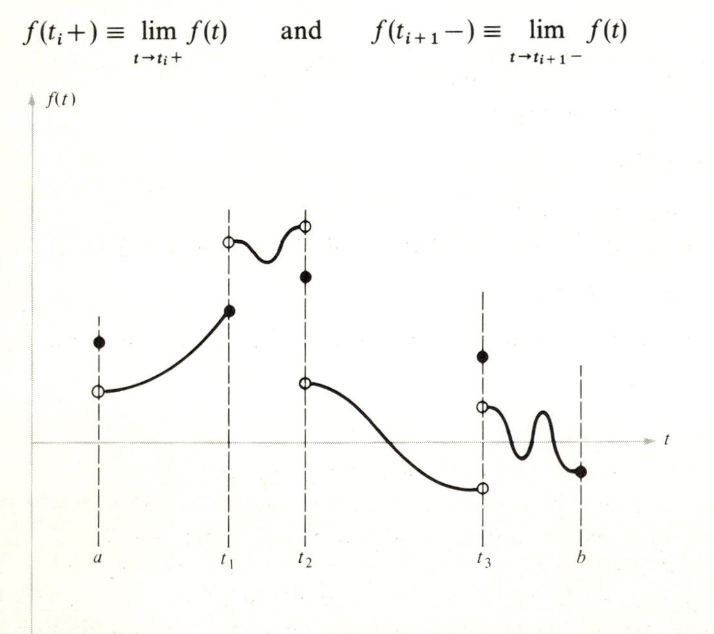

Figure 3.1 Piecewise continuous function.

exist for $i = 0, 1, \ldots, n-1$ (see Figure 3.1). Notice that a piecewise continuous function is continuous except possibly for a finite number of *jump* discontinuities. The function f is said to be of *exponential order* if there exist numbers M and σ such that

$$|f(t)| \leq Me^{\sigma t} \qquad \text{for all } t \geq 0$$

Clearly the functions $f(t) \equiv e^{at}$ and $f(t) \equiv \sin bt$ are of exponential order. To see that $f(t) \equiv t^n$ for $t \geq 0$ is of exponential order for each positive integer n, note that $t^n \leq (t^n e^{-t})e^t$. Since the maximum value of $t^n e^{-t}$ on $[0, \infty)$ occurs at $t = n$ (this is where its derivative is zero), we have $t^n e^{-t} \leq n^n e^{-n}$, and hence $t^n \leq (n^n e^{-n})e^t$ for all $t \geq 0$. A fundamental result that is sufficient for the existence of the Laplace transform is given by the following theorem:

Theorem 3.1-1

Suppose that the function f on $[0, \infty)$ is piecewise continuous on $[0, T]$ for each $T > 0$ and that f is of exponential order, say, $|f(t)| \leq Me^{\sigma t}$ for all $t \geq 0$. Then $\mathscr{L}\{f\}(s)$ exists for all $s > \sigma$.

As an indication of the validity of this theorem, note that if $T > 0$ and $s > \sigma$.

$$\int_0^T |e^{-st}f(t)|\, dt \leq \int_0^T e^{-st}Me^{\sigma t}\, dt = \int_0^T Me^{(\sigma - s)t}\, dt$$

$$= \left[\frac{M}{a - s} e^{(\sigma - s)t}\right]_0^T = \frac{M}{s - a}[1 - e^{(\sigma - s)T}]$$

Therefore, it follows that

$$\int_0^T |e^{-st}f(t)|\, dt \leq \frac{M}{s - a} \qquad \text{for all } T > 0,\ s > \sigma$$

and the convergence of

$$\int_0^\infty |e^{-st}f(t)|\, dt$$

follows. The convergence of

$$\int_0^\infty e^{-st}f(t)\, dt$$

(and hence the existence of the Laplace transform) can now be established using the comparison test for improper integrals.

As indicated in the illustrative example at the beginning of this section [see equation (1)], obtaining the Laplace transform of the derivative of a function in terms of the Laplace transform of the function itself is essential in the applications to differential equations. Since this method is applied mainly to second order equations, the Laplace transform of the second derivative also needs to be expressed in terms of the original function. These crucial formulas are contained in the next theorem.

Theorem 3.1-2

Suppose that f is continuous and of exponential order, say $|f(t)| \leq Me^{\sigma t}$ for $t \geq 0$, and that f' exists and is piecewise continuous on $[0, T]$ for each $T > 0$. Then both $\mathscr{L}\{f\}(s)$ and $\mathscr{L}\{f'\}(s)$ exist for $s > \sigma$, and the formula

(5) $$\mathscr{L}\{f'\}(s) = s\mathscr{L}\{f\}(s) - f(0)$$

is valid. If, in addition, f' is continuous and of exponential order, say, $|f'(t)| \leq Ke^{\rho t}$ for $t \geq 0$, and f'' exists and is piecewise continuous on $[0, T]$ for each $T > 0$, then $\mathscr{L}\{f''\}(s)$ exists for $s > \rho$. Moreover, if $s > \max\{\sigma, \rho\}$ the formula

(6) $$\mathscr{L}\{f''\}(s) = s^2\mathscr{L}\{f\}(s) - sf(0) - f'(0)$$

is valid.

When f' is continuous, formula (5) is a direct consequence of integration by parts since

$$\begin{aligned}\int_0^T e^{-st}f'(t)\,dt &= [e^{-st}f(t)]_0^T + s\int_0^T e^{-st}f(t)\,dt \\ &= e^{-sT}f(T) - f(0) + s\int_0^T e^{-st}f(t)\,dt\end{aligned}$$

for all $T > 0$. However, $e^{-sT}f(T) \to 0$ as $T \to \infty$ [recall that $|f(T)| \leq Me^{\sigma T}$, where $s > \sigma$ by assumption], and formula (5) follows by letting $T \to \infty$ in both sides of the preceding equation. Under the suppositions of the theorem, (6) can be obtained from (5) with f replaced by f':

$$\begin{aligned}\mathscr{L}\{f''\}(s) &= s\mathscr{L}\{f'\}(s) - f'(0) \\ &= s[s\mathscr{L}\{f\}(s) - f(0)] - f'(0)\end{aligned}$$

Therefore, the validity of (5) and (6) is indicated.

The final result of this section points out further basic properties of the Laplace transform.

Theorem 3.1-3

Suppose that f is piecewise continuous on $[0, T]$ for each $T > 0$ and that f is of exponential order, say $|f(t)| \leq Me^{\sigma t}$ for all $t > 0$ where $\sigma > 0$. Then each of the following formulas is valid:

(i) $\mathscr{L}\{tf(t)\}(s) = -\dfrac{d}{ds}\mathscr{L}\{f(t)\}(s)$ for all $s > \sigma$,

(ii) $\mathscr{L}\left\{\displaystyle\int_0^t f(r)\,dr\right\}(s) = \dfrac{1}{s}\mathscr{L}\{f(t)\}(s)$ for all $s > \sigma$, and

(iii) $\mathscr{L}\{e^{at}f(t)\}(s) = \mathscr{L}\{f(t)\}(s-a)$ for all $s > \sigma + a$ and each number a.

Assuming that differentiation under the integral sign is valid, we have

$$\begin{aligned}\frac{d}{ds}\mathscr{L}\{f(t)\}(s) &= \frac{d}{ds}\int_0^\infty e^{-st}f(t)\,dt \\ &= \int_0^\infty \frac{\partial}{\partial s}[e^{-st}f(t)]\,dt \\ &= -\int_0^\infty e^{-st}tf(t)\,dt \\ &= -\mathscr{L}\{tf(t)\}(s)\end{aligned}$$

and assertion (i) follows. Assertion (ii) follows from (5) in Theorem 3.1-2. For if

$$g(t) \equiv \int_0^t f(r)\,dr \qquad \text{for all } t \geq 0$$

then g is of exponential order [in fact, $|g(t)| \leq Me^{\sigma t}/\sigma$ for $t \geq 0$], $g'(t) = f(t)$ for $t \geq 0$, and

$$\mathscr{L}\{f(t)\}(s) = \mathscr{L}\{g'(t)\}(s) = s\mathscr{L}\{g(t)\}(s) - g(0)$$

by property (5). Assertion (ii) now follows since $g(0) = 0$. Assertion (iii) follows directly from the definition

$$\begin{aligned}\mathscr{L}\{e^{at}f(t)\}(s) &= \int_0^\infty e^{-st}e^{at}f(t)\,dt \\ &= \int_0^\infty e^{-(s-a)t}f(t)\,dt \\ &= \mathscr{L}\{f(t)\}(s-a)\end{aligned}$$

This establishes Theorem 3.1-3.

One of the main uses of the various properties of the Laplace transform is actually to compute the transform in some important circumstances. This is illustrated by the following examples.

EXAMPLE 3.1-2

By direct integration we have

$$\mathscr{L}\{1\}(s) = \int_0^\infty e^{-st} 1\, dt = \lim_{T \to \infty} \left[-\frac{e^{-st}}{s} \right]_{t=0}^{T} = \frac{1}{s}$$

and hence using (iii) in Theorem 3.1-3,

$$\mathscr{L}\{e^{at}\}(s) = \mathscr{L}\{e^{at} \cdot 1\}(s) = \mathscr{L}\{1\}(s-a) = \frac{1}{s-a}$$

This result was also obtained in the preceding example. Using (i) in Theorem 3.1-3 we obtain

$$\mathscr{L}\{te^{at}\}(s) = -\frac{d}{ds}\left(\frac{1}{s-a}\right) = \frac{1}{(s-a)^2}$$

In particular,

$$\mathscr{L}\{t\}(s) = \frac{1}{s^2} \qquad \text{for } s > 0$$

Using (ii) and the linear property (4), it follows that

$$\mathscr{L}\{t^2\}(s) = 2\mathscr{L}\left\{\int_0^t r\, dr\right\}(s) = 2 \cdot \frac{1}{s}\, \mathscr{L}\{t\}(s)$$

and hence

$$\mathscr{L}\{t^2\}(s) = \frac{2}{s^3} \qquad \text{for } s > 0$$

The Laplace transform of t^n for each positive integer n is listed in the short table at the end of this chapter (see also Problem 3.1-7).

EXAMPLE 3.1-3

Suppose that $\omega > 0$ and $f(t) = \sin \omega t$ for all $t \geq 0$. If $s > 0$, then

$$\mathscr{L}\{\sin \omega t\}(s) = \int_0^\infty e^{-st} \sin \omega t\, dt$$

by definition. Using integration by parts twice, we obtain

$$\int e^{-st} \sin \omega t \, dt = -e^{-st}\omega^{-1} \cos \omega t - \int se^{-st}\omega^{-1} \cos \omega t \, dt$$
$$= -e^{-st}\omega^{-1} \cos \omega t - se^{-st}\omega^{-2} \sin \omega t$$
$$- s^2\omega^{-2} \int e^{-st} \sin \omega t \, dt$$

and it follows that

$$\int e^{-st} \sin \omega t \, dt = -\frac{\omega^{-1}e^{-st} \cos \omega t + s\omega^{-2}e^{-st} \sin \omega t}{1 + s^2\omega^{-2}}$$

Since the expression on the right side of this equation has limit zero as $t \to \infty$, we have

$$\int_0^\infty e^{-st} \sin \omega t \, dt = \frac{\omega^{-1}}{1 + s^2\omega^{-2}}$$

Multiplying both the numerator and denominator by ω^2 shows that

$$\mathscr{L}\{\sin \omega t\}(s) = \frac{\omega}{s^2 + \omega^2} \qquad \text{for all } s > 0 \tag{7}$$

Since $\cos \omega t = d/dt\,(\omega^{-1} \sin \omega t)$, we have from (5) in Theorem 3.1-2 that

$$\mathscr{L}\{\cos \omega t\}(s) = \mathscr{L}\left\{\frac{d}{dt}(\omega^{-1} \sin \omega t)\right\}(s)$$
$$= s\mathscr{L}\{\omega^{-1} \sin \omega t\}(s) - \sin 0$$

From this it follows that

$$\mathscr{L}\{\cos \omega t\} = \frac{s}{s^2 + \omega^2} \qquad \text{for all } s > 0 \tag{8}$$

Using property (iii) of Theorem 3.1-3 along with (7) and (8), we obtain the formulas

(9)

(a) $\mathscr{L}\{e^{at} \sin \omega t\}(s) = \dfrac{\omega}{(s-a)^2 + \omega^2}$ for all $s > a$

(b) $\mathscr{L}\{e^{at} \cos \omega t\}(s) = \dfrac{s-a}{(s-a)^2 + \omega^2}$ for all $s > a$

These formulas are also listed in the table at the end of this chapter.

EXAMPLE 3.1-4

Using first part (i) of Theorem 3.1-3 and then formula (7), we see that

$$\mathscr{L}\{t \sin t\}(s) = -\frac{d}{ds}\mathscr{L}\{\sin t\}(s) = -\frac{d}{ds}\left(\frac{1}{s^2 + 1}\right) = -\frac{2s}{(s^2 + 1)^2}$$

Using part (1) of Theorem 3.1-3 twice shows that

$$\mathscr{L}\{t^2 \sin t\}(s) = -\frac{d}{ds}\mathscr{L}\{t \sin t\}(s) = \frac{d^2}{ds^2}\mathscr{L}\{\sin t\}(s) = \frac{2 - 2s^2}{(s^2+1)^3}$$

It is also of interest to compute the Laplace transform of the hyperbolic sine and cosine functions. Recall that

$$\sinh \omega t \equiv \frac{e^{\omega t} - e^{-\omega t}}{2} \qquad \text{and} \qquad \cosh \omega t \equiv \frac{e^{\omega t} + e^{-\omega t}}{2}$$

for each $\omega > 0$ and all numbers t. Using (2) along with linearity [see (4)], it is immediately seen that

$$\mathscr{L}\{\sinh \omega t\}(s) = \frac{1}{2}\mathscr{L}\{e^{\omega t}\}(s) - \frac{1}{2}\mathscr{L}\{e^{-\omega t}\}(s)$$

$$= \frac{1}{2}\frac{1}{s-\omega} - \frac{1}{2}\frac{1}{s+\omega} = \frac{\omega}{s^2 - \omega^2}$$

and

$$\mathscr{L}\{\cosh \omega t\}(s) = \frac{1}{2}\frac{1}{s-\omega} + \frac{1}{2}\frac{1}{s+\omega} = \frac{s}{s^2 - \omega^2}$$

Using (iii) in Theorem 3.1-3, we obtain the formulas

(10)

(a) $$\mathscr{L}\{e^{at} \sinh \omega t\}(s) = \frac{\omega}{(s-a)^2 - \omega^2}$$

(b) $$\mathscr{L}\{e^{at} \cosh \omega t\}(s) = \frac{s-a}{(s-a)^2 - \omega^2}$$

► **EXAMPLE 3.1-5**

Using (i) of Theorem 3.1-3 and (10a), we obtain

$$\mathscr{L}\{te^{3t} \sinh t\}(s) = -\frac{d}{ds}\mathscr{L}\{e^{3t} \sinh t\}(s)$$

$$= -\frac{d}{ds}\left[\frac{1}{(s-3)^2 + 1}\right]$$

$$= \frac{2s - 6}{(s^2 - 6s + 10)^2}$$

PROBLEMS

3.1-1. For each of the following functions f compute by direct evaluation of the integral the Laplace transform $\mathscr{L}\{f\}(s)$ for appropriate values of s.

(a) $f(t) = \sin 2t$

(b) $f(t) = te^{t/2}$

(c) $f(t) = t^2$

(d) $f(t) = \begin{cases} 1 & 0 \le t < 1 \\ 0 & t > 1 \end{cases}$

(e) $f(t) = \begin{cases} \sin (t) & 0 \le t \le \pi \\ 0 & t > \pi \end{cases}$

3.1-2. Use the formulas and properties for the Laplace transform developed in this section to compute $\mathscr{L}\{f\}(s)$ for each of the following:

(a) $f(t) = t^2 e^{-4t}$

(b) $f(t) = t \sin 2t$

(c) $f(t) = t \cos t$

(d) $f(t) = te^{-t} \sin t$

(e) $f(t) = \cos^2 t$

(f) $f(t) = \sinh 2t \cos t$

(g) $f(t) = te^{-t} - 3e^t \cosh 2t$

(h) $f(t) = t^2 \cos t$

(i) $f(t) = t^2 \sin 4t$

(j) $f(t) = \sin^2 3t$

***3.1-3.** Show that if $f(t) = 2te^{t^2} \cos e^{t^2}$ for $t \ge 0$, then f is not of exponential order. Show, however, that $\mathscr{L}\{f\}(a)$ exists for all $s > 0$. [*Hint:* Show that $f(t) = d/dt \sin e^{t^2}$ and apply Theorem 3.1-2.]

***3.1-4.** Show that if $f(t) = e^{t^2}$ for all $t \ge 0$, then $\mathscr{L}\{f\}(s)$ does not exist for any number s.

***3.1-5.** If f is defined on $[0, 1]$ by $f(0) = 0$ and $f(t) = \sin (1/t)$ for $t \in (0, 1]$, then f is continuous on $[0, 1]$ except at the point $t = 0$. Show, however, that f is not piecewise continuous.

***3.1-6.** Suppose that f is a continuous function on $[0, \infty)$ and is of exponential order: $|f(t)| \le Me^{at}$ for all $t \ge 0$. Show that $\mathscr{L}\{f\}(s) \to 0$ as $s \to \infty$ and further that $s\mathscr{L}\{f\}(s)$ is bounded as $s \to \infty$.

***3.1-7.** Show that

$$\mathscr{L}\{t^n\}(s) = \frac{n!}{s^{n+1}}$$

for all positive integers n and positive numbers s.

***3.1-8.** Suppose that n is a positive integer and f has n derivatives on $[0, \infty)$. Suppose further that $f, df/dt, \ldots, df^{n-1}/dt^{n-1}$ are each of exponential order and that d^nf/dt^n is piecewise continuous on $[0, T]$ for each $T > 0$. Show that $\mathscr{L}\{d^nf/dt^n\}$ exists and that the formula

$$\mathscr{L}\left\{\frac{d^nf}{dt^n}\right\}(s) = s^n \mathscr{L}\{f\}(s) - \sum_{j=0}^{n-1} \frac{s^j d^{n-1-j} f(0)}{dt^{n-1-j}}$$

is valid for all sufficiently large s. (*Hint:* Use Theorem 3.1-2 and mathematical induction on n.)

3.2 INVERSE TRANSFORMS

In our application of the Laplace transform to differential equations, the first step is to compute the transform of the solution directly from the given equation. The problem then is to determine the solution of the differential equation from its Laplace transform. Therefore, in this section we consider the question of determining a function from its transform—that is, can the Laplace transform be inverted? So suppose that F is a given function for $s > s_0$ and that f is a continuous function on $[0, \infty)$ such that

$$F(s) = \mathscr{L}\{f\}(s) \qquad \text{for all } s > s_0$$

Then f is called the *inverse Laplace transform* of F and is denoted $\mathscr{L}^{-1}\{F\}$:

$$\text{(1)} \qquad \begin{aligned} &\mathscr{L}^{-1}\{F\}(t) = f(t) \qquad \text{for } t \geq 0 \qquad \text{if and only if} \\ &F(s) = \mathscr{L}\{f\}(s) \qquad \text{for } s > s_0 \end{aligned}$$

The fact that definition (1) is well-defined when f is required to be continuous is not simple and depends on a type of uniqueness assertion for the Laplace transform known as *Lerch's theorem*: If f and g are continuous functions on $[0, \infty)$ such that $\mathscr{L}\{f\}(s) = \mathscr{L}\{g\}(s)$ for all $s > s_0$, then $f(t) \equiv g(t)$ for all $t \geq 0$.

According to Example 3.1-1, if σ is a number and $F(s) = 1/(s - \sigma)$ for all $s > \sigma$, then $\mathscr{L}^{-1}\{F\}(t) = e^{\sigma t}$ for all $t \geq 0$. However, note that if f is *any* function on $[0, \infty)$ such that $f(t) = e^{\sigma t}$ for all but a finite number of $t \geq 0$, then $\mathscr{L}\{f\}(s)$ is also equal to $1/(s - \sigma)$ for $s > \sigma$. This follows from the fact that the integral of a function over an interval is not altered by changing the function at any finite number of points in the interval. In order to avoid identification problems of this type, it is always assumed in this section that the inverse Laplace transform is a continuous function.

It is crucial to realize that each formula for the Laplace transform derived in Section 3.1 can also be used to compute inverse Laplace transforms. First, it is important to observe that the linear property of the Laplace transform implies the linear property of the inverse Laplace transform: If F and G are functions for $s > s_0$ and c_1 and c_2 are constants, then

$$\text{(2)} \qquad \mathscr{L}^{-1}\{c_1 F + c_2 G\}(t) = c_1 \mathscr{L}^{-1}\{F\}(t) + c_2 \mathscr{L}^{-1}\{G\}(t)$$

Formula (2) follows from the analogous property of $\mathscr{L}$ [see (4) in Section 3.1].

As a specific illustration, suppose that

$$\mathscr{L}\{f\}(s) = \frac{3}{s^2 + 4} \qquad \text{for all } s > 0$$

Rewriting in the form

$$\frac{3}{s^2 + 4} = \frac{3}{2} \cdot \frac{2}{s^2 + 2^2}$$

we see from formula (9) in the previous section that $f(t) = \frac{3}{2} \sin 2t$—that is,

$$\mathscr{L}^{-1}\left\{\frac{3}{s^2 + 4}\right\}(t) = \tfrac{3}{2} \sin 2t$$

EXAMPLE 3.2-1

Suppose that

$$\mathscr{L}\{f\}(s) = \frac{s}{s^2 - 2s + 3} \qquad \text{for all } s > 1$$

Completing the square of the denominator and rewriting the resulting fraction into terms of the form indicated by (9) in Section 3.1, we have

$$\begin{aligned}\frac{s}{s^2 - 2s + 3} &= \frac{s}{(s-1)^2 + 2} \\ &= \frac{s-1}{(s-1)^2 + (\sqrt{2})^2} + \frac{1}{\sqrt{2}} \frac{\sqrt{2}}{(s-1)^2 + (\sqrt{2})^2}\end{aligned}$$

From (9) in Section 3.1 and the linear properties it follows that

$$\begin{aligned}\mathscr{L}^{-1}\left\{\frac{s}{s^2 - 2s + 3}\right\}(t) &= \mathscr{L}^{-1}\left\{\frac{s-1}{(s-1)^2 + (\sqrt{2})^2}\right\}(t) \\ &\qquad + \frac{1}{\sqrt{2}} \mathscr{L}^{-1}\left\{\frac{\sqrt{2}}{(s-1)^2 + (\sqrt{2})^2}\right\}(t) \\ &= e^t \cos \sqrt{2}t + \frac{1}{\sqrt{2}} e^t \sin \sqrt{2}t\end{aligned}$$

Therefore,

$$f(t) = e^t \cos \sqrt{2}t + \frac{1}{\sqrt{2}} e^t \sin \sqrt{2}t$$

and

$$\mathscr{L}^{-1}\left\{\frac{s}{s^2 - 2s + 3}\right\}(t) = e^t \cos \sqrt{2}t + \frac{1}{\sqrt{2}} e^t \sin \sqrt{2}t$$

One of the most useful techniques for determining inverse tranforms is *partial fraction expansion.* The following examples illustrate this technique.

▶ **EXAMPLE 3.2-2**

Suppose that

$$\mathscr{L}\{f\}(s) = \frac{s-5}{s^2+6s+5} \qquad \text{for all } s > -1$$

This fraction has the partial fraction expansion of the form

$$\frac{s-5}{s^2+6s+5} = \frac{s-5}{(s+5)(s+1)} = \frac{d_1}{s+5} + \frac{d_2}{s+1}$$

for some constants d_1 and d_2. Multiplying by $(s+5)(s+1)$ we obtain

$$s - 5 = d_1(s+1) + d_2(s+5)$$

Setting $s = -5$ shows that $d_1 = \frac{5}{2}$, and setting $s = -1$ shows that $d_2 = -\frac{3}{2}$. Therefore,

$$\mathscr{L}^{-1}\left\{\frac{s-5}{s^2+6s+5}\right\}(t) = \tfrac{5}{2}\mathscr{L}^{-1}\left\{\frac{1}{s+5}\right\}(t) - \tfrac{3}{2}\mathscr{L}^{-1}\left\{\frac{1}{s+1}\right\}(t)$$

and it follows from Example 3.1-1 that

$$f(t) = \mathscr{L}^{-1}\left\{\frac{s-5}{s^2+6s+5}\right\}(t) = \tfrac{5}{2}e^{-5t} - \tfrac{3}{2}e^{-t}$$

for all $t \geq 0$.

 **EXAMPLE 3.2-3**

Suppose that

$$\mathscr{L}\{f\}(s) = \frac{s+1}{s^3+s} \qquad \text{for all } s > 0$$

Assuming the partial fraction expansion

$$\frac{s+1}{s(s^2+1)} = \frac{d_1}{s} + \frac{d_2 s + d_3}{s^2+1}$$

shows that

$$\frac{s+1}{s^3+s} = \frac{1}{s} + \frac{-s+1}{s^2+1}$$

Therefore, using the linear property of $\mathscr{L}^{-1}$,

$$\mathscr{L}^{-1}\left\{\frac{s+1}{s^3+s}\right\}(t) = \mathscr{L}^{-1}\left\{\frac{1}{s}\right\}(t) - \mathscr{L}^{-1}\left\{\frac{s}{s^2+1}\right\}(t) + \mathscr{L}^{-1}\left\{\frac{1}{s^2+1}\right\}(t)$$

and it follows from Example 3.1-1 and formula (9) in Section 3.1 that

$$f(t) = \mathscr{L}^{-1}\left\{\frac{s+1}{s^3+5}\right\}(t) = 1 - \cos t + \sin t$$

*3.2a Convolution Theorem

The convolution of two functions can also be a helpful tool for the application of the Laplace transform. Suppose that f and g are two piecewise continuous functions on $[0, \infty)$. The *convolution of f and g* is the function $f * g$ that is defined on $[0, \infty)$ by

$$f * g\,(t) = \int_0^t f(t-r)g(r)\,dr \qquad \text{for all } t \geq 0 \tag{3}$$

The convolution operation $*$ has several properties that resemble those of multiplication. For instance, convolution is a commutative operation:

$$f * g\,(t) \equiv g * f\,(t) \qquad \text{for all } t \geq 0 \tag{4}$$

To see that (4) holds observe that

$$\begin{aligned} f * g\,(t) &= \int_0^t f(t-r)g(r)\,dr = -\int_t^0 f(\tau)g(t-\tau)\,d\tau \\ &= \int_0^t g(t-\tau)f(\tau)\,d\tau = g * f\,(t) \end{aligned}$$

where the second equality was obtained by using the change of variables $\tau = t - r$. Similarly, the distributive and associative properties are also valid for convolution:

$$\begin{aligned} &\textbf{(a)}\quad f * (g + h) = f * g + f * h \\ &\textbf{(b)}\quad f * (g * h) = (f * g) * h \end{aligned} \tag{5}$$

The student is asked to establish these properties in Problem 3.2-5. One of the most important properties of convolution is the fact that the Laplace transform of the convolution of two functions is the product of the Laplace transform of the functions:

■ **Theorem 3.2-1**

Suppose that f and g are piecewise continuous on $[0, T]$ for each $T > 0$ and are of exponential order. Then there is a number s_0 such that

$$\mathscr{L}\{f * g\}(s) = \mathscr{L}\{f\}(s) \times \mathscr{L}(g\}(s) \tag{6}$$

for all $s > s_0$.

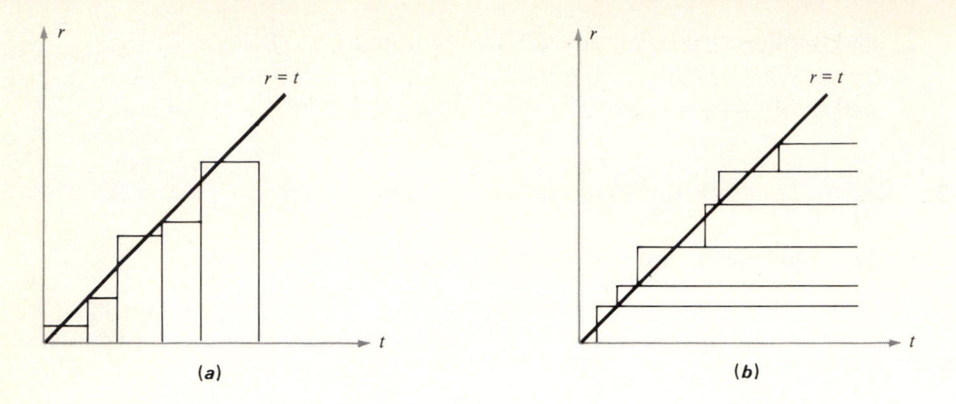

Figure 3.2 ***a, b*** Interchanging order of integration.

Formula (6) can be indicated easily by reversing the order of integration in the double integral that results from the definitions. Note first that

$$\mathscr{L}\{f * g\}(s) = \int_0^\infty e^{-st}\left[\int_0^t f(t-r)g(r)\,dr\right]dt$$

$$= \int_0^\infty \left[\int_0^t e^{-st} f(t-r)g(r)\,dr\right]dt$$

Reversing the order of integration (see Figure 3.2*a* and *b*), it follows that

$$\mathscr{L}\{f * g\}(s) = \int_0^\infty \left[\int_r^\infty e^{-st} f(t-r)g(r)\,dt\right]dr$$

$$= \int_0^\infty g(r)\left[\int_r^\infty e^{-st} f(t-r)\,dt\right]dr$$

Making the change of variables $t = \tau + r$ in the inside integral, it follows that

$$\mathscr{L}\{f * g\}(s) = \int_0^\infty g(r)\left[\int_0^\infty e^{-s(\tau+r)} f(\tau)\,d\tau\right]dr$$

$$= \left[\int_0^\infty e^{-sr} g(r)\,dr\right] \times \left[\int_0^\infty e^{-s\tau} f(\tau)\,d\tau\right]$$

and assertion (6) follows by the definition of the Laplace transform.

Theorem 3.2-1 is known as the *convolution theorem* for the Laplace transform. This result can also be used to compute inverse Laplace transforms, and it follows from (6) that if $F(s) \equiv \mathscr{L}\{f\}(s)$ and $G(s) \equiv \mathscr{L}\{g\}(s)$ for all $s > s_0$, then

$$\mathscr{L}^{-1}\{F(s)G(s)\}(t) = f * g\,(t) \qquad \text{for all } t \geq 0 \tag{7}$$

This technique is illustrated in the following example.

EXAMPLE 3.2-4

Suppose that

$$\mathscr{L}\{f\}(s) = \frac{1}{s^2(s^2+3)} \qquad \text{for all } s > 0$$

Since $\mathscr{L}^{-1}\{1/s^2\}(t) = t$ [see (4)] and since

$$\mathscr{L}^{-1}\left\{\frac{1}{s^2+3}\right\}(t) = \frac{1}{\sqrt{3}}\,\mathscr{L}^{-1}\left\{\frac{\sqrt{3}}{s^2+3}\right\}(t) = \frac{1}{\sqrt{3}} \sin \sqrt{3}t$$

it follows from (7) that $f = g * h$, where $g(t) \equiv t$ and $h(t) \equiv (1/\sqrt{3}) \sin \sqrt{3}t$ for $t \geq 0$. Therefore, using integration by parts,

$$\begin{aligned} f(t) &= \int_0^t (t-r)\frac{1}{\sqrt{3}} \sin \sqrt{3}r\, dr \\ &= [(t-r)(-\tfrac{1}{3}\cos\sqrt{3}r)]_{r=0}^t - \int_0^t (-1)(-\tfrac{1}{3}\cos\sqrt{3}r)\, dr \\ &= \frac{t}{3} - \left[\frac{1}{3\sqrt{3}}\sin\sqrt{3}r\right]_{r=0}^t \\ &= \frac{1}{3}t - \frac{1}{3\sqrt{3}}\sin\sqrt{3}t \end{aligned}$$

for all $t \geq 0$.

PROBLEMS

3.2-1. Determine the inverse Laplace transform of each of the following functions:

(a) $\dfrac{2s-1}{s^2+9}$ **(c)** $\dfrac{s}{s^2+2s+6}$

(b) $\dfrac{5}{(s-1)^3}$ **(d)** $\dfrac{1}{s^2+4s-5}$

(e) $\dfrac{1}{s^2(s^2+1)}$ **(h)** $\dfrac{2s-1}{s(s-1)^2}$

(f) $\dfrac{1}{(s-1)(s^2+1)}$ **(i)** $\dfrac{1}{s^3-8}$

(g) $\dfrac{1}{s^3-s}$ **(j)** $\dfrac{s^2+1}{s^3+3s^2+2s}$

3.2-2. Determine the inverse Laplace transform of each of the following functions:

(a) $\dfrac{(s+1)^2}{(s+2)^3}$ $\left(\textit{Hint: } \text{Rewrite in the form } \dfrac{d_1}{s+2} + \dfrac{d_2}{(s+2)^2} + \dfrac{d_3}{(s+2)^3}.\right)$

(b) $\dfrac{1}{(s^2+4)^2}$ $\left(\textit{Hint: } \text{Rewrite as } \dfrac{1}{s^2+4} \cdot \dfrac{1}{s^2+4}.\right)$

(c) $\dfrac{1}{s^4-1}$ $\left(\textit{Hint: } \text{Rewrite as } \dfrac{d_1}{s-1} + \dfrac{d_2}{s+1} + \dfrac{d_3 s + d_4}{s^2+1}.\right)$

(d) $\dfrac{s}{(s^2+1)^2}$

(e) $\dfrac{s^2}{(s^2+1)^2}$

3.2-3. Suppose that f is continuously differentiable on $[0, \infty)$ and that both f and f' are of exponential order.

(a) Show that $\lim_{s \to +\infty} s\mathscr{L}\{f\}(s) = f(0)$.

(b) If $\lim_{t \to \infty} f(t) = f(\infty)$ exists, show that $\lim_{s \to 0} s\mathscr{L}\{f\}(s) = f(\infty)$ [since $f(\infty)$ exists, f is bounded on $[0, \infty)$ and $\mathscr{L}\{f\}(s)$ exists for all $s > 0$].

(c) Discuss the validity of part b when $f(t) = e^t$, $f(t) = 1$, $f(t) = \sin t$, and $f(t) = e^{-t}$.

3.2-4. Suppose that $F(s) = \mathscr{L}\{f(t)\}(s)$ exists for all $s > s_0$ and that $\lambda > 0$ is given. Show that

$$\mathscr{L}\{f(\lambda t)\}(s) = \lambda^{-1}F(\lambda^{-1}s) \qquad \text{and} \qquad \mathscr{L}^{-1}\{F(\lambda s)\}(t) = \lambda^{-1}f(\lambda^{-1}t)$$

whenever $\lambda^{-1}s > s_0$ in the first equation and $\lambda s > s_0$ in the second.

***3.2-5.** Establish properties (5a) and (5b) for the convolution operation.

***3.2-6.** Suppose that f is continuous on $[0, \infty)$, of exponential order, and $\lim_{t \to 0^+} t^{-1}f(t)$ exists. If $F(s) = \mathscr{L}\{f\}(s)$ for $s > s_0$, show that

$$(*) \qquad \mathscr{L}\left\{\frac{f(t)}{t}\right\}(s) = \int_s^\infty F(r)\, dr \qquad \text{for } s > s_0$$

[Under these assumptions interchanging the order of integration is valid for the right side of equation (*).] Using equation (*) determine the Laplace transform of the following functions g:

(a) $g(t) = \dfrac{\sin t}{t}$

(b) $g(t) = \dfrac{1 - e^{-t}}{t}$

(c) $g(t) = \dfrac{\cos t - 1}{t}$

***3.2-7.** Suppose that f is of exponential order on $[0, \infty)$ and that $F(s) = \mathscr{L}\{f\}(s)$ for all $s > s_0$. Show that

$$f(t) = -\frac{1}{t}\,\mathscr{L}^{-1}\{F'(s)\}(t) \qquad \text{for all } t \geq 0$$

[see (i) of Theorem 3.2-1]. Use this result to determine the inverse transform of the following functions:

(a) $\arctan \dfrac{1}{s}$ **(b)** $\ln \dfrac{s+1}{s-1}$

***3.2-8.** Suppose that $T > 0$ and f is continuous and periodic with period T on $[0, \infty)$: $f(t + T) \equiv f(t)$ for all $t \geq 0$. Show that for $s > 0$,

$$\mathscr{L}\{f\}(s) = \int_0^T e^{-st}f(t)\,dt + \int_T^{2T} e^{-st}f(t)\,dt + \int_{2T}^{3T} e^{-st}f(t)\,dt + \cdots$$

and then deduce that

$$\mathscr{L}\{f\}(s) = \int_0^T e^{-st}f(t)\,dt\,[1 + e^{-sT} + e^{-2sT} + \cdots]$$

Conclude further that if $s > 0$,

$$\mathscr{L}\{f\}(s) = \frac{\int_0^T e^{-st}f(t)\,dt}{1 - e^{-sT}}$$

Apply this result to obtain the Laplace transform of the following functions:

(a) $f(t) = |\sin t|$

(b) $f(t) = \begin{cases} 1 & \text{if } 0 \leq t < 1 \\ -1 & \text{if } 1 \leq t < 2 \end{cases}$ and f is extended periodically of period 2 to $[0, \infty)$

(c) $f(t) = t$ if $0 \leq t < 1$ and f is extended periodically of period 1 to $[0, \infty)$.

Sketch the graph of each of the preceding functions.

3.3 NONHOMOGENEOUS EQUATIONS AND DISCONTINUITIES

One of the principal applications of the Laplace transform is in solving nonhomogeneous linear differential equations with constant coefficients. In the first part of this section a few routine equations are solved in order to illustrate the basic ideas. Subsequently, it is shown how this method can be applied when the nonhomogeneous term is only piecewise continuous. In fact, nonhomogeneous terms with jump discontinuities occur in several important physical models.

Assume initially that f is a continuous function on $[0, \infty)$ that is of exponential order, and consider the second order, nonhomogeneous initial value problem

$$ay'' + by' + cy = f(t) \qquad y(0) = y_0 \qquad y'(0) = y_1 \tag{1}$$

where a, b, and c are constants with $a \neq 0$. Although it is not proved in this text, it can be shown that the solution y to (1), as well as y' and y'', are of exponential order, and hence they each have a Laplace transform for s sufficiently large, say $s > s_0$. Using the properties developed in Section 3.1, the Laplace transform of the solution y to (1) can be obtained directly from (1). Taking the Laplace transform of each side of (1) and using linearity [see (4) in Section 3.1], it follows that

$$a\mathscr{L}\{y''\}(s) + b\mathscr{L}\{y'\}(s) + c\mathscr{L}\{y\}(s) = \mathscr{L}\{f\}(s)$$

Using the formulas for the transform of derivatives given in Theorem 3.1-2, along with the initial conditions in (1), we have

$$a[s^2\mathscr{L}\{y\}(s) - sy_0 - y_1] + b[s\mathscr{L}\{y\}(s) - y_0] + c\mathscr{L}\{y\}(s) = \mathscr{L}\{f\}(s)$$

and hence

$$[as^2 + bs + c]\mathscr{L}\{y\}(s) = \mathscr{L}\{f\}(s) + ay_1 + by_0 + asy_0$$

Since this holds for all s sufficiently large, it follows that there is a number s_0 such that

$$\mathscr{L}\{y\}(s) = \frac{\mathscr{L}\{f\}(s) + ay_1 + by_0 + ay_0 s}{as^2 + bs + c} \qquad \text{for all } s > s_0 \tag{2}$$

It is important to note that the denominator for the right-hand side of (2) is the auxiliary polynomial of the homogeneous equation associated with (1). Of course, once the transform of the solution is computed, the solution can be obtained from the inverse transform. There is a short table of Laplace transforms at the end of this chapter which can be used to help in these computations. The following two examples illustrate this procedure.

EXAMPLE 3.3-1

Consider the initial value problem

$$y'' + y' - 2y = \sin t \qquad y(0) = 0 \qquad y'(0) = 0 \tag{3}$$

According to (2) and the formula for $\mathscr{L}\{\sin t\}(s)$ in the table,

$$\mathscr{L}\{y\}(s) = \frac{\mathscr{L}\{\sin t\}(s)}{s^2 + s - 2} = \frac{1}{(s-1)(s+2)(s^2+1)} \qquad \text{for all } s > 1$$

Assuming the partial fraction expansion

$$\frac{1}{(s-1)(s+2)(s^2+1)} = \frac{d_1}{s-1} + \frac{d_2}{s+2} + \frac{d_3 s + d_4}{s^2+1}$$

it follows that $d_1 = \frac{1}{6}$, $d_2 = -\frac{1}{15}$, $d_3 = -\frac{1}{10}$, and $d_4 = -\frac{3}{10}$. Therefore,

$$\begin{aligned} y(t) &= \mathscr{L}^{-1}\left\{\frac{1}{(s-1)(s+2)(s^2+1)}\right\}(t) \\ &= \tfrac{1}{6}\mathscr{L}^{-1}\left\{\frac{1}{s-1}\right\}(t) - \tfrac{1}{15}\mathscr{L}^{-1}\left\{\frac{1}{s+2}\right\}(t) \\ &\qquad - \tfrac{1}{10}\mathscr{L}^{-1}\left\{\frac{s}{s^2+1}\right\}(t) - \tfrac{3}{10}\mathscr{L}^{-1}\left\{\frac{1}{s^2+1}\right\}(t) \\ &= \tfrac{1}{6}e^{t} - \tfrac{1}{15}e^{-2t} - \tfrac{1}{10}\cos t - \tfrac{3}{10}\sin t \end{aligned}$$

is the solution to the initial value problem (3).

EXAMPLE 3.3-2

Consider the initial value problem

$$y'' - y = t \qquad y(0) = -1 \qquad y'(0) = 1 \tag{4}$$

Using formula (2) and the formula $\mathscr{L}\{t\}(s) = s^{-2}$ from the table,

$$\mathscr{L}\{y\}(s) = \frac{1 - s + s^{-2}}{s^2 - 1} = \frac{s^2 - s^3 + 1}{s^2(s-1)(s+1)}$$

Assuming the partial fraction expansion

$$\frac{s^2 - s^3 + 1}{s^2(s-1)(s+1)} = \frac{d_1}{s} + \frac{d_2}{s^2} + \frac{d_3}{s-1} + \frac{d_4}{s+1}$$

leads to the equation

$$\mathscr{L}\{y\}(s) = -\frac{1}{s^2} + \frac{\frac{1}{2}}{s-1} - \frac{\frac{3}{2}}{s+1}$$

Therefore, taking the inverse transform of both sides,

$$y(t) = -t + \tfrac{1}{2}e^{t} - \tfrac{3}{2}e^{-t}$$

is the solution to (4).

3.3a Unit Step Functions

Now we consider the more subtle problem in which the nonhomogeneous term in (1) is only assumed to be piecewise continuous on each closed bounded interval. One needs to be careful in deciding on the appropriate notion of what is meant by a solution to (1) when f has discontinuities. It is required that y *and* y' *are continuous on* $[0, \infty)$, that y'' exists and is continuous on each open interval where f is continuous, and that equation (1) is satisfied on each open interval where f is continuous. With this definition of solution it is easy to check that y'' must be piecewise continuous and, in fact, has a discontinuity at precisely the same points where f is discontinuous.

As a simple example to illustrate this situation, consider the equation

$$(5) \qquad y''(t) = \begin{cases} 1 & \text{if } 0 \le t < 3 \\ 0 & \text{if } t \ge 3 \end{cases} \qquad y(0) = 0 \qquad y'(0) = 0$$

Using Newton's second law of motion, this initial value problem could describe the rectilinear motion of a unit mass (initially at rest) under a constant applied unit force up until time $t = 3$. At time $t = 3$ the force is instantaneously switched off (and hence the right-hand side of this equation jumps from 1 to 0 at time $t = 3$). Equation (5) can be solved easily. For if $0 \le t < 3$, then $y''(t) \equiv 1$, $y(0) = y'(0) = 0$, and hence $y(t) = \frac{1}{2}t^2$ for $0 \le t < 3$. Note that $y(3) = \frac{9}{2}$, $y'(3) = 3$, and so the equation $y''(t) \equiv 0$, $y(3) = \frac{9}{2}$, $y'(3) = 3$ must be satisfied for $t > 3$. Therefore, $y(t) = \frac{9}{2} + 3(t - 3)$ for all $t > 3$, and the solution to (5) is seen to be

$$y(t) = \begin{cases} \frac{1}{2}t^2 & \text{for } 0 \le t \le 3 \\ \frac{9}{2} + 3(t-3) & \text{for } t > 3 \end{cases}$$

It is easy to check that y and y' are continuous and that y'' is continuous except for a jump discontinuity at $t = 3$. It is also important to note that changing the value of the right-hand side of (5) at the single point $t = 3$ does not change the solution.

For convenience we introduce a notation for a basic family of step functions: for each $\sigma > 0$ the function u_σ is defined on $(-\infty, \infty)$ by

$$(6) \qquad u_\sigma(t) = \begin{cases} 0 & \text{if } t < \sigma \\ 1 & \text{if } t \ge \sigma \end{cases}$$

These functions are called *unit step functions.* The graph of u_σ is given in Figure 3.3.

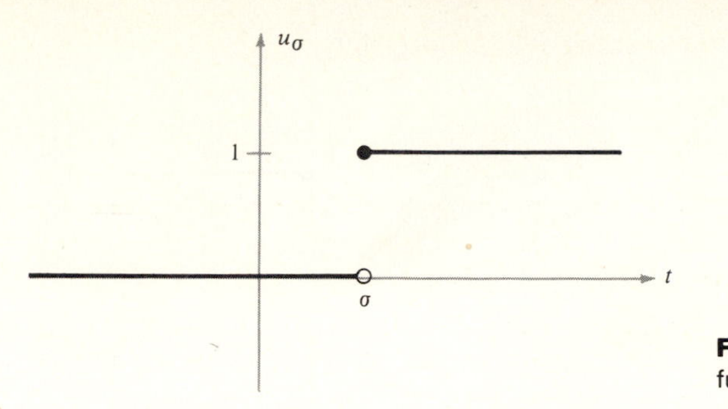

Figure 3.3 Graph of unit step function u_σ.

Since u_σ is piecewise continuous and bounded on $[0, \infty)$, it has a Laplace transform for all $s > 0$. Moreover,

$$\mathscr{L}\{u_\sigma\}(s) = \int_0^\infty e^{-st}u_\sigma(t)\,dt = \int_\sigma^\infty e^{-st}\,dt$$

$$= \lim_{T\to\infty}\left[-\frac{1}{s}e^{-st}\right]_{t=\sigma}^{T} = \frac{1}{s}e^{-s\sigma}$$

and so

$$\mathscr{L}\{u_\sigma\}(s) = \frac{e^{-\sigma s}}{s} \qquad \text{for all } s > 0 \text{ and } \sigma > 0 \tag{7}$$

The unit step functions can be used to represent and to help in the computation of the Laplace transform of several piecewise continuous functions and also right translations of functions. Note, for example, that the function defined by the right-hand side of equation (5) can be written in the form $1 - u_3(t)$ for all $t > 0$. Our first examples are simple step functions.

▶ **EXAMPLE 3.3-3**

Consider the function f defined on $[0, \infty)$ by

$$f(t) = \begin{cases} 1 & \text{if } 0 \le t < 1 \\ 2 & \text{if } 1 \le t < 3 \\ 4 & \text{if } 3 \le t < 4 \\ -2 & \text{if } 4 \le t \end{cases} \tag{8}$$

The graph of f is given in Figure 3.4*a*. Using the unit step function (6), f can be written in the form

$$f(t) = 1 + u_1(t) + 2u_3(t) - 6u_4(t) \qquad \text{for } t \ge 0$$

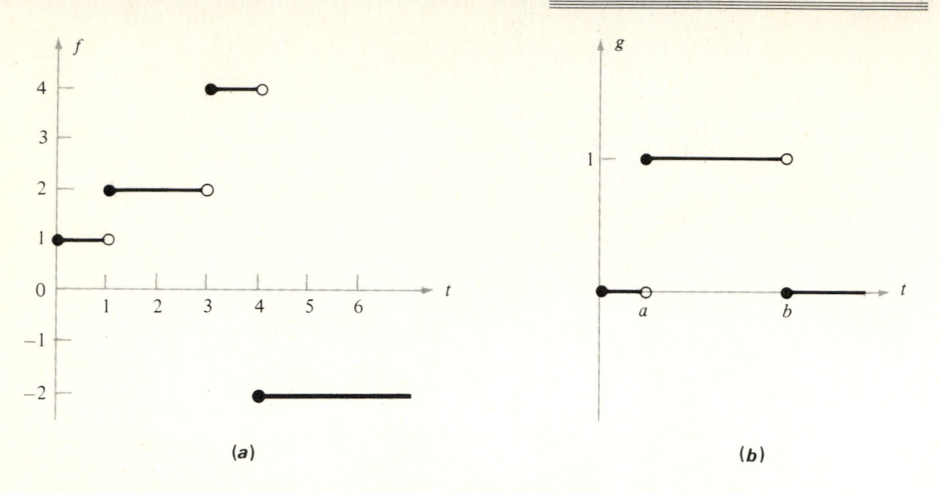

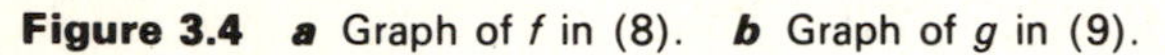
Figure 3.4 ***a*** Graph of f in (8). ***b*** Graph of g in (9).

Therefore, by the linear property of the Laplace transform and (7) it follows that

$$\mathscr{L}\{f\}(s) = \frac{1}{s} + \frac{e^{-s}}{s} + 2\,\frac{e^{-3s}}{s} - 6\,\frac{e^{-4s}}{s}$$

for all $s > 0$.

EXAMPLE 3.3-4

Suppose that $0 < a < b$ and that

$$g(t) = \begin{cases} 0 & \text{if } 0 \le t < a \\ 1 & \text{if } a \le t < b \\ 0 & \text{if } b \le t \end{cases} \tag{9}$$

This function is called a "square pulse," and its graph is given in Figure 3.4*b*. Since

$$g(t) = u_a(t) - u_b(t) \qquad \text{for } t \ge 0$$

it follows that

$$\mathscr{L}\{g\}(s) = \frac{e^{-as}}{s} - \frac{e^{-bs}}{s} = \frac{e^{-as} - e^{-bs}}{s}$$

for all $s > 0$.

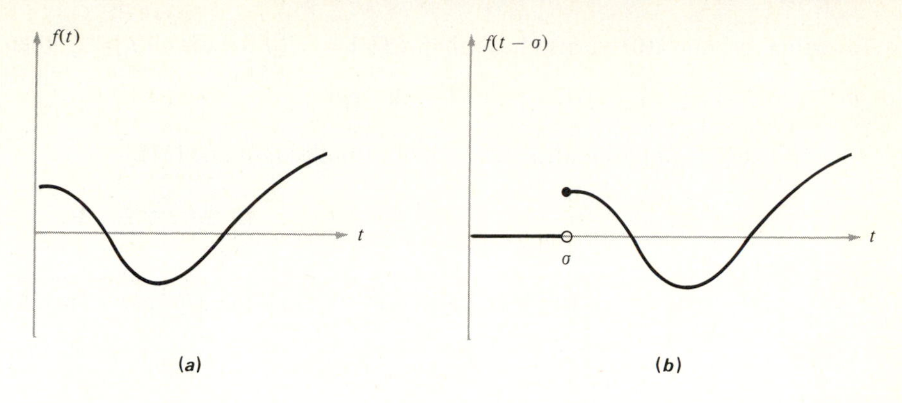

Figure 3.5 ***a*** Graph of $f(t)$. ***b*** Graph of $u_\sigma(t)\, f(t-\sigma)$.

Now suppose that f is defined for $t \geq 0$ and let σ be a positive number. Consider the function g obtained from f by moving the graph of f over σ units to the right and setting g equal to 0 to the left of σ: $g(t) = 0$ for $0 \leq t < \sigma$ and $g(t) = f(t-\sigma)$ for $t \geq \sigma$. A convenient representation of this function is

$$g(t) = u_\sigma(t) f(t-\sigma) \qquad \text{for all } t \geq 0$$

A graph indicating a typical shift is given in Figure 3.5.

There is a simple formula connecting the Laplace transform of a translation with the Laplace transform of the original function.

Theorem 3.3-1

Suppose that f is piecewise continuous and of exponential order and that $\sigma > 0$. Then

$$\mathscr{L}\{u_\sigma(t) f(t-\sigma)\}(s) = e^{-\sigma s}\mathscr{L}\{f(t)\}(s) \tag{10}$$

for all $s > s_0$.

Formula (10) follows easily from the definitions, with a simple change of variable:

$$\begin{aligned}\mathscr{L}\{u_\sigma(t) f(t-\sigma)\}(s) &= \int_\sigma^\infty e^{-st} f(t-\sigma)\, dt \\ &= \int_0^\infty e^{-s(\tau+\sigma)} f(\tau)\, d\tau \\ &= e^{-s\sigma}\mathscr{L}\{f(t)\}(s)\end{aligned}$$

Inverting formula (10) we also have that if $F(s) = \mathcal{L}\{f\}(s)$ for all $s > s_0$, then

$$\mathcal{L}^{-1}\{e^{-\sigma s}F(s)\}(t) = u_\sigma(t)f(t-\sigma) \qquad \text{for all } t \geq 0 \tag{11}$$

The following examples indicate the use of formulas (10) and (11).

EXAMPLE 3.3-5

Suppose that $f(t) = 1$ for $0 \leq t < \pi/6$ and $f(t) = \sin t$ for $\pi/6 \leq t$. Therefore, f can be represented in the form

$$f(t) = 1 + u_{\pi/6}(t)(\sin t - 1) \tag{12}$$

for all $t \geq 0$. In order to use (10) to compute the Laplace transform of the second term on the right side of (12), it is necessary for the function in parentheses to be expressed in terms of the variable $(t - \pi/6)$. Noting that

$$\begin{aligned}\sin t &= \sin\left[\left(t - \frac{\pi}{6}\right) + \frac{\pi}{6}\right] \\ &= \cos\frac{\pi}{6}\sin\left(t - \frac{\pi}{6}\right) + \sin\frac{\pi}{6}\cos\left(t - \frac{\pi}{6}\right) \\ &= \frac{\sqrt{3}}{2}\sin\left(t - \frac{\pi}{6}\right) + \frac{1}{2}\cos\left(t - \frac{\pi}{6}\right)\end{aligned}$$

it follows that

$$f(t) = 1 + u_{\pi/6}(t)\left[\frac{\sqrt{3}}{2}\sin\left(t - \frac{\pi}{6}\right) + \frac{1}{2}\cos\left(t - \frac{\pi}{6}\right) - 1\right]$$

It follows from (10) that

$$\mathcal{L}\{f\}(s) = \frac{1}{s} + e^{-\pi s/6}\left(\frac{\sqrt{3}}{2}\frac{1}{s^2+1} + \frac{1}{2}\frac{s}{s^2+1} - \frac{1}{s}\right)$$

for all $s > 0$.

EXAMPLE 3.3-6

Consider the initial value problem

$$y'' - 3y' - 4y = f(t) \equiv \begin{cases} e^t & \text{if } 0 \leq t < 2 \\ 0 & \text{if } 2 \leq t \end{cases} \qquad y(0) = 0 \qquad y'(0) = 0 \tag{13}$$

Using a unit step function we have that

$$f(t) = e^t - u_2(t)e^t = e^t - e^2u_2(t)e^{t-2}$$

for all $t \geq 0$, and it follows that

$$\mathscr{L}\{f(t)\}(s) = \frac{1}{s-1} - e^2 e^{-2s} \frac{1}{s-1}$$

Therefore, since $s^2 - 3s - 4 = (s+1)(s-4)$, we have by formula (2) that

$$\mathscr{L}\{y\}(s) = \frac{1}{(s-1)(s+1)(s-4)} - \frac{e^2 e^{-2s}}{(s-1)(s+1)(s-4)}$$

Since

$$\frac{1}{(s-1)(s+1)(s-4)} = \frac{-\frac{1}{6}}{s-1} + \frac{\frac{1}{10}}{s+1} + \frac{\frac{1}{15}}{s-4}$$

by partial fractions, it follows that the solution y to (13) has the form

$$y(t) = -\tfrac{1}{6}e^t + \tfrac{1}{10}e^{-t} + \tfrac{1}{15}e^{4t} - e^2 u_2(t)[-\tfrac{1}{6}e^{t-2} + \tfrac{1}{10}e^{-(t-2)} + \tfrac{1}{15}e^{4(t-2)}]$$

[see formula (11)]. Hence, the solution y can be written as

$$y(t) = -\tfrac{1}{6}e^t + \tfrac{1}{10}e^{-t} + \tfrac{1}{15}e^{4t} + u_2(t)[\tfrac{1}{6}e^t - \tfrac{1}{10}e^4 e^{-t} + \tfrac{1}{15}e^{-6}e^{4t}]$$

or as

$$y(t) = \begin{cases} -\frac{1}{6}e^t + \frac{1}{10}e^{-t} + \frac{1}{15}e^{4t} & 0 \leq t < 2 \\ \dfrac{1-e^4}{10}e^{-t} + \dfrac{1+e^{-6}}{15}e^{4t} & 2 \leq t \end{cases}$$

PROBLEMS

3.3-1. Determine a solution to each of the following equations by using Laplace transform techniques.

(a) $y'' - y = 1 \qquad y(0) = -1 \qquad y'(0) = 1$

(b) $y'' + 4y = \cos t \qquad y(0) = 0 \qquad y'(0) = 0$

(c) $y'' + y' - 2y = e^{3t} \qquad y(0) = 0 \qquad y'(0) = 2$

(d) $y'' - 6y' + 10y = 1 \qquad y(0) = 0 \qquad y'(0) = 0$

(e) $y'' + y' - 6y = e^{2t} \qquad y(0) = 0 \qquad y'(0) = 0$

(f) $y'' - 2y' + y = e^t \qquad y(0) = 0 \qquad y'(0) = 0$

3.3-2. Determine the Laplace transform of the following functions (u_σ is the unit step function).

(a) $f(t) = \begin{cases} 1 & 0 \leq t < 2 \\ -8 & 2 \leq t < \pi \\ 0 & \pi < t \end{cases}$ **(b)** $f(t) = \begin{cases} -2 & 0 \leq t < 1 \\ 2 & 1 \leq t < 2 \\ -2 & 2 \leq t \end{cases}$

(c) $f(t) = 1 + 2u_1(t) + 4u_3(t)$ **(d)** $f(t) = 6 - 2u_\pi(t) - 4u_{2\pi}(t)$

3.3-3. Determine the Laplace transform of each of the following functions (u_σ is the unit step function).

(a) $f(t) = \begin{cases} 1 & 0 \le t < 3 \\ t & 3 \le t \end{cases}$

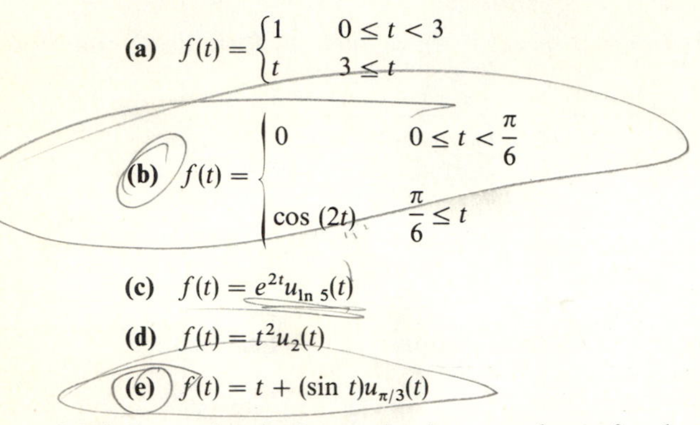

(b) $f(t) = \begin{cases} 0 & 0 \le t < \dfrac{\pi}{6} \\ \cos(2t) & \dfrac{\pi}{6} \le t \end{cases}$

(c) $f(t) = e^{2t}u_{\ln 5}(t)$

(d) $f(t) = t^2 u_2(t)$

(e) $f(t) = t + (\sin t)u_{\pi/3}(t)$

3.3-4. Determine the inverse Laplace transform of each of the following functions.

(a) $F(s) = \dfrac{e^{-s}}{s^2 + 4}$ **(c)** $F(s) = \dfrac{2^{-s}}{s(s^2 - 1)}$

(b) $F(s) = \dfrac{e^{-4s}}{s^2 - s - 2}$ **(d)** $F(s) = \dfrac{e^{-3s}}{(2s - 1)^3}$

3.3-5. The *Heaviside function* is the function H defined on $(-\infty, \infty)$ by

$$H(t) = \begin{cases} 0 & \text{if } t < 0 \\ 1 & \text{if } t \ge 0 \end{cases}$$

Show that if $\sigma > 0$ and u_σ is the unit step function defined by (6), then $u_\sigma(t) = H(t - \sigma)$ for all $t \ge 0$. Consider the graph of the function f in Figure 3.6. Using this graph, sketch the graph of each of the following functions:

(a) $g_1(t) = H(t - \sigma)f(t - \sigma)$ for $t > 0$

(b) $g_2(t) = H(t - \sigma)f(t)$ for $t \ge 0$

(c) $g_3(t) = H(\sigma - t)f(t)$ for $t \ge 0$

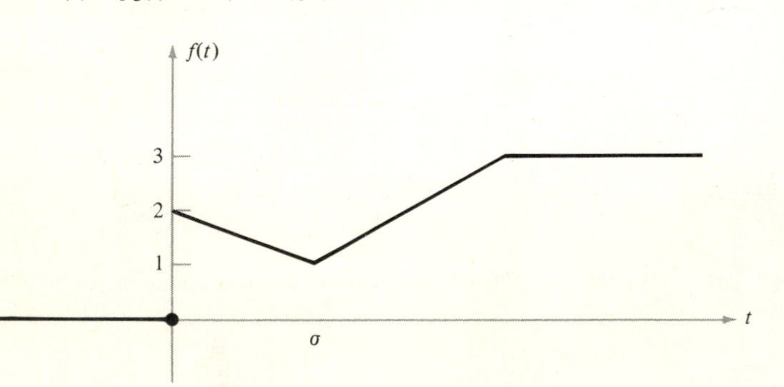

Figure 3.6 Graph of function $f(t)$ from Problem 3.3-5.

3.3-6. Determine a solution to each of the following equations, using Laplace transform methods.

(a) $y'' - y = f(t) \equiv \begin{cases} 0 & \text{if } 0 \le t < 2 \\ -1 & \text{if } 2 \le t \end{cases}$ $\quad y(0) = y'(0) = 0$

(b) $y'' + y = f(t) \equiv \begin{cases} 1 & \text{if } 0 \le t < 3 \\ 0 & \text{if } 3 \le t \end{cases}$ $\quad y(0) = y'(0) = 0$

(c) $y'' - y' - 2y = f(t) \equiv \begin{cases} e^{3t} & 0 \le t < 1 \\ e^{3t} + 1 & 1 \le t \end{cases}$ $\quad y(0) = y'(0) = 0$

(d) $y'' - y = f(t) \equiv \begin{cases} e^{t} & 0 \le t < 2 \\ 0 & 2 \le t \end{cases}$ $\quad y(0) = y'(0) = 0$

3.3-7. Define the function f on $[0, \infty)$ by $f(t) = 1$ if $t \in [n, n + 1)$ where n is even, and $f(t) = -1$ if $t \in [n, n + 1)$ where n is odd.

(a) Show that $f(t) = 1 - 2u_1(t) + 2u_2(t) - 2u_3(t) + \cdots$, that is,

$$f(t) \equiv 1 + 2 \sum_{n=1}^{\infty} (-1)^n u_n(t) \qquad \text{for all } t \ge 0$$

(b) Using the series in part *a*, compute the Laplace transform of $f(t)$, using term-by-term addition. (Compare your answer with the answer for Problem 3.2-8*b*.)

3.4 MECHANICAL VIBRATIONS AND IMPULSES

This section indicates some elementary applications of Laplace transforms to the study of forced vibrations in linear springs. Vibrations in linear springs have already been studied in Section 2.6. We now analyze this type of problem, using Laplace transforms and the techniques studied in the preceding three sections. For the derivation of the equation of motion for an externally forced linear spring, the reader is referred to Section 2.6. It is assumed that a mass m is attached to a linear spring having spring constant k and that there is an applied force $f(t)$ for each time $t \ge 0$. Assuming that *no damping* is present and that the mass is initially at equilibrium, the equation of motion is

$$my'' + ky = f(t) \qquad y(0) = y'(0) = 0 \tag{1}$$

where $y(t)$ is the position of the mass relative to equilibrium (see Figure 3.7). Taking the Laplace transform of each side of (1) and using the initial conditions along with the formula for the transform of the second derivative leads to the equation

$$ms^2\mathscr{L}\{y\}(s) + k\mathscr{L}\{y\}(s) = \mathscr{L}\{f\}(s)$$

Therefore,

$$\mathscr{L}\{y\}(s) = \frac{\mathscr{L}\{f\}(s)}{ms^2 + k} = \frac{1}{m}\frac{\mathscr{L}\{f\}(s)}{s^2 + k/m}$$

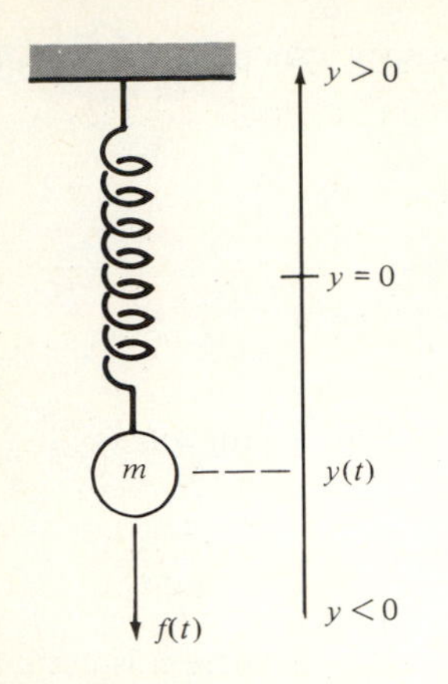

Figure 3.7 Mass-spring system with external forcing.

and it follows that

(2) $$\mathscr{L}\{y\}(s) = \frac{1}{m}\frac{\mathscr{L}\{f\}(s)}{s^2 + \omega_0^2} \qquad \text{where } \omega_0 = \sqrt{\frac{k}{m}}$$

Since the Laplace transform of the solution to (1) can be written as the product $m^{-1}(s^2 + \omega_0^2)^{-1} \times \mathscr{L}\{f\}(s)$ and since

$$\mathscr{L}\{m^{-1}\omega_0^{-1} \sin \omega_0 t\}(s) = m^{-1}(s^2 + \omega_0^2)^{-1}$$

it follows from the convolution theorem (see Theorem 3.2-1) that the solution y to (1) has the form

(3) $$y(t) = \frac{1}{m\omega_0}\int_0^t \sin\,[\omega_0(t - r)]\, f(r)\, dr \qquad \text{for all } t \geq 0$$

Although the form (3) is valid for general piecewise continuous functions f, we analyze the behavior of solutions to (1) in several special cases that are of interest.

Suppose first that the *forcing term is sinusoidal*, say $f(t) = A \cos \omega t$, where A and ω are constants with $\omega > 0$. In this case

$$\mathscr{L}\{y\}(s) = \frac{A}{m}\frac{s}{(s^2 + \omega_0^2)(s^2 + \omega^2)}$$

and if $\omega^2 \neq \omega_0^2$, then

$$\frac{A}{m}\frac{s}{(s^2 + \omega_0^2)(s^2 + \omega^2)} = \frac{A}{m(\omega_0^2 - \omega^2)}\left(\frac{s}{s^2 + \omega^2} - \frac{s}{s^2 + \omega_0^2}\right)$$

by a partial fraction expansion. Looking up the inverse Laplace transform from the table at the end of this chapter, we obtain

(4) $$y(t) = \frac{A}{m(\omega_0^2 - \omega^2)}(\cos \omega t - \cos \omega_0 t)$$

This is, of course, the same answer obtained in Section 2.6 [see (12) in Section 2.6]. In the case $\omega = \omega_0$, equation (4) is no longer valid, and we have

$$\mathscr{L}\{y\}(s) = \frac{A}{m}\frac{s}{(s^2 + \omega_0^2)^2}$$

Noting further, however, that

$$\mathscr{L}\{y\}(s) = \frac{A}{m}\frac{s}{(s^2 + \omega_0^2)^2} = -\frac{d}{ds}\left(\frac{A}{2m}\frac{1}{s^2 + \omega_0^2}\right)$$

it follows from Theorem 3.2-1 that

$$y(t) = t\mathscr{L}^{-1}\left\{\frac{A}{2m}\frac{1}{s^2 + \omega_0^2}\right\}(t) = t\mathscr{L}^{-1}\left\{\frac{A}{2m\omega_0}\cdot\frac{\omega_0}{s^2 + \omega_0^2}\right\}(t)$$

and hence

$$y(t) = \frac{A}{2m\omega_0} t \sin \omega_0 t \tag{5}$$

if $\omega = \omega_0$. This also agrees with the answer obtained in Section 2.6. Note in particular that the solution (5) is *unbounded* on $[0, \infty)$.

Now we want to consider the case when the forcing term is *constant over an interval* $[0, \sigma)$ and then zero beyond time $t = \sigma$, say $f(t) = A$ for $0 \leq t < \sigma$ and $f(t) = 0$ for $t \geq 0$. We use the convention here that $f(t) \equiv A$ on $[0, \infty)$ if $\sigma = \infty$. If $u_\sigma(t)$ is the unit step function [that is, $u_\sigma(t) = 0$ for $0 \leq t < \sigma$, $u_\sigma(t) = 1$ for $t \geq \sigma$], then $f(t) = A \cdot [1 - u_\sigma(t)]$ for all $t \geq 0$ [here $u_\infty(t) \equiv 0$ for all $t \geq 0$], and it follows from formula (7) in the preceding section and formula (2) above that

$$\mathscr{L}\{y\}(s) = \frac{A}{m}\frac{1 - e^{-\sigma s}}{s(s^2 + \omega_0^2)}$$

where $e^{-\infty s} \equiv 0$ for all $s > 0$. Using the partial fraction expansion

$$\frac{1}{s(s^2 + \omega_0^2)} = \frac{1}{\omega_0^2}\left(\frac{1}{s} - \frac{s}{s^2 + \omega_0^2}\right)$$

shows that

$$\mathscr{L}\{y\}(s) = \frac{A}{m\omega_0^2}\left(\frac{1}{s} - \frac{s}{s^2 + \omega_0^2} - \frac{e^{-\sigma s}}{s} + \frac{se^{-\sigma s}}{s^2 + \omega_0^2}\right)$$

Therefore,

$$\begin{aligned} y(t) &= \frac{A}{m\omega_0^2}(1 - \cos \omega_0 t) \qquad \text{if } \sigma = \infty \qquad \text{and} \\ y(t) &= \frac{A}{m\omega_0^2}(1 - \cos \omega_0 t) - u_\sigma(t)\frac{A}{m\omega_0^2}\{1 - \cos[\omega_0(t - \sigma)]\} \qquad \text{if } \sigma < \infty \end{aligned} \tag{6}$$

Rewriting we have

$$(7) \qquad y(t) = \begin{cases} \dfrac{A}{m\omega_0^2}(1 - \cos \omega_0 t) & \text{if } 0 \le t < \sigma \\ \dfrac{A}{m\omega_0^2}\{\cos [\omega_0(t - \sigma)] - \cos \omega_0 t\} & \text{if } \sigma \le t \end{cases}$$

Using the trigonometric identity

$$\cos(\phi - \psi) - \cos(\phi + \psi) = 2 \sin \psi \sin \phi$$

with $\phi = \omega_0 t - \frac{1}{2}\omega_0 \sigma$ and $\psi = \frac{1}{2}\omega_0 \sigma$, shows that

$$\cos [\omega_0(t - \sigma)] - \cos \omega_0 t = 2 \sin (\tfrac{1}{2}\omega_0 \phi) \sin \left[\omega_0\left(t - \frac{\sigma}{2}\right)\right]$$

Therefore, the solution y to (1) can be written

$$(7') \qquad y(t) = \begin{cases} \dfrac{A}{m\omega_0^2}(1 - \cos \omega_0 t) & \text{if } 0 \le t < \sigma \\ \dfrac{2A}{m\omega_0^2} \sin (\tfrac{1}{2}\omega_0 \sigma) \sin \left[\omega_0\left(t - \dfrac{\sigma}{2}\right)\right] & \text{if } t \ge \sigma \end{cases}$$

Writing $y(t)$ in the form (7′) leads to an interesting physical interpretation. The mass oscillates with the same frequency for $t < \sigma$ and for $t > \sigma$ [provided that $\sin(\omega_0 \sigma/2) \ne 0$]. However, the amplitude for $0 \le t < \sigma$ is $A/(m\omega_0^2)$, and the amplitude for $t \ge \sigma$ is $2A \sin(\omega_0 \sigma/2)/(m\omega_0^2)$. For example, if $\sigma = \omega_0^{-1}\pi$ then the amplitude increases from $A/(m\omega_0^2)$ to $2A/(m\omega_0^2)$ at $t = \sigma$; if $\sigma = \omega_0^{-1}\pi/3$ the amplitude remains $A/(m\omega_0^2)$ at $t = \sigma$; and if $\sigma = \omega_0^{-1}2\pi$, then the amplitude decreases from $A/(m\omega_0^2)$ to zero (i.e., the mass remains at equilibrium for all $t \ge \sigma$).

3.4a Unit Impulse Functions

The problem now is to analyze the problem associated with what would loosely be called an "instantaneous impulse" acting on the mass (for example, a hammer striking the mass initially). As a finite approximation, it is supposed that the force $f(t)$ has a "large" constant value for "small" times t and is zero elsewhere.

In order to discuss this case we introduce a new concept. For each $\varepsilon > 0$ define the function q_ε on $[0, \infty)$ by

$$q_\varepsilon(t) = \begin{cases} \dfrac{1}{\varepsilon} & \text{for } 0 \le t < \varepsilon \\ 0 & \text{for } t \ge \varepsilon \end{cases}$$

{that is, $q_\varepsilon(t) = (1/\varepsilon)[1 - u_\varepsilon(t)]$ for $t \ge 0$, where u_ε is the unit step function—see Figure 3.8}. The function q_ε is called an ε *pulse*.

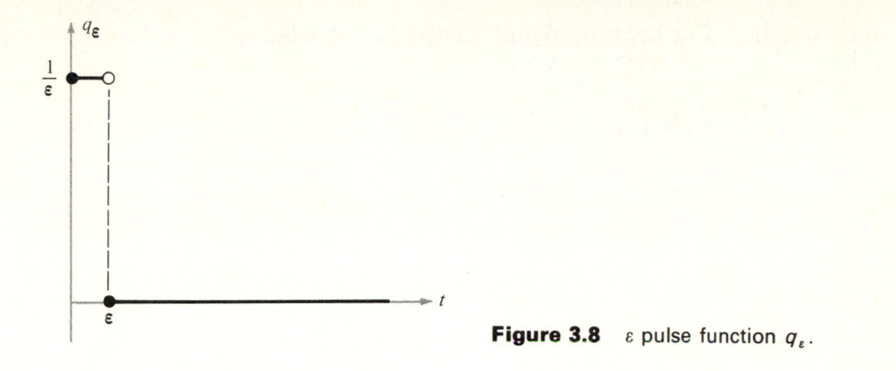

Figure 3.8 ε pulse function q_ε.

Since

$$\int_0^\infty q_\varepsilon(t)\,dt = \int_0^\varepsilon \varepsilon^{-1}\,dt = 1 \qquad \text{for all } \varepsilon > 0$$

it follows that if $q_\varepsilon(t)$ describes a force for $t \geq 0$, then q_ε has impulse unity {recall that if $f(t)$ describes a force for t in an interval $[a, b]$, then $\int_a^b f(t)\,dt$ is the *impulse* imparted by f}. Therefore, q_ε is called a *unit impulse force.* Observe that

$$\mathscr{L}\{q_\varepsilon\}(s) = \int_0^\varepsilon e^{-st}\varepsilon^{-1}\,dt = \frac{1-e^{-\varepsilon s}}{\varepsilon s}$$

for all $\varepsilon > 0$, and hence

$$\lim_{\varepsilon \to 0+} \mathscr{L}\{q_\varepsilon\}(s) = \lim_{\varepsilon \to 0+} \frac{1-e^{-\varepsilon s}}{\varepsilon s} = 1 \tag{8}$$

Note still further that if f is any continuous function on $[0, \infty)$ and

$$f_\varepsilon(t) \equiv \int_0^t f(r)q_\varepsilon(t-r)\,dr = f * q_\varepsilon(t) \qquad \text{for } t > 0$$

then

$$f_\varepsilon(t) = \int_{t-\varepsilon}^t f(r)\varepsilon^{-1}\,dr = \varepsilon^{-1}\int_0^\varepsilon f(t-\xi)\,d\xi$$

for all $t \geq \varepsilon > 0$. Since f is continuous, $f(t-\xi) \to f(t)$ as $\xi \to 0+$, and it follows from the mean value theorem for integrals that

$$f(t) = \lim_{\varepsilon \to 0+} \int_0^t f(r)q_\varepsilon(t-r)\,dr = \lim_{\varepsilon \to 0+} f * q_\varepsilon(t) \qquad \text{for all } t > 0 \tag{9}$$

Although the limit of the integral in (9) exists for every continuous function f, the function q_ε does not have a limit function: $[q_\varepsilon(t) \to 0$ as $\varepsilon \to 0+$ for $t > 0$ and $q_\varepsilon(0) \to \infty$ as $\varepsilon \to 0+]$. However, by convenient abuse of terminology, the "function" (or operation) δ is defined by the relation $f * \delta\ (t) = f(t)$ for all $t > 0$:

$$f * \delta\ (t) = \int_0^t f(r)\delta(t - r)\ dr = f(t)$$

for every continuous function f. Therefore, in this generalized sense, δ is the limit of the ε pulses q_ε as $\varepsilon \to 0+$: that is,

$$\int_0^t f(r)\delta(t - r)\ dr = \lim_{\varepsilon \to 0+} \int_0^t f(r)q_\varepsilon(t - r)\ dr = f(t)$$

for all continuous functions f on $[0, \infty)$. The operation δ is called the *Dirac delta function* (or *unit impulse function*). Note, however, that the operation δ is by no means a function in the usual sense [observe that $\delta(t - r) = 0$ if $0 \leq r < t$ and that $\int_0^t \delta(t - r)\ dr = 1$ for all $t \geq 0$]. From formula (8) we have that if $a \geq 0$ then

$$\begin{aligned}\mathscr{L}\{\delta(t - a)\}(s) &= \lim_{\varepsilon \to 0+} \mathscr{L}\{q_\varepsilon(t - a)\}(s) \\ &= \lim_{\varepsilon \to 0+} e^{-as}\mathscr{L}\{q_\varepsilon(t)\}(s) = e^{-as}\end{aligned}$$

and hence

$$\mathscr{L}\{\delta(t - a)\}(s) = e^{-as} \qquad \text{for all } a \geq 0 \tag{10}$$

It is the Laplace transform of the Dirac delta function that helps in describing the motion generated by the application of an impulsive force of short duration to the mass.

Suppose now that A is a nonzero constant and that equation (1) has the form

$$my'' + ky = A\delta(t) \qquad y(0) = y'(0) = 0 \tag{11}$$

Therefore, the force f is, loosely speaking, assumed to be applied for a short duration and with impulse A. As before, taking the Laplace transform of each side of (11) and using the appropriate properties and initial data, we obtain

$$\mathscr{L}\{y\}(s) = \frac{A}{m} \cdot \frac{1}{s^2 + \omega_0^2}$$

where $\omega_0 = \sqrt{k/m}$ [it is easy to check that $\mathscr{L}\{A\delta(t)\}(s) \equiv A$ for all s—see (10)]. By taking the inverse transform, the solution y to (11) is seen to be

$$y(t) = \frac{A}{m\omega_0} \sin \omega_0 t \tag{12}$$

It is interesting to note that the function y defined by (12) is in fact a solution to the homogeneous equation associated with (11) (that is, the equation $my'' + ky = 0$), except that the initial conditions are $y(0) = 0$ and $y'(0) = A/m$.

Therefore, this "instantaneous impulse" of magnitude A initiates the same motion as would starting the motion of the mass free of external forces with an initial velocity A/m.

▶ **EXAMPLE 3.4-1**

Suppose that a mass of 4 units is attached to a linear spring with spring constant 9. Suppose also that there is a resistance force proportional to the velocity, with proportionality constant 4. It is assumed that the mass is initially at equilibrium and that it is set into motion with an impulsive force of 4 units. Moreover, after 3 units of time a second impulsive force of 7 units is applied. The differential equation of motion is

$$4y'' + 4y' + 9y = 4\delta(t) + 7\delta(t-3) \qquad y(0) = y'(0) = 0 \tag{13}$$

Taking the Laplace transform of each side of (13), we obtain

$$\mathscr{L}\{y(t)\}(s) = \frac{4\mathscr{L}\{\delta(t)\}(s) + 7\mathscr{L}\{\delta(t-3)\}(s)}{4s^2 + 4s + 9}$$

$$= \frac{4 + 7e^{-3s}}{4(s+\frac{1}{2})^2 + 8}$$

Since

$$\mathscr{L}^{-1}\left\{\frac{1}{4(s+\frac{1}{2})^2 + 8}\right\}(t) = \frac{1}{4\sqrt{2}}\,\mathscr{L}^{-1}\left\{\frac{\sqrt{2}}{(s+\frac{1}{2})^2 + 2}\right\}(t)$$

$$= \frac{1}{4\sqrt{2}}\,e^{-t/2}\sin\sqrt{2}t$$

it follows that the solution y to (13) is

$$y(t) = \begin{cases} \dfrac{1}{\sqrt{2}}\,e^{-t/2}\sin\sqrt{2}t & \text{for } 0 \le t < 3 \\ \dfrac{1}{\sqrt{2}}\,e^{-t/2}\sin\sqrt{2}t + \dfrac{7}{4\sqrt{2}}\,e^{-(t-3)/2}\sin[\sqrt{2}(t-3)] & \text{for } t \ge 3 \end{cases}$$

Using the unit step function $u_3(t)$ the solution y can also be represented as

$$y(t) = \frac{1}{\sqrt{2}}\,e^{-t/2}\sin\sqrt{2}t + u_3(t)\,\frac{7}{4\sqrt{2}}\,e^{-(t-3)/2}\sin[\sqrt{2}(t-3)]$$

for all $t \ge 0$. Again it is of interest to note that y is a solution to the homogeneous equation associated with (13) on the interval $(0, 3)$ and the interval $(3, \infty)$. There is, however, a jump in the first derivative of y at the point $t = 3$.

PROBLEMS

3.4-1. Each of the following problems represents the forced vibration of an undamped or damped linear spring. Determine the solution, using Laplace transform techniques, and interpret your answer relative to the motion of the mass.

(a) $y'' + 4y = 3 + \cos t \qquad y(0) = y'(0) = 0$

(b) $y'' + 4y' + 3y = 3 + \sin t \qquad y(0) = y'(0) = 0$

(c) $y'' + y = 2 + \delta(t-1) \qquad y(0) = y'(0) = 0$

(d) $y'' + 2y' + y = \begin{cases} 2 & \text{if } 0 \le t < 5 \\ \delta(t-6) & \text{if } t \ge 5 \end{cases} \qquad y(0) = y'(0) = 0$

3.4-2. Using Laplace transform techniques, determine the solution to each of the following problems, having impulsive nonhomogeneous terms. When possible, relate your answer to a solution of the corresponding homogeneous equations, indicating all jumps in the first derivative.

(a) $y'' + 3y' + 2y = \delta(t) + 3\delta(t-2) \qquad y(0) = y'(0) = 0$

(b) $y'' + 2y' + 2y = 2\delta(t) - \delta(t-1) \qquad y(0) = y'(0) = 0$

(c) $y'' + y = \delta(t-2) \qquad y(0) = 0 \qquad y'(0) = 1$

(d) $y'' + 4y' + 4y = \delta(t) - \delta(t-1) \qquad y(0) = y'(0) = 0$

(e) $y'' + 3y' + 2y = \delta(t) \qquad y(0) = 1 \qquad y'(0) = -6$

Remarks Additional information on the Laplace transform can be found in Churchill [2], Hildebrand [6], Kreyszig [10], and Wylie [15].

Basic Properties of the Laplace Transform

Function	Laplace transform
$e^{\sigma t} f(t)$	$\mathscr{L}\{f(t)\}(s-\sigma)$
$f(\rho t)$	$\frac{1}{\rho}\mathscr{L}\{f(t)\}\left(\frac{s}{\rho}\right)$
$u_\sigma(t) f(t-\sigma)$	$e^{-\sigma s}\mathscr{L}\{f(t)\}(s)$
$f * g\,(t)$	$\mathscr{L}\{f(t)\}(s) \cdot \mathscr{L}\{g(t)\}(s)$
$\int_0^t f(r)\,dr$	$\frac{1}{s}\mathscr{L}\{f(t)\}(s)$
$tf(t)$	$-\frac{d}{ds}\mathscr{L}\{f(t)\}(s)$
$t^n f(t),\ n = 1, 2, \ldots$	$(-1)^n \frac{d^n}{ds^n}\mathscr{L}\{f(t)\}(s)$
$f'(t)$	$s\mathscr{L}\{f(t)\}(s) - f(0)$
$f''(t)$	$s^2\mathscr{L}\{f(t)\}(s) - sf(0) - f'(0)$
$f^{(n)}(t),\ n = 1, 2, \ldots$	$s^n\mathscr{L}\{f(t)\}(s) - s^{n-1}f(0) - \cdots - f^{n-1}(0)$

Short Table of Laplace Transforms

$f(t)$	$\mathscr{L}\{f(t)\}(s)$
1	$\dfrac{1}{s}$
$t^n,\ n = 1, 2, \ldots$	$\dfrac{n!}{s^{n+1}}$
e^{at}	$\dfrac{1}{s-a}$
$t^n e^{at},\ n = 1, 2, \ldots$	$\dfrac{n!}{(s-a)^{n+1}}$
$\sin \omega t$	$\dfrac{\omega}{s^2+\omega^2}$
$\cos \omega t$	$\dfrac{s}{s^2+\omega^2}$
$e^{at} \sin \omega t$	$\dfrac{\omega}{(s-a)^2+\omega^2}$
$e^{at} \cos \omega t$	$\dfrac{s-a}{(s-a)^2+\omega^2}$
$t \sin \omega t$	$\dfrac{2\omega s}{(s^2+\omega^2)^2}$
$t \cos \omega t$	$\dfrac{s^2-\omega^2}{(s^2+\omega^2)^2}$
$\sinh bt$	$\dfrac{b}{s^2-b^2}$
$\cosh bt$	$\dfrac{s}{s^2-b^2}$
$e^{at} \sinh bt$	$\dfrac{b}{(s-a)^2-b^2}$
$e^{at} \cosh bt$	$\dfrac{s-a}{(s-a)^2-b^2}$
$t \sinh \omega t$	$\dfrac{2\omega s}{(s^2-\omega^2)^2}$
$t \cosh \omega t$	$\dfrac{s^2+\omega^2}{(s^2-\omega^2)^2}$
$u_\sigma(t)$	$\dfrac{e^{-\sigma s}}{s}$
$\delta(t-a)$	e^{-as}

4 Power Series Methods

As opposed to the case for first order linear equations, there is no general procedure for the explicit computation of solutions to second order linear differential equations with nonconstant coefficients. The purpose of this chapter is to show that in some situations power series expansions can be used to obtain information about the approximation and form of solutions. The first section contains a review of elementary properties of power series, and the two basic power series methods are studied in Sections 4.2 and 4.3. The application of these techniques to some important equations from mathematical physics is indicated in Section 4.5.

4.1 FUNDAMENTAL CONCEPTS

The basic notations and definitions involved in solving second order linear differential equations by power series are discussed in this section. Suppose that t_0 is a number and that $a_0, a_1, a_2, \ldots$ are constants. For each pair k, N of nonnegative integers with $N \geq k$, the notation

$$\sum_{n=k}^{N} a_n(t - t_0)^n = a_k(t - t_0)^k + \cdots + a_N(t - t_0)^N$$

is used for each t. Observe that

$$\sum_{n=k}^{N} a_n(t - t_0)^n$$

is a polynomial in t of degree no larger than N (and exactly N if $a_N \neq 0$).

With this notation, for example,

$$\sum_{n=2}^{4} n^2 t^n = 4t^2 + 9t^3 + 16t^4$$

and

$$\sum_{n=0}^{3} (-1)^n \frac{n+1}{2^n} (t-1)^n = 1 - (t-1) + \tfrac{3}{4}(t-1)^2 - \tfrac{1}{2}(t-1)^3$$

In order to take into account sums that involve an infinite number of terms, we define

$$(1) \qquad \sum_{n=k}^{\infty} a_n(t-t_0)^n = \lim_{N\to\infty} \sum_{n=k}^{N} a_n(t-t_0)^n$$

The limit in (1) is called a *power series* and is said to be *expanded about* t_0. The series is said to *converge* at t whenever the limit in (1) exists and is said to *converge absolutely* if series

$$\sum_{n=k}^{\infty} |a_n(t-t_0)^n|$$

converges. Notice that the series in (1) always converges if $t = t_0$. In general, there is a ρ, $0 \le \rho \le +\infty$, such that the series converges (absolutely) if $|t - t_0| < \rho$ and diverges if $|t - t_0| > \rho$. The power series may converge at both, one, or none of the points $|t - t_0| = \rho$. The number ρ is called the *radius of convergence* of the power series. If $a_n \neq 0$ for all large n, then

$$\rho = \lim_{n\to\infty} \left| \frac{a_n}{a_{n+1}} \right|$$

provided this limit exists. This gives a convenient method for determining the radius of convergence for a large class of power series and is called the *ratio test*.

Consider, for example, the power series

$$(2) \qquad \sum_{n=1}^{\infty} \frac{3^n}{n} t^n = 3t + \frac{3^2}{2} t^2 + \frac{3^3}{3} t^3 + \cdots$$

Since

$$\lim_{n\to\infty} \frac{3^n}{n} \div \frac{3^{n+1}}{n+1} = \lim_{n\to\infty} \frac{3^n}{3^{n+1}} \cdot \frac{n+1}{n} = \frac{1}{3}$$

we see from the ratio test that the radius of convergence for (2) is $\rho = \frac{1}{3}$. Therefore, (2) converges if $|t| < \frac{1}{3}$ and diverges if $|t| > \frac{1}{3}$. The convergence or divergence of (2) when $|t| = \frac{1}{3}$ must be checked separately. If $t = \frac{1}{3}$ then (2) becomes

$$\sum_{n=1}^{\infty} \frac{3^n}{n} \left(\frac{1}{3}\right)^n = \sum_{n=1}^{\infty} \frac{1}{n}$$

which diverges (see Problem 4.1-4), and if $t = -\frac{1}{3}$ then (2) becomes

$$\sum_{n=1}^{\infty} \frac{3^n}{n} \left(-\frac{1}{3}\right)^n = \sum_{n=1}^{\infty} \frac{(-1)^n}{n}$$

which can be shown to converge. Thus the series converges if $t \in [-\frac{1}{3}, \frac{1}{3})$ and diverges otherwise.

Suppose now that the power series (1) has a positive radius of convergence ρ, and define the function

$$y(t) = \sum_{n=0}^{\infty} a_n(t - t_0)^n \qquad \text{for } |t - t_0| < \rho \tag{3}$$

Observe that the function y is defined for $t \in (t_0 - \rho, t_0 + \rho)$ by "substituting" t into the power series in (3). This function y is said to have a *power series expansion* about the point $t = t_0$. Any polynomial certainly has a power series expansion about $t = 0$ (and, in fact, about any t_0). For if p is the Nth-degree polynomial

$$p(t) = c_0 + c_1 t + c_2 t^2 + \cdots + c_N t^N$$

then p is of the form (3), with $a_n = c_n$ for $n = 0, 1, \ldots, N$, and $a_n = 0$ for $n = N + 1, N + 2, \ldots$. Thus any polynomial has a power series expansion with radius of convergence $\rho = \infty$. There are of course many important functions that have power series expansions and are not polynomials [see (6)]. As an elementary example, use the multiplication formula

$$(1 - t)(1 + t + t^2 + \cdots + t^N) = 1 - t^{N+1}$$

to obtain

$$\sum_{n=0}^{N} t^n = 1 + t + \cdots + t^N = \frac{1}{1 - t} - \frac{t^{N+1}}{1 - t}$$

If $|t| < 1$ then $t^{N+1} \to 0$ as $N \to \infty$, and it follows that

$$\lim_{N \to \infty} \sum_{n=0}^{N} t^n = \frac{1}{1 - t} - \lim_{N \to \infty} \frac{t^{N+1}}{1 - t} = \frac{1}{1 - t}$$

Therefore,

$$\frac{1}{1 - t} = \sum_{n=0}^{\infty} t^n \qquad \text{if } |t| < 1$$

and the function $y(t) = (1 - t)^{-1}$ has a power series expansion about $t = 0$ with radius of convergence $\rho = 1$.

Now suppose again that y has the power series expansion (3). Then this function y has derivatives of all orders on $(t_0 - \rho, t_0 + \rho)$, and the derivatives of y also have power series expansions about t_0 with the same radius of convergence ρ. Moreover, the power series for the derivatives of y can be computed from the power series of y by termwise differentiation. In particular, y' and y'' have the representations

$$\begin{aligned} y'(t) &= \sum_{n=0}^{\infty} n a_n (t - t_0)^{n-1} \qquad \text{and} \\ y''(t) &= \sum_{n=0}^{\infty} n(n-1) a_n (t - t_0)^{n-2} \end{aligned} \tag{4}$$

Note, for example, that $d/dt\,(1-t)^{-1} = (1-t)^{-2}$ and hence

$$\frac{1}{(1-t)^2} = \frac{d}{dt}\left(\sum_{n=0}^{\infty} t^n\right) = \sum_{n=1}^{\infty} nt^{n-1}$$

if $|t| < 1$. Using (3) and (4) it is easy to see that

$$a_0 = y(t_0) \qquad \text{and} \qquad a_1 = y'(t_0)$$

It is in fact easy to check that $a_n = y^{(n)}(t_0)/n!$ for all $n = 0, 1, 2, \ldots$. This formula is particularly important for second order linear equations, since the initial value problem designates $y(t_0)$ and $y'(t_0)$. Note further that if y has a power series expansion about t_0, then

$$y(t) = \sum_{n=0}^{\infty} \frac{y^{(n)}(t_0)}{n!}(t-t_0)^n \qquad \text{if } |t-t_0| < \rho$$

This series representation for y is known as the *Taylor series expansion* of y about t_0. These assertions are basic, but their proofs are beyond the scope of this text and are omitted.

The algebraic operations for power series can also be handled termwise. Suppose that $b_0, b_1, b_2, \ldots$ are real constants and that the power series

$$x(t) = \sum_{n=0}^{\infty} b_n(t-t_0)^n$$

has positive radius of convergence $\bar{\rho}$. If c and $\bar{c}$ are constants, then $cy + \bar{c}x$ has a power series expansion about t_0 with positive radius of convergence $\hat{\rho} \geq \min\{\rho, \bar{\rho}\}$, and

$$cy(t) + \bar{c}x(t) = \sum_{n=0}^{\infty} (ca_n + \bar{c}b_n)(t-t_0)^n$$

whenever $|t-t_0| < \min\{\rho, \bar{\rho}\}$. A crucial point in the addition of two power series is that the coefficients of like powers of $(t-t_0)$ are added together. For example, if $0 < j \leq k$ and

$$x(t) = \sum_{n=k}^{\infty} b_n(t-t_0)^{n-j}$$

then care must be taken, and it is convenient to change the summation index in order to obtain the power series for $y + x$. In particular, we have

$$\begin{aligned} y(t) + x(t) &= \sum_{n=0}^{\infty} a_n(t-t_0)^n + \sum_{n=k}^{\infty} b_n(t-t_0)^{n-j} \\ &= \sum_{n=j}^{\infty} a_{n-j}(t-t_0)^{n-j} + \sum_{n=k}^{\infty} b_n(t-t_0)^{n-j} \\ &= \sum_{n=k}^{\infty} (a_{n-j} + b_n)(t-t_0)^{n-j} + a_0 + \cdots + a_{k-j}(t-t_0)^{k-j} \end{aligned}$$

Another important property needed in these methods is the following:

(5) $$\sum_{n=k}^{\infty} a_n(t - t_0)^n \equiv 0$$ for all t in any open subinterval of $(t_0 - \rho, t_0 + \rho)$ only in case $a_n = 0$ for $n = k, k + 1, k + 2, \ldots$

The student should recall the power series for the following important functions:

(6) $$\begin{cases} e^t = \sum_{n=0}^{\infty} \dfrac{1}{n!} t^n \\ \sin t = \sum_{n=0}^{\infty} \dfrac{(-1)^n}{(2n+1)!} t^{2n+1} \\ \cos t = \sum_{n=0}^{\infty} \dfrac{(-1)^n}{(2n)!} t^{2n} \end{cases}$$

The radius of convergence for each of these power series is infinite, and hence these series are valid for all numbers t. It is easy to check that term-by-term differentiation of these power series gives the power series expansion for the derivative function. For example,

$$\begin{aligned} \frac{d}{dt} \sin t &= \sum_{n=0}^{\infty} \frac{d}{dt} \left[\frac{(-1)^n}{(2n+1)!} t^{2n+1} \right] \\ &= \sum_{n=0}^{\infty} \frac{(-1)^n(2n+1)}{(2n+1)!} t^{2n} \\ &= \sum_{n=0}^{\infty} \frac{(-1)^n}{(2n)!} t^{2n} = \cos t \end{aligned}$$

Also, combinations of these series give series representations for related functions. For example,

$$e^{3t} = \sum_{n=0}^{\infty} \frac{3^n}{n!} t^n \qquad e^{t^2} = \sum_{n=0}^{\infty} \frac{1}{n!} t^{2n}$$

and

$$\cos \sqrt{t} = \sum_{n=0}^{\infty} \frac{(-1)^n}{(2n)!} t^n$$

Now we want to consider the differential equation

(7) $$p(t)y'' + q(t)y' + r(t)y = 0$$

where

(8) $\begin{cases} p, q, \text{ and } r \text{ are polynomials and there is no} \\ \text{common factor for all three polynomials} \end{cases}$

We could be more general and require only that p, q, and r have power series expansions (see Problem 4.1-5). However, in most important applications, (8) is satisfied and the procedures are considerably less complicated. Even though p, q, and r are defined on all of $(-\infty, \infty)$, we assume throughout that I is an open interval and consider equation (7) to be defined on I. The purpose of this chapter is to investigate the possibilities of solutions to (7) having power series expansions about various points t_0 in I. These considerations separate into two cases. A point t_0 in I is said to be an *ordinary point of (7)* if $p(t_0) \neq 0$, and it is said to be a *singular point of (7)* if $p(t_0) = 0$. Singular points are divided into two classes: regular singular point and irregular singular point. A singular point t_0 is a *regular singular point* if either t_0 is a simple zero of p [that is, $p(t_0) = 0$ and $p'(t_0) \neq 0$] or if t_0 is a double root of p [that is, $p(t_0) = p'(t_0) = 0$ and $p''(t_0) \neq 0$] and a zero of q [that is, $q(t_0) = 0$]. It is easy to check that t_0 is a regular singular point of (7) only if (7) can be written in one of the following two forms: There are polynomials p_1 and q_1 such that $p_1(t_0) \neq 0$ and equation (7) has the form

(9)
(a) $(t - t_0)p_1(t)y'' + q(t)y' + r(t)y = 0$ or
(b) $(t - t_0)^2 p_1(t)y'' + (t - t_0)q_1(t)y' + r(t)y = 0$

If the singular point t_0 is not a regular singular point, then it is said to be an *irregular singular point.*

Consider, for example, the equation

$$t^2y'' + ty' + y = 0$$

The only singular point is $t = 0$, and since 0 is a double root of the coefficient of y'' and a simple root of the coefficient of y', it follows directly from the definition that 0 is a regular singular point. The point $t = 0$ is also a singular point of the equation

$$t^2y'' + (t + 2)y' + y = 0$$

since it is a double zero of the coefficient of y''. However, since 0 is not a zero of the coefficient of y', it follows that 0 is an irregular singular point.

EXAMPLE 4.1-1

Consider the equation $(t^3 - t^2)y'' + (t + 1)y' + ty = 0$ on $I = (-\infty, \infty)$. The singular points are 0 and 1 since $t^3 - t^2 = t^2(t - 1) = 0$ only if $t = 0, 1$. Since 1 is a simple zero, 1 is a regular singular point. Since 0 is a double zero of $t^3 - t^2$ and not a zero of $t + 1$, the singular point 0 is irregular.

The aim of this chapter is to determine circumstances when equation (7) has solutions that can be described with the aid of power series expansions, as well as to develop methods of computing power series solutions. As will be indicated, equation (7) has two linearly independent solutions that have a power series expansion about any ordinary point, but modifications must be made when trying to expand a solution about a singular point.

PROBLEMS

4.1-1. Use the ratio test to determine the radius of convergence for each of the following power series.

(a) $\sum_{n=0}^{\infty} 2^n t^n$ (d) $\sum_{n=0}^{\infty} \frac{n}{1+2^n} t^n$

(b) $\sum_{n=0}^{\infty} n^2 t^n$ (e) $\sum_{n=0}^{\infty} \frac{n+1}{(2n)!} t^{2n}$

(c) $\sum_{n=0}^{\infty} \frac{(-3)^n}{n!} t^n$ (f) $\sum_{n=0}^{\infty} \frac{2n+3}{5n+1} t^{2n+1}$

4.1-2. Recall the geometric series

$$\frac{1}{1-t} = \sum_{n=0}^{\infty} t^n$$

which converges for $|t| < 1$. Using this series along with those given in (6), express in terms of elementary functions the functions represented by the following series.

(a) $\sum_{n=0}^{\infty} t^{2n+1}$ (d) $\sum_{n=2}^{\infty} 2^n t^n$

(b) $\sum_{n=1}^{\infty} \frac{(-1)^n}{(2n)!} t^n \quad t > 0$ (e) $\sum_{n=0}^{\infty} \frac{(-1)^n}{(2n)!} t^{4n}$

(c) $\sum_{n=1}^{\infty} \frac{(-1)^{n+1}}{(n+1)!} t^n$ (f) $\sum_{n=1}^{\infty} \frac{1}{n!} t^{2n+1}$

4.1-3. Determine the ordinary points, the regular singular points, and the irregular singular points for each of the following equations.

(a) $(t^2 - 1)y'' + ty' + 6y = 0$

(b) $(t^4 - t^2)y'' + (t^2 + t)y' + y = 0$

(c) $(t^4 - t^2)y'' - (t + 1)y + y = 0$

(d) $t^3 y'' + t^2 y' + y = 0$

(e) $(t^2 - t - 2)y'' + (1 + t)^2 y' + ty = 0$

4.1-4. Suppose that $b_n > 0$ for $n = 0, 1, 2, \ldots$ and f is a continuous *nonincreasing* function on $[0, \infty)$ such that $f(n) = b_n$ for $n = 0, 1, 2, \ldots$. Show that

$$\int_n^{n+1} f(s)\,ds \le b_n \le \int_{n-1}^{n} f(s)\,ds \qquad \text{for } n = 1, 2, \ldots$$

(*Hint:* Deduce these inequalities from the area interpretation of the definite integral.)

(a) Show that if $\int_0^\infty f(s)\,ds < \infty$, then $\sum_{n=0}^{\infty} b_n$ converges.

(b) Show that if $\int_0^\infty f(s)\,ds = \infty$, then $\sum_{n=0}^{\infty} b_n$ diverges.

(c) Show that the series $\sum_{n=1}^{\infty} \frac{1}{n^p}$ converges if $p > 1$ and diverges if $p \le 1$.

The test for convergence or divergence of a series by using parts *a* and *b* is called the *integral test.*

4.1-5. Instead of assuming that p, q, and r are polynomials, consider equation (7), where p, q, and r have power series expansions about $t = t_0$. Then t_0 is an ordinary point if $p(t_0) \neq 0$ and a singular point if $p(t_0) = 0$. If t_0 is a singular point and if

$$\lim_{t \to t_0} \frac{(t - t_0)q(t)}{p(t)} \quad \text{and} \quad \lim_{t \to t_0} \frac{(t - t_0)^2 r(t)}{p(t)}$$

both exist, then t_0 is said to be a regular singular point.

(a) If p, q, and r are polynomials, show that these definitions agree with the ones in the paragraph preceding (9).

(b) Determine the singular points for the following equations and classify each as being a regular singular point or not.

(i) $(\sin t)y'' + y' + y = 0$

(ii) $(1 - \cos t)y'' + (\sin t)y' - y = 0$

(iii) $(e^t - 1 - t)y'' + e^{2t}y' + (\sin t)y = 0$

(iv) $(t^2 - t)(\sin \pi t)y'' + te^{2t}y' - y = 0$

4.1-6. Use the Taylor series expansion

$$y(t) = \sum_{n=0}^{\infty} \frac{y^{(n)}(t_0)}{n!}(t - t_0)^n$$

to obtain a power series expansion for the following polynomials about the given points.

(a) $y(t) = 2 + t + t^2$ about $t = 1$ and $t = -1$

(b) $y(t) = t^5$ about $t = 1$ and $t = 2$

(c) $y(t) = 4 - t^3$ about $t = 1$ and $t = -2$

4.2 ORDINARY POINTS: POWER SERIES SOLUTIONS

In this section we study a class of second order linear equations whose solutions can be expanded in a power series and indicate a procedure for determining the coefficients in the power series expansion. As in the previous section [see equation (7)] let p, q, and r be polynomials with no common factor, and consider the second order equation

$$p(t)y'' + q(t)y' + r(t)y = 0 \tag{1}$$

If $p(t_0) \neq 0$ [that is, t_0 is an ordinary point of (1)], then (1) has two linearly independent solutions that have power series expansions about t_0, and the purpose of this section is to develop a method for computing these power series.

In order to illustrate some of these ideas, consider the specific initial value problem

$$y'' - y = 0 \qquad y(0) = 1 \qquad y'(0) = 1 \tag{2}$$

Since this equation has constant coefficients, it is easy to see that $y(t) \equiv e^t$ is the solution (see Section 2.3). However, we want to obtain the power series expansion for e^t [see formula (6) in Section 4.1] by using equation (2) directly. So assume that the solution y to (2) has the form

$$y(t) = \sum_{n=0}^{\infty} a_n t^n$$

where the numbers a_n, $n = 0, 1, 2, \dots$, are to be determined. By termwise differentiation we have

$$y'(t) = \sum_{n=0}^{\infty} n a_n t^{n-1} \qquad \text{and} \qquad y''(t) = \sum_{n=0}^{\infty} n(n-1)a_n t^{n-2}$$

and then, by substituting in the equation in (2),

$$\sum_{n=0}^{\infty} n(n-1)a_n t^{n-2} - \sum_{n=0}^{\infty} a_n t^n = 0$$

Since the powers of t in these two series are not the same, we must relabel the summation index in order to combine the two series into one. Therefore, replacing n by $n - 2$ in the second series, we get

$$\sum_{n=0}^{\infty} n(n-1)a_n t^{n-2} - \sum_{n=2}^{\infty} a_{n-2} t^{n-2} = 0$$

(note that the summation now begins with $n = 2$ in the second series). Since the first two terms in the first series are 0, we can add these two series term-by-term to obtain

$$\sum_{n=2}^{\infty} [n(n-1)a_n - a_{n-2}]t^{n-2} = 0$$

Since this power series is the zero function, each of the coefficients must be zero, and it follows that

$$n(n-1)a_n - a_{n-2} = 0 \qquad \text{for } n = 2, 3, 4, \dots$$

and hence that

$$a_n = \frac{a_{n-2}}{n(n-1)} \qquad \text{for } n = 2, 3, 4, \ldots \tag{3}$$

Equation (3) is known as a recursion formula for the coefficients a_n. Since $a_0 = y(0)$ and $a_1 = y'(0)$, the initial conditions imply that $a_0 = a_1 = 1$. Now we can use the recursion formula (3) to compute the remaining coefficients:

$$a_2 = \frac{a_0}{2 \cdot 1} = \frac{1}{2!} \qquad a_3 = \frac{a_1}{3 \cdot 2} = \frac{1}{3!}$$

$$a_4 = \frac{a_2}{4 \cdot 3} = \frac{1}{4 \cdot 3(2!)} = \frac{1}{4!} \qquad a_5 = \frac{a_3}{5 \cdot 4} = \frac{1}{5 \cdot 4(3!)} = \frac{1}{5!}$$

Continuing, $a_n = 1/n!$ for $n = 0, 1, 2, \ldots$, and so

$$y(t) = \sum_{n=0}^{\infty} \frac{1}{n!} t^n = e^t$$

is the solution to (2). The student should verify that if the initial conditions are changed to $y(0) = 1$ and $y'(0) = -1$, then the power series solution obtained by this procedure is $y(t) = e^{-t}$ [that is, $a_n = (-1)^n/n!$ for $n = 0, 1, 2, \ldots$].

The procedure indicated for computing a power series solution to (2) applies analogously to the general equation (1) if we consider the power series expanded about an ordinary point of (1). Before considering further examples we first state the basic result. The proof of this theorem is beyond the scope of this book and is omitted.

Theorem 4.2-1

Suppose that t_0 is a number and that $p(t_0) \neq 0$ [that is, t_0 is an ordinary point of (1)]. Then (1) has two linearly independent solutions y_1 and y_2 having the form

$$y_1(t) = \sum_{n=0}^{\infty} a_n(t - t_0)^n \qquad \text{and} \qquad y_2(t) = \sum_{n=0}^{\infty} b_n(t - t_0)^n$$

and both series converge in some open interval I containing t_0.

The procedure for constructing the solutions y_1 and y_2 to (1) is similar to the above, and we outline it here. Assume that a solution y to (1) has the form

$$y(t) = \sum_{n=0}^{\infty} a_n(t - t_0)^n \tag{4}$$

Use the representations

$$y'(t) = \sum_{n=0}^{\infty} na_n(t - t_0)^{n-1} \qquad \text{and}$$

(5)

$$y''(t) = \sum_{n=0}^{\infty} n(n-1)a_n(t - t_0)^{n-2}$$

and substitute into equation (1). Since p, q, and r are polynomials, they can be expanded in a finite number of terms involving powers of $(t - t_0)$. Therefore, the multiplications $p(t)y''(t)$, $q(t)y'(t)$, and $r(t)y(t)$ can easily be computed. Combining each of the power series on the left side of (1) into one series (by changing the summation index) and equating the coefficients of this series to zero give formulas for computing the unknown coefficients a_n, $n \geq 0$. The formula describing the interdependence of the coefficients a_n is called the *recursion formula.* This procedure is further illustrated with specific examples.

EXAMPLE 4.2-1

Consider the equation $y'' - 2ty' - 2y = 0$. Since $t_0 = 0$ is an ordinary point, we assume that

$$y = \sum_{n=0}^{\infty} a_n t^n$$

is a solution. Using (5), with $t_0 = 0$, and substituting into the expression $y''(t) - 2ty'(t) - 2y(t)$, we have

$$\sum_{n=0}^{\infty} n(n-1)a_n t^{n-2} - 2t \sum_{n=0}^{\infty} na_n t^{n-1} - 2 \sum_{n=0}^{\infty} a_n t^n$$

$$= \sum_{n=2}^{\infty} n(n-1)a_n t^{n-2} - \sum_{n=0}^{\infty} 2na_n t^n - \sum_{n=0}^{\infty} 2a_n t^n$$

$$= \sum_{n=2}^{\infty} n(n-1)a_n t^{n-2} - \sum_{n=2}^{\infty} 2(n-2)a_{n-2} t^{n-2} - \sum_{n=2}^{\infty} 2a_{n-2} t^{n-2}$$

$$= \sum_{n=2}^{\infty} \{n(n-1)a_n - [2(n-2) + 2]a_{n-2}\}t^{n-2}$$

$$= \sum_{n=2}^{\infty} [n(n-1)a_n - 2(n-1)a_{n-2}]t^{n-2}$$

Note that following the second equal sign the summation index of two of the series was changed by replacing n by $n - 2$. This was done in order that the power of t be $n - 2$ in all three series so that the addition could be performed term-by-term. Since the coefficient of t^{n-2} must be zero for all $n \geq 2$, we have $n(n-1)a_n - 2(n-1)a_{n-2} = 0$ for all $n \geq 2$. Canceling $n - 1$ from each term it follows that the a_n's must satisfy the recursion formula

$$a_n = \frac{2}{n} a_{n-2} \qquad \text{for } n = 2, 3, \ldots \tag{6}$$

Observe that (6) does not restrict a_0 and a_1 [note that if the initial value problem at $t = 0$ associated with this equation is considered, then $a_0 = y(0)$ and $a_1 = y'(0)$, so a_0 and a_1 are specified by the initial values]. However, the recursion formula (6) implies that the remaining coefficients $a_2, a_3, a_4, \ldots$ are determined in terms of a_0 and a_1. In this particular example the coefficients subscripted with even integers depend on a_0 and those subscripted with odd integers depend on a_1. Using (6) for the even integers, we have

$$a_{2k} = \frac{2}{2k} a_{2k-2} = \frac{1}{k} a_{2(k-1)}$$

Hence

$$a_2 = a_0 \qquad a_4 = \frac{a_2}{2} = \frac{a_0}{2} \qquad a_6 = \frac{a_4}{3} = \frac{a_0}{3 \cdot 2} \qquad a_8 = \frac{a_6}{4} = \frac{a_0}{4 \cdot 3 \cdot 2}$$

and it follows easily that

$$a_{2k} = \frac{a_0}{k!} \qquad \text{for k} = 0, 1, 2, \ldots \tag{7}$$

Using (6) for odd integers we have

$$a_3 = \frac{2a_1}{3} \qquad a_5 = \frac{2a_3}{5} = \frac{2^2 a_1}{5 \cdot 3} \qquad a_7 = \frac{2a_5}{7} = \frac{2^3 a_1}{7 \cdot 5 \cdot 3}$$

and it follows easily that

$$a_{2k+1} = \frac{2^k a_1}{(2k+1)(2k-1) \cdots 5 \cdot 3} \qquad \text{for } k = 0, 1, 2, \ldots \tag{8}$$

Separating the sums defining $y(t)$ into those indexed by even integers and those indexed by odd integers, we see from (7) and (8) that $y(t)$ can be expressed as

$$\begin{aligned} y(t) &= \sum_{n=0}^{\infty} a_n t^n = \sum_{k=0}^{\infty} a_{2k} t^{2k} + \sum_{k=0}^{\infty} a_{2k+1} t^{2k+1} \\ &= a_0 \sum_{k=0}^{\infty} \frac{1}{k!} t^{2k} + a_1 \sum_{k=0}^{\infty} \frac{2^k}{(2k+1)(2k-1) \cdots 3 \cdot 1} t^{2k+1} \end{aligned}$$

Both of these power series have an infinite radius of convergence (this follows easily from the ratio test), and if

$$y_1(t) \equiv \sum_{k=0}^{\infty} \frac{1}{k!} t^{2k}$$

and

$$y_2(t) \equiv \sum_{k=0}^{\infty} \frac{2^k}{(2k+1)(2k-1)\cdots 3\cdot 1} t^{2k+1}$$

then y_1 and y_2 are linearly independent solutions to the equation in question. There seems to be no elementary expression for $y_2(t)$, but replacing t by t^2 in the power series for e^t, it is easily seen that $y_1(t) = e^{t^2}$ (it is routine to check that e^{t^2} is indeed a solution).

The recursion formula (6) was reasonably simple in this case, and we were able to express each a_n explicitly in terms of n [see equations (7) and (8)]. In many situations this is impossible and we must be satisfied only with the recursion formula itself and an explicit calculation of only a few of the a_n's.

▶ EXAMPLE 4.2-2

Consider the initial value problem

$$t^2y'' + 2y' + (t-1)y = 0 \qquad y(1) = 2 \qquad y'(1) = 3 \tag{9}$$

Since the initial conditions are at $t_0 = 1$, the solution y is expanded about the ordinary point $t_0 = 1$. Therefore

$$y(t) = \sum_{n=0}^{\infty} a_n(t-1)^n$$

and since

$$t^2 = (t-1)^2 + 2t - 1 = (t-1)^2 + 2(t-1) + 1$$

equation (9) can be written in the form

$$[(t-1)^2 + 2(t-1) + 1]y'' + 2y' + (t-1)y = 0$$

Computing y' and y'' and substituting into this equation, it follows that

$$\sum_{n=2}^{\infty} n(n-1)a_n(t-1)^n + 2\sum_{n=2}^{\infty} n(n-1)a_n(t-1)^{n-1} + \sum_{n=2}^{\infty} n(n-1)a_n(t-1)^{n-2}$$
$$+ 2\sum_{n=1}^{\infty} a_n n(t-1)^{n-1} + \sum_{n=0}^{\infty} a_n(t-1)^{n+1} \equiv 0$$

Changing notation so that all terms inside the summations contain the factor $(t-1)^{n-2}$, we have

$$\sum_{n=4}^{\infty}(n-2)(n-3)a_{n-2}(t-1)^{n-2}+2\sum_{n=3}^{\infty}(n-1)(n-2)a_{n-1}(t-1)^{n-2}$$
$$+\sum_{n=2}^{\infty}n(n-1)a_n(t-1)^{n-2}+2\sum_{n=2}^{\infty}(n-1)a_{n-1}(t-1)^{n-2}$$
$$+\sum_{n=3}^{\infty}a_{n-3}(t-1)^{n-2}\equiv 0$$

Simplifying and beginning the summation at $n=4$ so that the first two terms of the resulting power series are outside the summation sign, we obtain

$$\{2a_2+2a_1\}+\{8a_2+6a_3+a_0\}(t-1)$$
$$+\sum_{n=4}^{\infty}\{(n-2)(n-3)a_{n-2}+2(n-1)^2a_{n-1}+n(n-1)a_n+a_{n-3}\}(t-1)^{n-2}\equiv 0$$

Setting the coefficients of the powers of $(t-1)$ equal to zero, we obtain from the first two terms

$$2a_2+2a_1=0 \qquad \text{and} \qquad 8a_2+6a_3+a_0=0 \tag{10}$$

and we obtain the following recursion formula

$$a_n=-\frac{2(n-1)^2a_{n-1}+(n-2)(n-3)a_{n-2}+a_{n-3}}{n(n-1)} \qquad n\geq 4 \tag{11}$$

from the coefficients inside the summation sign. From the initial values we have

$$a_0=y(1)=2 \qquad \text{and} \qquad a_1=y'(1)=3$$

and then from (10) we have

$$a_2=-a_1=-3 \qquad a_3=\frac{-8a_2-a_0}{6}=\frac{11}{3}$$

Any given number of the remaining coefficients can now be computed in an iterative manner from the recursion formula (11). For example,

$$a_4=-\frac{18a_3+2a_2+a_1}{4\cdot 3}=-\frac{21}{4} \qquad \text{and}$$

$$a_5=-\frac{32a_4+6a_3+a_2}{5\cdot 4}=\frac{149}{20}$$

Therefore, neglecting terms of order $(t-1)^6$ and higher,

$$y(t)\cong 2+3(t-1)-3(t-1)^2+\tfrac{11}{3}(t-1)^3-\tfrac{21}{4}(t-1)^4+\tfrac{149}{20}(t-1)^5$$

is an "approximation" of the solution to (9).

PROBLEMS

4.2-1. Obtain a power series expansion about $t_0 = 0$ for each of the following initial value problems and compare with the exact solution.

(a) $y'' - y' - 2y = 0 \quad y(0) = 1 \quad y'(0) = 2$

(b) $y'' + 4y = 0 \quad y(0) = 0 \quad y'(0) = 2$

(c) $y'' - 9y = 0 \quad y(0) = -2 \quad y'(0) = -6$

(d) $y'' - 4y = 0 \quad y(0) = 1 \quad y'(0) = 0$

4.2-2. Assume that the solution y to each of the following equations has the form

$$y(t) = \sum_{n=0}^{\infty} a_n t^n$$

Determine the recursion formula for the a_n's and compute a_n explicitly in terms of a_0, a_1, and n $(n = 2, 3, \ldots)$.

(a) $y'' + ty' + 2y = 0$

(b) $y'' - 2ty' + 2y = 0$

(c) $(1 - t^2)y'' - 2ty' + 2y = 0$

(d) $(2 - t^2)y'' + 2ty' - 2y = 0$

(e) $(t^2 + 2t + 1)y'' - (2t + 2)y' + 2y = 0$

(f) $y'' - 2ty' + 5 = 0$

4.2-3. In each of the following equations assume that the solution y has a power series expansion about the given initial time t_0:

$$y(t) = \sum_{n=0}^{\infty} a_n(t - t_0)^n$$

Determine the recursion formula for the a_n's and then determine the first four nonzero coefficients in the expansion.

(a) $y'' + (3 - t)y = 0 \quad y(0) = 1 \quad y'(0) = 0$

(b) $y'' + (3 - t)y = 0 \quad y(0) = 0 \quad y'(0) = 1$

(c) $y'' + t^2 y = 0 \quad y(0) = 1 \quad y'(0) = 2$

(d) $(t - 3)y'' + y = 0 \quad y(1) = 1 \quad y'(1) = -3$

(e) $y'' - ty = 0 \quad y(-2) = 3 \quad y'(-2) = 4$

(f) $t^2 y'' + y' + y = 0 \quad y(1) = 2 \quad y'(1) = -1$

(g) $y'' - t^2 y' - y = 0 \quad y(0) = 1 \quad y'(0) = 0$

4.2-4. The method of power series can also be applied to solutions of nonhomogeneous equations. Consider the equation

$$p(t)y'' + q(t)y' + r(t)y = f(t) \qquad (*)$$

where f has a power series expansion about t_0:

$$f(t) = \sum_{n=0}^{\infty} d_n(t - t_0)^n$$

and $p(t_0) \neq 0$. Equation (*) has a particular solution y of the form

(**) $$y(t) = \sum_{n=2}^{\infty} c_n(t - t_0)^n$$

[c_0 and c_1 are set to be zero for simplicity—this corresponds to determining the solution to (*) such that $y(t_0) = y'(t_0) = 0$]. The procedure is to substitute the power series representations for y, y', and y'' into (*) and collect terms involving like powers of $(t - t_0)$. Then set the coefficient of $(t - t_0)^n$ on the left side of (*) equal to d_n. Each of the c_n's can then be determined by this recursion formula. In the following nonhomogeneous equations, assume that the solution y has the form (**) and determine the recursion formula involving the c_n's. Compute the first four nonzero c_n's from the recursion formula.

(a) $y'' - ty = \dfrac{1}{1-t} \qquad t_0 = 0 \qquad$ (See Problem 4.1-2.)

(b) $y'' - 2ty' - 2y = e^t \qquad t_0 = 0$

(c) $y'' - t^2y' + y = 1 - 3t^2 \qquad t_0 = 0$

4.2-5. Show that the equation

$$y'' + 2ty' - 2py = 0$$

has a polynomial solution for each positive integer p. Compute the polynomial solution explicitly for $p = 0, 1, 2$, and 3.

4.3 SINGULAR POINTS: FROBENIUS' METHOD

In this section the possibility of obtaining power series or "modified power series" expansions for solutions about a singular point is investigated. Euler's differential equation (see Section 2.7) illustrates the type of problem that can occur at a singular point. Suppose that α, β, and γ are constants and consider the equation

(1) $$\alpha t^2 y'' + \beta t y' + \gamma y = 0$$

According to the results in Section 2.7, if the solutions to (1) are assumed to be of the form $y = t^\lambda$, then λ satisfies

(2) $$\alpha\lambda^2 + (\beta - \alpha)\lambda + \gamma = 0$$

(See, in particular, Problem 2.7-8.) Taking $\alpha = 2$, $\beta = 5$, and $\gamma = 1$, the roots of (2) are $-\frac{1}{2}$ and -1, and so $y_1(t) = t^{-1/2}$ and $y_2(t) = t^{-1}$ are linearly independent solutions to (1) on $(0, \infty)$. Neither y_1 nor y_2 has a power series expansion about the singular point $t_0 = 0$. Taking $\alpha = 1$, $\beta = 2$, and $\gamma = 2$, the roots of (2) are 1 and 2, and so $y_1(t) = t$ and $y_2(t) = t^2$ are linearly independent solutions to (1) on $(0, \infty)$. Therefore, both y_1 and y_2 have a power series expansion about the singular point $t_0 = 0$. Taking $\alpha = \beta = 1$ and $\gamma = -1$ gives $y_1(t) = 1/t$ and $y_2(t) = t$ as solutions to (1): y_2 has a power series expansion about $t_0 = 0$ and y_1 does not.

Now consider on an open interval I the second order equation

$$p(t)y'' + q(t)y' + r(t) = 0 \tag{3}$$

where p, q, and r are polynomials. Also, let t_0 be in I and suppose that t_0 is a singular point of (3): $p(t_0) = 0$. Rather than search for solutions that have a power series expansion about t_0, it is assumed that σ is a constant; that a_0, $a_1, a_2, \ldots$ are constants; and that a solution y to (3) has the form

$$y(t) = (t - t_0)^\sigma \sum_{n=0}^{\infty} a_n(t - t_0)^n \tag{4}$$

The expression given in (4) is called a *Frobenius-type expansion* of y about t_0. Frobenius-type solutions include power series solutions (take $\sigma = 0$) and solutions that are fractional or negative powers of t (take $\sigma = -\frac{3}{2}$ and $a_n = 0$ for $n \geq 1$, for example).

In order to apply a Frobenius-type expansion to represent a solution to (3) about a singular point, it is convenient to rewrite (4) in the form

$$y(t) = \sum_{n=0}^{\infty} a_n(t - t_0)^{n+\sigma} \tag{4'}$$

The first and second derivatives of y can be computed termwise [use the product rule in (4) and then simplify], and so

$$\begin{aligned} y'(t) &= \sum_{n=0}^{\infty} (n + \sigma)a_n(t - t_0)^{n+\sigma-1} \qquad \text{and} \\ y''(t) &= \sum_{n=0}^{\infty} (n + \sigma)(n + \sigma - 1)a_n(t - t_0)^{n+\sigma-2} \end{aligned} \tag{5}$$

As opposed to the case when $\sigma = 0$, the summations in (5) must begin with $n = 0$, since the first term in y' and the first two terms in y'' are not zero. The procedure for determining conditions on σ, a_0, a_1, $a_2, \ldots$ so that y is a solution to (3) is similar to the power series method: expand the polynomials p, q, and r in terms of $(t - t_0)$; substitute the Frobenius-type expansions for y, y', and y'' [see (4′) and (5) above] into equation (3); multiply the polynomials with the series; change the summation index n so that the series can be added term by term; and combine the resulting series into one series. Now *choose σ so that the coefficient of the lowest power of $(t - t_0)$ is zero.* This equation is called the *indicial equation*, and the roots σ of this equation are called the *indicial roots.* Once the indicial roots are computed, they can be substituted for σ, and a recursion formula for the a_n's can be obtained by setting the coefficients of the remaining powers of $t - t_0$ equal to zero. This procedure is called the *method of Frobenius* and is illustrated by the following example.

EXAMPLE 4.3-1

Consider the equation $2ty'' + y' - 2y = 0$. Taking $t_0 = 0$ in (4′) and (5), it follows that

$$2ty''(t) + y' - 2y(t)$$

$$= 2\sum_{n=0}^{\infty}(n+\sigma)(n+\sigma-1)a_n t^{n+\sigma-1} + \sum_{n=0}^{\infty}(n+\sigma)a_n t^{n+\sigma-1} - 2\sum_{n=0}^{\infty} a_n t^{n+\sigma}$$

$$= \sum_{n=0}^{\infty}[2(n+\sigma)(n+\sigma-1) + (n+\sigma)]a_n t^{n+\sigma-1} - 2\sum_{n=1}^{\infty} a_{n-1} t^{n+\sigma-1}$$

$$= [2\sigma(\sigma-1) + \sigma]a_0 t^{\sigma-1} + \sum_{n=1}^{\infty}\{[2(n+\sigma)-1][n+\sigma]a_n - 2a_{n-1}\}t^{n+\sigma-1}$$

The lowest power of t is $t^{\sigma-1}$, and hence the indicial equation is $2\sigma(\sigma-1) + \sigma = (2\sigma-1)\sigma = 0$. Thus $\sigma = 0, \frac{1}{2}$ are the indicial roots. Before substituting the indicial roots, we compute the recursion formula for the a_n's in terms of σ. Setting the terms in the braces equal to zero,

$$[2(n+\sigma)-1][n+\sigma]a_n = 2a_{n-1} \qquad \text{for } n = 1, 2, \ldots$$

and hence

$$a_n = \frac{2a_{n-1}}{(2n+2\sigma-1)(n+\sigma)} \qquad \text{for } n = 1, 2, \ldots$$

Setting $\sigma = 0$ we obtain

$$a_n = \frac{2a_{n-1}}{(2n-1)(n)} = \frac{4a_{n-1}}{(2n-1)(2n)} \qquad n = 1, 2, \ldots$$

and it follows that a_0 is arbitrary and

$$a_1 = \frac{4a_0}{2!} \qquad a_2 = \frac{4a_1}{3\cdot 4} = \frac{4^2 a_0}{4!} \qquad a_3 = \frac{4a_2}{5\cdot 6} = \frac{4^3 a_0}{6!} \qquad \ldots$$

It follows now that $a_n = 4^n a_0/(2n)!$ for all $n \geq 0$. Therefore,

$$y_1(t) = \sum_{n=0}^{\infty}\frac{4^n}{(2n)!}t^n$$

$$= \sum_{n=0}^{\infty}\frac{1}{(2n)!}(2\sqrt{t})^{2n}$$

is a solution to this equation. Setting $\sigma = \frac{1}{2}$ we obtain

$$a_n = \frac{2a_{n-1}}{2n(n+\frac{1}{2})} = \frac{4a_{n-1}}{2n(2n+1)} \qquad n = 1, 2, \ldots$$

and it follows that a_0 is arbitrary and

$$a_1 = \frac{4a_0}{3!} \qquad a_2 = \frac{4a_1}{4 \cdot 5} = \frac{4^2 a_0}{5!} \qquad a_3 = \frac{4a_2}{6 \cdot 7} = \frac{4^3 a_0}{7!} \qquad \cdots$$

It follows now that $a_n = 4^n a_0/(2n+1)!$ for all $n \geq 0$. Therefore,

$$\begin{aligned} y_2(t) &= \sum_{n=0}^{\infty} \frac{4^n}{(2n+1)!} t^{n+1/2} \\ &= \frac{1}{2} \sum_{n=0}^{\infty} \frac{1}{(2n+1)!} (2\sqrt{t})^{2n+1} \end{aligned}$$

is a solution to this equation. From the series representation of y_1 and y_2 one can check that

$$y_1(t) = \cosh 2\sqrt{t} \qquad \text{and} \qquad y_2(t) = \tfrac{1}{2} \sinh 2\sqrt{t}$$

for all $t > 0$, and hence

$$y(t) = c_1 \cosh 2\sqrt{t} + c_2 \sinh 2\sqrt{t}$$

is the general solution to the given equation on $(0, \infty)$.

▶ **EXAMPLE 4.3-2**

Consider the equation $t^3 y'' - y = 0$. Assuming a Frobenius-type solution about $t_0 = 0$, we have

$$\begin{aligned} t^3 y''(t) - y(t) &= \sum_{n=0}^{\infty} (n+\sigma)(n+\sigma-1) a_n t^{n+\sigma+1} - \sum_{n=0}^{\infty} a_n t^{n+\sigma} \\ &= \sum_{n=1}^{\infty} [(n+\sigma-1)(n+\sigma-2) a_{n-1} - a_n] t^{n+\sigma} - a_0 t^{\sigma} \end{aligned}$$

and it follows that if y is a solution to this equation, then

$$a_0 = 0 \qquad \text{and} \qquad a_n = (n+\sigma-1)(n+\sigma-2) a_{n-1} \qquad \text{for } n = 1, 2, \ldots$$

Therefore, $a_n = 0$ for all $n \geq 0$ and the only solution of Frobenius-type expanded about $t_0 = 0$ is the trivial solution.

Example 4.3-2 shows that not all equations of the form (3) have a nontrivial Frobenius-type solution. However, if a singular point t_0 is a *regular singular point* [see equation (9) in Section 4.1], then at least one nontrivial Frobenius-type solution is guaranteed to exist, and we indicate a method of computing it here. Suppose now that equation (3) can be written in the form

$$t^2 p_0(t) y'' + t q_0(t) y' + r_0(t) y = 0 \tag{6}$$

where p_0, q_0, and r_0 are polynomials, with $p_0(0) \neq 0$. We have assumed here that $t_0 = 0$ is the singular point in question [if this is not the case, we can use the simple transformation from $(t - t_0)$ into t in order that our equation has the form (6)]. Since $p_0(0) \neq 0$ it is easy to check that $t_0 = 0$ *is a regular singular point for (6).* Conversely, if $t_0 = 0$ is a regular singular point for equation (3), then equation (3) can be written in the form (6) [notice that, since it has not been assumed that p_0, q_0, and r_0 have no common factors, equation (9*a*) in Section 4.1 can be written in the form (6) by taking $t_0 = 0$ and multiplying each side of the equation by t].

Suppose that a solution y to (6) has a Frobenius-type expansion about $t_0 = 0$. Substituting into (6) we have

$$\begin{aligned} t^2p_0(t)y'' + tq_0(t)y'(t) + r_0(t)y(t) &= t^2p_0(t)\sum_{n=0}^{\infty}(n+\sigma)(n+\sigma-1)a_n t^{n+\sigma-2} \\ &\quad + tq_0(t)\sum_{n=0}^{\infty}(n+\sigma)a_n t^{n+\sigma-1} + r_0(t)\sum_{n=0}^{\infty}a_n t^{n+\sigma} \\ &= \sum_{n=0}^{\infty}p_0(t)(n+\sigma)(n+\sigma-1)a_n t^{n+\sigma} + \sum_{n=0}^{\infty}q_0(t)(n+\sigma)t^{n+\sigma} \\ &\quad + \sum_{n=0}^{\infty}r_0(t)a_n t^{n+\sigma} \end{aligned}$$

The lowest power of t in this expression is t^σ, and since the constant terms in the polynomials p_0, q_0, and r_0 are $p_0(0)$, $q_0(0)$, and $r_0(0)$, respectively, the term with the lowest power of t is

$$[p_0(0)\sigma(\sigma-1) + q_0(0)\sigma + r_0(0)]a_0 t^\sigma$$

and hence

$$p_0(0)\sigma^2 + [q_0(0) - p_0(0)]\sigma + r_0(0) = 0 \tag{7}$$

is the *indicial equation for (6).* Since $p_0(0) \neq 0$, equation (7) is a quadratic equation and there always exist two indicial roots (or one double indicial root). The following two results describe the applicability of Frobenius' method to equation (6). We omit the proofs of these results, since they require techniques beyond the scope of this text.

Theorem 4.3-1

Suppose that the indicial roots σ_1 and σ_2 to equation (6) are distinct and that $\sigma_1 - \sigma_2$ is not an integer. Then (6) has linearly independent solutions y_1 and y_2 of the form

$$y_1(t) = t^{\sigma_1}\sum_{n=0}^{\infty}a_n t^n \quad \text{and} \quad y_2(t) = t^{\sigma_2}\sum_{n=0}^{\infty}b_n t^n \tag{8}$$

where $a_0 \neq 0$, $b_0 \neq 0$, and which are valid for t in some interval $(0, \delta)$ where $\delta > 0$.

Theorem 4.3-1 is valid for complex as well as real indicial roots (note that if two indicial roots σ_1 and σ_2 are complex, then they are conjugate, and so $\sigma_1 - \sigma_2$ is imaginary and Theorem 4.3-1 applies). However, this theorem is applied in this text only to equations with real indicial roots. The case when the indicial roots differ by an integer is somewhat more complicated.

Theorem 4.3-2

Suppose that the indicial roots σ_1 and σ_2 to equation (6) differ by an integer: say $\sigma_1 - \sigma_2 = k$, where $k = 0, 1, 2, \ldots$. Then corresponding to the larger root σ_1, there is a solution y_1 to (6) of the form

$$y_1(t) = t^{\sigma_1} \sum_{n=0}^{\infty} a_n t^n \tag{9}$$

where $a_0 \neq 0$. Also, corresponding to the smaller root σ_2, there is a solution y_2 to (6) of the form

$$y_2(t) = t^{\sigma_2} \sum_{n=0}^{\infty} b_n t^n + K y_1(t) \ln t \tag{10}$$

where K is some constant (if $\sigma_1 = \sigma_2$ then K is necessarily nonzero, but if $\sigma_1 > \sigma_2$ then K may or may not be zero). Moreover, y_1 and y_2 are defined on $(0, \delta)$ for some $\delta > 0$ and are linearly independent.

Note that if $p_0(t) \equiv \alpha$, $q_0(t) \equiv \beta$, and $r_0(t) \equiv \gamma$, then equation (6) is actually Euler's equation (1), and the indicial equation is the same as equation (2). Thus applying Theorem 4.3-2 to Euler's equation shows that if σ_1 is a double root of (7), then the logarithm term in (10) is necessary; and if the indicial roots σ_1 and σ_2 differ by a nonzero integer, then the logarithm term is not always necessary since $y_1(t) = t^{\sigma_1}$ and $y_2(t) = t^{\sigma_2}$ are linearly independent solutions. However, there are some cases when the indicial roots differ by a nonzero integer and the logarithm term must be retained in (10). This is shown by the following example.

EXAMPLE 4.3-3

Consider the equation $ty'' + 3y' - 2y = 0$. Multiplying each side of this equation by t shows that it has the form (6) so that Frobenius' method applies.

Assuming that

$$y(t) = \sum_{n=0}^{\infty} a_n t^{n+\sigma}$$

we have

$$t^2 y''(t) + 3ty' - 2ty(t)$$

$$= \sum_{n=0}^{\infty} [(n+\sigma)(n+\sigma-1) + 3(n+\sigma)]a_n t^{n+\sigma} - 2\sum_{n=0}^{\infty} a_n t^{n+\sigma+1}$$

$$= [\sigma(\sigma-1) + 3\sigma]a_0 t^{\sigma} + \sum_{n=1}^{\infty} [(n+\sigma)(n+\sigma+2)a_n - 2a_{n-1}]t^{n+\sigma}$$

Therefore, $\sigma(\sigma-1) + 3\sigma = \sigma(\sigma+2) = 0$ is the indicial equation, and $\sigma = 0$, -2 are the indicial roots. The recursion formula for the a_n is

$$(n+\sigma)(n+\sigma+2)a_n = 2a_{n-1} \qquad \text{for } n = 1, 2, \ldots \tag{11}$$

Setting $\sigma = -2$ in (11) we have that $(n-2)na_n = 2a_{n-1}$ for $n \geq 1$. Thus, $-a_1 = 2a_0$, $0 \cdot 2 = 2a_1$, and hence $a_0 = a_1 = 0$; and it follows that a Frobenius-type solution with leading term t^{-2} does not exist. Setting $\sigma = 0$ (the largest indicial root), we have from (11) that $a_n = 2a_{n-1}/[n(n+2)]$ for $n \geq 1$, and hence

$$a_1 = \frac{2a_0}{1 \cdot 3} = \frac{2^2 a_0}{3!} \qquad a_2 = \frac{2a_1}{2 \cdot 4} = \frac{2^3 a_0}{2!4!} \qquad a_3 = \frac{2a_2}{3 \cdot 5} = \frac{2^4 a_0}{3!5!} \qquad \cdots$$

An easy argument shows that $a_n = 2^{n+1}a_0/[n!(n+2)!]$ for all $n \geq 0$, and the function

$$y_1(t) = \sum_{n=0}^{\infty} \frac{2^{n+1}}{n!(n+2)!} t^n \tag{12}$$

is a solution to the equation. Taking $K = 1$ in (1) (we know that $K \neq 0$ since no Frobenius solution has leading term t^{-2}), we have from Theorem 4.3-2 that there is a second solution having the form

$$y(t) = \sum_{n=0}^{\infty} b_n t^{n-2} + y_1(t) \ln t \tag{13}$$

for t in some interval $(0, \delta)$ and $b_0 \neq 0$. The computations involved in actually computing the b_n's are quite involved. Although it is not the case in this example, if y_1 can be recognized as a series representing an elementary function, then reduction of order (see Section 2.5) may be an effective method of determining a second solution. Sometimes, if one requires only that a few of the b_n's be computed, one can assume y has the form (13) [or, in general, the form (10) in Theorem 4.3-2] and substitute this expression into the equation. These computations, however, can be tedious.

PROBLEMS

4.3-1. Determine the indicial equation and the indicial roots about each regular singular point for each of the following equations. Also indicate the form of the solutions in each case.

(a) $3t^2y'' + (7t - 7t^2)y' + (1 + t^3)y = 0$

(b) $t^2y'' + ty' + (t^2 - \frac{1}{4})y = 0$

(c) $(t + 2)y'' + \frac{1}{2}y' - \frac{1}{4}y = 0$

(d) $t(t - 1)y'' + (7t - 1)y' + y = 0$

(e) $(t + 1)y'' - ty' + 6y = 0$

4.3-2. Determine the solutions y to the following equations that have the form

$$y(t) = t^{\sigma} \sum_{n=0}^{\infty} a_n t^n$$

where σ is real, $a_0 \neq 0$, and t is in $(0, \delta)$ for some $\delta > 0$. Determine the indicial roots, the recursion formula for the a_n's, and, whenever possible, compute the a_n's explicitly.

(a) $4ty'' + 2y' - y = 0$

(b) $ty'' + y' - y = 0$

(c) $ty'' + y = 0$

(d) $t^2y'' + t(t + 1)y' - y = 0$

(e) $ty'' + y' + ty = 0$

(f) $t(t - 1)y'' + (3t - 1)y' + y = 0$

(g) $t(t - 1)y'' + (7t - 1)y' + 5y = 0$

(h) $(t - t^2)y'' - 3y' + 2y = 0$

4.3-3. Suppose that n is a positive integer and consider the equation

$$(*) \qquad ty'' + (1 - t)y' + ny = 0 \qquad t > 0$$

(a) Show that $t = 0$ is a regular singular point and that $\sigma = 0$ is a double indicial root.

(b) Show that (*) has a polynomial solution of degree n for all $n = 1, 2, 3, \ldots$.

(c) Let L_n denote the polynomial solution to (*) such that $L_n(0) = 1$ (L_n is called the nth *Laguerre polynomial*). Compute L_n for $n = 1, 2,$ and 3.

4.4 EXPANSIONS FOR LARGE TIMES

Sections 4.2 and 4.3 are concerned with expanding a solution to a second order linear equation about a point t_0 in a power series or a Frobenius-type series. The leading term (or terms) in these expansions determine the behavior of the solution $y(t)$ for t near t_0 (that is, as $t \to t_0$). The purpose of this section is to indicate that similar techniques can be used to study the behavior of solutions for large values of t (and, in particular, as $t \to \infty$). Again it is assumed that p, q, and r are polynomials and the equation

$$(1) \qquad p(t)y'' + q(t)y' + r(t)y = 0$$

is considered. Also, instead of analyzing directly the behavior of solutions to (1) for large values of t, we use the change of independent variable $s = 1/t$ (see Section 2.7) and study the behavior of solutions to the resulting equation for small values of s: that is, expand the solutions of the resulting equation about $s = 0$.

Setting $s = 1/t$ we have that $t = 1/s$ and also that

$$\frac{dy}{dt} = \frac{ds}{dt}\frac{dy}{ds} = -\frac{1}{t^2}\frac{dy}{ds} = -s^2\dot{y}$$

where $\dot{y} = dy/ds$ and

$$\frac{d^2y}{dt^2} = \frac{d}{dt}\left(-\frac{1}{t^2}\frac{dy}{ds}\right) = \frac{2}{t^3}\frac{dy}{ds} - \frac{1}{t^2}\left(-\frac{1}{t^2}\frac{d^2y}{ds^2}\right) = 2s^3\dot{y} + s^4\ddot{y}$$

where $\ddot{y} = d^2y/ds^2$. Substituting these expressions into equation (1), we obtain the following equation in the variable s:

(2) $$s^4 p\left(\frac{1}{s}\right)\ddot{y} + \left[2s^3 p\left(\frac{1}{s}\right) - s^2 q\left(\frac{1}{s}\right)\right]\dot{y} + r\left(\frac{1}{s}\right)y = 0$$

The technique in Sections 4.2 and 4.3 can be used to expand the solutions of this equation in a Frobenius-type series about the point $s = 0$. [Since p, q, and r are polynomials, each side of equation (2) can always be multiplied by some power of s so that the resulting equation has polynomial coefficients.] The types of singularities of (1) at $t = \infty$ are defined in terms of the types of singularities of (2) at $s = 0$. *Infinity is said to be an ordinary (or a singular) point of equation (1) if zero is an ordinary (or singular) point of equation (2).* Moreover, if $t = \infty$ is singular point of (1), then it is said to be a *regular singular point* or an *irregular singular point* according to whether $s = 0$ is a regular or irregular singular point of (2). Therefore, if $t = \infty$ is an ordinary point of equation (1), there are two linearly independent solutions to (2) of the form

$$y(s) = \sum_{n=0}^{\infty} a_n s^n$$

and hence (1) has two linearly independent solutions of the form

(3\` $$y(t) = \sum_{n=0}^{\infty} a_n \left(\frac{1}{t}\right)^n$$

Similarly, if $t = \infty$ is a regular singular point, then (1) has the least one nontrivial solution of the form

(4) $$y(t) = \left(\frac{1}{t}\right)^{\sigma} \sum_{n=0}^{\infty} a_n \left(\frac{1}{t}\right)^n$$

where σ is an indicial root of (2). Of course, if the indicial roots do not differ by an integer, there is a solution to (1) of the form (4) for each indicial root σ. The series in (3) and (4) can be considered either for $|t| > M$ or just for $t > M$ (where M is some constant).

EXAMPLE 4.4-1

Consider the equation $(1 - t^2)y'' + 2ty + 2y = 0$. The points $t = \pm 1$ are regular singular points, and all other finite points t are ordinary points. Comparing with equation (1), $p(t) = 1 - t^2$, $q(t) = 2t$, and $r(t) = 2$. Thus,

$$p\left(\frac{1}{s}\right) = 1 - \frac{1}{s^2} \qquad q\left(\frac{1}{s}\right) = \frac{2}{s} \qquad \text{and} \qquad r\left(\frac{1}{s}\right) = 2$$

and equation (2) is seen to have the form

$$s^2(s^2 - 1)\ddot{y} + s(2s^2 - 4)\dot{y} + 2y = 0 \tag{5}$$

Since this equation has a regular singular point at $s = 0$, the original equation has a regular singular point at $t = \infty$. From (7) in Section 4.3, $-\sigma^2 - 3\sigma + 2 = 0$ is the indicial equation for (5), and hence the original equation has solutions of the form

$$y(t) = \left(\frac{1}{t}\right)^{\sigma} \sum_{n=0}^{\infty} a_n \left(\frac{1}{t}\right)^n$$

where $\sigma = (-3 \pm \sqrt{17})/2$, the indicial roots of (5) at $s = 0$.

EXAMPLE 4.4-2

Consider the equation $t^4y'' + (2t^3 - t)y' + y = 0$. Setting $s = 1/t$ we have $p(1/s) = 1/s^4$, $q(1/s) = 2/s^3 - 1/s$, $r(1/s) = 1$, and so from (2) we have $\ddot{y} + sy + \dot{y} = 0$. Therefore, $s = 0$ is an ordinary point and so $t = \infty$ is an ordinary point for the original equation. Assuming

$$y(s) = \sum_{n=0}^{\infty} a_n s^n$$

leads to the equation

$$\ddot{y}(s) + s\dot{y}(s) + y(s) = \sum_{n=2}^{\infty} [n(n-1)a_n + (n-2)a_{n-2} + a_{n-2}]s^{n-2}$$

and hence the recursion formula $a_n = -a_{n-2}/n$ for $n = 2, 3, \ldots$. From this it follows that

$$a_{2k} = \frac{(-1)^k}{2^k k!} a_0 \quad \text{and} \quad a_{2k+1} = \frac{(-1)^k}{1 \cdot 3 \cdot 5 \cdot \cdots \cdot (2k+1)} a_1 \quad k = 1, 2, \ldots$$

and it follows that *every* solution to the given equation has the form

$$y(t) = a_0 \sum_{k=0}^{\infty} \frac{(-1)^k}{2^k k!} \left(\frac{1}{t}\right)^{2k} + a_1 \sum_{k=0}^{\infty} \frac{(-1)^k}{1 \cdot 3 \cdot 5 \cdot \cdots \cdot (2k+1)} \left(\frac{1}{t}\right)^{2k+1}$$

where a_0 and a_1 are constants.

PROBLEMS

4.4-1. Determine the solutions y to the following equations that have the form

$$y(t) = \left(\frac{1}{t}\right)^{\sigma} \sum_{n=0}^{\infty} a_n \left(\frac{1}{t}\right)^{n}$$

where σ is real, $a_0 \neq 0$, and t is in (M, ∞) for some $M > 0$.

(a) $t^3y'' + 2t^2y' - y = 0$

(b) $t^4y'' + (2t^3 + t)y' + 2y = 0$

(c) $t^3y'' + (t^2 + t)y' - y = 0$

4.4-2. Consider solutions y to the following equations that have the form

$$y(t) = \left(\frac{1}{t}\right)^{\sigma} \sum_{n=0}^{\infty} a_n \left(\frac{1}{t}\right)^{n}$$

where σ is real, $a_0 \neq 0$, and t is in (M, ∞) for some $M > 0$. For each value of σ, determine the recursion formula for the a_n's and compute a_1, a_2, a_3 in terms of a_0.

(a) $2t^3y'' + t^2y' + y = 0$

(b) $(1 - t^2)y'' - 2ty' + 6y = 0$

4.5 SPECIAL EQUATIONS

In this section the solutions to several important classes of second order homogeneous equations are studied, using the methods developed in this chapter. However, this analysis is just a basic introduction and describes only the most elementary properties of these functions.

4.5*a* Bessel's Equation

The first equation is one of the most important differential equations in applied mathematics and is called *Bessel's differential equation:*

$$t^2y'' + ty' + (t^2 - v^2)y = 0 \qquad t > 0 \tag{1}$$

where $v \geq 0$ is a given real parameter. It is easy to check that all nonzero numbers are ordinary points and that zero is a regular singular point for Bessel's equation (the point at ∞ is an irregular singular point—see Problem 4.5-1). Applying the method of Frobenius in order to determine expansions of solutions to (1) about $t = 0$, it is assumed that

$$y(t) = \sum_{n=0}^{\infty} a_n t^{n+\sigma}$$

and hence that

$$t^2y''(t) + ty'(t) + (t^2 - \nu^2)y(t)$$

$$= \sum_{n=0}^{\infty} [(n+\sigma)(n+\sigma-1)a_n + (n+\sigma)a_n - \nu^2 a_n]t^{n+\sigma} + \sum_{n=0}^{\infty} a_n t^{n+\sigma+2}$$

$$= [\sigma(\sigma-1) + \sigma - \nu^2]a_0 t^\sigma + [(1+\sigma)\sigma + (1+\sigma) - \nu^2]a_1 t^{\sigma+1}$$

$$+ \sum_{n=2}^{\infty} \{[(n+\sigma)^2 - \nu^2]a_n + a_{n-2}\}t^{n+\sigma}$$

Setting the coefficients of the powers of t to zero, we obtain

$$(\sigma^2 - \nu^2)a_0 = 0 \qquad \text{and} \qquad [(\sigma+1)^2 - \nu^2]a_1 = 0 \tag{2}$$

and the recursion formula

$$a_n = \frac{-a_{n-2}}{(n+\sigma+\nu)(n+\sigma-\nu)} \qquad \text{for } n = 2, 3, 4, \ldots \tag{3}$$

The first equation in (2) gives the indicial equation $\sigma^2 - \nu^2 = 0$, and hence $\sigma = \pm\nu$ are the indicial roots. Using the largest indicial root $\sigma = \nu$, the second equation in (2) implies that $a_1 = 0$, and hence $a_n = 0$ for all odd positive integers n. Therefore, with $\sigma = \nu$ and $n = 2k$, the recursion formula (3) becomes

$$a_{2k} = \frac{-a_{2(k-1)}}{4k(k+\nu)} \qquad \text{for } k = 1, 2, 3, \ldots$$

From this formula it follows that

$$a_2 = \frac{-a_0}{4(1+\nu)} \qquad a_4 = \frac{-a_2}{4 \cdot 2(2+\nu)} = \frac{a_0}{4^2 \cdot 2!(1+\nu)(2+\nu)}$$

and, in general,

$$a_{2k} = \frac{(-1)^k a_0}{4^k k!(1+\nu)(2+\nu)\cdots(k+\nu)} \qquad \text{for } k = 1, 2, \ldots$$

Conventionally, the arbitrary constant a_0 is chosen so that

$$a_0 2^\nu \int_0^\infty s^\nu e^{-s}\, ds = 1 \tag{4}$$

and we obtain the solution

$$J_\nu(t) = a_0 t^\nu \left[1 + \sum_{k=1}^{\infty} \frac{(-1)^k}{4^k k!(1+\nu)\cdots(k+\nu)} t^{2k}\right] \tag{5}$$

The solution J_v is the *Bessel function of the first kind of order v*. If $v = p$ is a nonnegative integer, it is routine to check from (4) that $a_0 = 1/2^p p!$ and hence

$$J_p(t) = \sum_{k=0}^{\infty} \frac{(-1)^k}{k!(k+p)!}\left(\frac{t}{2}\right)^{2k+p} \qquad \text{for } p = 0, 1, 2, \ldots \tag{6}$$

It is easy to check, using the ratio test, that J_v is defined for all $t > 0$. Note also that the series in (5) converges for $t \in (0, \infty)$ whenever v is not a negative integer. Thus

$$J_{-v}(t) = \bar{a}_0 t^{-v}\left[1 + \sum_{k=1}^{\infty} \frac{(-1)^k}{4^k k!(1-v)\cdots(k-v)} t^{2k}\right] \tag{7}$$

exists for all $v > 0$ with $v \neq 1, 2, \ldots$, and is also a solution to Bessel's equation that is defined for all $t > 0$. The constant $\bar{a}_0$ is defined as in (4) (with v replaced by $-v$) whenever $-1 < -v < 0$, but is defined in other terms when $-v < -1$ and $-v \neq -2, -3, \ldots$. Since this definition involves the use of the gamma function, we omit these ideas (see Problem 4.5-3).

By the method of Frobenius if v is neither an integer nor half an odd integer, then the difference of the indicial roots equals $2v$, which is not an integer, and hence J_v and J_{-v} are linearly independent solutions to (1). It is also the case that if v is half an odd integer, then J_v and J_{-v} are still linearly independent, and hence the family of functions y on $(0, \infty)$ defined by

$$y = c_1 J_v + c_2 J_{-v} \qquad v \geq 0 \qquad v \neq 0, 1, 2, \ldots$$

is the general solution to Bessel's equation. If $v = p \geq 0$ is an integer, then a second linearly independent solution has the form indicated by formula (10) in Theorem 4.3-2.

Several basic properties of Bessel functions can be obtained with elementary techniques. If the change of dependent variable $y = t^{-1/2}z$ is made in equation (1), it is routine to check that z must satisfy the equation

$$z'' + \left(1 + \frac{1 - 4v^2}{4t^2}\right)z = 0 \tag{8}$$

Taking $v = \pm\frac{1}{2}$ equation (8) becomes $z'' + z = 0$, and it follows easily that

$$J_{1/2}(t) = a_0 t^{-1/2} \sin t \qquad \text{and} \qquad J_{-1/2}(t) = \bar{a}_0 t^{-1/2} \cos t \tag{9}$$

It can be shown that $a_0 = \bar{a}_0 = \sqrt{2/\pi}$ in (9). As a further example observe that

$$\begin{aligned} J_0(t) &= 1 - \frac{t^2}{2^2(1!)^2} + \frac{t^4}{2^4(2!)^2} - \frac{t^6}{2^6(3!)^2} + \cdots \\ J_1(t) &= \frac{t}{2} - \frac{t^3}{2^3(1!)(2!)} + \frac{t^5}{2^5(2!)(3!)} - \frac{t^7}{2^7(3!)(4!)} + \cdots \end{aligned} \tag{10}$$

for all $t > 0$ [set $p = 0$ and $p = 1$ in (6)]. From these expansions it is easy to check that

$$J_0'(t) = -J_1(t) \qquad \text{and} \qquad [tJ_1(t)]' = tJ_0(t) \qquad \text{for all } t > 0 \tag{11}$$

The formulas in (11) imply an interesting property of the zeros of J_0 and J_1: *the zeros of J_0 and J_1 occur alternately.* This fact follows from (11) and Rolle's theorem. Since between two zeros of J_0 there is a zero of J_0', the first formula in (11) implies that between any two zeros of J_0 there is a zero of J_1. Similarly, from the second formula in (11), between any two zeros of $tJ_1(t)$ there is a zero of $tJ_0(t)$, and it is immediately seen that the zeros of J_0 and J_1 are interlaced.

▶ **EXAMPLE 4.5-1**

Consider the equation $t^2y'' + ty' + (\delta^2t^2 - \nu^2)y = 0$, where $\delta, \nu > 0$. Making the change of variable $s = \delta^{-1}t$, we have that

$$\frac{dy}{dt} = \delta^{-1}\frac{dy}{ds} \qquad \text{and} \qquad \frac{d^2y}{dt^2} = \delta^{-2}\frac{d^2y}{ds^2}$$

Since $t = \delta s$ the given equation is transformed into the equation

$$s^2\frac{d^2y}{ds^2} + s\frac{dy}{ds} + (s^2 - \nu^2)y = 0$$

which is Bessel's equation of order ν. Since $y(s) = J_\nu(s)$ is a solution to the transformed equation, $y(t) = J_\nu(\delta^{-1}t)$ for $t > 0$ is a solution to the original equation. Problem 4.5-5 indicates a few more equations that can be transformed into Bessel's equation.

4.5b Legendre's Equation

The second equation to be analyzed also occurs in the applications of mathematics to engineering and physics and is known as *Legendre's equation:*

$$(1 - t^2)y'' - 2ty' + \rho(\rho + 1)y = 0 \tag{12}$$

where $\rho > -\frac{1}{2}$ is a constant. [Note that if $\gamma < -\frac{1}{2}$ and $\rho = -\gamma - 1$, then $\rho > -\frac{1}{2}$ and $\rho(\rho + 1) = \gamma(\gamma + 1)$. Hence there is no loss in assuming that $\rho > -\frac{1}{2}$ in (12).] The solutions to equation (12) are called *Legendre's functions of order ρ.* It is easy to see that $t = \pm 1$ are regular singular points of (12) and all other finite points are ordinary points. The point $t = \infty$ is also a regular singular point (see Problem 4.5-1).

Since $t = 0$ is an ordinary point, Legendre's equation has two linearly independent solutions of the form

$$y(t) = \sum_{n=0}^{\infty} a_n t^n$$

Assuming that y has this form implies that

$$(1 - t^2)y''(t) - 2ty'(t) + \rho(\rho + 1)y(t)$$

$$= \sum_{n=0}^{\infty} n(n-1)a_n t^{n-2} - \sum_{n=0}^{\infty} [n(n-1)a_n + 2na_n - \rho(\rho+1)a_n]t^n$$

$$= \sum_{n=2}^{\infty} \{n(n-1)a_n - [(n-2)(n-3) + 2(n-2) - \rho(\rho+1)]a_{n-2}\}t^{n-2}$$

and since $(n-2)(n-3) + 2(n-2) - \rho(\rho+1) = (n-2-\rho)(n-1+\rho)$, it follows that the recursion formula is

$$(13) \qquad a_n = \frac{-(\rho + 2 - n)(\rho - 1 + n)}{n(n-1)} a_{n-2} \qquad n = 2, 3, 4, \ldots$$

and that a_0 and a_1 are arbitrary constants. It is evident from (13) that each even subscripted coefficient is a multiple of a_0 and each odd subscripted coefficient a multiple of a_1. Noting that

$$a_2 = -\frac{\rho(\rho+1)}{2!} a_0$$

and

$$a_4 = -\frac{(\rho-2)(\rho+3)}{4 \cdot 3} a_2 = \frac{(\rho-2)\rho(\rho+1)(\rho+3)}{4!} a_0$$

it follows by induction that

$$(14) \qquad a_{2k} = (-1)^k \frac{(\rho - 2k + 2) \cdots (\rho - 2)\rho(\rho+1) \cdots (\rho + 2k - 1)}{(2k)!} a_0$$

for $k = 1, 2, 3, \ldots$.

Noting that

$$a_3 = \frac{-(\rho-1)(\rho+2)}{3!} a_1$$

and

$$a_5 = -\frac{(\rho-3)(\rho+4)}{5 \cdot 4} a_3 = \frac{(\rho-3)(\rho-1)(\rho+2)(\rho+4)}{5!} a_1$$

it follows by induction that

$$(15) \qquad a_{2k+1} = (-1)^k \frac{(\rho - 2k + 1) \cdots (\rho - 1)(\rho + 2) \cdots (\rho + 2k)}{(2k + 1)!} a_1$$

for $k = 0, 1, 2, \ldots$

Therefore, Legendre's equation has two linearly independent solutions y_1 and y_2 having the form

$$(16) \qquad y_1(t) = 1 - \frac{\rho(\rho + 1)}{2!} t^2 + \frac{(\rho - 2)\rho(\rho + 1)(\rho + 3)}{4!} t^4 - \frac{(\rho - 4)(\rho - 2)\rho(\rho + 1)(\rho + 3)(\rho + 5)}{6!} t^6 + \cdots$$

and

$$(17) \qquad y_2(t) = t - \frac{(\rho - 1)(\rho + 2)}{3!} t^3 + \frac{(\rho - 3)(\rho - 1)(\rho + 2)(\rho + 4)}{5!} t^5 - \frac{(\rho - 5)(\rho - 3)(\rho - 1)(\rho + 2)(\rho + 4)(\rho + 6)}{7!} t^7 + \cdots$$

In particular, y_1 and y_2 are linearly independent, and the general solution of Legendre's equation [for t in $(-1, 1)$] is the family $y = c_1 y_1 + c_2 y_2$, where c_1 and c_2 are constants.

It is important to observe that if m is a nonnegative integer and $\rho = m$, then y_1 is a polynomial of degree m if m is even, and y_2 is a polynomial of degree m if m is odd. Therefore, if m is a nonnegative integer, *the polynomial solution* P_m *to equation (12) such that* $P_m(1) = 1$ *is called Legendre's polynomial of degree* m. If m is odd then P_m contains only terms with odd powers of t, and if m is even P_m contains only terms with even powers of t. It is easy to check that the first five Legendre polynomials are

$$(18) \qquad \begin{array}{lll} P_0(t) \equiv 1 & P_1(t) \equiv t & P_2(t) \equiv \dfrac{3t^2 - 1}{2} \\ P_3(t) \equiv \dfrac{5t^3 - 3t}{2} & \text{and} & P_4(t) \equiv \dfrac{35t^4 - 30t^2 + 3}{8} \end{array}$$

The Legendre polynomials can also be expressed in the form

$$(19) \qquad P_m(t) \equiv \frac{1}{2^m m!} \frac{d^m}{dt^m} [(t^2 - 1)^m] \qquad \text{for } m = 1, 2, 3, \ldots$$

The formula (19) is known as *Rodrigues' formula* and is easily verified for the polynomials $P_0 - P_4$ given in (18) (see also Problem 4.5-6).

4.5c Hypergeometric Equation

Suppose that α, β, and γ are real constants, and consider the second order homogeneous equation

$$(20) \qquad t(1 - t)y'' + [\gamma - (\alpha + \beta + 1)t]y' - \alpha\beta y = 0$$

This equation is known as the *hypergeometric equation* (or sometimes as *Gauss' differential equation*), and its solutions are called *hypergeometric functions.* It is easy to see that $t = 0$ and $t = 1$ are regular singular points and that all other finite points are ordinary points (see also Problem 4.5-2). Expanding the solution y in the Frobenius-type series

$$y(t) = \sum_{n=0}^{\infty} a_n t^{n+\sigma}$$

we obtain

$$\begin{aligned}(t - t^2)y''(t) &+ [\gamma - (\alpha + \beta + 1)t]y'(t) - \alpha\beta y(t)\\ &= \sum_{n=0}^{\infty} [(n + \sigma)(n + \sigma - 1) + \gamma(n + \sigma)]a_n t^{n+\sigma-1}\\ &\qquad - \sum_{n=0}^{\infty} [(n + \sigma)(n + \sigma - 1) + (\alpha + \beta + 1)(n + \sigma) + \alpha\beta]a_n t^{n+\sigma}\\ &= [\sigma(\sigma - 1) + \gamma\sigma]a_0 t^{\sigma-1}\\ &\quad + \sum_{n=1}^{\infty} \{[(n + \sigma)(n + \sigma - 1) + \gamma(n + \sigma)]a_n\\ &\quad -[(n + \sigma - 1)(n + \sigma - 2) + (\alpha + \beta + 1)(n + \sigma - 1) + \alpha\beta]a_{n-1}\}t^{n+\sigma-1}\end{aligned}$$

Therefore, the indicial equation is $\sigma(\sigma - 1) + \gamma\sigma = 0$, and the indicial roots are $\sigma = 0, 1 - \gamma$. The recursion formula for the a_n has the form

$$a_n = \frac{(n + \sigma + \alpha - 1)(n + \sigma + \beta - 1)}{(n + \sigma)(n + \sigma - 1 + \gamma)} a_{n-1} \qquad n = 1, 2, \ldots \tag{21}$$

Assuming that $\gamma \neq 0, -1, -2, \ldots$ and taking $\sigma = 0$ in the recursion formula (21), we obtain

$$a_1 = \frac{\alpha\beta}{1 \cdot \gamma} a_0 \qquad a_2 = \frac{(\alpha + 1)(\beta + 1)}{2(\gamma + 1)} a_1 = \frac{\alpha(\alpha + 1)(\beta(\beta + 1)}{2!\,\gamma(\gamma + 1)} a_0$$

$$a_3 = \frac{(\alpha + 2)(\beta + 2)}{3(\gamma + 2)} = \frac{\alpha(\alpha + 1)(\alpha + 2)\beta(\beta + 1)(\beta + 2)}{3!\,\gamma(\gamma + 1)(\gamma + 2)} \qquad \ldots$$

The series solution to (20) defined by

$$\begin{aligned}F(\alpha, \beta, \gamma; t) \equiv 1 + \frac{\alpha\beta}{1 \cdot \gamma} t + \frac{\alpha(\alpha + 1)\beta(\beta + 1)}{2!\,\gamma(\gamma + 1)} t^2\\ + \frac{\alpha(\alpha + 1)(\alpha + 2)\beta(\beta + 1)(\beta + 2)}{3!\,\gamma(\gamma + 1)(\gamma + 2)} t^3 + \cdots\end{aligned} \tag{22}$$

is known as the *hypergeometric series.* When γ is not an integer, a second linearly independent solution to (20) can be found by using the indicial root $\sigma = 1 - \gamma$ (see Problem 4.5-8).

Several elementary functions can be described in terms of the hypergeometric series F. Taking $\alpha = 1$ and $\beta = \gamma$ in (22), it is seen that

$$F(1, \beta, \beta; t) = 1 + t + t^2 + t^3 + \cdots$$

$$= \frac{1}{1 - t} \qquad |t| < 1$$

Note also that

$$\ln(1 - t) = -tF(1, 1, 2; t)$$

Other examples can be found in Problem 4.5-9.

PROBLEMS

4.5-1. Show that the point $t = \infty$ is an irregular singular point for Bessel's equation and a regular singular point for Legendre's equation.

4.5-2. Show that $t = \infty$ is an ordinary point for the hypergeometric equation when $\alpha = 0$, $\beta = 1$ and when $\alpha = 1$, $\beta = 0$. Show further that, in all other cases, $t = \infty$ is a regular singular point for the hypergeometric equation.

4.5-3. For each number $v > 0$ define

$$\Gamma(v) = \int_0^\infty t^{v-1} e^{-t}\, dt$$

The function Γ is called the *gamma function.*

(a) Show that $\Gamma(v + 1) = v\Gamma(v)$ for all $v > 0$. (*Hint:* Use integration by parts.)

(b) Show that $\Gamma(v + k) = v(v + 1) \cdots (v + k - 1)\Gamma(v)$ whenever $v > 0$ and k is a positive integer.

(c) Show that $\Gamma(k + 1) = k!$ for each positive integer k.

(d) Using formulas (4) and (5) as well as the preceding properties of Γ, show that

$$J_v(t) = \sum_{k=0}^{\infty} \frac{(-1)^k}{k!\,\Gamma(v + k + 1)} \left(\frac{t}{2}\right)^{2k+v}$$

for each $v > 0$.

4.5-4. Show that

$$\frac{d}{dt}[t^v J_v(t)] = t^v J_{v-1}(t) \qquad \text{for } v \geq 1 \text{ and } t > 0$$

4.5-5. In each of the following equations, show that the indicated change of variable transforms the given equation into Bessel's equation. Express at least one nontrivial solution in terms of an appropriate Bessel function.

(a) $t^2y'' + ty' + (36t^4 - 1)y = 0$ (Let $t = \sqrt{s}/\sqrt{3}$.)

(b) $4t^2y'' + 4ty' + (t - \frac{4}{9})y = 0$ (Let $t = s^2$.)

(c) $ty'' - 3y' + ty = 0$ [Let $y(t) = t^2v(t)$.]

(d) $y'' + ty = 0$ [Let $y(t) = t^{1/2}v(2t^{3/2}/3)$.]

(e) $ty'' - y' + 4t^3y = 0$ [Let $y(t) = tv(t^2)$.]

4.5-6. Using formulas (16) and (17) compute the Legendre polynomials P_5 and P_6, and compare the answer with Rodrigues' formula (19).

4.5-7. Show that the change of independent variable $s = t^2$ transforms Legendre's equation (12) into a hypergeometric equation [show, in fact, that one can select real numbers α and β such that $\alpha + \beta = \frac{1}{2}$ and $\alpha\beta = -\rho(\rho + 1)/4$].

4.5-8. Suppose that the parameter γ in the hypergeometric equation (20) is not an integer [and hence the indicial roots 0 and $1 - \gamma$ to (20) do not differ by an integer].

(a) Show that if $\sigma = 1 - \gamma$ in the recursion formula (21), then

$$a_n = \frac{(\alpha - \gamma + 1) \cdots (\alpha - \gamma + n)}{n!} \frac{(\beta - \gamma + 1) \cdots (\beta - \gamma + n)}{(2 - \gamma) \cdots (n + 1 - \gamma)} a_0$$

for $n = 1, 2, \ldots$.

(b) Show that the a_n's computed in part *a* can be obtained from the a_n's computed with $\sigma = 0$ [see formula (22)] with α, β, γ replaced by $\alpha - \gamma + 1$, $\beta - \gamma + 1$, $2 - \gamma$, respectively.

(c) Deduce from part *b* that whenever γ is not an integer,

$$y(t) = c_1F(\alpha, \beta, \gamma; t) + c_2t^{1-\gamma}F(\alpha - \gamma + 1, \beta - \gamma + 1, 2 - \gamma; t)$$

($|t| < 1$ and c_1 and c_2 constants) is the general solution to the hypergeometric equation (20).

4.5-9. Deduce the following formulas involving the hypergeometric series F [see (22)].

(a) $(1 - t)^{-\alpha} = F(\alpha, \beta, \beta; t)$ for $|t| < 1$

(b) $\ln(1 + t) = tF(1, 1, 2; t)$ for $|t| < 1$

(c) $\ln\left(\frac{1 + t}{1 - t}\right) = 2tF(\frac{1}{2}, 1, \frac{3}{2}; t^2)$ for $|t| < 1$

(d) $\frac{1}{1 + t} + \frac{1}{1 - t} = 2F(\frac{1}{2}, 1, \frac{1}{2}; t^2)$ for $|t| < 1$

(e) $\arctan t = tF(1, \frac{1}{2}, \frac{3}{2}; -t^2)$ for $|t| < 1$

4.5-10. Derive the formula

$$\frac{d}{dt} F(\alpha, \beta, \gamma; t) = \frac{\alpha\beta}{\gamma} F(\alpha + 1, \beta + 1, \gamma + 1; t)$$

for the hypergeometric series F.

4.5-11. Express at least one nontrivial solution to each of the following equations in terms of the hypergeometric series F.

(a) $8(t - t^2)y'' + (4 - 9t)y' - y = 0$

(b) $4(t - t^2)y'' + (6 - 12t)y' - y = 0$

(c) $(t - t^2)y'' + (2 - 4t)y' - 2y = 0$

Remarks Proofs of the convergence of power series solutions are given in Birkhoff and Rota [1]. The method of Frobenius is proved in Coddington and Levinson [4]. Additional properties of Bessel functions and Legendre polynomials can be found in Hildebrand [6] and Wylie and Barrett [15].

5 First Order Systems in the Plane

The aim of this chapter is to introduce some of the fundamental ideas involved in the study of first order differential equations in the plane. The two main topics are linear systems (Sections 5.2 and 5.3) and nonlinear systems (Sections 5.4 and 5.5). General properties of linear systems are discussed in Section 5.2, and methods for explicitly solving linear equations with constant coefficients are indicated in Section 5.3. The main emphasis for nonlinear systems is using elementary techniques to study the behavior of solutions when one cannot determine a solution explicitly. Section 5.4 indicates basic procedures for sketching phase plots and then explains how to use plots in analyzing the solutions. These methods are developed further in Section 5.5 in the study of periodic solutions. Also included in this chapter are a detailed analysis of phase plots for linear systems in Section 5.6 and an indication of linearization techniques for the study of solutions locally about a critical point in Section 5.7. Section 5.8 contains an analysis of the Volterra-Lotka equation modeling predator-prey–type interactions of populations, and Section 5.9 gives a detailed discussion of the periodic motions in undamped nonlinear vibrations.

5.1 INTRODUCTORY CONCEPTS

The basic type of equations studied in this chapter is a system of two differential equations that involves two unknown functions and their first derivatives. The unknown functions are denoted by x and y and the independent variable is denoted by t. In general the type of equations under consideration is of the form

$$\begin{aligned} x' &= F(t, x, y) \\ y' &= G(t, x, y) \end{aligned} \tag{1}$$

where F and G are continuous functions of three variables. A *pair* x, y of real-valued, differentiable functions on an interval I is said to be a *solution to* (1) *on* I if $(t, x(t), y(t))$ is in the domain of both F and G and if

$$x'(t) = F(t, x(t), y(t)) \qquad \text{and} \qquad y'(t) = G(t, x(t), y(t))$$

for all t in I. It is important to note that such a system is "coupled" in the functions x and y but not in the derivatives x' and y'. Therefore, the only derivative in the first equation is x' and the only derivative in the second equation is y'.

Equations of the form (1) occur frequently in engineering and physical sciences as well as in biological modeling. Suppose, for example, that $x(t)$ and $y(t)$ denote quantities at time t of two elements or species that *interact*. The instantaneous rate of change of x at time t might be dependent on the amount of y present as well as the amount of x. Similarly, the rate of change of y could depend on both x and y. The modeling of this type of interaction between elements or species leads naturally to a system of differential equations having the form (1). Specific applications of such systems are given throughout this chapter.

Associated with the differential equation (1) is a corresponding initial value problem. Suppose that t_0 is in an interval I and x_0, y_0 are numbers such that (t_0, x_0, y_0) is in the domains of F and G. A pair of functions x, y on I is said to be a solution to the *initial value problem*

$$\begin{aligned} x' &= F(t, x, y) \qquad x(t_0) = x_0 \\ y' &= G(t, x, y) \qquad y(t_0) = y_0 \end{aligned} \tag{2}$$

if x, y is a solution to the differential equation (1) such that $x(t_0) = x_0$ and $y(t_0) = y_0$. The pair of differential equations in (1) is called a *first order, two-dimensional differential system*, and the pair of initial value problems (2) is called a *first order, two-dimensional initial value problem.*

▶ **EXAMPLE 5.1-1**

Consider the initial value system

$$\begin{aligned} x' &= x + y \qquad x(0) = 3 \\ y' &= 3x - y \qquad y(0) = -5 \end{aligned}$$

If c_1 and c_2 are constants and $x(t) = c_1e^{2t} + c_2e^{-2t}$ and $y(t) = c_1e^{2t} - 3c_2e^{-2t}$, then x, y is a solution to the corresponding differential equation on $(-\infty, \infty)$. To see that this is so note that

$$\begin{aligned} x'(t) &= 2c_1e^{2t} - 2c_2e^{-2t} = x(t) + y(t) \qquad \text{and} \\ y'(t) &= 2c_1e^{2t} + 6c_2e^{-2t} = 3x(t) - y(t) \end{aligned}$$

In order to determine a solution that satisfies the given initial conditions, we solve the algebraic system

$$x(0) = c_1 + c_2 = 3$$
$$y(0) = c_1 - 3c_2 = -5$$

to determine that $c_1 = 1$ and $c_2 = 2$. Thus, $x(t) = e^{2t} + 2e^{-2t}$, $y(t) = e^{2t} - 6e^{-2t}$ is a solution to the given initial value problem.

▶ **EXAMPLE 5.1-2**

Consider the initial value system

$$x' = -x^3 \qquad x(0) = 1$$
$$y' = xy \qquad y(0) = 4$$

Since the first equation only involves x, we use the methods for solving first order equations with one unknown function by first solving for x in the first equation and then substituting for x in the second equation in order to solve for y. Since the variables separate in the first equation, we have that $dx = -x^3\,dt$ or $x^{-3}\,dx = -dt$. Antidifferentiating, $-x^{-2}/2 = -t + c$ or $x^2 = (2t - 2c)^{-1}$. Since $x(0) = 1$, it follows that $c = -\frac{1}{2}$ and hence

$$x(t) = (2t + 1)^{-1/2}$$

Substituting this expression for x in the second equation we see that

$$y' = (2t + 1)^{-1/2}y \qquad y(0) = 4$$

This is a first order, homogeneous linear equation (see Section 1.2), and the general solution to the differential equation is

$$y(t) = ce^{\int (2t+1)^{-1/2}\,dt} = ce^{(2t+1)^{1/2}}$$

The initial condition $y(0) = ce = 4$ implies that $c = 4e^{-1}$, and hence the pair $x(t) \equiv (2t + 1)^{-1/2}$, $y(t) \equiv 4e^{(2t+1)^{1/2} - 1}$ is a solution to the given initial value problem.

A very important class of differential equations that can be written in the form (1) or (2) is the class of *second order equations* (see Chapter 2). Suppose that H is a continuous function of three variables, and consider the second order differential equation

$$z'' = H(t, z, z') \tag{3}$$

and the second order initial value problem

$$z'' = H(t, z, z') \qquad z(t_0) = z_0 \qquad z'(t) = v_0 \tag{4}$$

The following observation is important and can be easily established: *A function z on an interval I is a solution to (3) if and only if $z = x$ and $z' = y$ on I, where x and y is a solution to (1) on I, with $F(t, x, y) \equiv y$ and* $G(t, x, y) \equiv H(t, x, y)$. Therefore, equation (3) is equivalent to the two-dimensional system

$$\begin{aligned} x' &= y \\ y' &= H(t, x, y) \end{aligned} \tag{3'}$$

For assume that x, y is a solution to (3′) and set $z = x$. Then $z' = x' = y$ by the first equation in (3′), and the second equation implies

$$z'' = (z')' = y' = H(t, x, y) = H(t, x, z')$$

Thus z is a solution to (3). It is also easy to see that if z is a solution to (3), then $x = z$, $y = z'$ is a solution to (3′).

EXAMPLE 5.1-3

Consider the system

$$\begin{aligned} x' &= y \qquad & x(0) &= 3 \\ y' &= 3x - 2y \qquad & y(0) &= 1 \end{aligned} \tag{5}$$

and note that it is of the form (3′), with $H(t, x, y) = 3x - 2y$. By the equivalence of (3) and (3′), if $z = x$ then $z' = y$ and

$$z'' + 2z' - 3z = 0$$

Since $\lambda = -1, 3$ are the auxiliary roots for this equation, $z = c_1 e^{-t} + c_2 e^{3t}$ is the general solution, and

$$x = c_1 e^{-t} + c_2 e^{3t} \qquad y = -c_1 e^{-t} + 3c_2 e^{3t}$$

is a family of solutions to the differential equation in (5). From the initial data, $x(0) = c_1 + c_2 = 3$ and $y(0) = -c_1 + 3c_2 = 1$. Thus $c_1 = 2$, $c_2 = 1$, and $x = 2e^{-t} + e^{3t}$, $y = -2e^{-t} + 3e^{3t}$ is the solution to (5).

PROBLEMS

5.1-1. Determine a solution to each of the following initial value systems by solving for x in the first equation and then substituting into the second equation in order to determine y.

(a) $x' = 2x \qquad x(0) = 3$

$y' = x - y \qquad y(0) = 0$

(b) $x' = -x^2 \qquad x(0) = 1$

$y' = x + 2 \qquad y(0) = 3$

(c) $x' = \dfrac{1}{t}x + t \qquad x(1) = 2$

$y' = y + x + 1 \qquad y(1) = 0$

(d) $x' = -x^3 \qquad x(0) = 1$

$y' = -(1 + x^2)y^2 \qquad y(0) = 3$

5.1-2. Verify that the given family of pairs of functions x, y is a solution to the corresponding system, and then determine the constants c_1 and c_2 so that the given initial conditions are satisfied.

(a) $x' = y \qquad x(0) = 3 \qquad \begin{Bmatrix} x(t) = c_1 e^t + c_2 e^{2t} \\ y(t) = c_1 e^t + 2c_2 e^{2t} \end{Bmatrix}$
$\quad\; y' = -2x + 3y \qquad y(0) = 4$

(b) $x' = y \qquad x(0) = -3 \qquad \begin{Bmatrix} x(t) = c_1 e^t + c_2 te^t \\ y(t) = c_1 e^t + c_2(1 + t)e^t \end{Bmatrix}$
$\quad\; y' = -x + 2y \qquad y(0) = 1$

(c) $x' = 2y \qquad x\left(\frac{\pi}{6}\right) = 1 \qquad \begin{Bmatrix} x(t) = c_1 \cos 2t + c_2 \sin 2t \\ y(t) = -c_1 \sin 2t + c_2 \cos 2t \end{Bmatrix}$

$\quad\; y' = -2x \qquad y\left(\frac{\pi}{6}\right) = 2$

(d) $x' = 3y \qquad x(0) = 0 \qquad \begin{Bmatrix} x(t) = c_1 e^{3t} + c_2 e^{-3t} \\ y(t) = c_1 e^{3t} - c_2 e^{-3t} \end{Bmatrix}$
$\quad\; y' = 3x \qquad y(0) = 2$

5.1.3. Solve the following linear initial value problems by using (3) and (3′) to write them as a second order equation.

(a) $x' = y \qquad x(0) = 1$
$\quad\; y' = 4x \qquad y(0) = 0$

(b) $x' = y \qquad x(0) = -3$
$\quad\; y' = -9x \qquad y(0) = 2$

(c) $x' = y \qquad x(0) = 0$
$\quad\; y' = 3x + 2y \qquad y(0) = 2$

(d) $x' = y \qquad x(0) = 1$
$\quad\; y' = -5x - 2y \qquad y(0) = 0$

5.2 ALGEBRAIC PROPERTIES OF LINEAR SYSTEMS

In this section some of the properties of solutions to systems of two first order linear differential equations are discussed. Suppose that I is an interval and that a_1, a_2, b_1, and b_2 are continuous real-valued functions on I. We consider the system

$$\begin{aligned} x' &= a_1(t)x + b_1(t)y \\ y' &= a_2(t)x + b_2(t)y \end{aligned} \tag{1}$$

as well as the system

$$\begin{aligned} x' &= a_1(t)x + b_1(t)y + f_1(t) \\ y' &= a_2(t)x + b_2(t)y + f_2(t) \end{aligned} \tag{2}$$

where f_1 and f_2 are continuous functions on I. Of course, equation (1) is a special case of equation (2), with $f_1(t) \equiv f_2(t) \equiv 0$. Both of these systems are called *first order, two-dimensional linear systems.* The functions f_1 and f_2 in (2) are called *nonhomogeneous* (or *inhomogeneous*) terms and the pair of equations in (2) is called a *nonhomogeneous system.* The pair of equations (1) is called a *homogeneous system.*

This terminology for (1) and (2) corresponds with the previous use of these terms in Chapter 2. For consider the second order linear equation

$$a(t)z'' + b(t)z' + c(t)z = f(t) \tag{3}$$

where a, b, c, and f are continuous functions on I, with $a(t) \neq 0$ for all $t \in I$. Setting $x = z$ and $y = z'$, it follows that equation (3) is equivalent to the system

$$\begin{aligned} x' &= y \\ y' &= -\left[\frac{c(t)}{a(t)}\right]x - \left[\frac{b(t)}{a(t)}\right]y + \frac{f(t)}{a(t)} \end{aligned} \tag{3'}$$

[see equations (3) and (3′) in Section 5.1]. Note that the system (3′) is homogeneous (or nonhomogeneous) in the sense of the preceding paragraph if and only if the second order equation (3) is homogeneous (or nonhomogeneous) in the sense described in Section 2.2. Many of the properties of second order linear equations have natural extensions to two-dimensional linear systems discussed in Section 2.2, and the purpose of this section is to indicate some of these basic properties.

Suppose that x_1, y_1 is a solution to the nonhomogeneous system (2) and x_2, y_2 is a solution to the homogeneous system (1). If $x = x_1 + x_2$ and $y = y_1 + y_2$, then

$$\begin{aligned} x' &= [a_1(t)x_1 + b_1(t)y_1 + f_1(t)] + [a_1(t)x_2 + b_1(t)y_2] \\ &= a_1(t)(x_1 + x_2) + b_1(t)(y_1 + y_2) + f_1(t) \\ &= a_1(t)x + b_1(t)y + f_1(t) \end{aligned}$$

and similarly

$$y' = a_2(t)x + b_2(t)y + f_2(t)$$

and the following assertion is seen to be true:

(4) $\left\{\right.$ if x_1, y_1 is a solution to (2) and x_2, y_2 is a solution to (1), then $x = x_1 + x_2$, $y = y_1 + y_2$ is a solution to (2).

Taking $f_1(t) \equiv f_2(t) \equiv 0$ it follows from (4) that if x_1, y_1 and x_2, y_2 are both solutions to the homogeneous equation (1), then so is $x_1 + x_2$, $y_1 + y_2$. It is also easy to check that if x_1, y_1 is a solution to the homogeneous system (1) and c_1 is a constant, then c_1x_1, c_1y_1 is also a solution to (1). Therefore, we also have the following result:

(5) $\left\{\right.$ If the pairs x_1, y_1 and x_2, y_2 are solutions to the homogeneous system (1) and c_1 and c_2 are constants, then the pair

$$x = c_1x_1 + c_2x_2 \qquad y = c_1y_1 + c_2y_2$$

is also a solution to (1).

As in the second order case we consider the possibility of "generating" all the solutions to the homogeneous equation (1) by determining as few solutions to (1) as necessary. The set of all solutions to (1) or (2) is called the *general solution*, and our concern is to represent the general solution to (1) and (2) in as simple a manner as possible. In order to develop such a representation we need the concepts of linear independence and dependence for "pairs" of functions. For clarity we sometimes insert parentheses around a pair of functions. Thus the pair x, y is sometimes written (x, y).

Two pairs x_1, y_1 and x_2, y_2 on an interval I are said to be *linearly independent on I* if whenever c_1 and c_2 are constants such that both

$$\begin{aligned} c_1 x_1(t) + c_2 x_2(t) &\equiv 0 \\ c_1 y_1(t) + c_2 y_2(t) &\equiv 0 \end{aligned} \qquad \text{for all } t \in I \tag{6}$$

then c_1 and c_2 must be zero. If x_1, y_1 and x_2, y_2 are not linearly independent [i.e., if there exist c_1 and c_2, not both zero, such that (6) holds], these are said to be *linearly dependent on I*. These concepts are completely analogous to linear independence and dependence of functions discussed in Section 2.2. It is important to realize that in order for the two *pairs* of functions x_1, y_1 and x_2, y_2 to be linearly independent or dependent on I, both equations in (6) must be considered *simultaneously* for the same constants c_1 and c_2. Note, for example, if *either* the functions x_1 and x_2 or the functions y_1 and y_2 are linearly independent on I, then the pairs x_1, y_1 and x_2, y_2 are also linearly independent on I. However, the converse implication is *not* true. For example, the functions $x_1(t) \equiv e^t$ and $x_2(t) \equiv 0$ as well as $y_1(t) \equiv 0$ and $y_2(t) \equiv e^t$ are linearly dependent on any interval, yet the pairs $e^t, 0$ and $0, e^t$ are linearly independent on any interval (verify this).

▶ EXAMPLE 5.2-1

Consider the pairs of functions $(x_1, y_1) = (2 \sin t, -\sin t)$, $(x_2, y_2) = (\sin t, \sin t)$ on $(-\infty, \infty)$. In order to check the independence or dependence of these pairs we must consider the system

$$c_1 x_1 + c_2 x_2 = 2c_1 \sin t + c_2 \sin t \equiv 0$$

$$c_1 y_1 + c_2 y_2 = -c_1 \sin t + c_2 \sin t \equiv 0$$

Subtracting the second equation from the first, we obtain $3c_1 \sin t \equiv 0$, and hence $c_1 = 0$. Substituting into the first equation it follows that $c_2 = 0$ as well, so the pairs x_1, y_1 and x_2, y_2 are linearly independent.

Since equation (6) involves solving two linear equations simultaneously, a simple criterion for checking linear independence can be obtained using determinants: if $t_0 \in I$ and if

(7) $$\det\begin{pmatrix} x_1(t_0) & x_2(t_0) \\ y_1(t_0) & y_2(t_0) \end{pmatrix} = x_1(t_0)y_2(t_0) - y_1(t_0)x_2(t_0) \neq 0$$

then x_1, y_1 and x_2, y_2 are linearly independent on I. This criterion is immediate since (7) implies that the system

$$c_1 x_1(t_0) + c_2 x_2(t_0) = 0$$
$$c_1 y_1(t_0) + c_2 y_2(t_0) = 0$$

has only the trivial solution $c_1 = c_2 = 0$. As in the second order case, this determinant plays a crucial role, and the determinant of two pairs of functions x_1, y_1 and x_2, y_2 indicated by (7) is called the *wronskian* and is denoted $W[(x_1, y_1), (x_2, y_2); t]$:

(8) $$W[(x_1, y_1), (x_2, y_2); t] \equiv \det\begin{pmatrix} x_1(t) & x_2(t) \\ y_1(t) & y_2(t) \end{pmatrix}$$

EXAMPLE 5.2-2

Consider the pairs $-e^t, e^{-t}$ and $e^{2t}, 3e^{2t}$. Since

$$W[(-e^t, e^{-t}), (e^{2t}, 3e^{2t}); t] = \det\begin{pmatrix} -e^t & e^{2t} \\ e^{-t} & 3e^{2t} \end{pmatrix} = -3e^{3t} - e^t \neq 0$$

for all t, we have that $-e^t, e^{-t}$ and $e^{2t}, 3e^{2t}$ are linearly independent on any interval I.

The definition of the wronskian for two pairs of functions has a direct correspondence to the definition of the wronskian for two real-valued functions given in Section 2.2. Recall that if z_1 and z_2 are continuously differentiable functions on I, then

$$W[z_1, z_2; t] \equiv \det\begin{pmatrix} z_1(t) & z_2(t) \\ z_1'(t) & z_2'(t) \end{pmatrix} = z_1(t)z_2'(t) - z_1'(t)z_2(t)$$

Comparing this formula with formula (8), it follows that the wronskian for the pairs z_1, z_1' and z_2, z_2' is precisely the wronskian for z_1 and z_2:

$$W[z_1, z_2; t] \equiv W[(z_1, z_1'), (z_2, z_2'); t]$$

Recall that the relationship between the solutions to the second order equation (3) and the first order system (3′) is exactly this [that is, z is a solution to (3) only in case $x = z$, $y = z'$ is a solution to (3′)]. In particular, two solutions z_1 and z_2 to (3) are linearly independent if and only if z_1, z_1' and z_2, z_2' are linear independent solutions to (3′).

EXAMPLE 5.2-3

Consider the second order homogeneous equation

$$z'' + 2z' - 3z = 0 \tag{9}$$

Setting $x = z$ and $y = z'$ we obtain the equivalent system

$$\begin{aligned} x' &= y \\ y' &= 3x - 2y \end{aligned} \tag{9'}$$

The auxiliary equation for (9) is $\lambda^2 + 2\lambda - 3 = (\lambda - 1)(\lambda + 3) = 0$, so $\lambda = 1$, -3 are the auxiliary roots, and it follows that

$$z = c_1 e^t + c_2 e^{-3t} \tag{10}$$

is the general solution of (9). Since

$$\frac{d}{dt}(e^t) = e^t \qquad \text{and} \qquad \frac{d}{dt}(e^{-3t}) = -3e^{-3t}$$

we have that the pairs e^t, e^t and e^{-3t}, $-3e^{-3t}$ are linearly independent solutions to the system (9′). If z is a solution to (9) it follows from (10) that there are constants c_1 and c_2 such that

$$\begin{aligned} z &= c_1 e^t + c_2 e^{-3t} \qquad \text{and} \\ z' &= c_1 e^t - 3c_2 e^{-3t} \end{aligned} \tag{10'}$$

Since every solution of (9′) must have the form (10′), we see that every solution to (9′) is a "linear combination" of the two pairs e^t, e^t and e^{-3t}, $-3e^{-3t}$:

$$x = c_1 e^t + c_2 e^{-3t} \qquad y = c_1 e^t + c_2(-3e^{-3t})$$

Note in particular that the general solution to (9′) is generated by two linearly independent solutions.

If the pairs x_1, y_1 and x_2, y_2 are solutions to (1) on I and if

$$W(t) \equiv W[(x_1, y_1), (x_2, y_2); t] = x_1(t)y_2(t) - x_2(t)y_1(t)$$

then, suppressing the variable t, using equation (1), and simplifying,

$$\begin{aligned} W' &= x_1'y_2 + x_1y_2' - x_2'y_1 - x_2y_1' \\ &= a_1(x_1y_2 - x_2y_1) + b_2(x_1y_2 - x_2y_1) \\ &= (a_1 + b_2)W \end{aligned}$$

This equation is first order linear and

$$\exp\left(-\int_0^t [a_1(s) + b_2(s)]\, ds\right)$$

is an integrating factor. Therefore, it follows that

$$W[x_1, y_1), (x_2, y_2); t] \equiv W[(x_1, y_1), (x_2, y_2); 0]e^{\int_0^t (a_1 + b_2)\, dt} \tag{11}$$

for all t, and hence the wronskian of solutions to (1) is either never zero or identically zero.

The formula (11) should be compared with Abel's identity (Problem 2.2-3). The fundamental connection between linear independence and the general solution to (1) is included in the following theorem (compare with Theorem 2.2-2).

■ **Theorem 5.2-1**

Suppose that the pairs x_1, y_1 and x_2, y_2 are solutions to the homogeneous system (1). Then any one of the following statements implies the other two.

(i) The general solution of (1) is the family

$$x = c_1x_1 + c_2x_2 \qquad y = c_1y_1 + c_2y_2$$

where c_1 and c_2 are arbitrary constants.

(ii) The pairs x_1, y_1 and x_2, y_2 are linearly independent.

(iii) The wronskian $W[(x_1, y_1), (x_2, y_2); t] \neq 0$ for some t (and hence for all t).

The proof of this important theorem depends on the existence and uniqueness to solutions of the initial value problem associated with (1) and is beyond the scope of this text. There is also an analogous result for solutions to the nonhomogeneous equation (2), which is also stated without proof.

■ **THEOREM 5.2-2**

Suppose that x_1, y_1 and x_2, y_2 are linearly independent solutions to the homogeneous equation (1) and that x_p, y_p is a particular solution to the nonhomogeneous equation (2). Then the general solution to (2) is the family

$$x = c_1x_1 + c_2x_2 + x_p \qquad y = c_1y_1 + c_2y_2 + y_p$$

where c_1 and c_2 are arbitrary constants.

As an illustration of these two results consider the system

$$\begin{aligned} x' &= x + y \\ y' &= 3x - y \end{aligned} \tag{12}$$

It is easy to check that e^{2t}, e^{2t} and e^{-2t}, $-3e^{-2t}$ are each solutions to (12) (take $c_1 = 1$, $c_2 = 0$ and then $c_1 = 0$, $c_2 = 1$ in Example 5.1-1). Since

$$\det\begin{pmatrix} e^{2t} & e^{-2t} \\ e^{2t} & -3e^{-2t} \end{pmatrix} = -3 - 1 = -4 \neq 0$$

these two solutions are linearly independent, and it follows from Theorem 5.2-1 that the family of pairs

$$x = c_1 e^{2t} + c_2 e^{-2t} \qquad y = c_1 e^{2t} - 3c_2 e^{-2t}$$

where c_1 and c_2 are constants, is the general solution to (12). Now consider a nonhomogeneous system corresponding to (12):

$$\begin{aligned} x' &= x + y - 2t \\ y' &= 3x - y + 2 - 6t \end{aligned} \tag{13}$$

If $x_p = 2t$, $y_p = 2$, then

$$x_p + y_p - 2t = 2 = x_p' \qquad \text{and} \qquad 3x_p - y_p + 2 - 6t = 0 = y_p'$$

Hence x_p, y_p is a solution to (13), and by Theorem 5.2-2,

$$x = c_1 e^{2t} + c_2 e^{-2t} + 2t \qquad y = c_1 e^{2t} - 3c_2 e^{-2t} + 2$$

is the general solution to (13).

PROBLEMS

5.2-1. Determine the linear independence or linear dependence of each of the following pairs of functions.

(a) $(e^{4t}, -e^{4t})$ $(-e^{4t}, 2e^{4t})$

(b) $(\sin t, \cos t)$ $(\cos t, \sin t)$

(c) $(-e^{2t}, e^{-2t})$ $(2e^{2t}, -2e^{-2t})$

(d) $(e^t \sin t, 1)$ $(e^t \cos t, 0)$

5.2-2. Write each of the following second order equations as a two-dimensional first order system, and determine the general solution to the resulting system.

(a) $z'' + 4z = 3$

(b) $z'' + 2z' + 2z = e^{2t}$

(c) $z'' - 4z' + 4z = 0$

(d) $z'' - 2z = t + 3$

5.3 LINEAR SYSTEMS WITH CONSTANT COEFFICIENTS

The general solution to two-dimensional, homogeneous, linear differential equations with constant coefficients can be explicitly determined by elementary means, and the main object of this section is to indicate a method for computing the solutions. Suppose that a_1, a_2, b_1, and b_2 are *constants*, and consider the homogeneous linear system

$$\begin{aligned} x' &= a_1 x + b_1 y \\ y' &= a_2 x + b_2 y \end{aligned} \tag{1}$$

The procedure for solving (1) is similar to the method developed for solving second order homogeneous equations with constant coefficients in Section 2.3. In fact, as was indicated in the preceding section, such second order equations can be written in the form of the system (1).

In order to illustrate the basic method consider the system

$$\begin{aligned} x' &= 4x - 2y \\ y' &= 3x - y \end{aligned} \tag{2}$$

As was done for homogeneous second order equations in Section 2.3, it is *assumed* that a solution to (2) is an exponential function: suppose that λ is a number, that α and β are constants, and that

$$x(t) = \alpha e^{\lambda t} \qquad y(t) = \beta e^{\lambda t} \tag{3}$$

is the solution to (2). Since $x'(t) = \alpha\lambda e^{\lambda t}$ and $y'(t) = \beta\lambda e^{\lambda t}$, we have, by substituting (3) into (2),

$$\begin{aligned} \alpha\lambda e^{\lambda t} &= 4\alpha e^{\lambda t} - 2\beta e^{\lambda t} \\ \beta\lambda e^{\lambda t} &= 3\alpha e^{\lambda t} - \beta e^{\lambda t} \end{aligned}$$

Canceling $e^{\lambda t}$ from each side of both equations, we see that x, y in (3) is a solution to (2) only in case

$$\begin{aligned} (4 - \lambda)\alpha - 2\beta &= 0 \\ 3\alpha + (-1 - \lambda)\beta &= 0 \end{aligned} \tag{4}$$

This algebraic equation has a *nontrivial* solution α, β if and only if the determinant of its coefficients is zero:

$$\begin{aligned} \det \begin{pmatrix} 4 - \lambda & -2 \\ 3 & -1 - \lambda \end{pmatrix} &= (4 - \lambda)(-1 - \lambda) - 3(-2) \\ &= \lambda^2 - 3\lambda + 2 \\ &= (\lambda - 1)(\lambda - 2) \\ &= 0 \end{aligned}$$

Therefore, the number λ in (3) must be either 1 or 2, and the coefficients α and β must satisfy (4) for the given value of λ. Taking $\lambda = 1$ we have from (4) that

$$\begin{aligned} 3\alpha - 2\beta &= 0 \\ 3\alpha - 2\beta &= 0 \end{aligned}$$

and hence $\alpha = 2$, $\beta = 3$ is a nontrivial solution, and

$$x(t) = 2e^t \qquad y(t) = 3e^t$$

is a corresponding solution to (2). Taking $\lambda = 2$ we have from (4) that

$$2\alpha - 2\beta = 0$$
$$3\alpha - 3\beta = 0$$

and hence $\alpha = 1$, $\beta = 1$ is a nontrivial solution, and

$$x(t) = e^{2t} \qquad y(t) = e^{2t}$$

is a corresponding solution to (2). Since these two solutions to (2) are linearly independent (verify this), it follows that

$$x = 2c_1 e^t + c_2 e^{2t} \qquad y = 3c_1 e^t + c_2 e^{2t}$$

is the general solution to (2). [The student should verify directly that these functions are solutions to (2).]

The procedure indicated in solving the specific system (2) gives a method for determining the general solution to (1) as well. Assuming that $x(t) = \alpha e^{\lambda t}$, $y(t) = \beta e^{\lambda t}$ is a solution to (1) implies by substitution directly into (1) that

$$\lambda\alpha e^{\lambda t} = a_1 \alpha e^{\lambda t} + b_1 \beta e^{\lambda t}$$
$$\lambda\beta e^{\lambda t} = a_2 \alpha e^{\lambda t} + b_2 \beta e^{\lambda t}$$

Canceling $e^{\lambda t}$ from each equation, the constants α and β must satisfy

$$(5) \qquad \begin{aligned} (a_1 - \lambda)\alpha + b_1 \beta &= 0 \\ a_2 \alpha + (b_2 - \lambda)\beta &= 0 \end{aligned}$$

Since this algebraic system has nontrivial solutions α, β only in case the determinant of the coefficients is zero, we see that the equation

$$\det \begin{pmatrix} a_1 - \lambda & b_1 \\ a_2 & b_2 - \lambda \end{pmatrix} = (a_1 - \lambda)(b_2 - \lambda) - a_2 b_1 = 0$$

should be satisfied. If λ is chosen so that this determinant is zero, then the nontrivial α and β that satisfy (5) generate a nontrivial solution to (1). The quadratic polynomial

$$(6) \qquad p(\lambda) = \det \begin{pmatrix} a_1 - \lambda & b_1 \\ a_2 & b_2 - \lambda \end{pmatrix} = \lambda^2 - (a_1 + b_2)\lambda + (a_1 b_2 - a_2 b_1)$$

is called the *characteristic polynomial* of (1), and the zeros of this polynomial are called the *characteristic roots*. As in the second order case, whether the characteristic roots are real and distinct, double, or conjugate complex determines the form of the solutions to (1) (see Section 2.3).

Consider the second order equation with constant coefficients that has the characteristic polynomial $p(\lambda)$ as its auxiliary polynomial:

$$(7) \qquad z'' - (a_1 + b_2)z' + (a_1 b_2 - a_2 b_1)z = 0$$

By the methods of Section 2.3 the general solution to this equation can be obtained directly from the roots of the auxiliary polynomial. Now suppose that the pair x, y is any solution to (1). If $b_1 \neq 0$ we have $y = b_1^{-1}(x' - a_1 x)$ from the first equation in (1), and hence $y' = b_1^{-1}(x'' - a_1 x')$. Substituting for y and y' in the second equation of (1) shows that

$$b_1^{-1}(x'' - a_1 x') = a_2 x + b_2 b_1^{-1}(x' - a_1 x)$$

Simplifying shows that $z = x$ must also be a solution to (7). If $b_1 = 0$ then $x' = a_1 x$, and so $x = c_1 e^{a_1 t}$, and one can again show that $z = x$ is a solution to (7) since $\lambda = a_1$ is an auxiliary root of (7) when $b_1 = 0$. In a similar manner one can show that $z = y$ must also be a solution to (7) whenever x, y is a solution to (1) (consider the cases $a_2 \neq 0$ and $a_2 = 0$). Therefore, we have the following fundamental result: *If the pair x, y is a solution to (1), then necessarily $z = x$ and $z = y$ are solutions to (7).* So in determining the solutions to (1) we need only consider pairs x, y where both x and y are solutions to (7). Once the form of x and y is determined from equation (7), we must still compute a relationship between x and y by substituting directly into (1). The following procedure may be easily applied:

(8)

(a) Determine two linearly independent solutions z_1 and z_2 to the second order equation (7) [z_1 and z_2 can be obtained *directly* from the characteristic roots of (1)].

(b) Assume that $x = c_1 z_1 + c_2 z_2$ and $y = d_1 z_1 + d_2 z_2$ are solutions to (1), and substitute into (1) in order to obtain the relationships among c_1, c_2, d_1, and d_2.

Once the relationships among the four constants in part b are obtained, there should be *exactly two arbitrary constants* remaining. This procedure is illustrated with the following two examples.

EXAMPLE 5.3-1

Consider the initial value system

$$\begin{aligned} x' &= x + 2y \qquad x(0) = 1 \\ y' &= 6x - 3y \qquad y(0) = -1 \end{aligned} \tag{9}$$

The characteristic polynomial [see (6)] is

$$p(\lambda) = \det \begin{pmatrix} 1 - \lambda & 2 \\ 6 & -3 - \lambda \end{pmatrix} = (1 - \lambda)(-3 - \lambda) - 12 = \lambda^2 + 2\lambda - 15$$

and since $\lambda^2 + 2\lambda - 15 = (\lambda - 3)(\lambda + 5)$, we see that $\lambda = 3, -5$ are the characteristic roots. Therefore, the general solution to (9) has the form

$$x = c_1 e^{3t} + c_2 e^{-5t} \qquad y = d_1 e^{3t} + d_2 e^{-5t}$$

In this case we can solve for y in the first equation in (9) to obtain

$$y = \tfrac{1}{2}(x' - x) = \tfrac{1}{2}(3c_1 e^{3t} - 5c_2 e^{-5t} - c_1 e^{3t} - c_2 e^{-5t})$$
$$= c_1 e^{3t} - 3c_2 e^{-5t}$$

Therefore, $d_1 = c_1$, $d_2 = -3c_2$, and

$$x = c_1 e^{3t} + c_2 e^{-5t} \qquad y = c_1 e^{3t} - 3c_2 e^{-5t}$$

is the general solution to the differential equation in (9). In order to satisfy the initial conditions, the constants c_1 and c_2 should be selected so that

$$x(0) = c_1 + c_2 = 1 \qquad \text{and} \qquad y(0) = c_1 - 3c_2 = -1$$

From this it follows that $c_1 = c_2 = \frac{1}{2}$, and hence

$$x(t) = \tfrac{1}{2}e^{3t} + \tfrac{1}{2}e^{-5t} \qquad y(t) = \tfrac{1}{2}e^{3t} - \tfrac{3}{2}e^{-5t}$$

is the solution to the initial value problem (9).

EXAMPLE 5.3-2

Consider the initial value system

$$\begin{aligned} x' &= -x + 3y \qquad & x(0) &= 2 \\ y' &= -3x - y \qquad & y(0) &= -3 \end{aligned} \tag{10}$$

The characteristic polynomial is

$$\det \begin{pmatrix} -1-\lambda & 3 \\ -3 & -1-\lambda \end{pmatrix} = (-1-\lambda)^2 + 9 = \lambda^2 + 2\lambda + 10$$

and by the quadratic formula the characteristic roots are

$$\lambda = \frac{-2 \pm \sqrt{4 - 40}}{2} = -1 \pm 3i$$

Therefore [see (13) in Section 2.3] the solutions x, y to (10) have the form

$$x = c_1 e^{-t} \cos 3t + c_2 e^{-t} \sin 3t$$
$$y = d_1 e^{-t} \cos 3t + d_2 e^{-t} \sin 3t$$

Since $x' = (-c_1 + 3c_2)e^{-t} \cos 3t - (3c_1 + c_2)e^{-t} \sin 3t$, it follows from the first equation in (10) that

$$y = \tfrac{1}{3}(x' + x) = c_2 e^{-t} \cos 3t - c_1 e^{-t} \sin 3t$$

(that is, $d_1 = c_2$ and $d_2 = -c_1$). Selecting c_1 and c_2 so that $x(0) = c_1 = 2$ and $y(0) = c_2 = -3$, it follows that

$$x = 2e^{-t} \cos 3t - 3e^{-t} \sin 3t$$

$$y = -3e^{-t} \cos 3t - 2e^{-t} \sin 3t$$

is the solution to the given initial value problem (10).

If the nonhomogeneous terms are simple, the method of undetermined coefficients can be applied easily to nonhomogeneous systems with constant coefficients. The form of the particular solution is essentially the same as in the second order case (see Section 2.4), and we indicate this procedure with a specific example.

EXAMPLE 5.3-3

Consider the nonhomogeneous initial value system

$$\begin{aligned} x' &= -x + 4y + e^{3t} \qquad & x(0) &= 0 \\ y' &= -x + 3y - 1 \qquad & y(0) &= 0 \end{aligned} \tag{11}$$

The characteristic polynomial for the corresponding homogeneous system is

$$\det\begin{pmatrix} -1-\lambda & 4 \\ -1 & 3-\lambda \end{pmatrix} = (-1-\lambda)(3-\lambda) + 4 = \lambda^2 - 2\lambda + 1$$

and since $\lambda^2 - 2\lambda + 1 = (\lambda - 1)^2$, $\lambda = 1$ is a double characteristic root, and the solutions x_H, y_H to the corresponding homogeneous equation have the form

$$x_H = c_1 e^t + c_2 te^t \qquad y_H = d_1 e^t + d_2 te^t$$

Substituting back into the homogeneous equation, we obtain

$$\begin{aligned} y_H &= \tfrac{1}{4}(x_H + x_H') = \tfrac{1}{4}(c_1 e^t + c_2 te^t + c_1 e^t + c_2 e^t + c_2 te^t) \\ &= \left(\frac{2c_1 + c_2}{4}\right)e^t + \frac{c_2}{2} te^t \end{aligned}$$

and it follows that

$$x_H = c_1 e^t + c_2 te^t \qquad y_H = \left(\frac{2c_1 + c_2}{4}\right)e^t + \frac{c_2}{2} te^t \tag{12}$$

is the general solution to the corresponding homogeneous equation. According to the types of the nonhomogeneous terms, we look for a particular solution x_p, y_p to (11) of the form

$$x_p = d_1 + d_2 e^{3t} \qquad y_p = d_3 + d_4 e^{3t}$$

Even though e^{3t} appears only in the first equation of (11), we must assume that *both* x_p and y_p contain e^{3t} terms. Similarly, both x_p and y_p should contain constant terms. Substituting $x_p' = 3d_2 e^{2t}$, $y_p' = 3d_4 e^{3t}$ into (11), we find that

$$\begin{aligned} 3d_2 e^{3t} &= (-d_1 + 4d_3) + (-d_2 + 4d_4 + 1)e^{3t} \\ 3d_4 e^{3t} &= (-d_1 + 3d_3 - 1) + (-d_2 + 3d_4)e^{3t} \end{aligned}$$

Equating appropriate coefficients (the constant terms give two equations for d_1 and d_3, and the coefficients of e^{3t} give two equations in d_2 and d_4), we see that the systems

$$\begin{aligned} -d_1 + 4d_3 &= 0 \\ -d_1 + 3d_3 &= 1 \end{aligned} \qquad \text{and} \qquad \begin{aligned} -4d_2 + 4d_4 &= -1 \\ -d_2 &= 0 \end{aligned}$$

must be satisfied. The first system implies that $d_1 = -4$, $d_3 = -1$ and the second implies $d_2 = 0$, $d_4 = -\frac{1}{4}$. Therefore,

$$x_p = -4 \qquad y_p = -1 - \frac{e^{3t}}{4}$$

is a particular solution to (11) and

$$x = c_1 e^t + c_2 te^t - 4 \qquad y = \left(\frac{2c_1 + c_2}{4}\right)e^t + \frac{c_2}{2} te^t - 1 - \frac{e^{3t}}{4}$$

is the general solution. Selecting c_1 and c_2 so that

$$x(0) = c_1 - 4 = 0 \qquad \text{and} \qquad y(0) = \frac{2c_1 + c_2}{4} - 1 - \tfrac{1}{4} = 0$$

it follows that $c_1 = 4$ and $c_2 = -3$. Therefore,

$$x = 4e^t - 3te^t - 4 \qquad y = \frac{5e^t - 6te^t - 4 - e^{3t}}{4}$$

is the solution to the given initial value problem.

As a further illustration, a particular solution x_p, y_p to

$$\begin{aligned} x' &= -x + 4y + \sin 2t \\ y' &= -x + 3y \end{aligned}$$

should be assumed of the form

$$x_p = d_1 \sin 2t + d_2 \cos 2t \qquad y_p = d_3 \sin 2t + d_4 \cos 2t$$

5.3a Mixing in Interconnected Tanks

An illustrative application of a two-dimensional linear system is a liquid mixture problem in two interconnected tanks (see Section 1.3 for an example involving one tank). Suppose that there are two tanks T_1 and T_2 that initially contain V_1 and V_2 gallons of a brine mixture, respectively. Suppose also that the tanks are interconnected with pipes, both tanks are stirred continuously,

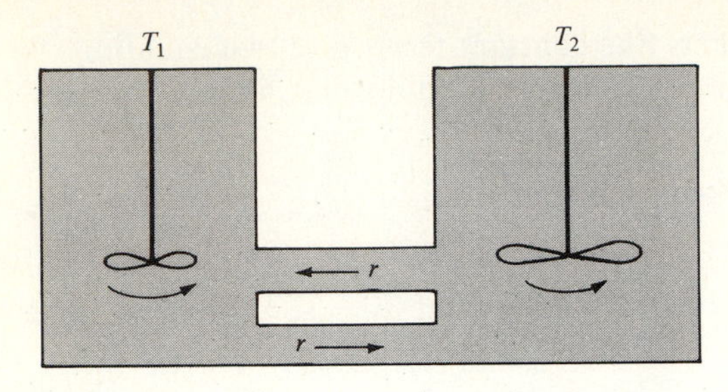

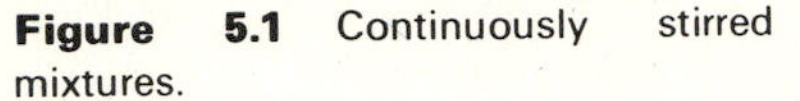
Figure 5.1 Continuously stirred mixtures.

and the mixture is pumped from T_1 to T_2 and from T_2 to T_1 at a constant rate of r gal/min (see Figure 5.1). It is assumed that the amount of the mixture in the pipes and the time that the mixture remains in the pipes are both negligible.

Assume that initially there is A_1 pounds of salt in T_1 and A_2 pounds in T_2. For each $t \geq 0$ let $Q_1(t)$ and $Q_2(t)$ denote the number of pounds of salt in tanks T_1 and T_2. From the principle indicated in Section 1.3 it follows that

$$\begin{aligned} Q_1'(t) &= \text{rate of salt into } T_1 - \text{rate of salt out of } T_1 \\ &= \text{rate of salt out of } T_2 - \text{rate of salt out of } T_1 \\ &= \frac{rQ_2(t)}{V_2} - \frac{rQ_1(t)}{V_1} \end{aligned}$$

and similarly,

$$Q_2'(t) = \frac{rQ_1(t)}{V_1} - \frac{rQ_2(t)}{V_2}$$

Therefore, since the quantities of salt are initially A_1 and A_2, the system

$$\begin{aligned} Q_1' &= -rV_1^{-1}Q_1 + rV_2^{-1}Q_2 \qquad Q_1(0) = A_1 \\ Q_2' &= rV_1^{-1}Q_1 - rV_2^{-1}Q_2 \qquad Q_2(0) = A_2 \end{aligned} \tag{13}$$

must be satisfied. This is a two-dimensional, homogeneous, linear initial value problem with constant coefficients, and hence the solution can be explicitly computed.

As a specific illustration, suppose that tank T_1 holds initially 50 gallons of pure water (that is, $V_1 = 50$, $A_1 = 0$) and that tank T_2 holds initially 100 gallons of brine with $\frac{3}{4}$ pound of salt per gallon (that is, $V_2 = 100$, $A_2 = 75$). If the rate r that the mixtures are pumped between the tanks is 5 gal/min, the quantities $Q_1(t)$ and $Q_2(t)$ of the pounds of salt in T_1 and T_2, respectively, must satisfy the initial value system

$$\begin{aligned} Q_1' &= -\tfrac{1}{10}Q_1 + \tfrac{1}{20}Q_2 \qquad Q_1(0) = 0 \\ Q_2' &= \tfrac{1}{10}Q_1 - \tfrac{1}{20}Q_2 \qquad Q_2(0) = 75 \end{aligned} \tag{14}$$

The characteristic polynomial for this differential equation is

$$\det\begin{pmatrix} -\frac{1}{10}-\lambda & \frac{1}{20} \\ \frac{1}{10} & -\frac{1}{20}-\lambda \end{pmatrix} = \left(\lambda+\frac{1}{10}\right)\left(\lambda+\frac{1}{20}\right)-\frac{1}{10}\cdot\frac{1}{20}$$

$$= \lambda\left(\lambda+\frac{1}{10}+\frac{1}{20}\right)$$

Hence $\lambda = 0, -\frac{3}{20}$ are the characteristic roots, so the solution to (14) has the form

$$Q_1 = c_1 + c_2 e^{-3t/20} \qquad Q_2 = d_1 + d_2 e^{-3t/20}$$

Since $Q_2 = 20Q_1' + 2Q_1$ it follows that

$$Q_1 = c_1 + c_2 e^{-3t/20} \qquad Q_2 = 2c_1 - c_2 e^{-3t/20}$$

is the general solution. The initial conditions

$$Q_1(0) = c_1 + c_2 = 0 \qquad \text{and} \qquad Q_2(0) = 2c_1 - c_2 = 75$$

imply that $c_1 = 25$, $c_2 = -25$, and hence

$$Q_1(t) = 25 - 25e^{-3t/20} \qquad Q_2(t) = 50 + 25e^{-3t/20}$$

is the solution to (14). Note that $Q_1(t) \to 25$ and $Q_2(t) \to 50$ as $t \to +\infty$ (does this agree with what you believe would happen from the physical interpretation?). Actually it is not difficult to explicitly solve the system (13) in general (see Problem 5.3-7).

PROBLEMS

5.3-1. Determine the general solution to each of the following systems and, when given, find a solution that satisfies the initial values.

(a) $x' = 2y$, $y' = -x$

(b) $x' = 2y$, $y' = x$

(c) $x' = 5x + y$, $y' = -x + 7y$

(d) $x' = x - 4y$, $y' = 4x + y$

(e) $x' = x + 2y$, $y' = 2x + y$ $\quad x(0) = 1$, $y(0) = -1$

(f) $x' = y$, $y' = y$ $\quad x(0) = -3$, $y(0) = 2$

(g) $x' = -x + 4y$, $y' = -2x + 3y$ $\quad x(0) = -2$, $y(0) = 1$

(h) $x' = -x + y$, $y' = -x - 3y$ $\quad x(0) = 5$, $y(0) = -7$

5.3-2. Determine the general solution to each of the following systems and, when given, find a solution that satisfies the initial values.

(a) $x' = 2y + 1$, $y' = -x + 3$ (See Problem5.3-1*a*.)

(b) $x' = 2y + t$ (See Problem 5.3-1*b*.)
$y' = x$

(c) $x' = x + y + 1$ (*Hint:* Assume the particular solution to be
$y' = x + y$ $x_p = d_1 + d_2 t,\ y_p = d_3 + d_4 t$.)

(d) $x' = x + 2y + e^t,\quad x(0) = 0$ (See Problem 5.3-1*e*.)
$y' = 2x + y + 1\quad y(0) = 0$

(e) $x' = y + \sin t\quad x(0) = 2$
$y' = x + 2\cos t\quad y(0) = 0$

5.3-3. Determine *all* initial values x_0, y_0 such that the solution x, y to the initial value problem

$$x' = x + 3y \qquad x(0) = x_0$$
$$y' = 2x + 2y \qquad y(0) = y_0$$

satisfies $\lim_{t\to+\infty} x(t) = \lim_{t\to+\infty} y(t) = 0$.

5.3-4. Determine *all* initial values x_0, y_0 such that the solution x, y to the initial value problem

$$x' = x + y + 5 \qquad x(0) = x_0$$
$$y' = 4x + y - 7 \qquad y(0) = y_0$$

has a limit as $t \to +\infty$ [that is, $\lim_{t\to\infty} x(t)$ and $\lim_{t\to\infty} y(t)$ both exist]. What is the limit of such solutions?

***5.3-5.** Suppose that a_1, b_1 and a_2, b_2 are constants and that $p(\lambda)$ is the characteristic polynomial of (1) [see (6)].

(a) Show that *every* solution (x, y) to (1) has limit (0, 0) as $t \to +\infty$ [that is, $\lim_{t\to\infty} x(t) = \lim_{t\to\infty} y(t) = 0$] if and only if the real part of the characteristic roots are negative.

(b) Show that the real parts of the characteristic roots are negative if and only if $a_1 + b_2 < 0$ and $a_1 b_2 - a_2 b_1 > 0$.

5.3-6. Consider the interconnecting tanks described in Figure 5.1 and equation (13). Suppose that initially there is 50 pounds of salt in T_1 and in T_2 (that is, $A_1 = A_2 = 50$) and that the brine is pumped between the tanks at a rate of $r = 10$ gal/min.

(a) Determine $Q_1(t)$ and $Q_2(t)$ for all $t \geq 0$ if $V_1 = 100$ gallons and $V_2 = 300$ gallons. Compute $\lim_{t\to+\infty} Q_1(t)$ and $\lim_{t\to+\infty} Q_2(t)$.

(b) Determine $Q_1(t)$ and $Q_2(t)$ for all $t \geq 0$ if $V_1 = 50$ gallons and $V_2 = 350$ gallons. Compute $\lim_{t\to+\infty} Q_1(t)$ and $\lim_{t\to+\infty} Q_2(t)$. Compare and interpret these limits with the analogous ones in part *a*.

***5.3-7.** Show that

$$Q_1(t) = \frac{V_1(A_1 + A_2) + (V_2 A_1 - V_1 A_2) \exp(-(V_2^{-1} + V_1^{-1})rt)}{V_1 + V_2}$$

and

$$Q_2(t) = \frac{V_2(A_1 + A_2) - (V_2 A_1 - V_1 A_2) \exp(-(V_2^{-1} + V_1^{-1})rt)}{V_1 + V_2}$$

is the solution to the initial value problem (13). Show that

$$Q_1(\infty) = \lim_{t \to +\infty} Q_1(t) \qquad \text{and} \qquad Q_2(\infty) = \lim_{t \to +\infty} Q_2(t)$$

both exist and that $Q_1(\infty)Q_2(\infty)^{-1} = V_1 V_2^{-1}$. Interpret these results physically.

***5.3-8.** Consider the tanks T_1 and T_2 in Figure 5.2*a*, where initially T_1 and T_2 hold V_1 and V_2 gallons of a mixture containing A_1 and A_2 pounds of salt, respectively. Brine containing β pounds of salt per gallon enters T_1 at a rate r gal/min. The well-stirred mixture leaves T_1, enters T_2, and then leaves T_2 at the

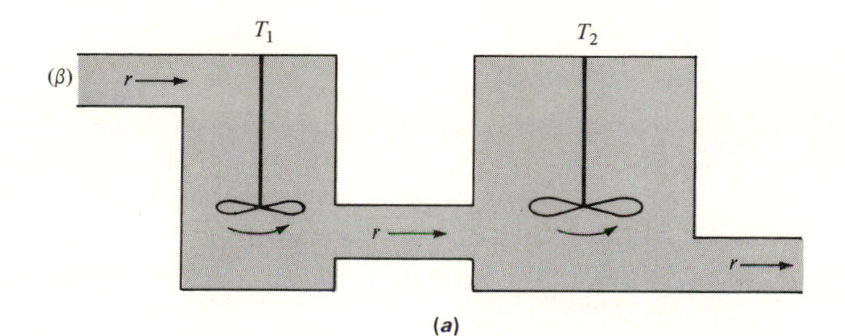

(*a*)

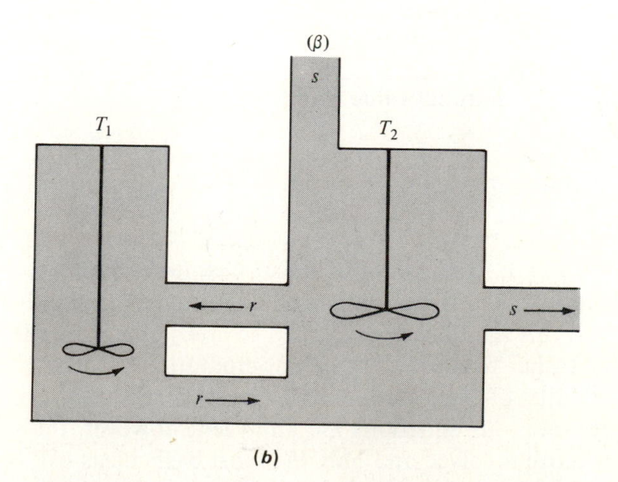

(*b*)

Figure 5.2 (*a*) Tank interconnection for Problem 5.3-8. (*b*) Tank interconnection for Problem 5.3-9.

same rate r (see Figure 5.2*a*). Determine the amounts $Q_1(t)$ and $Q_2(t)$ of pounds of salt in T_1 and T_2, respectively, for all $t \geq 0$. Do $Q_1(t)$ and $Q_2(t)$ have a limit as $t \to +\infty$?

***5.3-9.** Consider the interconnecting tanks in Figure 5.2*b*, where the brine is pumped between the tanks at a rate of r gal/min. Assume also that brine containing β pounds of salt per gallon enters T_2 at a rate of s gal/min and that the overflow leaves at the same rate (see Figure 5.2*b*). Assume initially that there are V_1 and V_2 gallons of mixture containing A_1 and A_2 pounds of salt, respectively, and let $Q_1(t)$ and $Q_2(t)$ denote the pounds of salt in T_1 and T_2, respectively, for each time $t \geq 0$. Show that Q_1, Q_2 is a solution to

$$Q_1' = -rV_1^{-1}Q_1 + rV_2^{-1}Q_2 \qquad Q_1(0) = A_1$$
$$Q_2' = rV_1^{-1}Q_1 - (r+s)V_2^{-1}Q_2 + s\beta \qquad Q_2(0) = A_2$$

Determine $Q_1(t)$ and $Q_2(t)$ explicitly and discuss their behavior as $t \to +\infty$ in each of the following two cases:

(a) $V_1 = V_2 = 30$, $r = 5$, $s = 1$, $\beta = 1$, $A_1 = 50$, and $A_2 = 28$

(b) $V_1 = 10$, $V_2 = 20$, $r = s = 1$, $\beta = \frac{1}{5}$, $A_1 = 10$, $A_2 = 4$

5.4 NONLINEAR SYSTEMS: CURVE SKETCHES

An important class of two-dimensional systems is that in which the equations do not depend explicitly on time. Such systems are called *autonomous* (see also Sections 1.4 and 1.5). Throughout this section it is supposed that f and g are continuous real-valued functions of two variables and that the first partial derivatives of f and g exist and are also continuous. Basic properties of solutions to the two-dimensional system

$$\begin{aligned} x' &= f(x, y) \\ y' &= g(x, y) \end{aligned} \tag{1}$$

as well as the associated initial value problem

$$\begin{aligned} x' &= f(x, y) \qquad x(0) = x_0 \\ y' &= g(x, y) \qquad y(0) = y_0 \end{aligned} \tag{2}$$

are considered. Unless stated otherwise the independent variable t is assumed to be in $[0, \infty)$. Except in the linear case (see the preceding section) it is rare that solutions to (1) can actually be computed explicitly. Therefore, the techniques in this section are developed to study the properties and behavior of solutions rather than their explicit computation. There are similarities between the results of this section and those of Section 1.5, where one-dimensional autonomous equations are analyzed; however, the situation here is considerably more involved and only the most basic ideas will be discussed.

A solution to (1) on an interval I is a pair x, y of functions on I such that $x'(t) = f(x(t), y(t))$ and $y'(t) = g(x(t), y(t))$ for all $t \in I$. Although it is not established in this book, for each given initial value x_0, y_0, equation (2) has exactly one solution $x = x(t)$ and $y = y(t)$. Moreover, this solution is defined for all $t \geq 0$ unless either $x(t)$ or $y(t)$ becomes arbitrarily large in finite time: there is a $T > 0$ such that

$$\lim_{t \to T-} x(t)^2 + y(t)^2 = +\infty$$

In order to study the "qualitative behavior" of the solutions to (1) and (2), *the procedure is to sketch the set of points* $\{(x(t),\ y(t)):\ t > 0\}$ *in the xy plane.* These ideas are related to some of those arising in the study of parametric equations: for each solution x, y to (1) on an interval I, sketch the curve in the xy plane described by the parametric equations

$$x = x(t) \qquad y = y(t) \qquad t \in I \tag{3}$$

The plane curve defined by the parametric equations (3) is called the *trajectory* of the solution x, y to (1) (sometimes, in place of trajectory, the term *path* or *orbit* is used). As will soon be indicated, formulas for trajectories of solutions to (1) are considerably easier to deduce than the solutions themselves. In fact, determining a trajectory can be reduced to solving a one-dimensional differential equation, and so the techniques from Chapter 1 can be applied.

When a trajectory is a single point, the corresponding solution is constant, and, as in the one-dimensional equation, such solutions play an important role. A point (x_0, y_0) in the plane is said to be a *critical point* of (1) if the simultaneous equations

$$f(x_0, y_0) = 0 \qquad \text{and} \qquad g(x_0, y_0) = 0 \tag{4}$$

are satisfied. If (x_0, y_0) is a critical point, then $x(t) \equiv x_0$, $y(t) \equiv y_0$ is a solution to (1), and this constant solution is called an *equilibrium solution* of (1). The set of all critical points to (1) is called the *critical point set* and is often denoted by CP. Since the determination of critical points involves solving two nonlinear equations simultaneously, the critical point set may be difficult to compute in some cases.

▶ **EXAMPLE 5.4-1**

Consider the linear system

$$\begin{aligned} x' &= y \\ y' &= -x \end{aligned} \tag{5}$$

It is easy to check that the origin is the only critical point for this system, and since this equation is homogeneous linear, the techniques in the previous section can be used to show that the general solution is

$$x(t) = c_1 \cos t + c_2 \sin t \qquad y(t) = -c_1 \sin t + c_2 \cos t$$

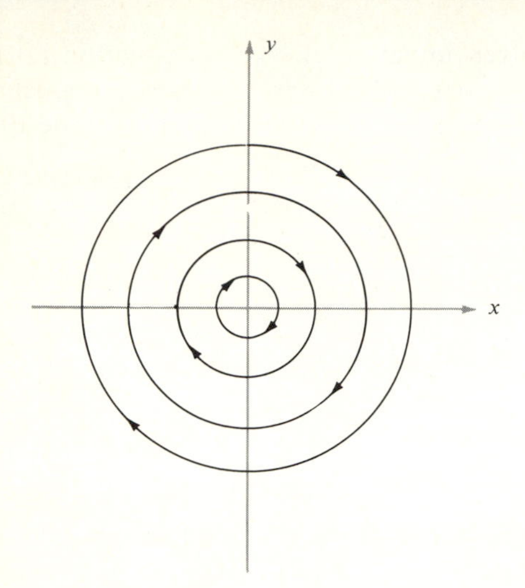

Figure 5.3 Representative sketch of trajectories for (5).

By "eliminating the parameter t" it follows that

$$x(t)^2 + y(t)^2 = c_1^2 \cos^2 t + c_2^2 \sin^2 t + (-c_1)^2 \sin^2 t + c_2^2 \cos^2 t$$
$$= c_1^2 + c_2^2$$

for all $t \geq 0$ and constants c_1 and c_2. Noting that $c_1 = x(0)$ and $c_2 = y(0)$, it follows that every trajectory of (5) lies on the circle with center at the origin and through the initial point $(x(0), y(0))$ [i.e., with radius equal $\sqrt{x(0)^2 + y(0)^2}$]. A sketch of these trajectories along with the direction of the flow is given in Figure 5.3.

In Example 5.4-1, the equation of the trajectory of a solution is obtained from the solution. This is, of course, not the usual situation. The basic problem is to *determine the equation of the trajectories directly from equation (1).* Since the trajectories of nonconstant solutions form essentially a family of curves in the plane, we try to determine a function $U(x, y)$ such that each trajectory of (1) lies on some curve $U(x, y) = c$, where c is a constant [in Example 5.4-1 such a function is $U(x, y) = x^2 + y^2$]. We indicate now how this function $U(x, y)$ can be determined directly from equation (5) in Example 5.4-1 instead of first solving equation (5) and then eliminating the parameter t.

▶ **EXAMPLE 5.4-2**

Consider the same system as in Example 5.4-1:

$$x' = y$$
$$y' = -x \tag{5}$$

Using the chain rule for differentiation it follows that if $x = x(t)$, $y = y(t)$ is a solution to (5), then

$$\frac{dy}{dx} = \frac{dy/dt}{dx/dt} = \frac{y'}{x'} = \frac{-x}{y}$$

and it follows that x, y must satisfy the total differential equation

$$x\,dx + y\,dy = 0$$

(see Section 1.7). The variables are separated in this equation, and direct integration shows that $U(x, y) = \frac{1}{2}x^2 + \frac{1}{2}y^2$ is a solution. Therefore, any solution to (5) satisfies

$$\tfrac{1}{2}x(t)^2 + \tfrac{1}{2}y(t)^2 \equiv \text{constant}$$

Substituting $t = 0$ in order to determine the constant, it is easy to see that

$$x(t)^2 + y(t)^2 \equiv x(0)^2 + y(0)^2$$

which is exactly the result obtained in Example 5.4-1.

Before considering the general case we look at a second specific illustrative example. Consider the system

$$\begin{aligned} x' &= y - 1 \\ y' &= 4x \end{aligned} \tag{6}$$

Setting $y - 1 = 0$ and $4x = 0$ shows that (0, 1) is the only critical point. If $x = x(t)$, $y = y(t)$ is any solution to (6) for $t \geq 0$, then so long as $x'(t) \neq 0$ [that is, $y(t) \neq 1$], we have

$$\frac{dy}{dx} = \frac{dy/dt}{dx/dt} = \frac{y'}{x'} = \frac{4x}{y - 1}$$

Therefore, considering y as a function of x we see that y is a solution to a single first order differential equation whose variables separate (see Section 1.4). Thus,

$$(y - 1)\,dy = 4x\,dx$$

and upon antidifferentiation,

$$\tfrac{1}{2}(y - 1)^2 = 2x^2 + c$$

Multiplying each side by -2 and labeling the new constant k, it follows that

$$4x^2 - (y - 1)^2 = k \tag{7}$$

The family of curves (7) in the plane are called *integral curves of* (6): that is, if x, y is a solution to (6), then there is a constant k such that

$$4x(t)^2 - [y(t) - 1]^2 \equiv k$$

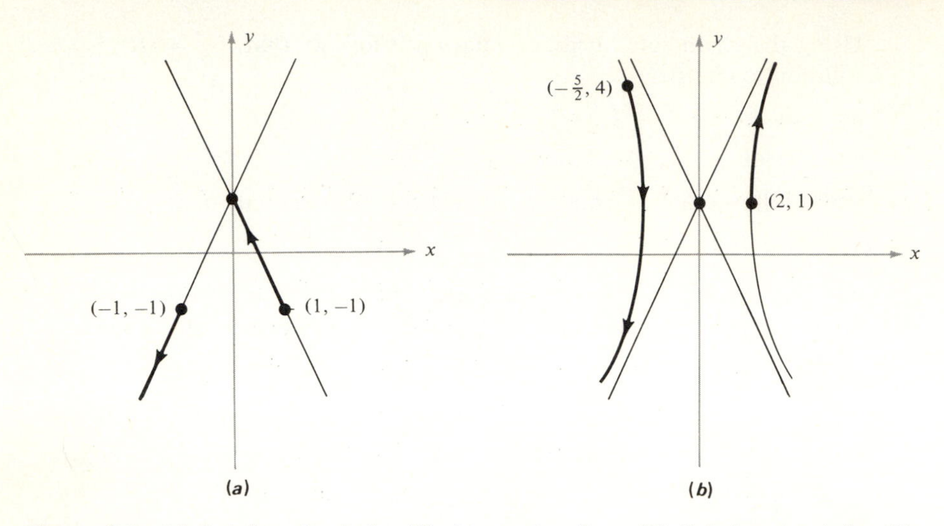

Figure 5.4 (*a*) Solution sketch for (6): Intersecting lines. (*b*) Solution sketch for (6): hyperbolic curves.

{in fact, $k = 4x(0)^2 - [y(0) - 1]^2$}. Therefore, solution curves $x = x(t)$, $y = y(t)$ to (6) must lie on one of the integral curves (7) in the plane.

Suppose now that x, y is the solution to (6) that satisfies $x(0) = 1$, $y(0) = -1$. Substituting $x = 1$, $y = -1$ in (7), we obtain $k = 0$, and hence the solution x, y satisfies

$$4x(t)^2 - [y(t) - 1]^2 \equiv 0 \qquad \text{for } t \geq 0$$

The equation $4x^2 - (y - 1)^2 = 0$ is a degenerate hyperbola and its graph is the intersecting lines $y - 1 = \pm 2x$. Therefore, the solution $x(t)$, $y(t)$ begins at the point $(1, -1)$ and must move along the line $y - 1 = -2x$ (see Figure 5.4*a*). Since $y'(t) = 4x(t) > 0$ so long as $x(t) > 0$, this solution moves upward along the line $y - 1 = -2x$ from the point $(1, -1)$ toward the point $(0, 1)$ as t increases. Since $(0, 1)$ is a critical point, this solution cannot reach $(0, 1)$ in finite time (this follows since the solutions are unique and cannot cross). Therefore, the trajectory of the solution beginning at $(1, -1)$ is the line segment joining $(1, -1)$ to $(0, 1)$ [including $(1, -1)$ but not including $(0, 1)$]. If the initial value of a solution is $(-1, -1)$, then $k = 0$ again, but in this case the solution is on the line $y - 1 = 2x$. Since $y' = 4x < 0$ for $x < 0$, the solution moves downward along the line, and the trajectory of this solution is the ray emanating from $(-1, -1)$ pointing downward along the line with slope 2. The graphs of these two trajectories and the direction of the motion are indicated in Figure 5.4*a*.

If the initial conditions are either $x(0) = 2$, $y(0) = 1$ or $x(0) = -\frac{5}{2}$, $y(0) = 4$, then the constant k in (7) equals 16, and hence the integral curve for these solutions is the hyperbola

$$4x^2 - (y - 1)^2 = 16$$

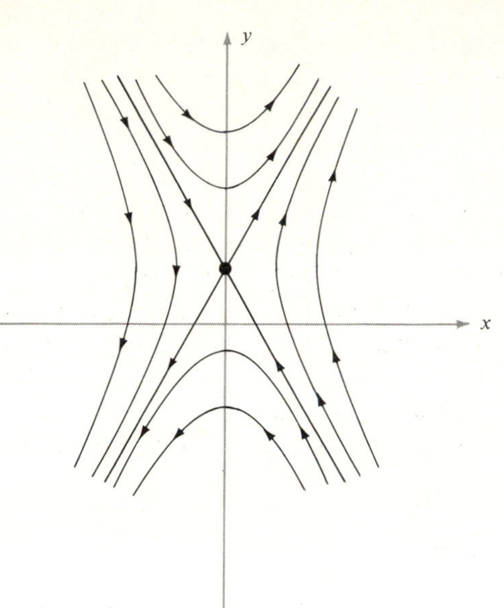

Figure 5.5 Phase plot for (6).

This hyperbola opens left and right and has the lines $y - 1 = \pm 2x$ as asymptotes (see Figure 5.4*b*). Again since $y' > 0$ if $x > 0$ and $y' < 0$ if $x < 0$, the motion of these two solutions along this hyperbola is as indicated by the arrows and dark lines in Figure 5.4*b*.

One can easily sketch a representative number of the curves (7) for various values of the constant k. If $k = 0$ then the graph of (7) is the two intersecting lines $y - 1 = \pm 2x$; if $k > 0$ then the graph of (7) is a hyperbola opening right and left with $y - 1 = \pm 2x$ as its asymptotes; and if $k < 0$ the graph of (7) is a hyperbola opening up and down with the same asymptotes. These are sketched in Figure 5.5. The arrows indicate the direction of the flow and can be determined directly from (6). For if $x > 0$ then $y' > 0$, and if $x < 0$ then $y' < 0$.

5.4a First Integrals and Phase Plots

Suppose that D is an open rectangle in the xy plane [that is, $D = \{(x, y): \alpha_1 < x < \beta_1, \alpha_2 < y < \beta_2\}$] and that $U = U(x, y)$ is a real-valued function defined on D, with U and its first partial derivatives U_x and U_y continuous on D. It is also assumed that U_x and U_y are not both zero at any point in D except possibly at critical points of (1) in D [this ensures that outside the critical point set the equation $U(x, y) = c$, c a constant, defines locally a curve in D]. Then U is called a *first integral of (1) in D* if

(8) $\quad U(x(t), y(t)) \equiv$ constant whenever x, y is a solution to (1) that remains in D

The *level curves* $U(x, y) = c$ of U are called the *integral curves* of (1). Since $U(x(t), y(t)) \equiv U(x(0), y(0))$ for a solution x, y to (1) that remains in D, the trajectory of x, y is a subset (perhaps a proper subset) of the level curve $U(x, y) = U(x(0), y(0))$. Thus each trajectory is a subset of an integral curve, but it is not necessarily equal to an integral curve. However, if a first integral of (1) on D can be found, *a sketch of the level curves $U(x, y) = c$ for a representative number of constants c gives a good picture of the behavior of the solutions to (1).*

An equation to use in determining a first integral for (1) can be easily found. Using the chain rule and equation (1), it follows that if x, y is a solution to (1) then

$$\frac{dy}{dx} = \frac{y'}{x'} = \frac{g(x, y)}{f(x, y)} \quad \text{if } x' \neq 0 \tag{9}$$
$$\frac{dx}{dy} = \frac{x'}{y'} = \frac{f(x, y)}{g(x, y)} \quad \text{if } y' \neq 0$$

Comparing (9) with equations (4) and (5) in Section 1.7, we see that the relationship between x and y is determined by the total differential equation

$$g(x, y)\,dx - f(x, y)\,dy = 0 \tag{10}$$

Therefore, we have the following result:

● **Proposition 1**

If $U(x, y)$ is a first integral on D to the total differential equation (10), then U is a first integral of (1) on D.

The methods in Section 1.7 can be applied to integrate equation (10), and the principal aim of this section is to develop results and techniques to obtain information about the behavior of solutions to (1) by studying the integral curves of (1). One of the most helpful procedures is to actually sketch a representative number of the integral curves of (1) in the xy plane. Such a sketch is called a *phase plot of (1)*, and there are essentially three main ingredients in the sketching of a phase plot:

(11)

- **(a)** Determine and sketch the critical point set CP.
- **(b)** Determine a first integral U and sketch a representative family of level curves $U(x, y) \equiv$ constant.
- **(c)** Determine the direction of the flow along integral curves.

Two examples are given to illustrate this procedure.

EXAMPLE 5.4-3

Consider the system

$$x' = -x \qquad (12)$$
$$y' = -x^2$$

Setting the right-hand side of each equation to zero, it follows that $x = 0$ is the equation of the critical point set, and hence the set of critical points for (12) is the y axis. Since

$$\frac{dy}{dx} = \frac{-x^2}{-x} = x \qquad \text{for } x \neq 0$$

direct integration shows that the integral curves of (12) is the family of parabolas $y = \frac{1}{2}x^2 + c$, where c is a constant. This family of curves is sketched along with the critical points in Figure 5.6. The direction of the flow is indicated with arrowheads. Since $x' < 0$ if $x > 0$ and $x' > 0$ if $x < 0$ (or, since $y' < 0$ if $x \neq 0$), it is easy to check that the solutions do indeed move along the integral curve in the indicated directions. Note, for example, if $x(0) = 2$ and $y(0) = 3$, then the solution lies on the parabola $y = \frac{1}{2}x^2 + 1$ (substituting $x = 2$, $y = 3$ into $y = \frac{1}{2}x^2 + c$ implies $c = 1$). From the direction of flow in Figure 5.6 it follows that the trajectory of this solution is $\{(x, y)\colon y = \frac{1}{2}x^2 + 1, 0 < x \leq 2\}$.

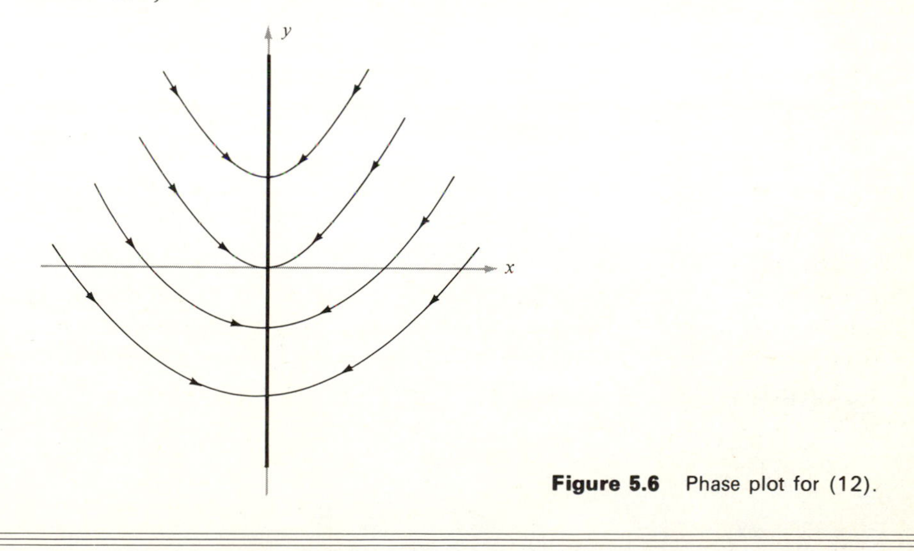

Figure 5.6 Phase plot for (12).

It is important to realize that a nonconstant solution to (1) cannot reach a critical point in finite time. This follows from the uniqueness of solutions to (1), for if x, y is a solution to (1) and (x_c, y_c) is a critical point of (1) [that is, $f(x_c, y_c) = g(x_c, y_c) = 0$] and there is a $T \geq 0$ such that $x(T) = x_c$, $y(T) = y_c$, then the solution x, y to equation (1) agrees with the constant solution x_c, y_c at a point, and hence by uniqueness of solutions they must be equal for all times $t \geq 0$ (and hence x, y is the equilibrium solution). This observation can be helpful in determining the behavior of solutions. For example, if an integral curve to (1) contains a critical point, then a solution to (1) that flows along this curve toward the critical point can never reach or move past this critical point.

In many cases once the integral curves and the critical points have been sketched, it is easy to see that certain solutions must remain bounded and continue to move in a certain direction as $t \to +\infty$, and also that they then must have a limit as $t \to +\infty$. Our next result can often be applied and asserts that if a solution converges as t approaches ∞, then it must converge to a critical point.

Proposition 2

Suppose that x, y is a solution to (1) and $x_0 = \lim_{t \to +\infty} x(t)$ and $y_0 = \lim_{t \to +\infty} y(t)$ both exist. Then (x_0, y_0) is a critical point of (1).

Since f and g are continuous,

$$\lim_{t \to \infty} x'(t) = \lim_{t \to \infty} f(x(t), y(t)) = f(x_0, y_0) \qquad \text{and}$$

$$\lim_{t \to \infty} y'(t) = \lim_{t \to \infty} g(x(t), y(t)) = g(x_0, y_0)$$

both exist. Since these two functions and their derivatives have a limit as $t \to \infty$, the limit of the derivatives must be zero (see Problem 1.5-6). Thus $f(x_0, y_0) = g(x_0, y_0) = 0$, and (x_0, y_0) must be a critical point of (1).

EXAMPLE 5.4-4

Consider the system

$$\begin{aligned} x' &= 4y(y - 2x) \\ y' &= (2 - x)(y - 2x) \end{aligned} \tag{13}$$

Considering the simultaneous equations

$$4y(y - 2x) = 0 \qquad \text{and} \qquad (2 - x)(y - 2x) = 0$$

we see that $\mathrm{CP} = \{(x, 2x) : -\infty < x < \infty\} \cup \{(2, 0)\}$: that is, the critical point set of (13) is the line $y = 2x$ and the point $(2, 0)$. This is indicated with the

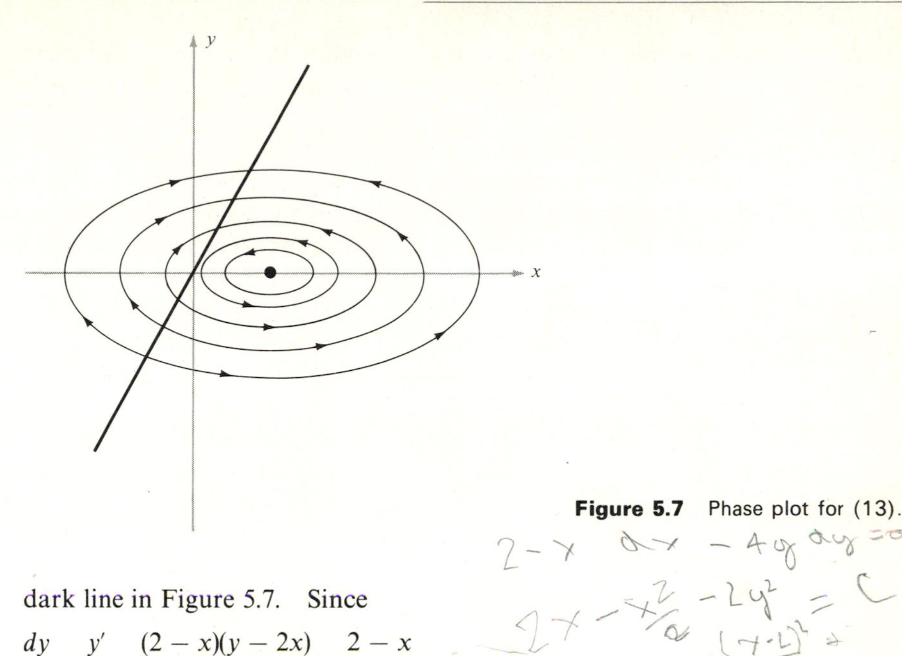

Figure 5.7 Phase plot for (13).

dark line in Figure 5.7. Since

$$\frac{dy}{dx} = \frac{y'}{x'} = \frac{(2-x)(y-2x)}{4y(y-2x)} = \frac{2-x}{4y}$$

if $y \neq 2x$, a solution to the total differential equation $(x-2)\,dx + 4y\,dy = 0$ is a first integral of (13). Since the variables are separated, direct integration shows that $U(x, y) = \frac{1}{2}(x-2)^2 + 2y^2$ is a first integral, and hence

$$\frac{(x-2)^2}{4} + y^2 = k \qquad k > 0 \text{ a constant} \tag{14}$$

is a family of integral curves for (13). The integral curves in (14) is the set of all ellipses with center (2, 0) and with horizontal major axis twice as long as its vertical minor axis. A phase plot for (13) is given in Figure 5.7. Since $x' = 4y(y-2x)$, we have that if $y > 2x$ and $y > 0$ or if $y < 2x$ and $y < 0$, then $x' > 0$, and the x direction of the flow is to the right. Similarly, if $y > 2x$ and $y < 0$ or if $y < 2x$ and $y > 0$, then $x' < 0$, and the x direction of the flow is to the left. This is indicated by the arrows in the phase plot for (13). As an illustration of Proposition 2, consider the solution to (13) that satisfies $x(0) = 4$, $y(0) = -1$. Substituting into (14), $(4-2)^2/4 + (-1)^2 = 2$, so $k = 2$, and

$$\frac{(x-2)^2}{4} + y^2 = 2$$

is the integral curve of this solution. The intersection points of this integral curve with the critical point set can be determined by setting $y = 2x$ and solving

$$\frac{(x-2)^2}{4} + 4x^2 = 2$$

Upon simplification, $17x^2 - 4x - 4 = 0$, and it follows from the quadratic formula that

$$x = \frac{2}{17} \pm \frac{12\sqrt{2}}{17} \qquad \text{and hence} \qquad y = \frac{4}{17} \pm \frac{24\sqrt{2}}{17}$$

Since the flow is counterclockwise on this side of the ellipse, it follows that the solution to (13) with initial value $(4, -1)$ must converge to the critical point $x_c = (2 + 12\sqrt{2})/17$, $y_c = (4 + 24\sqrt{2})/17$.

5.4*b* Model of an Epidemic

As a reasonably elementary application of the methods in this section, we consider a nonlinear system that is used to model the spread of a disease through a population. It is supposed that within a population there are two groups at each time $t \geq 0$: a number $x(t)$ that is *susceptible* to contracting an infectious disease and a number $y(t)$ that is actually *infected* with the disease. The contagiousness of the disease is represented by the number of contacts of the susceptible population $x(t)$ with the infected population $y(t)$ and is assumed to be proportional to the product $x(t)y(t)$. A member of the susceptible class in the population either remains susceptible or contracts the disease and moves from the susceptible class into the infected class (in particular, a susceptible person cannot become immune to the disease without contracting the disease). Therefore the rate out of the susceptible class equals the rate into the infected class. An equation modeling spread of this disease is the system

$$\begin{aligned} x' &= -\alpha xy \qquad & x(0) &= x_0 \geq 0 \\ y' &= \alpha xy - \beta y \qquad & y(0) &= y_0 \geq 0 \end{aligned} \tag{15}$$

where α and β are positive constants. The coefficient $\alpha > 0$ indicates how contagious the disease is, and the term $-\beta y$ indicates how rapidly the people that have the disease are cured (or how fast they die!). Also, once a person has the disease, that person is no longer susceptible (he or she either becomes immune or dies), and so there is no feedback into the susceptible class once a person leaves this class by becoming infected. The differential equation in (15) is nonlinear and autonomous, and so we sketch a phase plot for this system in the first quadrant (note that the physical interpretation requires that the solutions remain nonnegative). Since the simultaneous equations

$$-\alpha xy = 0 \qquad \text{and} \qquad (\alpha x - \beta)y = 0$$

are satisfied only in case $y = 0$, we see that the x axis is the critical point set for (15). Moreover,

$$\frac{dy}{dx} = \frac{y'}{x'} = \frac{\alpha xy - \beta y}{-\alpha xy} = -1 + \frac{\beta}{\alpha x}$$

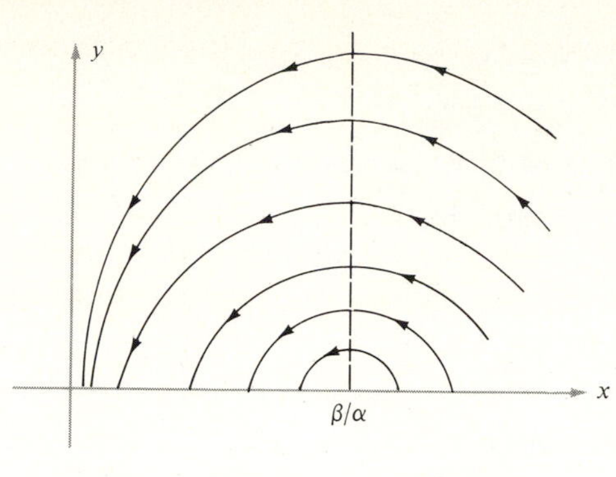

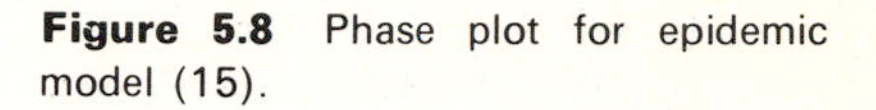
Figure 5.8 Phase plot for epidemic model (15).

and by direct integration the integral curves of (15) are

$$y = -x + \frac{\beta}{\alpha} \ln x + k \tag{16}$$

Since $dy/dx < 0$ if $x > \beta/\alpha$ and $dy/dx > 0$ if $x < \beta/\alpha$, and since $y(x) \to -\infty$ as $x \to 0+$, a phase plot for (15) has the form indicated in Figure 5.8 (note that $x' < 0$ in the first quadrant so the flow is to the left). It also follows (from the phase plot) that if x_0, y_0 are both positive, then the solution x, y to (15) satisfies

$$\lim_{t \to +\infty} y(t) = 0 \qquad \text{and} \qquad \lim_{t \to +\infty} x(t) = x(\infty) > 0 \tag{17}$$

There are several interesting interpretations of the preceding results. If the initial number of susceptibles x_0 is less than β/α, then the infected population $y(t)$ never increases and gradually dies out (and hence there is *no epidemic*). However, if $x_0 > \beta/\alpha$ then the infectives $y(t)$ grow until the susceptibles reach the number β/α and then die out (in particular, there is an epidemic, and the bigger x_0 is than β/α, the more severe the epidemic). This crucial value β/α for the number of susceptibles is called the *threshold value* of this disease. Finally, we point out that since $x(\infty) > 0$, this disease does not strike every susceptible person during an epidemic, and the epidemic subsides because there are not enough susceptibles for the disease to spread.

PROBLEMS

5.4-1. Determine the integral curves and sketch a phase plot for each of the following plane systems (indicate clearly all critical points and the direction of the flow).

(a) $x' = y + 1$, $y' = -4x + 8$

(b) $x' = 2y - 6$, $y' = x + 2$

(c) $x' = 2xy$, $y' = -x$

(d) $x' = e^x - 1$, $y' = ye^x$

(e) $x' = x + y$
$y' = 2x + 2y$

(f) $x' = y - x^2y - y^3$
$y' = x^2 + y^2 - 1$

(g) $x' = 2xy$
$y' = x + xy^2$

(h) $x' = 2x^2y$
$y' = (y^2 + 1)x$

5.4-2. Consider the initial value problem

$$(*) \quad \begin{aligned} x' &= y(y - x) & x(0) &= x_0 \\ y' &= (1 - x)(y - x) & y(0) &= y_0 \end{aligned}$$

(a) Sketch a phase plot for this system.

(b) Determine all initial values (x_0, y_0) such that the solution x, y to (*) has the property that $\lim_{t \to +\infty} x(t) = \lim_{t \to +\infty} y(t) = 2$.

(c) What is $\lim_{t \to +\infty} x(t)$ and $\lim_{t \to +\infty} y(t)$ if $x_0 = 3$ and $y_0 = 2$? if $x_0 = -1$ and $y_0 = 0$?

5.4-3. Consider the initial value problem

$$(**) \quad \begin{aligned} x' &= y & x(0) &= x_0 \\ y' &= 2xy & y(0) &= y_0 \end{aligned}$$

(a) Sketch a phase plot for this system.

(b) Determine $\lim_{t \to +\infty} x(t)$ and $\lim_{t \to +\infty} y(t)$ for each of the following initial values: $x_0 = 0$, $y_0 = -1$; $x_0 = -2$, $y_0 = 3$; and $x_0 = 2$, $y_0 = -1$.

(c) Determine all initial values x_0, y_0 such that the solution x, y to (**) satisfies

$$\lim_{t \to \infty} x(t) = \lim_{t \to \infty} y(t) = 0$$

(d) Describe the trajectory $\{(x(t), y(t)) : t \geq 0\}$ for each of the following initial values: $x_0 = -1$, $y_0 = 1$; $x_0 = 0$, $y_0 = 1$; $x_0 = 0$, $y_0 = -1$.

5.4-4. Sketch and compare phase plots for the following systems:

(a) $x' = -x$
$y' = -\frac{1}{2}y$

(b) $x' = -x$
$y' = -y$

(c) $x' = -x$
$y' = -2y$

(d) $x' = -x$
$y' = y$

(e) $x' = x$
$y' = y$

(f) $x' = x$
$y' = -y$

5.4-5. Suppose that the function f satisfies the conditions assumed in Section 1.5. Relate the sketch of a phase-time plot for the equation

$$y' = f(y)$$

to a phase-plot sketch of the system

$$\begin{aligned} x' &= 1 \\ y' &= f(y) \end{aligned}$$

5.4-6. Instead of the system (15), suppose that the susceptibles $x(t)$ and the infectives $y(t)$ satisfy the system

$$x' = -\alpha x^2 y \qquad x(0) = x_0 > 0$$
$$y' = \alpha x^2 y - \beta y \qquad y(0) = y_0 > 0$$

Sketch a phase plot for this system in the first quadrant and interpret the sketches physically. Compute, if possible,

$$\lim_{t\to\infty} x(t) = x(\infty) \qquad \text{and} \qquad \lim_{t\to\infty} y(t) = y(\infty)$$

in terms of α, β, x_0, and y_0.

5.4-7. Let $x(t)$ for $t \geq 0$ denote the position of an object of mass m moving vertically above the earth's surface ($x = 0$ at the earth's surface and $x > 0$ above the surface). Also let $y(t)$ for $t \geq 0$ denote the velocity of this object.

(a) Assuming only gravitational attraction ($x'' = -g$), write the equation of motion as a first order system in the plane, sketch a phase plot for this system in the first and fourth quadrants, and interpret the results.

(b) Repeat part *a*, assuming both a gravitational attraction and a resistance force proportional to the velocity ($x'' = -g - \beta x'/m$).

5.4-8. Sketch a phase plot for the linear system

$$x' = -rV_1^{-1}x + rV_2^{-1}y \qquad x(0) = x_0 \geq 0$$
$$y' = rV_1^{-1}x - rV_2^{-1}y \qquad y(0) = y_0 \geq 0$$

which describes the amount of salt in interconnected tanks that is continuously stirred [see equation (13) in Section 5.3].

5.5 PERIODIC SOLUTIONS IN NONLINEAR SYSTEMS

The concept of a periodic solution plays an important role in the analysis of autonomous systems, and some of the most basic ideas are discussed in this section. As in the preceding section, the autonomous system

$$\begin{aligned} x' &= f(x, y) \\ y' &= g(x, y) \end{aligned} \tag{1}$$

is considered, where both f and g are continuous and have continuous partial derivatives. A solution x, y to (1) is said to be *periodic* if there is a number $T > 0$ such that $x(t + T) \equiv x(t)$ and $y(t + T) \equiv y(t)$ for $t \geq 0$: that is, both x and y are periodic with the *same* period T. Any such number $T \geq 0$ is called a *period* of the solution. Normally, it is assumed in addition that periodic solutions are not equilibrium (constant) solutions (and hence have positive period T).

For a pair x, y that is not a solution to an autonomous system, it is certainly not necessarily the case that if $x(t_0 + T) = x(t_0)$ and $y(t_0 + T) = y(t_0)$ for *some* $t_0 \geq 0$ and $T > 0$, then x, y is periodic [for example, if

$x(t) \equiv y(t) \equiv t \sin t$, then $x(n\pi + 2\pi) = x(n\pi)$ and $y(n\pi + 2\pi) = y(n\pi)$ for $n = 0, 1, 2, \dots$, but x, y is not periodic]. However, if x, y is a solution to (1) this is indeed the case:

(2) If x, y is a solution to (1) and $x(t_0 + T) = x(t_0)$ and $y(t_0 + T) = y(t_0)$ for *some* $t_0 \geq 0$ and $T > 0$, then $x(t + T) \equiv x(t)$ and $y(t + T) \equiv y(t)$ for *all* $t \geq 0$.

Assertion (2) follows from the uniqueness of solutions to (1), along with the fact that translation of solutions of (1) are also solutions of (1). So assume that $x(t_0 + T) = x(t_0)$, $y(t_0 + T) = y(t_0)$, and define $\bar{x}(t) = x(t + T)$, $\bar{y}(t) = y(t + T)$ for all $t \geq 0$. Since $\bar{x}, \bar{y}$ is a translation of x, y, it must be a solution of (1), and since $\bar{x}(t_0) = x(t_0 + T) = x(t_0)$ and $\bar{y}(t_0) = y(t_0 + T) = y(t_0)$, $\bar{x}, \bar{y}$ and x, y agree at t_0, and hence agree for all $t \geq 0$ [that is, $x(t + T) \equiv x(t)$ and $y(t + T) \equiv y(t)$]. This establishes assertion (2).

A curve C in the plane is said to be a *closed curve* if there are continuous functions $x, y : [a, b] \to \mathrm{R}$ such that $x(a) = x(b)$, $y(a) = y(b)$, and

$$C = \{(x(t), y(t)) : a \leq t \leq b\}$$

The pair x, y is said to be a *parameterization* of C. If in addition, x and y can be chosen so that if $a \leq t < s < b$ then either $x(s) \neq x(t)$ or $y(s) \neq y(t)$, then the closed curve C is said to be a *simple closed curve*. A circle or an ellipse in the plane is a typical example of a simple closed curve. A figure eight is an example of a closed curve that is not simple. If x, y is any periodic solution to (1), with minimal period $T > 0$, then the trajectory of x, y is a simple closed curve. This follows easily using the periodicity of x, y. For if C is the trajectory, then

$$C = \{(x(t), y(t)) : t \geq 0\} = \{(x(t), y(t)) : 0 < t \leq T\}$$

Moreover, since T is the minimal period of x, y, we must have that if $0 \leq t < s < T$ then either $x(t) \neq x(s)$ or $y(t) \neq y(s)$. Thus the trajectory C is a simple closed curve by definition. The most important result of this section gives a very useful criterion for using the first integral (and the phase plot) to determine the existence and location of periodic solutions to (1):

Proposition 1

Suppose that $U = U(x, y)$ is a first integral of (1) and that the level curve $C = \{(x, y) : U(x, y) = k\}$ is a closed curve for some constant k. Suppose further that C contains no critical points of (1). Then every solution x, y to (1) with initial value on C is periodic and the level curve C is its trajectory (and, in particular, C is a simple closed curve).

The proof of this proposition is omitted. Geometrically, the fact that x, y must be periodic is clear. For as $t \geq 0$ increases, the pair $(x(t), y(t))$ moves in one direction around C and cannot "turn around" and move in the other direction on C (why?). If $(x(t), y(t))$ doesn't reach its starting point $(x(0), y(0))$ in finite time, then the pair $(x(t), y(t))$ must converge to a critical point as $t \to +\infty$ (see Proposition 2 in the preceding section). Since C contains no critical points, this cannot be the case, and we conclude that $x(T) = x(0)$ and $y(T) = y(0)$ at some finite time $T > 0$. Thus the solution x, y is periodic by (2).

We now study an example to illustrate the effectiveness of Proposition 1 and phase plots in analyzing the behavior of solutions to (1). Consider the system

$$\begin{aligned} x' &= -y(x^2 + y^2 - 4) \\ y' &= 4(x - 1)(x^2 + y^2 - 4) \end{aligned} \tag{3}$$

Since the right-hand sides are simultaneously zero when $x = 1$, $y = 0$, or when $x^2 + y^2 = 4$, it follows that

$$\text{CP} = \{(1, 0)\} \cup \{(x, y) : x^2 + y^2 = 4\}$$

In order to determine a first integral for (3) we have

$$\frac{dy}{dx} = \frac{y'}{x'} = \frac{4(x - 1)(x^2 + y^2 - 4)}{-y(x^2 + y^2 - 4)} = -\frac{4(x - 1)}{y}$$

Therefore $4(x - 1)\,dx + y\,dy = 0$, and it follows that $U(x, y) = 2(x - 1)^2 + \frac{1}{2}y^2$ is a first integral of (3). By multiplying by 2 we find that

$$4(x - 1)^2 + y^2 = k \qquad k > 0 \tag{4}$$

is a family of integral curves for (3) for each $k > 0$. Each of the integral curves are simple closed curves, and a phase plot for (3) is given in Figure 5.9. Using Proposition 1 above, Proposition 2 in the preceding section, and the phase plot in Figure 5.9, we can easily deduce information about the solutions to (3).

Specifically, suppose that x, y is a solution to (3) and

$$k = 4[x(0) - 1]^2 + y(0)^2 > 0$$

Then precisely one of the following holds:

(a) If $x(0)^2 + y(0)^2 = 4$, then x, y is an equilibrium solution.

(b) If the ellipse $4(x - 1)^2 + y^2 = k$ does not intersect the circle $x^2 + y^2 = 4$, then x, y is periodic and $\{(x, y) : 4(x - 1)^2 + y^2 = k\}$ is the trajectory of x, y.

(c) If the ellipse $4(x - 1)^2 + y^2 = k$ intersects the circle $x^2 + y^2 = 4$, then x, y converges to one of the intersection points as $t \to \infty$.

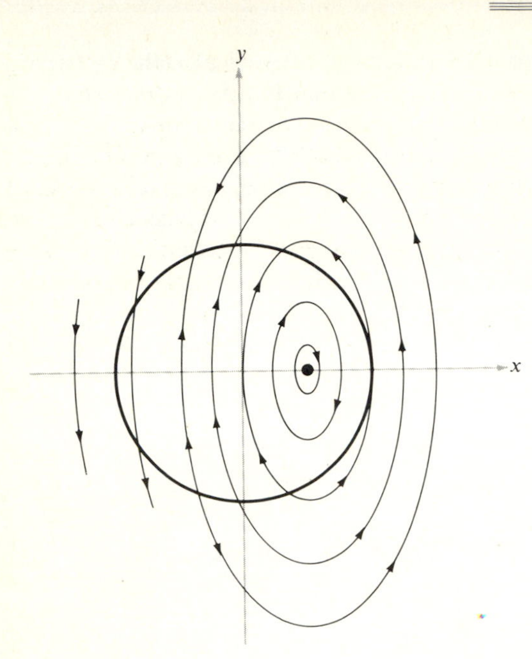

Figure 5.9 Phase plot for (3).

When part c is valid, the particular critical point that attracts the solution can be determined by solving the simultaneous equations $4(x-1)^2 + y^2 = k$ and $x^2 + y^2 = 4$ and using the phase plot in Figure 5.9. Suppose, for example, that $x(0) = \frac{3}{2}$ and $y(0) = \pm\sqrt{3}$. Then $k = 4$, and the intersection of the circle $x^2 + y^2 = 4$ with the ellipse $4(x-1)^2 + y^2 = 4$ are the three points $(\frac{2}{3}, 4\sqrt{2}/3)$, $(\frac{2}{3}, -4\sqrt{2}/3)$, and $(2, 0)$. Thus if $y_0 = \sqrt{3}$, then $x(t) \to \frac{2}{3}$ and $y(t) \to 4\sqrt{2}/3$ as $t \to \infty$; and if $y_0 = -\sqrt{3}$, then $x(t) \to 2$ and $y(t) \to 0$ as $t \to \infty$.

An analysis of the motion of an undamped mass-spring system gives an excellent illustrative application of these techniques. A detailed study of such systems is presented in Section 6.2 and so we only indicate a simple situation here. Suppose first that a unit mass is attached to a linear spring having spring constant $k > 0$. Using the results from Section 2.6 it follows that if $z(t)$ is the position of the mass relative to equilibrium, then z is a solution to the second order linear equation

$$z'' + kz = 0 \tag{5}$$

Setting $x = z$ and $y = z'$ equation (5) is equivalent to the system

$$\begin{aligned} x' &= y \\ y' &= -kx \end{aligned} \tag{6}$$

It is easy to check that the origin is the only critical point of (6) and that the equations

$$kx^2 + y^2 = c \tag{7}$$

are the integral curves of (6). These curves are ellipses centered at the origin, and hence all nontrivial solutions to (5) and (6) are periodic.

Instead of assuming that the spring is linear (i.e., exerts a force proportional to the distance displaced), it is sometimes necessary to assume that the force exerted by the spring is proportional to some nonlinear function of its distance to equilibrium. One of the simplest situations is to assume that the force F_s due to the spring is proportional to the distance from equilibrium cubed: $F_s = -kz^3$, where $k > 0$ is constant. In this case the equation of motion for a unit mass is

$$z'' + kz^3 = 0 \tag{8}$$

Setting $x = z$ and $y = z'$, we obtain the system

$$\begin{aligned} x' &= y \\ y' &= -kx^3 \end{aligned} \tag{9}$$

Again the origin is the only critical point, and since

$$\frac{dy}{dx} = \frac{-kx^3}{y} \qquad \text{or} \qquad kx^3\,dx + y\,dy = 0$$

it follows that

$$kx^4 + 2y^2 = c \qquad c > 0 \tag{10}$$

is the family of integral curves for (9). From the phase plot in Figure 5.10 we see that the integral curves are closed, and since the origin is the only critical point, all motions of the spring are periodic. Therefore, even though equation (8) cannot be solved explicitly, a great deal about the motion can be determined from the phase plot and the techniques associated with sketching it.

5.5a Pendulum Oscillations

One of the most basic applications of nonlinear vibrations is the analysis of the nonlinear oscillations in the motion of a pendulum. Suppose that a mass

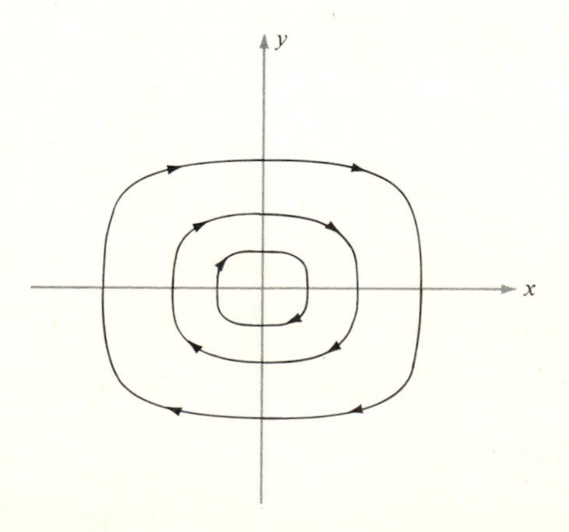

Figure 5.10 Phase plot for (9).

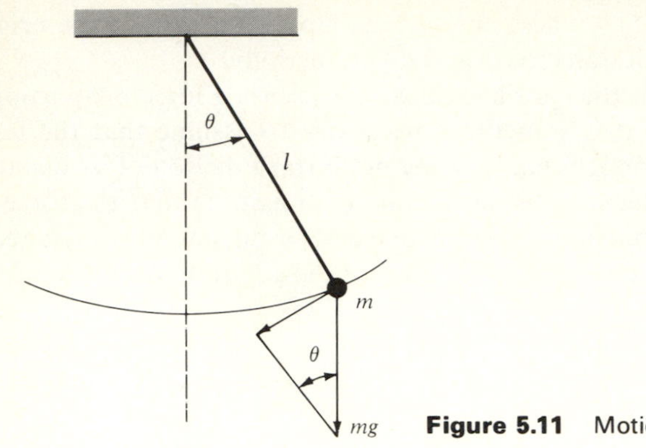

Figure 5.11 Motion of a pendulum.

m is attached to the end of a thin rod of length l that is suspended from a support (see Figure 5.11). For each time $t \geq 0$ let $\theta(t)$ denote the radian angle that the rod makes to a vertical line, with the positive direction counterclockwise. Neglecting any forces of resistance and the weight of the rod, it follows that if $s(t)$ denotes the length of arc along the path of motion of the mass (with $s = 0$ vertical, $s > 0$ counterclockwise measurement, and $s < 0$ clockwise measurement), then according to Newton's law the total force F_T acting on the mass is

$$F_T(t) = ms''(t)$$

and it follows from Figure 5.11 that

$$F_T(t) = mg \sin \theta(t)$$

Since $s(t) = l\theta(t)$ for all $t \geq 0$, it follows that

$$ml\theta''(t) = -mg \sin \theta(t)$$

and hence the equation of motion for the undamped pendulum is

$$\theta'' + \frac{g}{l} \sin \theta = 0 \tag{11}$$

where g is the acceleration due to gravity. Setting $x = \theta$ and $y = \theta'$, equation (11) is equivalent to the system

$$\begin{aligned} x' &= y \\ y' &= -\frac{g}{l} \sin x \end{aligned} \tag{12}$$

Since

$$\frac{dy}{dx} = \frac{(-g \sin x)/l}{y} \qquad \text{or} \qquad y\,dy + \frac{g}{l} \sin x\,dx = 0$$

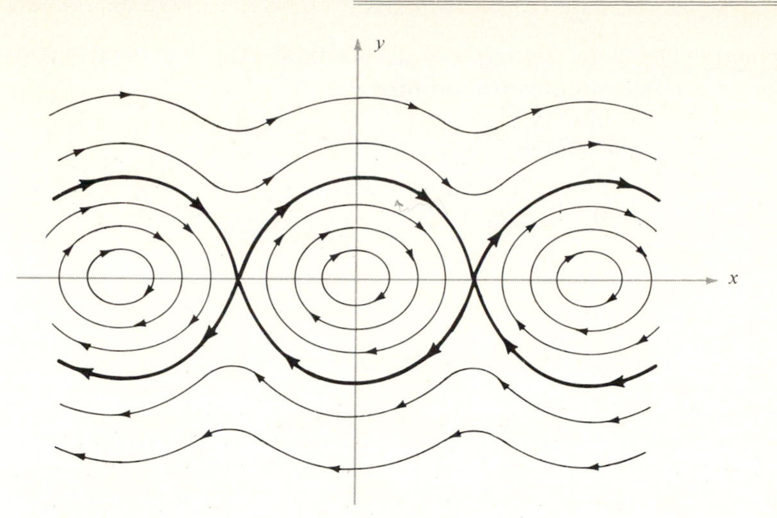

Figure 5.12 Phase plot for pendulum oscillations.

a first integral of (12) is easily seen to be

$$\frac{1}{2}y^2 + \frac{g}{l}\int_0^x \sin s\, ds = \frac{1}{2}y^2 + \frac{g}{l}(1 - \cos x)$$

Therefore, it follows that the trajectories of (12) lie on the level curves

$$\frac{1}{2}y^2 + \frac{g}{l}(1 - \cos x) = E \tag{13}$$

where $E \geq 0$ is a constant (in fact, mE is the total energy of the system). The phase plot for (12) is sketched in Figure 5.12. The critical points for this system are $y = 0$ and $x = 0, \pm\pi, \pm 2\pi, \ldots$. Also, since the left side of (13) is 2π periodic in x, it suffices to sketch the plot in the strip $-\pi < x < \pi$, and then extend periodically for all x. For small values of E in (13) we see from Proposition 1 that the trajectories correspond to periodic orbits and are simple closed curves that surround the origin. In this case it is easy to check that the solutions are periodic if $E < 2g/l$ and not periodic (in fact, unbounded) if $E > 2g/l$. When $E = 2g/l$ equation (13) becomes

$$y^2 = \frac{2g}{l}(1 + \cos x) \tag{14}$$

The solutions with these trajectories are not periodic or unbounded; they converge to one of the critical points. These trajectories are indicated with a somewhat darker curve in Figure 5.12. If $0 < E < 2g/l$, the *amplitude* A of the periodic motion is the intersection of the integral curve with the positive x axis (note that this is the maximum distance that the pendulum gets from equilibrium). Therefore, setting $x = A$, $y = 0$ in (13), we obtain that the amplitude A is related to E by the formula

$$\cos A = 1 - \frac{lE}{g} \tag{15}$$

where $0 < A < \pi$ and $0 < E < 2g/l$.

PROBLEMS

5.5-1. Sketch a phase plot for each of the following systems, indicating clearly all critical points. Also, classify all initial points that determine periodic solutions and all initial points that determine solutions having a limit as $t \to +\infty$.

(a) $x' = y - xy$

$y' = x^2 - x$

(b) $x' = 6(y + 1)(y - x^2)$

$y' = -x(y - x^2)$

(c) $x' = -x(y + 1)$

$y' = 3x^3(y + 1)$

5.5-2. Each of the following second order equations model the motion of a mass-spring system. Write each of these as a first order system and sketch and compare the phase plots. Moreover, compute the maximum velocity reached as a function of the amplitude $A > 0$ (the amplitude is the x coordinate of the intersection of the integral curve with the positive x axis).

(a) $z'' + z^5 = 0$

(b) $z'' + z|z| = 0$

(c) $z'' + \dfrac{z}{1 + z^2} = 0$

5.5-3. Use the half-angle formula for cosine to show that the curve defined by the equation in (14) is also defined by the equation

$$y = \pm 2\sqrt{\frac{g}{l}} \cos \tfrac{1}{2}x$$

5.5-4. For very small oscillations of the pendulum it is often sufficiently accurate to use the approximation $\sin x \cong x$ to describe the motion. Therefore, replace $\sin x$ by x in (12) to obtain the linear system

$$x' = y$$

$$y' = -\frac{gx}{l} \tag{*}$$

(a) Determine a first integral $\bar{U}$ for (*) and sketch the level curves $\bar{U}(x, y) = k$.

(b) Replace $\cos x$ by $1 - x^2/2$ in equation (13) [the level curves of (12)] and compare the resulting family of curves with the family $\bar{U}(x, y) = k$, where k is a constant and $\bar{U}$ is as in part *a*.

(c) Determine the general solution to (*) explicitly and show that all nontrivial solutions are periodic with the same period T, and compute T in terms of g and l. Does T increase or decrease if l is increased?

*5.6 PHASE PLOTS FOR LINEAR SYSTEMS

Even though solutions to linear systems with constant coefficients can be explicitly determined, it is still convenient and helpful to describe the forms of phase plots that can occur. Suppose that a_1, a_2, b_1 and b_2 are real *constants*, and consider the linear system

$$x' = a_1 x + b_1 y \qquad y' = a_2 x + b_2 y \tag{1}$$

as well as the associated initial value problem

$$x' = a_1 x + b_1 y \quad x(0) = x_0 \qquad y' = a_2 x + b_2 y \quad y(0) = y_0 \tag{2}$$

According to the results of Section 5.3 [see, in particular, (7) in Section 5.3] the solutions to (1) are determined by the roots λ_1 and λ_2 of the characteristic polynomial of (1):

$$\lambda^2 - (a_1 + b_2)\lambda + (a_1 b_2 - a_2 b_1) = 0 \tag{3}$$

The principal objective of this section is to indicate methods for sketching phase plots for (1) from the characteristic roots. This discussion is divided into four separate cases, each of which has various subcases that must be taken into account.

5.6*a* The Characteristic Roots Are Real, Distinct, and Nonzero

The model example in this case is the equation

$$x' = \lambda_1 x \qquad y' = \lambda_2 y \tag{4}$$

Since

$$\frac{dy}{dx} = \frac{\lambda_2}{\lambda_1}\frac{y}{x}$$

we have

$$\frac{dy}{y} = \frac{\lambda_2}{\lambda_1}\frac{dx}{x} \qquad \text{and hence} \qquad \ln|y| = \frac{\lambda_2}{\lambda_1}\ln|x| + c$$

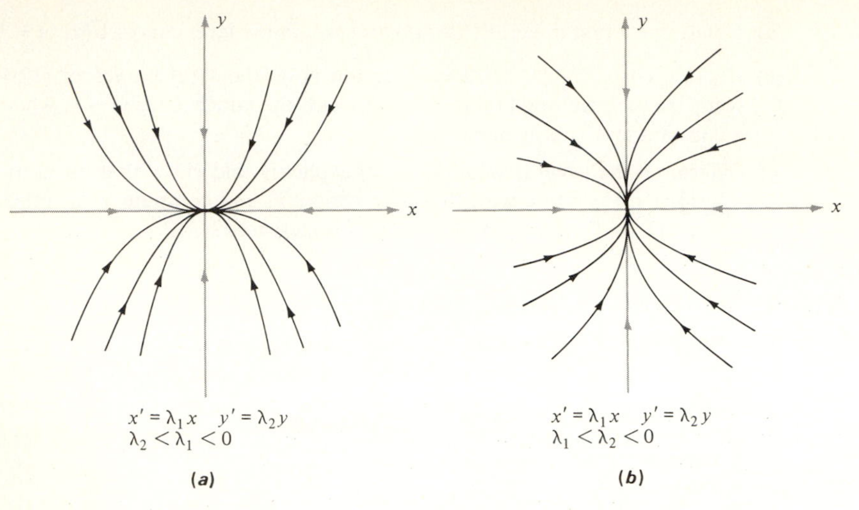

Figure 5.13 Nodal points.

From this it follows that the integral curves of (4) have the form

(4′) $$|y| = k|x|^{\lambda_2/\lambda_1} \qquad \text{where } k \geq 0 \text{ is a constant}$$

The two main distinctions are whether λ_1 and λ_2 have the same or the opposite sign. When λ_1 and λ_2 have the same sign, the origin [which is the only critical point of (4) since both λ_1 and λ_2 are nonzero] is called a *node*, and the curves resemble a family of parabolas that are tangent at the origin. A sketch of the integral curves (4′) in this case is given in Figure 5.13 The flow is toward the origin since λ_1 and λ_2 are both negative. Reversing the inequalities in Figure 5.13*a* and *b* leads to the same integral curves, except that the flow is in the opposite direction. When λ_1 and λ_2 are negative the node is said to be *asymptotically stable*, and when λ_1 and λ_2 are positive the node is said to be *unstable*. When λ_1 and λ_2 are of opposite sign, the integral curves (4′) resemble a family of hyperbolas with the axes as asymptotes (see Figure 5.14), and the origin is called a *saddle point*.

In general, when the characteristic roots are real, distinct, and nonzero the phase plot of equation (1) resembles a distortion of the corresponding plots for equation (4). A procedure that aids in sketching a phase plot for (1) in this case is to first determine which integral curves are straight lines through the origin by assuming that there is a number m such that $y = mx$ in the equation

$$\frac{dy}{dx} = \frac{a_2 x + b_2 y}{a_1 x + b_1 y}$$

Then $dy/dx = m$ and we obtain

(5) $$m = \frac{a_2 + b_2 m}{a_1 + b_1 m} \qquad \text{and hence} \qquad b_1 m^2 + (a_1 - b_2)m - a_2 = 0$$

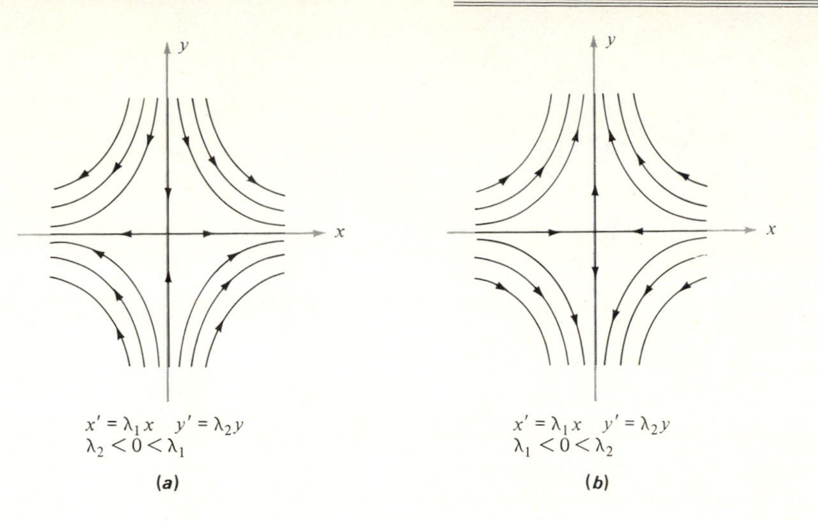

Figure 5.14 Saddle points.

(Note that if $b_1 = 0$ then $x = 0$ is an integral curve.) It can be shown that in this case one can always obtain *two distinct real values of m* (using the convention that $m = \infty$ if $b_1 = 0$), and so two lines through the origin are always obtained. These two lines take the place of the axes in Figures 5.13 and 5.14. The remaining problem is to determine to which line the "parabolas" are tangent when λ_1 and λ_2 are of the same sign (see Figure 5.13*a* and *b*) and the direction of the flow when λ_1 and λ_2 are of opposite sign (see Figure 5.14*a* and *b*). This information can be readily obtained by exhibiting explicitly the forms of the solutions in terms of λ_1 and λ_2 {if $b_1 \neq 0$, for example, then $x(t) = c_1 e^{\lambda_1 t} + c_2 e^{\lambda_2 t}$ and $y(t) = [x'(t) - a_1 x(t)]/b_1$}.

▶ **EXAMPLE 5.6.1**

Consider the system

$$x' = -x + y \qquad (6)$$
$$y' = 4x - y$$

The characteristic equation for this system is

$$\det\begin{pmatrix} -1-\lambda & 1 \\ 4 & -1-\lambda \end{pmatrix} = (\lambda + 1)^2 - 4 = 0$$

and hence $\lambda_1 = 1$ and $\lambda_2 = 3$ are the characteristic roots. Therefore the origin is a nodal point. Setting $y = mx$ in the equation

$$\frac{dy}{dx} = \frac{4x - y}{-x + y}$$

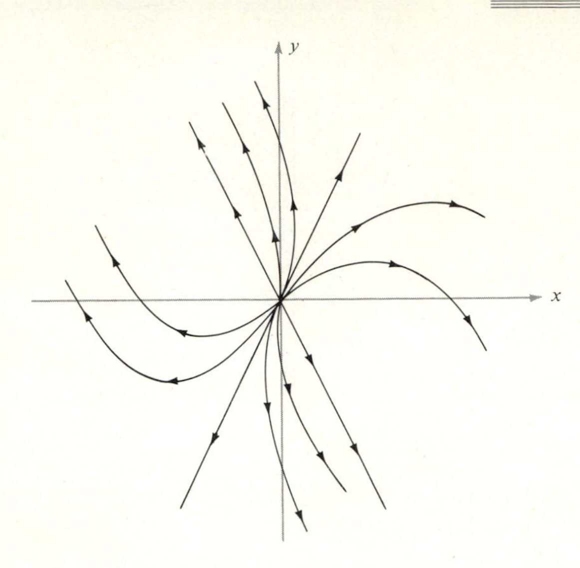

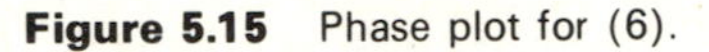

Figure 5.15 Phase plot for (6).

one obtains

$$m = \frac{4 - m}{-1 + m} \qquad \text{and hence} \qquad m^2 - 4 = 0$$

Therefore, the lines $y = \pm 2x$ are integral curves for (6). Setting $x(t) = c_1 e^t + c_2 e^{3t}$ and using the first equation in (6), it follows that

$$y(t) = x'(t) + x(t) = 2c_1 e^t + 4c_2 e^{3t}$$

Since $x(t)$, $y(t) \to 0$ as $t \to -\infty$, the slopes of the integral curves at the origin are the same as

$$\lim_{t \to -\infty} \frac{y(t)}{x(t)} = \lim_{t \to -\infty} \frac{2c_1 e^t + 4c_2 e^{3t}}{c_1 e^t + c_2 e^{3t}} = 2$$

whenever $c_1 \neq 0$. Thus the integral curves are tangent to the line $y = 2x$ at the origin, and the phase plot for (6) has the form indicated in Figure 5.15.

5.6*b* There Is a Nonzero, Real Double Root of the Characteristic Equation

There are two model examples that illustrate this case:

$$\text{(7)} \qquad \begin{aligned} x' &= \lambda_0 x \\ y' &= \lambda_0 y \end{aligned}$$

and

$$\text{(8)} \qquad \begin{aligned} x' &= \lambda_0 x \\ y' &= \lambda_0 x + \lambda_0 y \end{aligned}$$

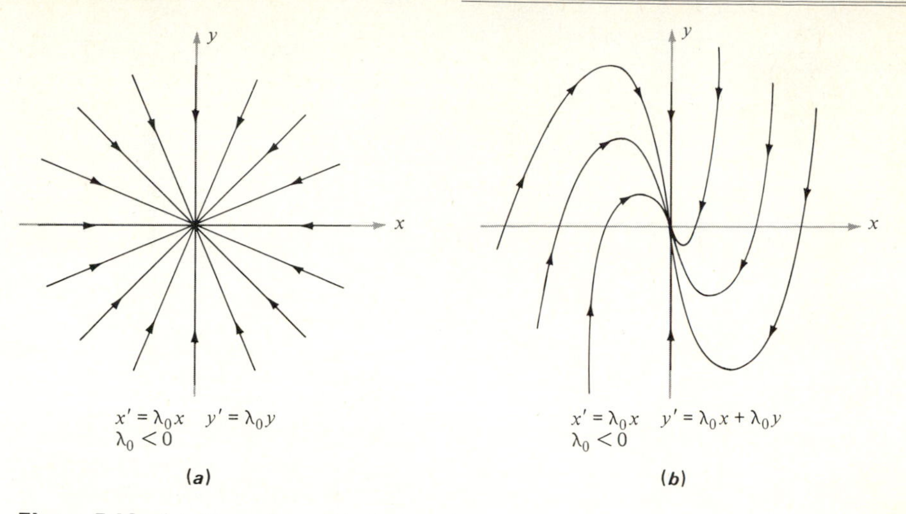

Figure 5.16 (*a*) Star point. (*b*) Nodal point.

where $\lambda_0 \neq 0$. Since $dy/dx = y/x$ in (7) and since $dy/dx = 1 + y/x$ in (8) and these are first order linear equations (see Section 1.2), it follows easily that a family of first integrals for (7) and (8) are

(7′) $\quad y = kx \qquad k$ a constant

and

(8′) $\quad y = kx + x \ln |x| \qquad k$ a constant

respectively (we use the convention that these curves are the y axis when $k = \infty$). Therefore, the phase plots for (7) and (8) resemble those in Figure 5.16*a* and *b*. The direction of the flow is toward the origin since $\lambda_0 < 0$, and in this case the origin is said to be *asymptotically stable*. If $\lambda_0 > 0$ the phase plots are exactly the same except that the flow is away from the origin (when $\lambda_0 > 0$ the origin is said to be *unstable*). The origin is a star point for (1) only if (1) has the form (7). Thus if λ_0 is a nonzero double characteristic root of (1) and (1) is coupled [i.e., not of the form (7)], then the origin is a single tangent nodal point and its phase plot resembles Figure 5.16*b*, except the straight line integral curve is not necessarily vertical. If $b_1 \neq 0$, the integral curve that is a line through the origin can be found as in the preceding case by noting that its equation is $y = mx$, where m is a solution to (5) [in this case (5) has exactly one double root].

▶ **EXAMPLE 5.6-2**

Consider the system

(9)
$$x' = y$$
$$y' = -x + 2y$$

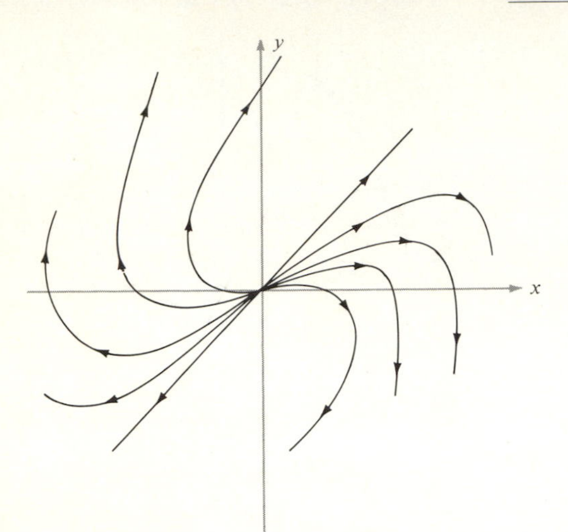

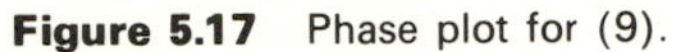

Figure 5.17 Phase plot for (9).

The characteristic equation is

$$\det\begin{pmatrix} -\lambda & 1 \\ -1 & 2-\lambda \end{pmatrix} = -\lambda(2-\lambda) + 1 = \lambda^2 - 2\lambda + 1 = 0$$

and it follows that $\lambda_0 = 1$ is a double root. Since (9) is coupled and $b_1 \neq 0$, the phase plot is similar to Figure 5.16*b*. Set $y = mx$ to obtain

$$m = \frac{dy}{dx} = \frac{-x + 2mx}{mx} = -\frac{1}{m} + 2$$

and hence $m^2 - 2m + 1 = 0$. Thus $m = 1$ and the straight line integral curve is $y = x$. Noting that $dy/dx = -x/y + 2$, we see that $dy/dx = 0$ whenever $y = \frac{1}{2}x$, and since the phase plot for (9) resembles Figure 5.16*b*, it is easy to see that the phase plot for (9) has the form in Figure 5.17. The direction of flow is away from the origin since $\lambda_0 = 1 > 0$.

5.6*c* The Roots of the Characteristic Equation Are Conjugate Complex

The model example to illustrate this case is

$$\begin{aligned} x' &= \alpha x + \beta y \\ y' &= -\beta x + \alpha x \end{aligned} \tag{10}$$

where α and β are real numbers with $\beta > 0$. It is easy to check that the characteristic roots of (10) are $\lambda = \alpha \pm \beta i$. The form of the phase plot for (10) becomes apparent when *polar coordinates* are introduced. For if x, y is a solution to (1) and $p(t) \equiv x(t)^2 + y(t)^2$, then (suppressing the variable t)

$$\begin{aligned} p' &= 2xx' + 2yy' = 2x(\alpha x + \beta y) + 2y(-\beta x + \alpha y) \\ &= 2\alpha(x^2 + y^2) = 2\alpha p \end{aligned}$$

Since $x^2 + y^2$ is a solution p of the equation $p' = 2\alpha p$, it follows that $p(t) = p(0)e^{2\alpha t}$ and hence

$$x(t)^2 + y(t)^2 = [x(0)^2 + y(0)^2]e^{2\alpha t} \qquad \text{for all } t \tag{11}$$

Let $\theta(t)$ denote the angle that the line segment from (0, 0) to $(x(t), y(t))$ makes with the positive x axis [with counterclockwise being the positive direction of $\theta(t)$]. Then so long as $x(t) \neq 0$ it follows that $\tan \theta(t) = y(t)/x(t)$, and differentiating each side of this equation with respect to time t, we obtain from (10) that

$$(\sec^2 \theta)\theta' = x^{-2}(xy' - yx') = -x^{-2}\beta(x^2 + y^2)$$

Since $\sec^2 \theta = (x^2 + y^2)/x^2$, it follows that $(x^2 + y^2)\theta' = -\beta(x^2 + y^2)$, and hence $\theta'(t) = -\beta$ so long as $x(t) \neq 0$. Since θ can be selected so that it is continuous, it follows that

$$\theta(t) = \theta(0) - \beta t \qquad \text{for all } t \tag{12}$$

Since $\beta > 0$ we see immediately that the solutions move around the origin in a clockwise direction as t tends to $+\infty$. When $\alpha = 0$ we have from (11) that the solutions remain on a circle with center at the origin. Moreover, if $\alpha < 0$ the solutions spiral to the origin as $t \to +\infty$, and if $\alpha > 0$ the solutions spiral away from the origin as $t \to +\infty$. Sketches of the phase plots in these three cases are indicated in Figure 5.18*a*, *b*, and *c*.

Note that with the polar coordinate variables $r = \sqrt{x^2 + y^2}$ and $\theta = \arctan (y/x)$, we obtain from (12) that $t = (\theta_0 - \theta)/\beta$, and using equation (11) it follows that the integral curves of (10) can be described by the polar coordinate equations

$$r = ke^{-\alpha\theta/\beta} \qquad \text{where } k > 0 \text{ is a constant} \tag{13}$$

In general, when the characteristic roots of (1) are the conjugate complex numbers $\alpha \pm i\beta$, where α and β are real and $\beta > 0$, the phase plot for (1) resembles those in Figure 5.18. When $\alpha \neq 0$, the phase plots are spirals that

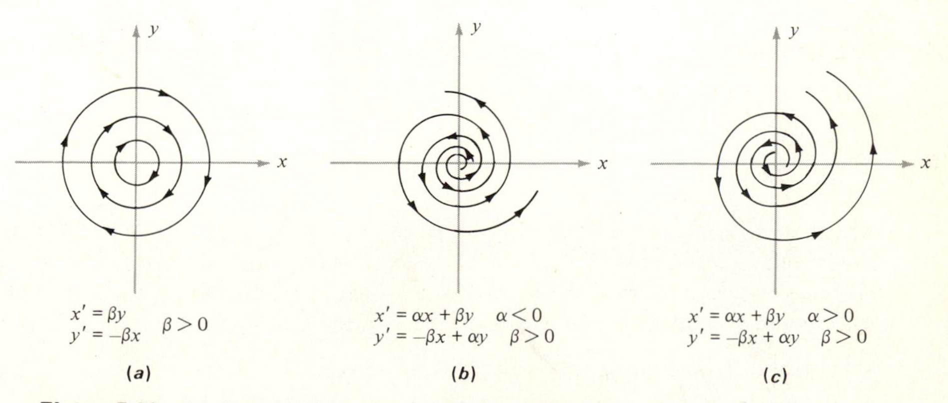

Figure 5.18 (*a*) Center point. (*b*) Focal point. (*c*) Focal point.

spiral toward the origin of $\alpha < 0$ (in this case the focus is said to be *asymptotically stable*) and away from the origin if $\alpha > 0$. If x, y is a solution to (1), then $y' = a_2 x$ when $y = 0$, and it follows that the spirals are clockwise if $a_2 > 0$ and counterclockwise if $a_2 < 0$. When $\alpha = 0$ the phase plot is a family of ellipses centered at the origin. A first integral U of (1) can be found readily in this case. Recall that a first integral is a solution to the total differential equation

$$(a_2 x + b_2 y)\,dx - (a_1 x + b_1 y)\,dy = 0 \tag{14}$$

Since the roots of the characteristic equation (3) are $\pm i\beta$ in this case (that is, $\alpha = 0$) it is immediately seen from the quadratic formula that $a_1 + b_2 = 0$, which is precisely the condition that (14) is exact (see Section 1.7). Therefore a first integral can be constructed immediately by determining a solution to (14).

EXAMPLE 5.6-3

Consider the system

$$\begin{aligned} x' &= x - y \\ y' &= 5x - y \end{aligned} \tag{15}$$

Since

$$\det\begin{pmatrix} 1-\lambda & -1 \\ 5 & -1-\lambda \end{pmatrix} = (\lambda - 1)(\lambda + 1) + 5 = \lambda^2 + 4$$

the characteristic roots are $\lambda = \pm 2i$, and hence the origin is a center point. A solution of the exact total differential equation

$$(5x - y)\,dx - (x - y)\,dy = 0 \qquad \text{is} \qquad U(x, y) = \tfrac{5}{2}x^2 - xy + \tfrac{1}{2}y^2$$

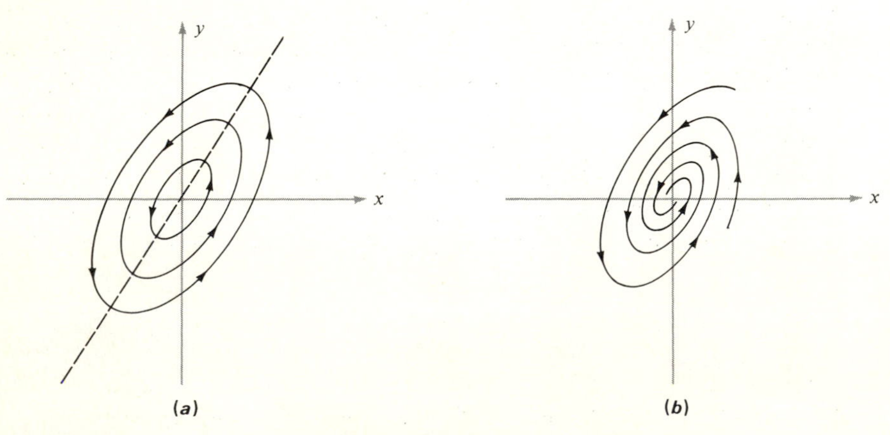

Figure 5.19 (*a*) Phase plot for (15). (*b*) Phase plot for (16).

and hence

$$5x^2 - 2xy + y^2 = k \qquad k \text{ a constant}$$

is the family of integral curves of (15). This is a family of ellipses centered at the origin (see Figure 5.19*a*). In this case the student should review rotational transformations in the plane—that is, the transformation of general second order equations into normal form (in this example one should rotate through an angle θ so that $\cot 2\theta = -2$, and hence $\theta \cong 1.33$ radians or 76°).

Even in the most complicated cases where first integrals cannot be easily computed or a polar coordinate transformation does not simplify the situation, one can still easily obtain the essential character of the phase plot. As an illustration consider the equation

$$\begin{aligned} x' &= -y \\ y' &= 5x - 2y \end{aligned} \tag{16}$$

Since

$$\det\begin{pmatrix} -\lambda & -1 \\ 5 & -2-\lambda \end{pmatrix} = \lambda(2+\lambda) + 5 = \lambda^2 + 2\lambda + 5$$

it follows that $\lambda = -1 \pm 2i$ are the characteristic roots of (16). Since these roots are complex conjugates and have negative real parts, we know that the solutions spiral toward the origin. Using the first equation in (16) we see that the flow in the x direction is negative in the upper half plane and positive in the lower half plane. From this observation it follows that the solutions must spiral in a counterclockwise direction. Therefore, Figure 5.19*b* gives a rough sketch of the phase plot for (16).

5.6*d* Zero Is a Root of the Characteristic Equation

The model examples in this case are the equations

$$\begin{aligned} x' &= \lambda_1 x \\ y' &= 0 \end{aligned} \tag{17}$$

where $\lambda_1 \neq 0$, and

$$\begin{aligned} x' &= y \\ y' &= 0 \end{aligned} \tag{18}$$

Equation (17) corresponds to zero being a simple root, and equation (18) corresponds to zero being a double root [we are excluding the trivial case when the right-hand side of (1) is identically zero]. Since $x(t) = c_1 e^{\lambda_1 t}$ and $y(t) = c_2$ in (17) and $x(t) = c_2 t + c_1$ and $y(t) = c_1$ in (18), it is easy to see that

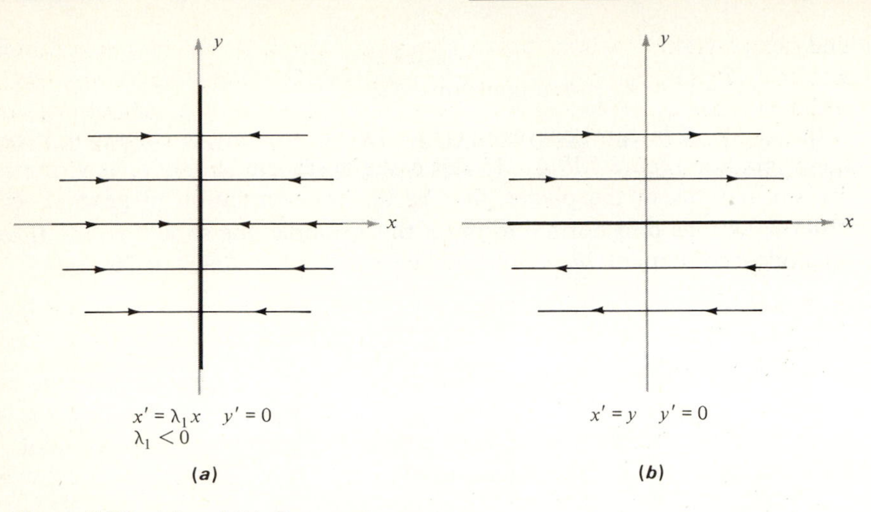

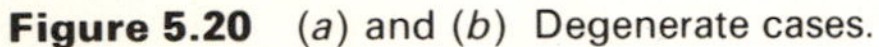

Figure 5.20 (*a*) and (*b*) Degenerate cases.

the phase plots for these equations have the forms indicated in Figure 5.20*a* and *b*. In the general case when zero is a characteristic root [and at least one coefficient in (1) is nonzero], the integral curves of (1) are straight lines and there is a line of critical points through the origin. These assertions follow easily from the fact that solutions x, y to (1) have the form $x(t) = c_1 + c_2 e^{\lambda_1 t}$, $y(t) = d_1 + d_2 e^{\lambda_1 t}$ if zero is a simple root and $x(t) = c_1 + c_2 t$ and $y(t) = d_1 + d_2 t$ if zero is a double root.

EXAMPLE 5.6-4

Consider the system

$$x' = x + 2y \qquad (19)$$
$$y' = x + 2y$$

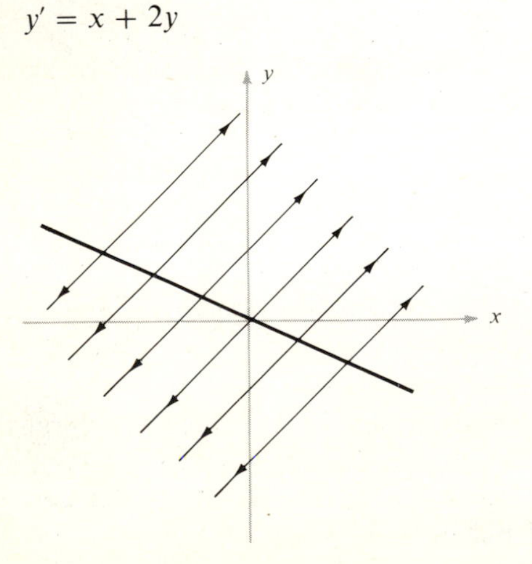

Figure 5.21 Phase plot of (19).

The characteristic roots of this equation are $\lambda = 0, 3$, and it is easy to check that the critical point set is $\{(x_c, y_c) : y_c = -x_c/2\}$. Since $dy/dx \equiv 1$ the integral curves are the lines $y = x + c$, where c is a constant. A phase-plot sketch is indicated in Figure 5.21. The direction of flow is away from the line of critical points since the second characteristic root is positive.

PROBLEMS

5.6-1. Sketch a phase plot for each of the following linear systems. In each case describe what type of critical point the origin is and indicate whether or not the origin is asymptotically stable.

(a) $x' = -2y$, $y' = 6x$

(b) $x' = y$, $y' = -2x - 3y$

(c) $x' = 2x - 3y$, $y' = 2x - 2y$

(d) $x' = 4x + 2x$, $y' = -x + y$

(e) $x' = -x + y$, $y' = -4x - y$

(f) $x' = -x + 3y$, $y' = x + y$

(g) $x' = x - y$, $y' = x - y$

(h) $x' = -4x - 4y$, $y' = x$

5.6-2. A unit mass attached to a linear spring having spring constant $k = 1$ has

$$x' = y$$
$$y' = -x - \mu y$$

as its equations of motion, where $\mu \geq 0$ is the resistance coefficient, $x = x(t)$ is the position of the mass relative to equilibrium, and $y = y(t)$ is the velocity of the mass. Sketch phase plots for this system for various values of μ and interpret these sketches relative to the motion of the mass.

5.6-3. In each of the following problems η is a real parameter. Sketch phase plots and describe the type of critical point the origin is relative to the values of η in $(-\infty, \infty)$.

(a) $x' = -x + \eta y$, $y' = x - y$

(b) $x' = -\eta x + y$, $y' = -x - y$

(c) $x' = y$, $y' = -x + \eta y$

5.6-4. Write each of the following second order linear equations as two first order systems by setting $x = z$ and $y = z'$. Sketch a phase plot and indicate the type of critical point of the origin.

(a) $z'' + 4z = 0$

(b) $z'' + 3z' + 4z = 0$

(c) $z'' + 5z' + 4z = 0$

(d) $z'' - 3z' + 2z = 0$

(e) $z'' + z' - 2z = 0$

(f) $z'' + 4z' + 4z = 0$

***5.6-5.** Suppose that λ_1 and λ_2 are real and distinct characteristic roots of (1). Show that if x, y is a solution to (1) and

$$\bar{x}(t) = a_2 x(t) + (\lambda_1 - a_1)y(t)$$
$$\bar{y}(t) = a_2 x(t) + (\lambda_2 - a_1)y(t)$$

then $\bar{x}$, $\bar{y}$ is a solution to the system

$$\bar{x}' = \lambda_1 \bar{x}$$
$$\bar{y}' = \lambda_2 \bar{y}$$

***5.6-6.** Show that the total differential equation (14) is exact if and only if $a_1 + b_2 = 0$. Show further that if $a_1 = -b_2 = \delta$, then the integral curves of (1) are

$$a_2 x^2 - 2\delta xy - b_1 y^2 = k \qquad (*)$$

where k is a constant.

(a) Show that if $\delta^2 - a_2 b_1 < 0$ then the curve defined by (*) is a hyperbola and that the origin is a saddle point.

(b) Show that if $\delta^2 - a_2 b_1 > 0$ then the curve defined by (*) is an ellipse and that the origin is a focal point.

*5.7 LOCAL BEHAVIOR ABOUT A CRITICAL POINT

In the preceding section it was shown how to sketch a phase plot for a linear system with constant coefficients, using the characteristic roots. The purpose of this section is to indicate that a sketch of the phase plot of a nonlinear system in the neighborhood of a critical point can be approximated by these linear techniques. Assume that f and g are smooth functions in the plane, and consider the autonomous system

$$\begin{aligned} x' &= f(x, y) \\ y' &= g(x, y) \end{aligned} \qquad (1)$$

Assume further that (x_c, y_c) is a critical point of (1):

$$f(x_c, y_c) = g(x_c, y_c) = 0 \qquad (2)$$

We show that in certain situations the behavior of solutions to (1) in a neighborhood of the critical point (x_c, y_c) is essentially the same as that of a linear system that can be obtained directly from (1) and the critical point (x_c, y_c).

The proofs of these results and in fact a precise statement of these results is beyond the scope of this book; however, an indication of what is valid and how to actually apply the results should be understandable and helpful to the reader, and so this is the aim of this section.

Since f and g are "smooth" functions, each of the partial derivatives $f_x(x, y)$, $f_y(x, y)$, $g_x(x, y)$, and $g_y(x, y)$ exists. Evaluating these partial derivatives at the critical point (x_c, y_c), we may consider the linear system

$$\begin{aligned} x' &= f_x(x_c, y_c)x + f_y(x_c, y_c)y \\ y' &= g_x(x_c, y_c)x + g_y(x_c, y_c)y \end{aligned} \tag{3}$$

This system is called the *linearization of (1) about the critical point* (x_c, y_c). As a simple illustration, consider the system

$$\begin{aligned} x' &= x^2 - x + \sin 2y \\ y' &= e^y - 1 + 3x \end{aligned} \tag{4}$$

Here $f(x, y) = x^2 - x + \sin 2y$, $g(x, y) = e^y - 1 + 3x$, and so

$$f_x(x, y) = 2x - 1 \qquad f_y(x, y) = 2 \cos 2y$$
$$g_x(x, y) = 3 \qquad g_y(x, y) = e^y$$

Since (0, 0) is a critical point and

$$f_x(0, 0) = -1 \qquad f_y(0, 0) = 2 \qquad g_x(0, 0) = 3 \qquad g_y(0, 0) = 1$$

we see that

$$\begin{aligned} x' &= -x + 2y \\ y' &= 3x + y \end{aligned}$$

is the linearization of (4) about (0, 0).

As an indication of how equation (3) arises from equation (1), recall that using the first order terms of the Taylor expansion for f about (x_c, y_c), we have

$$f(x_c + x, y_c + y) \cong f(x_c, y_c) + f_x(x_c, y_c)x + f_y(x_c, y_c)y$$

and, similarly,

$$g(x_c + x, y_c + y) \cong g(x_c, y_c) + g_x(x_c, y_c)x + g_y(x_c, y_c)y$$

Since $f(x_c, y_c) = g(x_c, y_c) = 0$, we see that the right-hand side of (1) in the neighborhood of (x_c, y_c) is approximated by the right-hand side of (3) in a neighborhood of the origin.

The results of the preceding section show that the characteristic roots of (3) determine the behavior of the solutions to (3), and since

$$\det\begin{pmatrix} f_x(x_c, y_c) - \lambda & f_y(x_c, y_c) \\ g_x(x_c, y_c) & g_y(x_c, y_c) - \lambda \end{pmatrix}$$
$$= [\lambda - f_x(x_c, y_c)][\lambda - g_y(x_c, y_c)] - g_x(x_c, y_c)f_y(x_c, y_c)$$

we have that

$$\lambda^2 - [f_x(x_c, y_c) + g_y(x_c, y_c)]\lambda + [f_x(x_c, y_c)g_y(x_c, y_c) - g_x(x_c, y_c)f_y(x_c, y_c)] = 0 \tag{5}$$

is the characteristic equation of the linearization of (1) about (x_c, y_c). Three important types of behavior for the solutions to (3) are indicated in Figure 5.22.

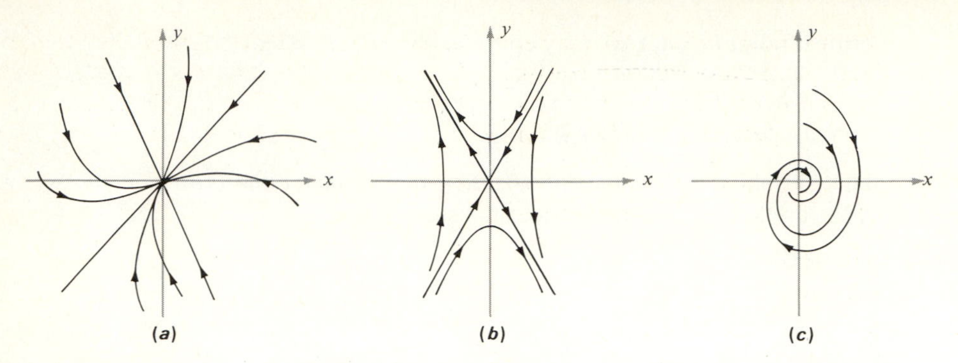

Figure 5.22 (*a*) Nodal point: Characteristic roots distinct and negative. (*b*) Saddle point: Characteristic roots real and opposite sign. (*c*) Spiral point: Characteristic roots conjugate complex and negative real parts.

There are possible behavior patterns that are analogous for the nonlinear equation (1) in the neighborhood of a critical point. We will not give precise definitions describing this behavior but will indicate the fundamental properties. Suppose that (x_c, y_c) is a critical point of (1). The three main types of behavior are indicated in Figure 5.23. *An asymptotically stable double-tangent node* (Figure 5.23*a*) has the property that there are two distinct lines L_1 and L_2 through the critical point such that the solutions on one integral curve approach the critical point in one of the directions tangent to L_1, and the solutions on every other integral curve beginning sufficiently near the critical point approach it in one of the directions tangent to L_2. A *saddle point* (Figure 5.23*b*) has the property that there are two distinct lines L_1 and L_2 through the critical point such that the solutions on one integral curve approach the critical point in one of the directions tangent to L_1, and the solutions on another integral curve tangent to L_2 move away from the critical point. In fact, there is a disk D centered at the critical point such that all solutions beginning in D, except those on the integral curve tangent to L_1, exit

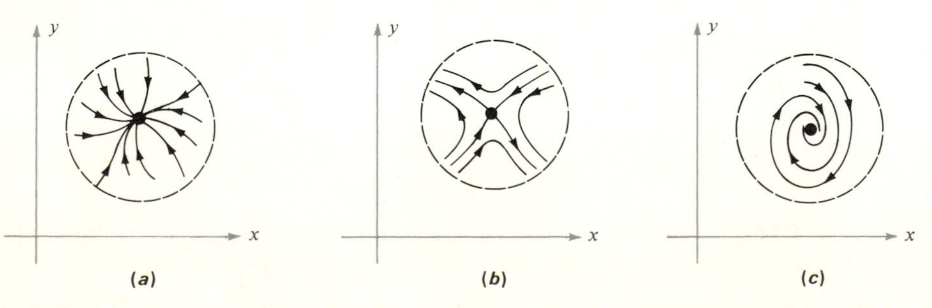

Figure 5.23 (*a*) Asymptotically stable double-tangent node. (*b*) Saddle point (unstable). (*c*) Asymptotically stable spiral point.

from D after a finite time. An *asymptotically stable spiral point* has the property that every solution starting sufficiently near the critical point continually spirals around the critical point and approaches it as $t \to +\infty$ (Figure 5.23c).

A fundamental and important result is the following theorem:

Theorem 5.7-1

Suppose that (x_c, y_c) is a critical point of (1) and that λ_1 and λ_2 are the characteristic roots of the linearized equation (3) of (1) about (x_c, y_c) [that is, λ_1 and λ_2 are the roots of (5)].

(a) If λ_1 and λ_2 are real with $\lambda_1 < \lambda_2 < 0$, then (x_c, y_c) is an asymptotically stable double-tangent node.

(b) If λ_1 and λ_2 are real with $\lambda_1 < 0 < \lambda_2$, then (x_c, y_c) is a saddle point.

(c) If λ_1 and λ_2 are conjugate complex and have negative real parts, then (x_c, y_c) is an asymptotically stable spiral point.

Linearization techniques give a detailed analysis of the behavior of trajectories in a neighborhood of the critical point (x_c, y_c). In fact, a *sketch of the phase plane of the linearized equation translated to the critical point* (x_c, y_c) *gives a good approximation of the phase plot for (1) locally about* (x_c, y_c). For example, if the linearized trajectories spiral clockwise (or counterclockwise) about the origin, then the trajectories of (1) spiral clockwise (or counterclockwise) about (x_c, y_c). Also, the tangent lines L_1 and L_2 mentioned in the description of the asymptotically stable double-tangent node and the saddle point are the straight line integral curves of the linearized equation translated to the critical point. As an indication of how one can apply Theorem 5.7-1, consider the following examples.

EXAMPLE 5.7-1

Consider the system

$$x' = -2x + y + xy - y^2 + 2$$

$$y' = x - 3y - 1$$

In order to determine the critical points use the second equation to obtain $x = 3y + 1$, and then substitute this expression for x in the first equation to obtain

$$-2(3y + 1) - y + (3y + 1)y - y^2 + 2 = 0$$

Therefore, upon simplification, $2y^2 - 6y = 0$ and so $y = 0, 3$, and it follows that (1, 0) and (10, 3) are the critical points of (6). In this particular case,

$$f_x(x, y) = -2 + y \qquad f_y(x, y) = -1 + x - 2y$$
$$g_x(x, y) = 1 \qquad g_y(x, y) = -3$$

and it follows that the linearization of (6) about the critical point (1, 0) is

$$x' = -2x$$
$$y' = x - 3y$$

Since the characteristic equation of this system is $(\lambda + 2)(\lambda + 3) = 0$, the characteristic roots are -2 and -3, and we have from part *a* of Theorem 5.7-1 that (1, 0) is an asymptotically stable double-tangent node. The linearization of (6) about the critical point (10, 3) is

$$x' = x + 3y$$
$$y' = x - 3y$$

The characteristic equation is $(1 - \lambda)(-3 - \lambda) - 3 = 0$; it follows that $\lambda^2 + 2\lambda - 6 = 0$, and hence $\lambda = -1 \pm \sqrt{7}$ are the characteristic roots. From part *b* of Theorem 5.7-1 we know that (10, 3) is a saddle point.

EXAMPLE 5.7-2

Suppose that α is a real parameter and consider the nonlinear system

$$\begin{aligned} x' &= -x + e^{\alpha y} - 1 \\ y' &= \sin(x - y) \end{aligned} \tag{7}$$

We study the behavioral properties of the solution near the origin relative to the values of the parameter α. Since

$$f_x(x, y) = -1 \qquad f_y(x, y) = \alpha e^{\alpha y}$$
$$g_x(x, y) = \cos(x - y) \qquad g_y(x, y) = -\cos(x - y)$$

it follows that the linearization of (7) about the critical point (0, 0) is the system

$$x' = -x + \alpha y$$
$$y' = x - y$$

The characteristic equation of this system is

$$(-1 - \lambda)(-1 - \lambda) - \alpha = 0$$

and so the characteristic roots are

$$\lambda = \begin{cases} -1 \pm \sqrt{\alpha} & \text{if } \alpha \geq 0 \\ -1 \pm \sqrt{-\alpha}\, i & \text{if } \alpha < 0 \end{cases}$$

Therefore, if $\alpha < 0$, the origin is an asymptotically stable spiral point; if $0 < \alpha < 1$, the origin is an asymptotically stable double-tangent node; and if $\alpha > 1$, the origin is a saddle point. Theorem 5.7-1 does not apply if $\alpha = 0$ or $\alpha = 1$ (why?).

It is important to realize that if the characteristic roots λ_1 and λ_2 of the linearized equation (3) are real, with $\lambda_1 > \lambda_2 > 0$, then the phase plot in a neighborhood of (x_c, y_c) looks exactly like the asymptotically stable double-tangent node (Figure 5.23*a*), except that the flow is *away* from (x_c, y_c) instead of toward (x_c, y_c). Similarly, if λ_1 and λ_2 are conjugate complex with positive real parts, then the solutions starting sufficiently close to (x_c, y_c) *spiral away* instead of spiraling toward (x_c, y_c). Therefore, this more detailed analysis applies whenever the characteristic roots of linearization of (1) about (x_c, y_c) are distinct and have nonzero real parts.

PROBLEMS

5.7-1. Determine all the critical points for each of the following systems. Determine also the linearization about each of these critical points, and, when possible, discuss the behavior in the neighborhood of the critical points.

(a) $x' = 2y - 4$, $y' = x + 3$

(b) $x' = x^2 - 2x$, $y' = x^2 - y$

(c) $x' = -2x + y - 1$, $y' = -3x + xy$

(d) $x' = y$, $y' = -x - x^3 - y$

(e) $x' = y$, $y' = -x + x^3 - y$

(f) $x' = y$, $y' = -\sin x - y$

5.7-2. In each of the following systems α is a real parameter and the origin is a critical point. Whenever possible discuss, relative to the values of α in $(-\infty, \infty)$, the stability properties of the origin and the behavior near the origin.

(a) $x' = -\alpha x + \sin y$, $y' = -x - y$

(b) $x' = -x + \sin \alpha y$, $y' = x - y + x^2$

(c) $x' = y$, $y' = -\sin x - \alpha y$

(d) $x' = \alpha x + e^y - 1$, $y' = e^{-x} + \alpha y - 1$

***5.7-3.** Suppose that h is a continuously differentiable function of two variables and consider the second order nonlinear equation

(*) $\quad z'' + h(z, z') = 0$

Set $x = z$ and $y = z'$ and write equation (*) as a first order system in the plane. Show that critical points of this system are precisely the points $(z_c, 0)$ where $h(z_c, 0) = 0$. Suppose that $(z_c, 0)$ is a critical point, and linearize the system about this critical point. Show further that the linearized system is equivalent to the second order equation

(**) $\quad z' + h_y(z_c, 0)z' + h_x(z_c, 0)z = 0$

Equation (**) is called the linearization of (*) about the constant solution $z(t) = z_c$.

***5.7-4.** If $z = z(t)$ denotes the measure of the angle that a swinging pendulum of unit mass makes with the vertical, then z obeys the second order equation

$$z'' + \mu z' + k \sin z = 0$$

where the constant $k > 0$ is inversely proportional to the length and $\mu \geq 0$ is the resistance coefficient. (If $\mu = 0$ the behavior of z is as described in Section 5.5*a*) Describe and interpret the behavior of z in a neighborhood of the origin for appropriate value ranges for $\mu > 0$ by using the linearization technique indicated in the preceding problem.

*5.8 NONLINEAR MODEL FOR INTERACTING POPULATIONS

In Section 1.8 some nonlinear differential equations that modeled the growth of a "homogeneous" population were studied. In this section a basic population model is introduced that involves *interaction between different species.* Only two species are considered, and throughout this section $p(t)$ and $q(t)$ are used to denote the number of individuals at time t in two species of a population. The principal concern is to indicate how interactions between these two species can affect the growth of the population.

As in Section 1.8 the functions p and q are assumed to be continuously differentiable functions of time t, and hence $p'(t)$ and $q'(t)$ are the "instantaneous" rate of change of the growth of these species. Also, the rate of interaction between the p species and the q species will be assumed proportional to the product of the sizes of these populations:

(1) The rate of interaction of the p species with the q species at time t is proportional to the product $p(t)q(t)$, where the constant of proportionality is independent of time t.

The product pq can be interpreted as an indication of the possibility of an "encounter" of an individual of the p species with one of the q species. Note in particular that if either p or q is "small," then the interaction between them is weak. Also, the measurement of this interaction has the same form as the measurement of the possible contacts between susceptibles and infectives in the epidemic model described in Section 5.4*b*.

Our basic model is a type of *predator-prey interaction* and has the form

$$p' = \alpha p - \beta pq \qquad (2)$$
$$q' = -\gamma q + \delta pq$$

where α, β, γ, and δ are positive constants. Equation (2) is also known as the *Volterra-Lotka equation.* The p species is the prey and the q species the predator (one can typically think of p representing the number of rabbits in a region and q the number of foxes). The prey p is assumed to have an abundant food supply, and in the absence of the predator q the prey species p increases according to a malthusian growth law ($p' = \alpha p$—see Section 1.8). However, whenever the predator q is also in the region, the rate of growth of the prey is adversely affected by interaction or encounters with the predators. This is reflected by the $-\beta pq$ term in the first equation of (2) [see assumption (1)]. The more adept the prey species p is in avoiding or eluding a predator is reflected by the smallness of the coefficient β. Thus if the prey is well camouflaged and has effective defenses, the coefficient β would be small. The predator q is assumed to depend upon the species p for existence. Hence, in the absence of the prey p the predator population q will die out (according to the law $q' = -\gamma q$). In contrast with the prey, the predator benefits from the interaction or encounters with the prey, and this is reflected by the δpq term in the second equation of (2). The more effective the predator is the larger the coefficient δ.

The problem now is to analyze the behavior of solutions to (2) and relate the results to the study of interacting populations in the predator-prey case. Since equation (2) is autonomous, the results in Sections 5.4 and 5.5 can be applied. The critical points can be found by solving the equations

$$0 = \alpha p - \beta pq = p(\alpha - \beta q)$$
$$0 = -\gamma q + \delta pq = q(-\gamma + \delta p)$$

and it is immediately seen that

$$(0, 0), \left(\frac{\gamma}{\delta}, \frac{\alpha}{\beta}\right) \qquad \text{are the critical points of (2)} \qquad (3)$$

Note that if $q(0) = 0$, then $q(t) \equiv 0$ and $p(t) \equiv p(0)e^{\alpha t}$ for all $t \geq 0$; and if $p(0) = 0$, then $p(t) \equiv 0$ and $q(t) = q(0)e^{-\gamma t}$ for all $t \geq 0$. Thus solutions initially on the positive p or q axis remain there for all time, and since trajectories cannot intersect, it follows that *solutions initially in the first quadrant remain there for all positive time* (this is certainly to be expected from the physical interpretation). Therefore, we study only the behavior of solutions to (2) in the first quadrant. Figure 5.24*a* indicates the critical points and the direction of the flow in certain parts of the first quadrant. [Note that $p' = p(\alpha - \beta q)$ and so $p' > 0$ if $q > \alpha/\beta$ and $p' < 0$ if $q > \alpha/\beta$. Similarly, $q' > 0$ if $p > \gamma/\delta$ and $q' < 0$ if $p < \gamma/\delta$.]

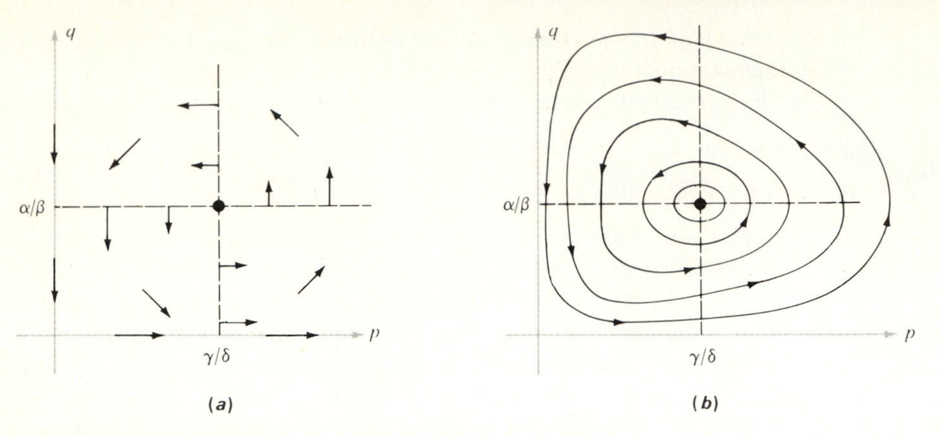

Figure 5.24 (*a*) Direction field of (2). (*b*) Phase plot of (2).

In order to determine a first integral for (2) use the fact that

$$\frac{dq}{dp} = \frac{-\gamma q + \delta pq}{\alpha p - \beta pq} = \frac{q(-\gamma + \delta p)}{p(\alpha - \beta q)}$$

and that the variables separate to obtain

$$\left(\delta - \frac{\gamma}{p}\right) dp + \left(\beta - \frac{\alpha}{q}\right) dq = 0$$

From this it follows that

$$U(p, q) = \delta p + \beta q - \gamma \ln p - \alpha \ln q \tag{4}$$

is a first integral for (2) in the first quadrant. We want to indicate that the curves $U(p, q) \equiv$ constant define a family of closed curves in the first quadrant that surround the critical point $(\gamma/\delta, \alpha/\beta)$. Taking the exponential of each side

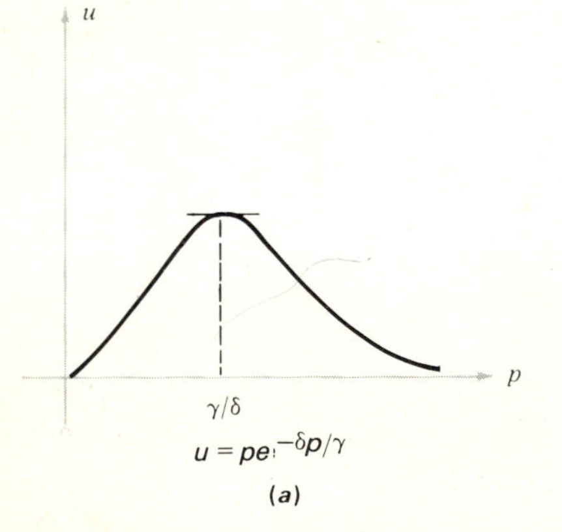

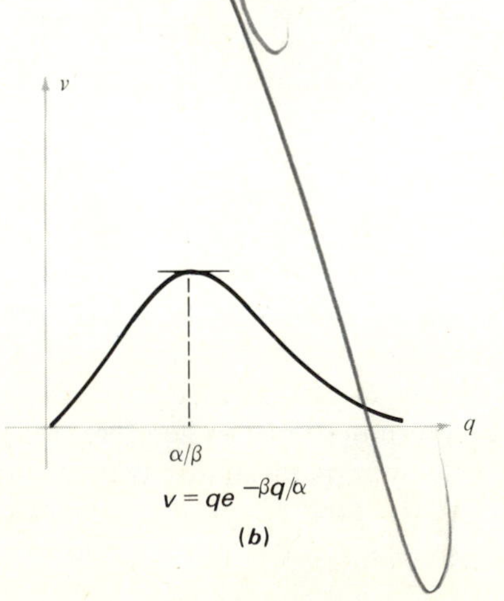

Figure 5.25 (*a*) $u = pe^{-\delta p/\gamma}$. (*b*) $v = qe^{-\beta q/\alpha}$.

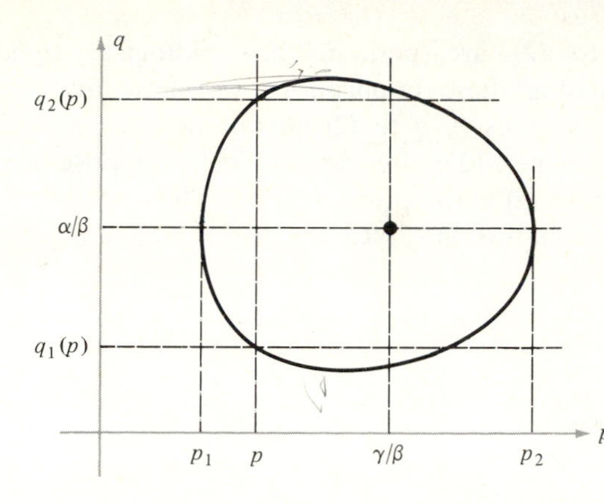

Figure 5.26 Sketch of a predator-prey curve.

of the equation $U(p, q) = c$, we obtain

$$e^c = e^{\delta p + \beta q + \ln p^{-\gamma} + \ln q^{-\alpha}} = e^{\delta p} e^{\beta q} p^{-\gamma} q^{-\alpha}$$

Taking the reciprocal of each side of this equation it follows that

$$(pe^{-\delta p/\gamma})^{\gamma}(qe^{-\beta q/\alpha})^{\alpha} = k \tag{5}$$

where $k = e^{-c}$. Sketches of the graphs of the functions $u: p \to pe^{-\delta p/\gamma}$ and $v: q \to qe^{-\beta q/\alpha}$ are given in Figure 5.25. In particular, both functions are bounded on the right half-line [$u(p) = pe^{-\delta p/\gamma}$ reaches a maximum at $p = \gamma/\delta$, and $v(q) = qe^{-\beta q/\alpha}$ reaches a maximum at $q = \alpha/\beta$] and have limit 0 at infinity. It now follows easily from (5) that the integral curves are bounded. Suppose that $0 < k < u(\gamma/\delta)^{\gamma} v(\alpha/\beta)^{\alpha}$ [if $k = u(\gamma/\delta)^{\gamma} v(\alpha/\beta)^{\alpha}$, then (5) defines the critical point $(\gamma/\delta, \alpha/\beta)$], and let p_1 and p_2 be such that $u(p_i)^{\gamma} v(\alpha/\beta)^{\alpha} = k$ for $i = 1, 2$. Note that $0 < p_1 < \gamma/\delta < p_2$, and that if $p_1 < p < p_2$ there are exactly two numbers $q_1(p)$ and $q_2(p)$ such that $0 < q_1(p) < \alpha/\beta < q_2(p)$ and

$$u(p)^{\gamma} v(q_i(p))^{\alpha} = k \qquad i = 1, 2$$

[i.e., the points $(p, q_1(p))$ and $(p, q_2(p))$ are on the integral curve (5)]. Moreover, for $i = 1, 2$, $q_i(p) \to \alpha/\beta$ as $p \to p_1+$ and $p \to p_2-$ (see Figure 5.26). The fact that (5) defines a closed curve surrounding the critical point $(\gamma/\delta, \alpha/\beta)$ now easily follows. Combining these ideas with Proposition 1 in Section 5.5 gives the following result.

Theorem 5.8-1

The predator-prey system (2) has a unique, strictly positive critical point $(p_c, q_c) = (\gamma/\delta, \alpha/\beta)$. Also, if p, q is a solution to (2) such that $p(0), q(0) > 0$ and $(p(0), q(0)) \neq (\gamma/\delta, \alpha/\beta)$, then $p(t), q(t) > 0$ for all $t \geq 0$. Moreover, the solution p, q is periodic and its trajectory is a simple closed curve that surrounds the critical point $(\gamma/\delta, \alpha/\beta)$.

The fact that solutions to (2) are periodic has biological significance. However, before discussing these implications we show that even though nonconstant periodic solutions p, q to (2) cannot be computed explicitly, the average value over one period can be determined. Suppose p, q is a nonconstant periodic solution to (2) with period $T > 0$. Then $p(t)$, $q(t) > 0$ for all $t \geq 0$, and from the first equation in (2) we see that $\alpha - \beta q = p'/p$, and hence

$$\int_0^T \frac{p'(s)}{p(s)}\, ds = \alpha T - \beta \int_0^T q(s)\, ds$$

Since $p(T) = p(0)$ we have also that

$$\int_0^T \frac{p'(s)}{p(s)}\, ds = \ln p(T) - \ln p(0) = 0$$

and it follows that

$$\bar{q} \equiv \frac{1}{T} \int_0^T q(s)\, ds = \frac{\alpha}{\beta} \tag{6}$$

Applying an analogous procedure to the second equation in (2),

$$\bar{p} \equiv \frac{1}{T} \int_0^T p(s)\, ds = \frac{\gamma}{\delta} \tag{6'}$$

and it follows that the integral average of each periodic solution over its period is the same as the critical point. Since $(\bar{p}, \bar{q})$ represents the average value of the population over one period, *the average population of each species is independent of the period and the initial population.*

The periodic fluctuations in predator-prey populations do exist in nature. Suppose, for example, in a fox-rabbit system that there is an abundance of rabbits. Then the foxes have plenty to eat and very little competition for food, and so the fox population increases rapidly. However, as the fox population becomes larger, they consume more and more rabbits, which causes the rabbit population to diminish. After the rabbit population diminishes, the larger fox population no longer has an adequate food supply and hence diminishes. This causes the rabbit population to grow again and one cycle is completed.

The predator-prey model applies to insect populations as well. An illustration is the interaction between the cottony cushion scale (an insect that is a threat to citrus trees) and its natural predator, the ladybug beetle. When both are located in a region, the scale population tends to remain at a tolerable level. However, if poison is applied in order to further reduce the scale population, a larger scale population usually appears later (as predicted by the Volterra-Lotka model). Predator-prey systems are found in certain fish populations as well. In fact, the study of various percentages of predator fish (sharks) relative to prey fish (sole) caught in the Adriatic Sea motivated Volterra to develop and analyze predator-prey models.

Now we investigate the effect of *harvesting* on the behavior of solutions to (2). It is assumed that the harvesting removes the amounts of prey p and predator q at a *rate proportional to the number present* (this is different from the assumption of a constant rate of harvesting considered in Section 1.8*b*). Therefore, assume there are numbers $\eta_1, \eta_2 > 0$ such that p and q satisfy the equation

$$\begin{aligned} p' &= \alpha p - \beta pq - \eta_1 p \qquad [=(\alpha - \eta_1)p - \beta pq] \\ q' &= -\gamma q - \delta pq - \eta_2 q \qquad [= -(\gamma + \eta_2)q - \delta pq] \end{aligned} \tag{7}$$

If $\eta_1 < \alpha$ and $\eta_2 < \gamma$, then (7) is precisely the same as (2), with α replaced by $\alpha - \eta_1$ and γ by $\gamma + \eta_2$. From equations (6) and (6′), the average values $\tilde{p}$ and $\tilde{q}$ for solutions p and q to (7) are

$$\tilde{p} = \frac{\gamma + \eta_2}{\delta} \qquad \text{and} \qquad \tilde{q} = \frac{\alpha - \eta_1}{\beta} \tag{8}$$

Therefore, $\tilde{p} > \bar{p}$ and $\tilde{q} < \bar{q}$, and it follows that a reasonable amount of harvesting (that is, $\eta_1 < \alpha$ and $\eta_2 < \gamma$) actually *increases* the average number of prey (and decreases the average number of predators).

PROBLEMS

5.8-1. In order to study the behavior of solutions to the predator-prey model in the vicinity of the critical point $(\gamma/\delta, \alpha/\beta)$, assume that

$$p = \bar{p} + \frac{\gamma}{\delta} \qquad \text{and} \qquad q = \bar{q} + \frac{\alpha}{\beta} \tag{*}$$

where $\bar{p}$ and $\bar{q}$ are "small." Neglecting terms that involve the product $\bar{p}\bar{q}$, show that if p, q is a solution to (2) then $\bar{p}, \bar{q}$ satisfies

$$\bar{p}' = -\frac{\beta\gamma}{\delta}\bar{q} \qquad \bar{q}' = \frac{\delta\alpha}{\beta}\bar{p} \tag{**}$$

[(**) is the linearization of (*) about $(\gamma/\delta, \alpha/\beta)$—see Section 5.7]. Show that the solutions to (**) form elliptical trajectories centered at the origin [and hence the periodic solutions to (2) that are very near $(\gamma/\delta, \alpha/\beta)$ are the trajectories that are approximately ellipses with center $(\gamma/\delta, \alpha/\beta)$].

(a) Determine $\bar{p}$ and $\bar{q}$ explicitly and show that the period of the solutions to (**) is $2\pi/\sqrt{\alpha\gamma}$.

(b) Determine the equation of the elliptical integral curves of equation (**).

(c) Use the substitution (*) in the integral curve

$$\delta p + \beta q - \gamma \ln p - \alpha \ln q = k \tag{***}$$

of equation (2) [see (4)] and the expansion $\ln(1 + x) \cong x - x^2/2$ to approximate the equation (***). Compare with the equation in part *b*.

5.8-2. Suppose that a disease is introduced into a population that has a number of susceptibles $x(t)$ and a number of infectives $y(t)$ [see the discussion preceding equation (15) in Section 5.4*b*]. In addition to the assumptions made concerning equation (15) in Section 5.4*b*, assume that the number $x(t)$ of susceptibles increases at a rate proportional to the number present, and hence that x and y satisfy the system

$$x' = -\alpha xy + \gamma x$$
$$y' = \alpha xy - \beta y$$

Relate this equation to the predator-prey model (2) and interpret the behavior of the solutions.

*5.9 UNDAMPED NONLINEAR VIBRATIONS

In this section an important class of nonlinear second order differential equations is analyzed. Typical phenomena that are described by the class of equations studied here are the motions induced by a "nonlinear spring" and the motions induced by a pendulum. The class of equations to be studied is the second order initial value problem

$$z'' + h(z) = 0 \qquad z(0) = x_0 \qquad z'(0) = y_0 \tag{1}$$

where the function h has the following two properties:

(*h*1) h is twice continuously differentiable on $(-\infty, \infty)$; and

(*h*2) $h(0) = 0$ and there is an r, $0 < r \leq +\infty$, such that $zh(z) > 0$ whenever $0 < |z| < r$.

If $z(t)$ denotes the position of an object moving in a straight line for each time $t \geq 0$, then according to Newton's second law (see Section 1.3) the total forces acting on z at time t is $mz''(t)$, where m is the mass of the object (which is assumed to be constant). From equation (1) it now follows that $mh(z(t))$ is the force acting on z at time t. The property (h2) of h implies that this force is always toward the equilibrium position $z = 0$ so long as $-r < z(t) < r$. See Section 2.6 for the analysis when h is linear and Section 5.5*a* for the analysis of pendulum motion.

In order to analyze the behavior of solutions, equation (1) is transformed into a two-dimensional first order system so that the methods of the preceding sections can be applied. Setting $x = z$ and $y = z'$, the second order equation (1) becomes the first order system

$$\begin{aligned} x' &= y & x(0) &= x_0 \\ y' &= -h(x) & y(0) &= y_0 \end{aligned} \tag{2}$$

Since $h(0) = 0$ by (h2) it follows that (0, 0) is a critical point of (2). Also, *all* critical points to (2) lie on the x axis and in fact are precisely the points $(x^*, 0)$, where $h(x^*) = 0$ [since $h(x) \neq 0$ for $0 < |x| < r$ by ($h2$), the critical point (0, 0) is "isolated"]. A first integral U of (2) can readily be computed since $dy/dx = -h(x)/y$ and hence the variables can be separated. Therefore, $y\,dy + h(x)\,dx = 0$, and it follows that

$$U(x, y) = \tfrac{1}{2} y^2 + \int_0^x h(s)\,ds \qquad \text{for all } x, y \tag{3}$$

is a first integral of (2). Therefore, solutions to (2) lie on the level curves

$$\tfrac{1}{2} y^2 + \int_0^x h(s)\,ds = E \qquad E \geq 0 \text{ a constant} \tag{4}$$

It is important to realize that the integral $\int_0^x h(s)\,ds$ in (3) [and (4)] is the *potential energy* of the system and the term $\frac{1}{2}y^2$ [$=\frac{1}{2}(z')^2$] is the *kinetic energy* [it is assumed here that equations (1) and (2) describe the motion of an object having unit mass]. Thus $U(x, y)$ is the *total energy* of the system, and the fact that the trajectories of solutions lie on the level curves (4) of U can be interpreted as the conservation of energy (since the system is undamped, there is no energy loss due to friction).

The first result indicates that locally about the critical point (0, 0) solutions to (2) [and hence to (1)] are periodic.

Theorem 5.9-1

If conditions ($h1$) and ($h2$) are satisfied, then there is a number $R > 0$ such that whenever $0 < x_0^2 + y_0^2 < R$, the solution to (2) is periodic. Moreover, its trajectory is the level curve

$$\tfrac{1}{2} y^2 + \int_0^x h(s)\,ds = \tfrac{1}{2} y_0^2 + \int_0^{x_0} h(s)\,ds$$

which is a simple closed curve that surrounds the origin.

In order to establish this theorem select $R > 0$ so that if $x_0^2 + y_0^2 < R$ then

$$\tfrac{1}{2} y_0^2 + \int_0^{x_0} h(s)\,ds < \min\left\{\int_0^{-r} h(s)\,ds, \int_0^{r} h(s)\,ds\right\} \tag{5}$$

[the right-hand side of this inequality is positive by virtue of (2)]. So let $0 < x_0^2 + y_0^2 < R$ and recall that the integral curve of (3) is

$$\tfrac{1}{2} y^2 + \int_0^x h(s)\,ds = E \qquad \text{where } E = \tfrac{1}{2} y_0^2 + \int_0^{x_0} h(s)\,ds \tag{6}$$

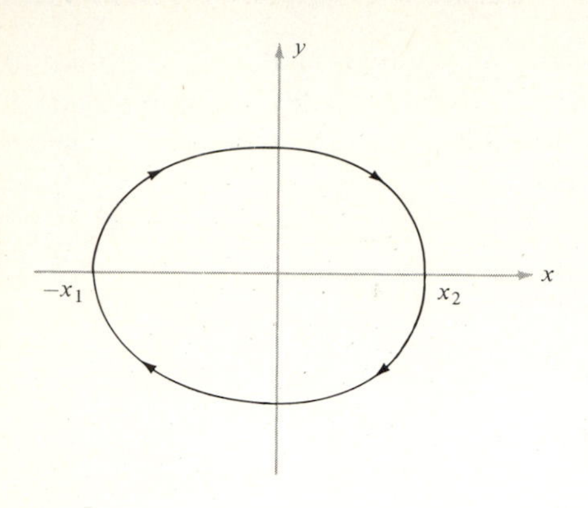

Figure 5.27 Closed curve depicting nonlinear oscillations.

Since this equation is unchanged, with y replaced by $-y$, the graph of (6) is symmetric with respect to the x axis. From the inequality (5) there are numbers $x_1, x_2 > 0$ such that $-r < -x_1 < x_2 < r$,

$$\int_0^{-x_1} h(s)\, ds = \int_0^{x_2} h(s)\, ds = E$$

and

$$\int_0^{x} h(s)\, ds < E \qquad \text{for all } -x_1 < x < x_2$$

From this it follows that the curve

$$y = \sqrt{2E - 2\int_0^{x} h(s)\, ds} \qquad \text{for } -x_1 < x < x_2$$

increases from $-x_1$ to 0 and decreases from 0 to x_2. Since (6) is a reflection of the curve about the x axis, it now follows that the level surface (6) is indeed a simple closed curve surrounding the origin (see Figure 5.27). Since there are no critical points on this curve, the fact that the solution is periodic is a direct consequence of Proposition 1 in Section 5.5.

5.9*a* Amplitude and Evaluation of the Period

In addition to the properties ($h1$) and ($h2$) in (2) for the function h it is now also assumed that h is odd:

(h3) $h(-z) = -h(z)$ for all z

Since h is odd it is easy to check that the potential energy is even:

$$\text{If } V(x) \equiv \int_0^{x} h(s)\, ds \qquad \text{then } V(-x) = V(x) \qquad \text{for all } x \tag{7}$$

To see that (7) holds note that by changing variables

$$V(-x) = \int_0^{-x} h(s)\, ds = \int_0^{x} h(-t)\,(-dt)$$

$$= \int_0^{x} h(t)\, dt = V(x)$$

Given an energy level $E > 0$, the level curve

$$\tfrac{1}{2} y^2 + V(x) = E \tag{8}$$

is symmetric with respect to both axes and the origin. If the energy level E is such that (8) defines a simple closed curve surrounding the origin (this is always the case for E sufficiently small—see Theorem 5.9-1), then the curve intersects the positive x axis exactly once, say at $x = A$. Since the curve is symmetric it intersects the negative x axis at $x = -A$, and since the x component of the solution to (2) remains between $-A$ and A, this number A is called the *amplitude of the system* corresponding to the energy level E. Since $y = 0$ when $x = A$, we have from (8) and the definition of V that for all $E > 0$ such that (8) is a simple closed curve surrounding the origin, the amplitude A corresponding to this energy is the least number such that

$$\int_0^{A} h(s)\, ds = E \tag{9}$$

EXAMPLE 5.9-1

Consider the linear system $z'' + kz = 0$, where $k > 0$. The level surfaces for the corresponding two-dimensional system $x = z, y = z'$ are the curves $\frac{1}{2}y^2 + \frac{1}{2}kx^2 = E$, which is a family of ellipses with center (0, 0) (see Figure 5.28). For a given energy level E, the amplitude $A = \sqrt{2E/k}$ [see (9)].

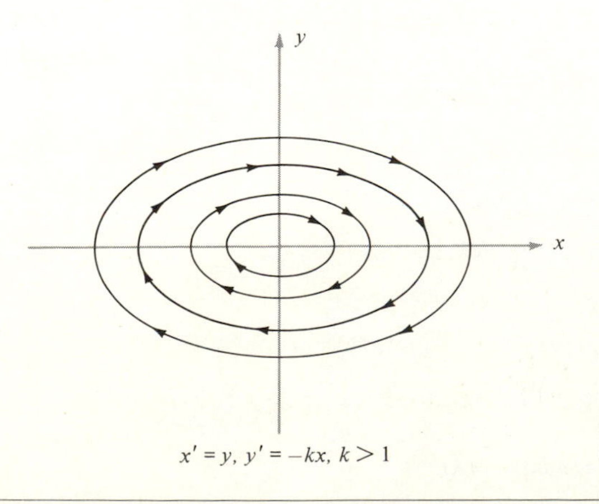

Figure 5.28 A family of ellipses with center (0, 0) for the system $z'' + kz = 0$ for $k > 0$.

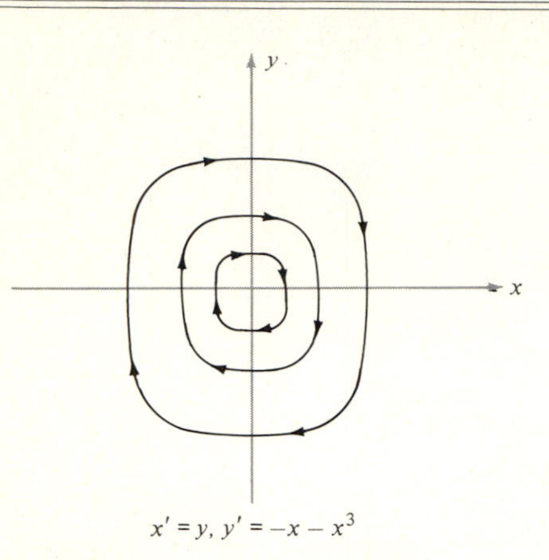

Figure 5.29 Phase plot of the system $z'' + kz + lz^3 = 0$ for $k, l > 0$.

EXAMPLE 5.9-2

Consider the system $z'' + kz + lz^3 = 0$, where $k, l > 0$. The level surfaces (with $x = z$, $y = z'$) are $\frac{1}{2}y^2 + \frac{1}{2}kx^2 + \frac{1}{4}lx^4 = E$. As in the linear case, the origin is the only critical point and all nontrivial solutions are periodic (see Theorem 5.9-1). A phase plot is given in Figure 5.29.

EXAMPLE 5.9-3

Consider the system $z'' + kz - lz^3 = 0$, where $k, l > 0$. The level surfaces (with $x = z$, $y = z'$) are $\frac{1}{2}y^2 + \frac{1}{2}kx^2 - \frac{1}{4}lx^4 = E$. In this example there are

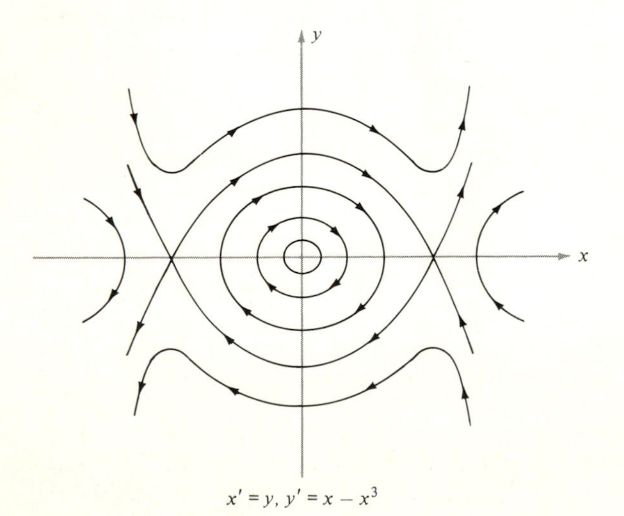

Figure 5.30 Phase plot of the system $z'' + kz - lz^3 = 0$ for $k = l = 1$.

three critical points: $(0, 0)$, $(\pm\sqrt{k/l}, 0)$. A sketch of a phase plot for $k = l = 1$ is given in Figure 5.30. Notice that the solutions to this system are not periodic (and even unbounded) for large values of E.

The formula

$$E = \int_0^A h(s)\, ds$$

[see (9)] gives a relationship between the energy E and the amplitude A, and in many cases A can be computed directly in terms of E (for example, $A = \sqrt{2E/k}$ in the linear case described in Example 5.9-1). One can also determine a formula for the period of a periodic solution. In the linear case in Example 5.9-1, the solutions can be explicitly computed, and the period equals $2\pi/\sqrt{k}$ for every nontrivial solution and hence is independent of the amplitude [note, however, that the period decreases as the stiffness $S(z) \equiv k$ increases].

In order to find a relationship between the amplitude A and the period T of a periodic solution, we let

$$V(x) \equiv \int_0^x h(s)\, ds$$

denote the potential energy and use the level curve (8), along with the fact that $y(t) = x'(t)$, to obtain

$$x' = \sqrt{2}\sqrt{E - V(x)}$$

so long as $x(t) \geq 0$. Since this is a first order autonomous equation with dependent variable x, we can separate variables (see Section 1.4) to obtain

$$\frac{x'(t)}{\sqrt{E - V(x(t))}} = \sqrt{2}$$

and it follows that if $x(0) = 0$ then

$$\int_0^x \frac{dx}{\sqrt{E - V(x)}} = \sqrt{2}\, t$$

As x goes from 0 to the amplitude A, one-fourth of the curve is traced (this follows from symmetry) and we see that $t = T/4$ when $x = A$, and hence

$$T = 2\sqrt{2} \int_0^A \frac{dx}{\sqrt{E - V(x)}} \tag{10}$$

is the period [the integral in (10) is actually improper since $V(A) = E$, but it converges so long as $h(A) \neq 0$, and hence $(A, 0)$ is not a critical point]. As opposed to the linear case the period is amplitude-dependent in the nonlinear case; however, these ideas are beyond the scope of this text.

PROBLEMS

5.9-1. Write each of the following second order equations as a first order system in the plane, with $x = z$ and $y = z'$. Determine the integral curves and sketch a phase plot. Also, for appropriate values of the energy E, determine the relationship between E and the amplitude A.

(a) $z'' + z^3 = 0$ **(d)** $z'' + z^5 = 0$

(b) $z'' + \dfrac{z}{1 + z^4} = 0$ **(e)** $z'' + ze^{-z^2} = 0$

(c) $z'' + z|z| = 0$

5.9-2. Determine the amplitude A explicitly in terms of k, l, and E in Examples 5.9-2 and 5.9-3.

5.9-3. Sketch and compare the phase plots (with $x = z$, $y = z'$) for each of the following second order systems.

(a) $z'' + z = 0$ **(b)** $z'' + z + z^3 = 0$ **(c)** $z'' + z + z^3 + z^5 = 0$

5.9-4. Verify formula (10) for the period T in the case that (1) is linear [that is, $h(z) = kz$].

Remarks A more advanced discussion of both linear and nonlinear systems in the plane can be found in Birkhoff and Rota [1], Petrovski [12], and Struble [14]. The interested reader should also consult Hirsch and Smale [7].

6 Numerical Methods

With the development of high-speed computers numerical methods have become an important technique in analyzing solutions to differential equations. In this chapter several basic methods for the approximation of solutions to initial value problems are developed. Only an introduction to these ideas is presented, and the purpose of this chapter is to make the student aware of some basic methods that are available and to indicate the procedures that are involved in implementing these methods. The technique developed in Section 6.1 is the use of Taylor's formula to obtain approximate solutions to initial value problems. Discrete variable methods are discussed in Sections 6.2, 6.3, and 6.4 (one-step methods are presented in Sections 6.2 and 6.3 and multistep methods in Section 6.4).

6.1 CONTINUOUS APPROXIMATIONS: TAYLOR'S EXPANSION

If exact solutions to differential equations cannot be computed, it is often necessary to obtain approximations of the solutions. In this section we indicate a procedure that can be used to obtain "continuous approximations" of solutions. By a continuous approximation of a solution we mean an approximating function for this solution, where the independent variable ranges over the entire time interval of interest. This is opposed to discrete approximations, where the interest is in estimating values of the solution at certain discrete values of the independent variable. Typical examples of continuous approximations are power series solutions, which are discussed in Chapter 4. The technique discussed in this section is the use of Taylor's formula, and some of the ideas involved are important in discrete approximations as well.

Suppose that $[a, b)$ is an interval and that f is a continuous function on $[a, b) \times (-\infty, \infty)$ such that the partial derivatives of all orders of f exist and are continuous. Consider the initial value problem

(1) $$y' = f(t, y) \qquad y(a) = y_0 \qquad t \in [a, b)$$

where y_0 is a given initial value. The fundamental existence and uniqueness theorem (see Section 1.4*b*) guarantees that there is some $c \in [a, b)$ such that (1) has a unique solution y on $[a, c]$:

(2) $$y'(t) = f(t, y(t)) \qquad \text{for } t \in [a, c] \qquad \text{and} \qquad y(a) = y_0$$

The fundamental problem is to consider techniques for constructing a function ϕ on $[a, c]$ that is "close to" the actual solution y on $[a, c]$. The meaning of the term *close to* depends on the accuracy that is needed. Therefore, not only is it important to have a procedure to construct the approximation ϕ, but it is also necessary to be able to estimate the error term ε on $[a, c]$:

(3) $$\varepsilon(t) = y(t) - \phi(t) \qquad \text{for } t \in [a, c]$$

The first procedure is to show that Taylor's formula (with remainder) can be used to construct an approximation ϕ.

Recall that if n is a positive integer and y is a function on $[a, c]$ that has at least $n + 1$ derivatives on $[a, c]$, then y can be expressed in the form

(4) $$y(t) = y(a) + y'(a)(t - a) + \cdots + \frac{y^{(n)}(a)}{n!}(t - a)^n + R_{n+1}(t)$$

for all $t \in [a, c]$, where the remainder term R_{n+1} is of the form

(5) $$R_{n+1}(t) = \frac{y^{(n+1)}(\xi_t)}{(n + 1)!}(t - a)^{n+1} \qquad \text{where } a \le \xi_t \le t$$

Formula (4) is known as *Taylor's theorem* and the expression for the remainder in (5) is known as *Lagrange's form* of the remainder. Therefore, if the values $y(a), y'(a), \ldots, y^{(n)}(a)$ of the solution y to equation (1) are known, it is convenient to take the polynomial

(6) $$\phi_n(t) = y(a) + y'(a)(t - a) + \cdots + \frac{y^{(n)}(a)}{n!}(t - a)^n$$

as an approximation, and ϕ_n is called *Taylor's expansion of order $n + 1$ to y on* $[a, c]$. An estimation of the remainder term in (5) is then an estimation of the approximation error ε in (3).

In many situations it is easy to compute a reasonable number of the terms $y(a), y'(a), \ldots, y^{(n)}(a)$. As an illustration consider the equation

(7) $$y' = t - y^3 \qquad y(0) = 1$$

Then $y(0) = 1$ and substituting into (7) gives $y'(0) = 0 - 1^3 = -1$. Differentiating each side of (7) we have

$$y'' = 1 - 3y^2 \cdot y'$$

and hence $y''(0) = 1 - 3(1)^2(-1) = 4$. Similarly

$$y''' = -6y \cdot y' - 3y^2 y''$$

and so $y'''(0) = -6(1)(-1) - 3(1)(4) = -6$. This procedure works (in theory) for general functions f, and one can proceed as follows: The initial value gives $y(a) = y_0$, and setting $t = a$ in equation (2) gives $y'(a) = f(a, y(a))$. Differentiating each side of the equation in (2) with respect to t and setting $t = a$ in the resulting equation shows that

$$y''(a) = f_t(a, y(a)) + f_y(a, y(a))y'(a)$$

Applying this procedure again leads to

$$y'''(a) = f_{tt}(a, y(a)) + 2f_{ty}(a, y(a))y'(a) + f_y(a, y(a))y''(a) + f_{yy}(a, y(a))[y'(a)]^2$$

This process can be continued so long as the computations are not too complicated; two examples are given for further illustration.

EXAMPLE 6.1-1

Consider the simple linear equation $y' = y$, $y(0) = 1$, and note that the exact solution is $y(t) = e^t$. In this case we have directly from the equation that

$$y(0) = 1 \qquad y'(0) = y(0) = 1 \qquad y''(0) = y'(0) = 1 \qquad y'''(0) = y''(0) = 1$$

and it easily follows that $y^{(k)}(0) = 1$ for each positive integer k. From equation (6) we have that Taylor's expansion ϕ_n has the form

$$\phi_n(t) = 1 + t + \frac{1}{2!}t^2 + \cdots + \frac{1}{n!}t^n$$

Since

$$e^t = \sum_{k=0}^{\infty} \frac{1}{k!} t^k$$

is the power series expansion for e^t about $t = 0$, we see that ϕ_n is a good approximation for the solution for sufficiently large values of n in a neighborhood of $t = 0$.

EXAMPLE 6.1-2

Consider the nonlinear equation $y' = t^2 - y^2$, $y(0) = \frac{1}{2}$, which does not have an elementary solution. Then

$$y'' = 2t - 2yy' \qquad y''' = 2 - 2(y')^2 - 2yy''$$

and

$$y^{(4)} = -6y'y'' - 2yy'''$$

and it follows that

$$y(0) = \tfrac{1}{2} \qquad y'(0) = -\tfrac{1}{4} \qquad y''(0) = \tfrac{1}{4} \qquad y'''(0) = \tfrac{13}{8} \qquad y^{(4)}(0) = -\tfrac{5}{4}$$

Therefore, Taylor's expansion of order 5 has the form

$$\phi_4(t) = \tfrac{1}{2} - \tfrac{1}{4}t + \tfrac{1}{8}t^2 + \tfrac{13}{48}t^3 - \tfrac{5}{96}t^4$$

and approximates the solution in a neighborhood of $t = 0$.

6.1*a* Taylor's Expansion for Systems

Taylor's formula can also be applied to systems of equations. Suppose that f and g are smooth functions in the plane and consider the two-dimensional autonomous system

$$x' = f(x, y) \qquad x(0) = x_0$$
$$y' = g(x, y) \qquad y(0) = y_0$$

(see Section 5.4). In this case we use the polynomial approximations ψ_n, ϕ_n where

$$\psi_n(t) = x(0) + x'(0)t + \cdots + \frac{x^{(n)}(0)}{n!}t^n$$
$$\phi_n(t) = y(0) + y'(0)t + \cdots + \frac{y^{(n)}(0)}{n!}t^n \tag{8}$$

for all t in some interval $[0, c]$ where $c > 0$. Note that $x(0) = x_0$, $y(0) = y_0$, $x'(0) = f(x_0, y_0)$, $y'(0) = g(x_0, y_0)$, and, using the chain rule,

$$x''(0) = f_x(x_0, y_0)x'(0) + f_y(x_0, y_0)y'(0)$$
$$y''(0) = g_x(x_0, y_0)x'(0) + g_y(x_0, y_0)y'(0)$$

Further values can be similarly computed.

▶ **EXAMPLE 6.1-3**

Consider the system

$$x' = x - xy \qquad x(0) = \tfrac{1}{2}$$
$$y' = y - xy \qquad y(0) = \tfrac{1}{4}$$

The equation has the form of the model for the competitive interaction of two species. The initial conditions give $x(0) = \frac{1}{2}$, $y(0) = \frac{1}{4}$, and substituting back into the equation gives

$$x'(0) = \tfrac{1}{2} - \tfrac{1}{2} \cdot \tfrac{1}{4} = \tfrac{3}{8} \qquad \text{and} \qquad y'(0) = \tfrac{1}{4} - \tfrac{1}{2} \cdot \tfrac{1}{4} = \tfrac{1}{8}$$

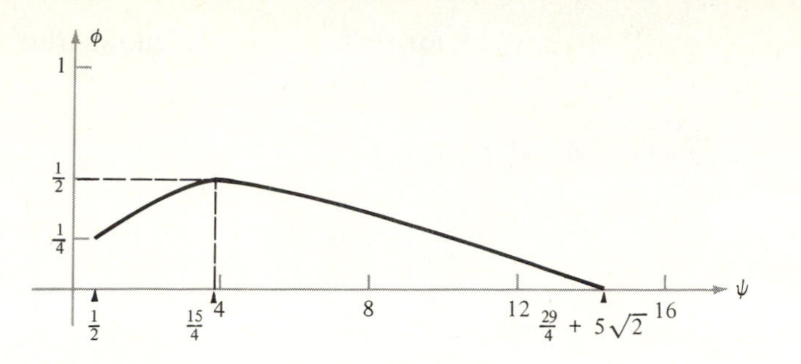

Figure 6.1 Approximate curve sketch for (ψ_2, ϕ_2).

Differentiating each side of both of the differential equations, we obtain

$$x'' = x' - x'y - xy' \qquad \text{and} \qquad y'' = y' - x'y - xy'$$

Therefore,

$$x''(0) = \tfrac{3}{8} - \tfrac{3}{8} \cdot \tfrac{1}{4} - \tfrac{1}{2} \cdot \tfrac{1}{8} = \tfrac{7}{32} \qquad \text{and} \qquad y''(0) = \tfrac{1}{8} - \tfrac{3}{8} \cdot \tfrac{1}{4} - \tfrac{1}{2} \cdot \tfrac{1}{8} = -\tfrac{1}{32}$$

and it follows from the formulas in (8) that

$$\psi_2(t) = \tfrac{1}{2} + \tfrac{3}{8}t + \tfrac{7}{64}t^2 \qquad \phi_2(t) = \tfrac{1}{4} + \tfrac{1}{8}t - \tfrac{1}{64}t^2$$

are approximations for the given system. Note that ϕ_2 increases on $[0, 4]$ and then decreases for $t \geq 4$. Also, $\phi_2(4 + 4\sqrt{2}) = 0$, and so the approximation is certainly no longer "good" for $t \geq 4 + 4\sqrt{2}$ [since it is known that $y(t)$ remains positive for all $t \geq 0$]. Clearly ψ_2 and all its derivatives are nonnegative. Since $\psi_2(4) = \frac{15}{4}$, $\phi_2(4) = \frac{1}{2}$ and $\psi_2(4 + 4\sqrt{2}) = \frac{29}{4} + 5\sqrt{2} \cong 14.31$, $\phi_2(4 + 4\sqrt{2}) = 0$, a rough sketch of the curve in the $\psi\phi$ plane whose parameteric equation is $\psi = \psi_2(t)$, $\phi = \phi_2(t)$, $0 \leq t \leq 4 + 4\sqrt{2}$ is given in Figure 6.1.

6.1*b* Error Estimation

In order to estimate the error term for Taylor's approximation of order $n + 1$ of the solution to (1) on $[a, c]$, it is sufficient to obtain a bound for the $(n + 1)$st derivative of y on $[a, c]$—see the expression for the remainder term in (5). The first step is to obtain bounds for y on $[a, c]$, say $y_- \leq y(t) \leq y_+$ for all $t \in [a, c]$. Bounds for y' on $[a, c]$ can then be obtained from equation (2), say $y'_- \leq y'(t) \leq y'_+$, for all $t \in [a, c]$, and since

$$y''(t) = f_t(t, y(t)) + f_y(t, y(t))y'(t)$$

we can estimate y'' on $[a, c]$, using the bounds for y and y', say $y''_- \leq y''(t) \leq y''_+(t)$ for $t \in [a, c]$. Continuing in this manner, numbers $y_-^{(n+1)}$ and $y_+^{(n+1)}$ can be found so that $y_-^{(n+1)} \le y^{(n+1)}(t) \le y_+^{(n+1)}$ for $t \in [a, c]$, and it follows from (5) that

$$|y(t) - \phi_n(t)| \le \frac{N(t-a)^{n+1}}{(n+1)!} \qquad \text{for all } t \in [a, c] \tag{9}$$

where N is the maximum of $|y_-^{(n+1)}|$ and $|y_+^{(n+1)}|$. In fact,

$$\frac{y_-^{(n+1)}(t-a)^{n+1}}{(n+1)!} \le y(t) - \phi_n(t) \le \frac{y_+^{(n+1)}(t-a)^{n+1}}{(n+1)!}$$

Of course, the computations and estimations can become very complicated, especially if it is necessary to use large values of n.

EXAMPLE 6.1-4

Consider the same equation as in Example 6.1-2:

$$y' = t^2 - y^2 \qquad y(0) = \tfrac{1}{2}$$

Taylor's expansion of order 3 has the form

$$\phi_2(t) = \tfrac{1}{2} - \tfrac{1}{4}t + \tfrac{1}{8}t^2$$

We show that this equation has a solution on [0, 1] and give an estimate of the error $|y(t) - \phi_2(t)|$ for $t \in [0, 1]$. First, bounds for y on [0, 1] are obtained. If $t > 0$ and $y(t) = 0$, then $y'(t) = t^2 > 0$, and it is immediately seen that $y(t) \ge 0$ on [0, 1]. Since $y'(t) = t^2 - y(t)^2 \le t^2$, we also see that

$$y(t) = \tfrac{1}{2} + \int_0^t y'(s)\,ds \le \tfrac{1}{2} + \int_0^t s^2\,ds = \tfrac{1}{2} + \tfrac{1}{3}t^3$$

and hence $y(t)$ exists for all $t \ge 0$, and we have the estimate $0 \le y(t) \le \frac{5}{6}$ for $t \in [0, 1]$. Also,

$$y'(t) = t^2 - y(t)^2 \le t^2 \le 1 \qquad y'(t) \ge t^2 - \left(\frac{5}{6}\right)^2 \ge -\frac{25}{36}$$

$$y''(t) = 2t - 2y(t)y'(t) \le 2 - 2\left(\frac{5}{6}\right)\left(-\frac{25}{36}\right) = \frac{341}{108}$$

$$y''(t) \ge -2\left(\frac{5}{6}\right)(1) = -\frac{5}{3}$$

$$y'''(t) = 2 - 2[y'(t)]^2 - 2y(t)y''(t) \le 2 - 2\left(\frac{5}{6}\right)\left(-\frac{5}{3}\right) = \frac{43}{18}$$

$$y'''(t) \ge 2 - 2(1)^2 - 2\left(\frac{5}{6}\right)\left(\frac{341}{108}\right) = -\frac{1705}{324}$$

Therefore, the estimate

$$\left| y(t) - \frac{1}{2} - \frac{1}{4}t + \frac{1}{8}t^2 \right| \leq \frac{1705}{1944}t^3 \qquad \text{for all } t \in [0, 1]$$

is valid. Note, for example, if t is restricted to the interval $[0, \frac{1}{10}]$, then the Taylor's third order approximation is accurate to within three decimal places (in fact, $|y(t) - \phi_2(t)| \leq 0.001$ if $0 \leq t \leq \frac{1}{10}$).

Finally, the reader should be aware that care must be taken in determining whether or not the solution actually exists on the interval in question. For one does not know and cannot easily determine the radius of convergence of the Taylor series expansion of a solution. As an illustration, the equation

$$y' = 1 + y^2 \qquad y(0) = 0 \tag{10}$$

has $y(t) = \tan t$ as its solution (this can be verified directly or obtained by separation of variables). Thus, the solution to (10) exists only for $-\pi/2 < t < \pi/2$. It is easy to check that $y(0) = 0$, $y'(0) = 1$, $y''(0) = 0$, and $y'''(0) = 2$. Thus,

$$\phi_3(t) = t + \frac{2t^3}{6}$$

is the Taylor's fourth order approximation for (10). This approximation gives no clue that the solution actually "blows up" at $t = \pi/2$.

PROBLEMS

6.1-1. Determine a Taylor's expansion of order 4 for the solution of each of the following initial value problems:

(a) $y' = t - y^3 \qquad y(0) = \frac{1}{2}$

(b) $y' = t^2 - y^2 \qquad y(0) = 0$

(c) $y' = e^{-y} + t \qquad y(0) = 0$

(d) $y' = -ty^2 + t^2 \qquad y(1) = 0$

6.1-2. Determine a Taylor's expansion of order 3 for the solution of the following two-dimensional systems:

(a) $x' = x - xy \qquad x(0) = \frac{1}{4}$
$y' = y - xy \qquad y(0) = \frac{1}{2}$

(b) $x' = x - xy \qquad x(0) = 1$
$y' = -y + xy \qquad y(0) = \frac{1}{2}$

6.1-3. Show that the method of using Taylor's formula can be adapted easily to apply to second order equations, and determine a Taylor's expansion of order 5 to each of the following second order initial value problems:

(a) $y'' + y - y^3 = 0 \qquad y(0) = \frac{1}{2} \qquad y'(0) = 0$

(b) $y'' + ty = 0 \qquad y(0) = 1 \qquad y'(0) = -3$

(c) $y'' + y' + \sin y = 0 \qquad y(0) = 0 \qquad y'(0) = \frac{1}{4}$

6.1-4. In each of the following initial value problems, compute Taylor's expansion of order 2 and determine a "reasonable" bound for the possible error over the given interval.

(a) $y' = 1 - y^3 \qquad y(0) = \frac{1}{2} \qquad t \in [0, \frac{1}{2}]$

(b) $y' = t + \sin y \qquad y(0) = 0 \qquad t \in [0, 1]$

(c) $y' = e^{-y} + t^2 \qquad y(0) = 1 \qquad t \in [0, 4]$

6.2 DISCRETE APPROXIMATIONS: ONE-STEP METHODS

The emphasis in the use of discrete methods is to concentrate on approximating the exact solution to an initial value problem at only a finite number of points in the interval. As in the previous section f is a continuous function with continuous partial derivatives of all orders on $[a, b) \times (-\infty, \infty)$, and the initial value problem

$$y' = f(t, y) \qquad y(a) = y_0 \qquad t \in [a, b) \tag{1}$$

is considered. It is also supposed that $a < c < b$ and that y is a solution to (1) on $[a, c]$. The problem under investigation is to approximate the solution y at a discrete set of points $a < t_1 < t_2 < \cdots < t_n \leq c$. Once estimates for y at these points are obtained, they can be "connected" (by straight line segments, for example) to provide an approximation for y over the entire interval. However, the emphasis in discrete methods is to obtain approximations for y at the points $t_1, t_2, \ldots, t_n$.

Although it is not always necessary, it is assumed for simplicity that the *mesh points* $t_1, t_2, \ldots, t_n$ where y is to be approximated are equidistant. Therefore, setting $t_0 = a$ there is a number $h > 0$ such that $t_i - t_{i-1} = h$ for all $i = 1, \ldots, n$. The positive number h is called the *stepsize* or simply the *step* of the mesh points. It follows easily that

$$t_0 = a \qquad \text{and} \qquad t_i = t_{i-1} + h = a + ih \qquad \text{for } i = 1, 2, \ldots, n \tag{2}$$

(It is always tacitly assumed that $a + nh \leq c$.) In order to conform with usual notation, the approximation of $y(t_i)$ is denoted by y_i [note that y_0 is indeed the initial value of (1)]. Therefore, given equation (1) and the stepsize h, the problem is to construct approximations $y_1, y_2, \ldots, y_n$ of the solution y at the

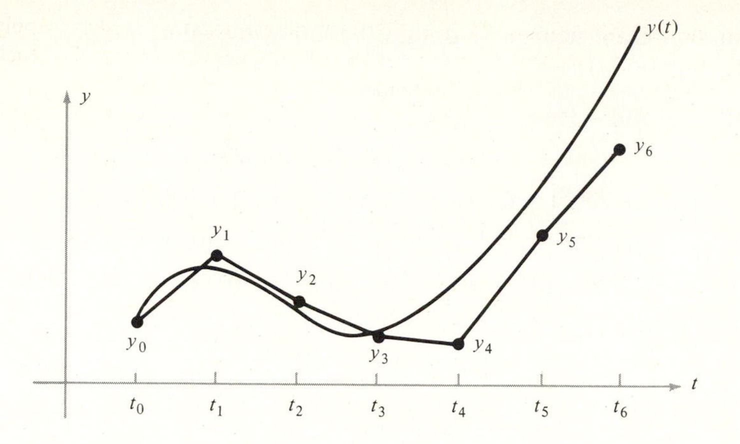

Figure 6.2 Illustrative sketch of Euler approximation.

times $t_1, t_2, \ldots, t_n$, respectively. If the computation of y_i depends only on y_{i-1} (and, of course, f, t_{i-1}, and h), then the method is called a *one-step method.*

6.2*a* Euler's Method

The most basic one-step method is known as *Euler's method.* Geometrically, in order to compute y_i from y_{i-1} it is assumed that the approximation is a line segment with slope equal to $f(t_{i-1}, y_{i-1})$. By the point-slope formula, the equation of the line through (t_{i-1}, y_{i-1}) with slope $f(t_{i-1}, y_{i-1})$ is

$$y - y_{i-1} = f(t_{i-1}, y_{i-1})(t - t_{i-1})$$

Setting $t = t_i$, $y = y_i$, and noting that $t_i - t_{i-1} = h$, it follows that Euler's method has the form

$$y_i = y_{i-1} + hf(t_{i-1}, y_{i-1}) \qquad \text{for } i = 1, 2, \ldots, n \tag{3}$$

A graph indicating a typical situation is given in Figure 6.2. Although Euler's method is not very accurate, it is easy to use and provides a good method for illustration. Therefore, a couple of simple examples are given.

EXAMPLE 6.2-1

Consider the equation $y' = -y + 2e^t$, $y(0) = 0$. The exact solution to this equation is $y(t) = e^t - e^{-t}$ (see the method developed in Section 1.2). We employ Euler's method [formula (3)] to approximate y on [0, 1] at the points 0.2, 0.4, 0.6, 0.8, and 1 (i.e., $h = 0.2$ and $n = 5$). In this case (3) has the form

$$y_i = y_{i-1} + 0.2(-y_{i-1} + 2e^{0.2(i-1)}) \qquad i = 1, 2, \ldots, 5$$

Carrying four significant figures we obtain the following values for $y_1, \ldots, y_5$ and $y(t_1), \ldots, y(t_5)$:

$$y_0 = 0 \qquad \text{and} \qquad y(0) = 0$$

$$y_1 = 0 + 0.2(-0 + 2e^0) = 0.4 \qquad \text{and} \qquad y(0.2) = 0.4027$$

$$y_2 = 0.4 + 0.2(-0.4 + 2e^{0.2}) = 0.8086 \qquad \text{and} \qquad y(0.4) = 0.8215$$

$$y_3 = 0.8086 + 0.2(-0.8086 + 2e^{0.4}) = 1.244 \qquad \text{and} \qquad y(0.6) = 1.273$$

$$y_4 = 1.244 + 0.2(-1.244 + 2e^{0.6}) = 1.724 \qquad \text{and} \qquad y(0.8) = 1.776$$

$$y_5 = 1.724 + 0.2(1.724 + 2e^{0.8}) = 2.269 \qquad \text{and} \qquad y(1) = 2.350$$

EXAMPLE 6.2-2

Consider the equation $y' = t^2 - y^2$, $y(0) = \frac{1}{2}$. Taking $h = 0.1$ and $n = 7$, we have from (3) that Euler's method has the form

$$y_i = y_{i-1} + 0.1[0.01(i-1)^2 - y_{i-1}^2] \qquad i = 1, \ldots, 7$$

Therefore, carrying four significant figures we obtain the following values for $y_1, \ldots, y_7$:

$$y_1 = 0.5 + 0.1[0 - (0.5)^2] = 0.4750$$

$$y_2 = 0.4750 + 0.1[0.01 - (0.4750)^2] = 0.4534$$

$$y_3 = 0.4534 + 0.1[0.04 - (0.4534)^2] = 0.4369$$

$$y_4 = 0.4369 + 0.1[0.09 - (0.4369)^2] = 0.4278$$

$$y_5 = 0.4278 + 0.1[0.16 - (0.4278)^2] = 0.4247$$

$$y_6 = 0.4247 + 0.1[0.25 - (0.4247)^2] = 0.4315$$

$$y_7 = 0.4315 + 0.1[0.36 - (0.4315)^2] = 0.4489$$

6.2*b* Higher-Order Taylor's Expansions

Comparing Euler's method with Taylor's expansion studied in the preceding section shows that the computation of y_i from y_{i-1} [see formula (3)] is in fact the value at $t = t_i$ of Taylor's expansion ϕ_2 of order 2 for the solution y on $[t_{i-1}, t_i]$ to the initial value problem $y' = f(t, y)$, $y(t_{i-1}) = y_{i-1}$. [See formula (6) in Section 6.1, with $n = 2$ and with a and t replaced by t_{i-1} and t_i, respectively.] Therefore, higher-order Taylor's expansions can also be used to compute y_i from y_{i-1}, and this should improve the accuracy of the approximation. Using Taylor's expansion of order 3, for example, leads to the following scheme:

$$
\begin{aligned}
y'_{i-1} &= f(t_{i-1}, y_{i-1}) \\
y''_{i-1} &= f_t(t_{i-1}, y_{i-1}) + f_y(t_{i-1}, y_{i-1})y'_{i-1} \qquad i = 1, \ldots, n \\
y_i &= y_{i-1} + hy'_{i-1} + \frac{h^2}{2!} y''_{i-1}
\end{aligned} \tag{4}
$$

Although higher-order Taylor's expansions lead to more accurate methods, they are also considerably more complicated. Also, the computations usually take much longer since they can involve computing complicated derivatives of f.

EXAMPLE 6.2-3

Consider the equation $y' = 1 + y^2$, $y(0) = 0$, whose solution is $y(t) = \tan t$ for $0 \le t < \pi/2$. The method in (4) will be used to approximate y at the points 0.1, 0.2, 0.3. Therefore, we take $h = 0.1$ and $n = 3$ to obtain the values of y_1, y_2, and y_3. The values of y at 0.1, 0.2, and 0.3 are also recorded.

$i = 1$: $y'_0 = 1 + y_0^2 = 1 \qquad y''_0 = 2y_0 y'_0 = 0$

$y_1 = y_0 + 0.1y'_0 + 0.005y''_0 = 0.1 \qquad y(0.1) = 0.0997$

$i = 2$: $y' = 1 + y_1^2 = 1.01 \qquad y''_1 = 2y_1 y'_1 = 0.202$

$y_2 = y_1 + 0.1y'_1 + 0.005y''_1 = 0.2020 \qquad y(0.2) = 0.1974$

$i = 3$: $y'_2 = 1 + y_2^2 = 1.041 \qquad y''_2 = 2y_2 y'_2 = 0.4110$

$y_3 = y_2 + 0.1y'_2 + 0.005y''_2 = 0.3082 \qquad y(0.3) = 0.2915$

The computational procedures (i.e., the flow of the computations) are easy to set up for one-step methods involving Taylor's expansion. The simple examples given can be done easily with a basic hand calculator, and the flow of the calculations is similar to that for a high-speed computer if one needs to consider more complicated equations or situations. For notational convenience define

$$
\begin{aligned}
f^{(0)}(t, y) &= f(t, y) \\
f^{(1)}(t, y, y') &= f_t(t, y) + f_y(t, y)y' \\
f^{(2)}(t, y, y', y'') &= f_{tt}(t, y) + 2f_{ty}(t, y)y' + f_{yy}(t, y)(y')^2 + f_y(t, y)y''
\end{aligned}
$$

and, inductively, for each positive integer j,

$$
f^{(j)}(t, y, \ldots, y^{(j)}) = \frac{d}{dt}\,[f^{(j-1)}(t, y, \ldots, y^{(j-1)})]
$$

Now suppose that the one-step method generated by Taylor's expansion of order k is to be used. The flow of the computations can be set up as follows:

1. Read in a, h, n, y_0 and set $t_0 = a$ and $i = 1$.
2. Set $y_{i-1}^{(0)} = y_{i-1}$.
3. Compute $y_{i-1}^{(j)} = f^{(j)}(t_{i-1}, y_{i-1}^{(0)}, \ldots, y_{i-1}^{(j-1)})$ for $j = 1, \ldots, k$.
4. Compute $y_i = y_{i-1}^{(0)} + \dfrac{h}{1!} y_{i-1}^{(1)} + \cdots + \dfrac{h^k}{k!} y_{i-1}^{(k)}$.
5. If $i < n$ replace i by $i + 1$ and return to 2. If $i = n$, the computations are complete.

The preceding discussions and examples indicate that, unless the function f is extremely complicated, one-step methods generated by Taylor's expansions can be routinely set up for computations and iterations. However, estimating the error and the time involved in these computations has not been considered. Although most of the techniques for error estimates are beyond this book's scope, we mention a few of the basic ideas in analyzing the effectiveness of numerical methods. The fundamental problem is to determine a "reasonable" estimate of the difference between the approximation y_i and the exact solution $y(t_i)$:

$$\varepsilon_i \equiv y_i - y(t_i) \qquad \text{for } i = 1, \ldots, n \tag{5}$$

The error ε_i is the *accumulated error* after i steps. Normally it is more effective to analyze the *local error* introduced at each step and then determine how the local error propagates in order to estimate ε_i. In fact, since we have already found error estimates for Taylor's expansions [see (9) in the preceding section], we have an estimate in this case. If the one-step method is generated by a Taylor's expansion of order k at each step, then the local error introduced at each step is, in absolute value, no larger than Nh^{k+1}, where N is a constant. By definition, the local error at the ith step is the difference $\bar{y}_i(t_i) - y_i$, where $\bar{y}_i$ is the solution to the initial value problem $\bar{y}_i' = f(t, \bar{y}_i)$, $\bar{y}_i(t_{i-1}) = y_{i-1}$. When the local error is bounded in absolute value by Nh^{k+1} (where the constant N is independent of h and i), we say that the *local error is of order h^{k+1} and write $O(h^{k+1})$*. If there is a constant $M > 0$ such that the absolute value of the accumulated error ε_i is no larger than Mh^k, then we say that the *accumulated error is of order h^k and write $O(h^k)$*. We have the following important result:

■ **Theorem 6.2-1**

Suppose that k is a positive integer and that the one-step method for approximating the solution to (1) is generated by Taylor's expansion of order $k + 1$. Then the accumulated error is of order h^k.

The proof of this theorem is omitted since it is beyond the scope of this book. Note that the larger the order k the faster the error estimates improve as h becomes smaller (for example, if the stepsize h is halved, the error estimation is reduced by a factor of $\frac{1}{2}^k$). However, the larger the order k the more complicated (and time-consuming) the computations are, since each step involves evaluating each of the derivatives of f up to order $k - 1$. Similarly, even though the error estimates improve as the stepsize h decreases, the number of steps needed to cover the same interval increases as h decreases. Therefore, one has to try to balance accuracy with the number of computations necessary. Moreover, if too many computations are involved, then the error introduced by rounding off can become significant. *Roundoff error* always exists in reality, since only a finite number of significant figures can be used in arithmetic operations. Again, an analysis of roundoff error is beyond the scope of this book and is omitted.

6.2c Systems in the Plane

It is important to note that the one-step methods are easily adaptable to systems of equations. Suppose that f and g are "smooth" functions in the plane, and consider the two-dimensional autonomous system

$$\begin{aligned} x' &= f(x, y) \qquad x(0) = x_0 \\ y' &= g(x, y) \qquad y(0) = y_0 \end{aligned} \tag{6}$$

Euler's method applied to equation (6) has the form

$$\begin{aligned} x_i &= x_{i-1} + hf(x_{i-1}, y_{i-1}) \\ y_i &= y_{i-1} + hg(x_{i-1}, y_{i-1}) \end{aligned} \qquad i = 1, 2, \ldots, n \tag{7}$$

One can also use higher-order Taylor's expansions to generate one-step procedures for equation (6). For example, by using Taylor's expansion of order 3 [see formula (8) in the preceding section], the following process is obtained:

$$\begin{aligned} x'_{i-1} &= f(x_{i-1}, y_{i-1}) \qquad y'_{i-1} = g(x_{i-1}, y_{i-1}) \\ x''_{i-1} &= f_x(x_{i-1}, y_{i-1})x'_{i-1} + f_y(x_{i-1}, y_{i-1})y'_{i-1} \\ y''_{i-1} &= g_x(x_{i-1}, y_{i-1})x'_{i-1} + g_y(x_{i-1}, y_{i-1})y'_{i-1} \\ x_i &= x_{i-1} + hx'_{i-1} + \frac{h^2}{2!} x''_{i-1} \\ y_i &= y_{i-1} + hy'_{i-1} + \frac{h^2}{2!} y''_{i-1} \end{aligned} \tag{8}$$

It is clear that for large systems of equations and for large values of the order k for the Taylor's expansion involved, the computations become unwieldy and, therefore, other procedures should be investigated.

PROBLEMS

6.2-1. Apply Euler's method, with $h = 0.1$ and $n = 5$, in order to obtain approximations for the solution to each of the following initial value problems at $t = 0.1, 0.2, 0.3, 0.4, 0.5$. Using an appropriate method from Chapter 1, determine the exact solutions and compare values with the approximations. (Use four significant figures in the computations.)

(a) $y' = -4y \quad y(0) = 1$ **(d)** $y' = -2ty^2 \quad y(0) = 1$

(b) $y' = -2y \quad y(0) = 1$ **(e)** $y' = 1 + y^2 \quad y(0) = 0$

(c) $y' = 2y - 1 \quad y(0) = 1$ **(f)** $y' = y - y^2 \quad y(0) = \frac{1}{2}$

6.2-2. Compute the value of the exact solution to each of the following equations at $t = \frac{1}{2}$. Then use Euler's method with $h = \frac{1}{8}$ and the one-step method generated by Taylor's expansion of order 3 with $h = \frac{1}{4}$ to obtain approximations of the exact solution at $t = \frac{1}{2}$. Repeat with $h = \frac{1}{8}$ and $h = \frac{1}{4}$ changed to $h = \frac{1}{12}$ and $h = \frac{1}{6}$, respectively.

(a) $y' = y^2 \quad y(0) = 1$ **(b)** $y' = -y^2 \quad y(0) = 1$

6.2-3. Consider the equation $y' = ay$, $y(0) = 1$, where a is a constant. For each positive integer n apply Euler's method with $h = 1/n$ to approximate the solution and show that

$$y_n = \left(1 + \frac{1}{n}a\right)^n$$

What is $\lim_{n \to \infty} y_n$?

6.2-4. Take $h = 0.1$ and $n = 2$ in the one-step method generated by Taylor's expansion of order 4, and then compute the approximations y_1 and y_2 to each of the following equations. (Use five significant figures in the computations.)

(a) $y' = 2ty \quad y(0) = 1$ **(b)** $y' = t^2 - y^2 \quad y(0) = \frac{1}{2}$

6.2-5. Apply Euler's method, with $h = 0.1$ and $n = 2$, in order to obtain approximations for solutions to each of the following two-dimensional systems at $t = 0.1, 0.2$. Repeat, using the one-step method generated by Taylor's expansion of order 3.

(a) $x' = x + 2y \quad x(0) = 1$ **(b)** $x' = x - \frac{1}{2}xy \quad x(0) = 1$

$y' = 6x - 3 \quad y(0) = -1$ $y' = y - \frac{1}{4}xy \quad y(0) = 3$

Compute the exact solution for part a and compare with your approximations.

*6.3 RUNGE-KUTTA–TYPE METHODS

The direct use of Taylor's expansion to generate one-step methods of rather high order was shown in the preceding section to involve the computation and evaluation of several of the derivatives of the function f. Since the computations of these derivatives are generally complex and time-consuming, we indicate here how one can generate one-step methods of higher order that involve only computations with f and no computations with the derivatives of f. As in the previous sections, the initial value problem

$$y' = f(t, y) \qquad y(a) = y_0 \tag{1}$$

is considered, where f is a "smooth" function on $[a, b) \times (-\infty, \infty)$. The problem under investigation is to obtain a discrete approximation of the solution y to (1). In particular, we indicate techniques for developing one-step methods of order larger than 1 that do not involve computing any derivatives of f.

These techniques are illustrated by the derivation of the second order methods. Using Taylor's expansion of order 3, the one-step method

$$y_i = y_{i-1} + hf(t_{i-1}, y_{i-1}) + \frac{h^2}{2}[f_t(t_{i-1}, y_{i-1}) + f_y(t_{i-1}, y_{i-1})f(t_{i-1}, y_{i-1})] \tag{2}$$

is obtained [see (4) in the preceding section]. In order to obtain an "equivalent" procedure that does not involve the computation of f_t and f_y, consider the one-step procedure

$$y_i = y_{i-1} + h[\sigma_1 f(t_{i-1}, y_{i-1}) + \sigma_2 f(t_{i-1} + \rho_1 h, y_{i-1} + \rho_2 hf(t_{i-1}, y_{i-1}))] \tag{3}$$

where σ_1, σ_2, ρ_1, and ρ_2 are constants to be determined. The general idea is to adjust the constants in (3) so that the expression for y_i in (3) equals the expression for y_i in (2) for all powers of h in terms up to and including h^2. Setting these expressions for y_i equal to each other, canceling y_{i-1}, and dividing each side of the resulting equation by h, we obtain

$$\begin{aligned} f(t, y) &+ \frac{h}{2}[f_t(t, y) + f_y(t, y)f(t, y)] \\ &= \sigma_1 f(t, y) + \sigma_2 f(t + \rho_1 h, y + \rho_2 hf(t, y)) + O(h^2) \end{aligned} \tag{4}$$

where the subscript $i - 1$ is omitted throughout. Using Taylor's formula for a function of two variables

$$f(t + \rho_1 h, y + \rho_2 hf(t, y)) = f(t, y) + \rho_1 hf_t(t, y) + \rho_2 hf(t, y)f_y(t, y) + O(h^2)$$

Substituting this expression into the right-hand side of (4),

$$\begin{aligned} f(t, y) &+ \frac{h}{2}[f_t(t, y) + f_y(t, y)f(t, y)] \\ &= (\sigma_1 + \sigma_2)f(t, y) + h[\sigma_2 \rho_1 f_t(t, y) + \sigma_2 \rho_2 f(t, y)f_y(t, y)] + O(h^2) \end{aligned}$$

Equating the constant terms and the coefficients of h the three equations

$$\sigma_1 + \sigma_2 = 1 \qquad \sigma_2 \rho_1 = \tfrac{1}{2} \qquad \text{and} \qquad \sigma_2 \rho_2 = \tfrac{1}{2}$$

must be satisfied. The solutions to these equations are

$$\sigma_1 = 1 - \gamma \qquad \sigma_2 = \gamma \qquad \text{and} \qquad \rho_1 = \rho_2 = \frac{1}{2\gamma} \tag{5}$$

where γ is a real parameter, $\gamma \neq 0$. Therefore, *for every nonzero value of the parameter* γ, *the method*

$$(6) \qquad y_i = y_{i-1} + h(1-\gamma)f(t_{i-1}, y_{i-1}) + h\gamma f\left(t_{i-1} + \frac{h}{2\gamma}, y_{i-1} + \frac{h}{2\gamma} f(t_{i-1}, y_{i-1})\right)$$

is of second order. Of course, this method does not involve the computation of any derivatives of f and, in fact, involves only two computations of f at each step [one at (t_{i-1}, y_{i-1}) and another at $(t_{i-1} + h/2\gamma, y_{i-1} + (h/2\gamma)f(t_{i-1}, y_{i-1}))$]. The procedure defined by (6) is known as a *simplified Runge-Kutta method of order 2.* For the particular case of (6) when $\gamma = \frac{1}{2}$ we obtain the method

$$(7) \qquad y_i = y_{i-1} + \frac{h}{2}[f(t_{i-1}, y_{i-1}) + f(t_{i-1} + h, y_{i-1} + hf(t_{i-1}, y_{i-1}))]$$

which is known as the *improved Euler method* or the *Heun method.*

EXAMPLE 6.3-1

Consider the equation $y' = 1 + y^2$, $y(0) = 0$. Taking $h = 0.1$ and $n = 4$ the improved Euler method (7) can be written in the form

$$k_1 = 1 + y_{i-1}^2 \qquad k_2 = 1 + (y_{i-1} + 0.1k_1)^2 \qquad i = 1, 2, 3, 4$$
$$y_i = y_{i-1} + 0.05[k_1 + k_2]$$

Therefore, the approximations y_1, y_2, y_3, and y_4 can be computed as follows:

$i = 1$: $k_1 = 1 \qquad k_2 = 1 + (0.1)^2 = 1.01$

$y_1 = 0.05[1 + 1.01] = 0.1005$

$i = 2$: $k_1 = 1 + (0.1005)^2 = 1.010 \qquad k_2 = 1 + (0.1005 + 0.101)^2 = 1.041$

$y_2 = 0.1005 + 0.05[1.01 + 1.041] = 0.2031$

$i = 3$: $k_1 = 1 + (0.1031)^2 = 1.041 \qquad k_2 = 1 + (0.2031) + 0.1041)^2 = 1.094$

$y_3 = 0.2031 + 0.05[1.041 + 1.094] = 0.3099$

$i = 4$: $k_1 = 1 + (0.3099)^2 = 1.096 \qquad k_2 = 1 + (0.3099 + 0.1096)^2 = 1.176$

$y_4 = 0.3099 + 0.05[1.096 + 1.094] = 0.4194$

The solution to this equation was also approximated in Example 6.2-3, with the second order one-step method generated by Taylor's expansion of order 3. The exact solution to this equation is $y(t) = \tan t$ [$y(0.1)$, $y(0.2)$, $y(0.3)$ are computed in Example 6.2-3], and checking the error indicates that for this example the Taylor's expansion method is slightly more accurate.

Higher-order Runge-Kutta–type methods can be obtained using analogous techniques. Although these techniques use only elementary ideas, the algebra involved becomes complicated. The best-known and most used one-step method is the *classical Runge-Kutta method*, and it has the form

$$
\begin{aligned}
&k_1 = f(t_{i-1}, y_{i-1}) \qquad k_2 = f(t_{i-1} + \tfrac{1}{2}h, y_{i-1} + \tfrac{1}{2}hk_1) \\
&k_3 = f(t_{i-1} + \tfrac{1}{2}h, y_{i-1} + \tfrac{1}{2}hk_2) \qquad k_4 = f(t_{i-1} + h, y_{i-1} + hk_3) \\
&y_i = y_{i-1} + \frac{h}{6}[k_1 + 2k_2 + 2k_3 + k_4] \qquad i = 1, 2, \ldots, n
\end{aligned}
\tag{8}
$$

This method is of order 4, and because it is simple to program and quite accurate, it is one of the most frequently used methods for approximating solutions to initial value problems. Note in particular that there are four evaluations of the function f at each step, and no evaluations of derivatives of f are required.

▶ **EXAMPLE 6.3-2**

Again consider the equation $y' = 1 + y^2$, $y(0) = 0$. Using the Runge-Kutta method (8), with $h = 0.2$ and $n = 2$, we obtain the following procedure:

$$
\begin{aligned}
&k_1 = 1 + y_{i-1}^2 \qquad k_2 = 1 + (y_{i-1} + 0.1k_1)^2 \\
&k_3 = 1 + (y_{i-1} + 0.1k_2)^2 \qquad k_4 = 1 + (y_{i-1} + 0.2k_3)^2 \\
&y_i = y_{i-1} + \frac{0.1}{3}[k_1 + 2k_2 + 2k_3 + k_4] \qquad i = 1, 2
\end{aligned}
$$

Therefore, carrying four significant figures,

$$
\begin{aligned}
i = 1: \quad & k_1 = 1 \qquad k_2 = 1 + (0.1)^2 = 1.01 \\
& k_3 = 1 + (0.101)^2 = 1.010 \\
& k_4 = 1 + (0.2 \cdot 1.012)^2 = 1.041 \\
& y_1 = \tfrac{1}{30}[1 + 2 \cdot 1.01 + 2 \cdot 1.010 + 1.041] = 0.2027 \\
i = 2: \quad & k_1 = 1 + (0.2027)^2 = 1.041 \qquad k_2 = 1 + (0.2027 + 0.1041)^2 = 1.094 \\
& k_3 = 1 + (0.2027 + 0.1094)^2 = 1.097 \\
& k_4 = 1 + (0.2027 + 0.2 \cdot 1.097)^2 = 1.178 \\
& y_2 = 0.2027 + \tfrac{1}{30}[1.041 + 2 \cdot 1.094 + 2 \cdot 1.097 + 1.178] = 0.4227
\end{aligned}
$$

and it follows that the accuracy here is about the same as that in Example 6.3-1, but the stepsize h is twice as large in this case.

6.3a Systems in the Plane

Runge-Kutta–type methods can also be applied to systems of equations. Suppose that f and g are smooth functions in the plane and consider the initial value system

$$(9)\qquad \begin{aligned} x' &= f(x, y) & x(0) &= x_0 \\ y' &= g(x, y) & y(0) &= y_0 \end{aligned}$$

The computations needed for the derivation of Runge-Kutta–type methods for the system (9) are elementary but complicated. However, the classical Runge-Kutta fourth order formula for (9) is analogous to the one-dimensional method and has the following form:

$$(10)\qquad \begin{aligned} k_1 &= f(x_{i-1}, y_{i-1}) & l_1 &= g(x_{i-1}, y_{i-1}) \\ k_2 &= f(x_{i-1} + \tfrac{1}{2}hk_1, y_{i-1} + \tfrac{1}{2}hl_1) & l_2 &= g(x_{i-1} + \tfrac{1}{2}hk_1, y_{i-1} + \tfrac{1}{2}hl_1) \\ k_3 &= f(x_{i-1} + \tfrac{1}{2}hk_2, y_{i-1} + \tfrac{1}{2}hl_2) & l_3 &= g(x_{i-1} + \tfrac{1}{2}hk_2, y_{i-1} + \tfrac{1}{2}hl_2) \\ k_4 &= f(x_{i-1} + hk_3, y_{i-1} + hl_3) & l_4 &= g(x_{i-1} + hk_3, y_{i-1} + hl_3) \\ x_i &= x_{i-1} + \frac{h}{6}[k_1 + 2k_2 + 2k_3 + k_4] & i &= 1, \ldots, n \\ y_i &= y_{i-1} + \frac{h}{6}[l_1 + 2l_2 + 2l_3 + l_4] \end{aligned}$$

The flow of the computations in (10) is quite easy to follow and can be readily programmed for use in a computer.

PROBLEMS

6.3-1. Determine approximations for the solution to each of the following equations, using the improved Euler method (7), with $h = 0.1$ and $n = 4$.

(a) $y' = -4y \quad y(0) = 1$ **(c)** $y' = y - y^2 \quad y(0) = \frac{1}{2}$

(b) $y' = -2y \quad y(0) = 1$ **(d)** $y' = t^2 + y^2 \quad y(0) = 0$

6.3-2. Approximate the solution y to each of the following equations at $t = \frac{1}{2}$ by using the classical Runge-Kutta method (8) successively by taking $h = \frac{1}{4}$ and then $h = \frac{1}{6}$.

(a) $y' = y^2 \quad y(0) = 1$ **(b)** $y' = -y^2 \quad y(0) = 1$

Compare the approximations with the exact solutions and also with the approximations obtained in Problem 6.2-2.

6.3-3. Set $\gamma = 1$ in the simplified Runge-Kutta method (6) to obtain the method

$$(11)\qquad y_i = y_{i-1} + hf(x_{i-1} + \tfrac{1}{2}h, y_{i-1} + \tfrac{1}{2}hf(x_{i-1}, y_{i-1}))$$

This is known as the *modified Euler method* or the *improved polygon method.*

Approximate the solutions to each of parts *a*, *b*, *c*, and *d* in Problem 6.3-1, using this method with $h = 0.1$ and $n = 4$. Compare the results with those using the improved Euler method.

***6.3-4.** Expand the expression for y_i in (8), using Taylor's formula for the function f of two variables, and compare with the Taylor's expansion

$$y_i = y_{i-1} + hy'_{i-1} + \frac{h^2}{2!} y''_{i-1} + \frac{h^3}{3!} y'''_{i-1} + \frac{h^4}{4!} y^{(iv)}_{i-1}$$

Show that these expansions agree up to $O(h^5)$, and hence that the classical Runge-Kutta method is of order 4 (and local order 5).

6.3-5. Show that if the function f in equation (1) is independent of y, then the Runge-Kutta formula (8) gives Simpson's rule for approximating an integral:

$$\int_a^{a+h} f(t)\, dt \cong \frac{h}{6}\left[f(a) + 4f\left(a + \frac{h}{2}\right) + f(a + h)\right]$$

6.3-6. Derive a simplified Runge-Kutta method of order 2 [see formula (6)] for the system (9).

6.3-7. Use the method (10), with $h = 0.1$ and $n = 2$, to approximate the solution to the following systems at $t = 0.1$ and $t = 0.2$.

(a) $x' = x + 2y \qquad x(0) = 1$
$\quad\;\; y' = 6x - 3 \qquad y(0) = -1$

(b) $x' = x - \frac{1}{2}xy \qquad x(0) = 1$
$\quad\;\; y' = y - \frac{1}{4}xy \qquad y(0) = 3$

Compute the exact solution for part *a* and compare with your approximations (see also Problem 6.2-5).

6.3-8. Write each of the following second order initial value problems as a system of two first order equations, and then apply both the improved Euler method (with $h = 0.1$ and $n = 4$) and the classical Runge-Kutta method (with $h = 0.2$ and $n = 2$) to obtain approximations of the solutions.

(a) $x'' + 4x = 0 \qquad x(0) = 0 \qquad x'(0) = 1$

(b) $x'' + \sin x = 0 \qquad x(0) = 0.1 \qquad x'(0) = 0$

(c) $x'' + x' + x - x^3 = 0 \qquad x(0) = 0.1 \qquad x'(0) = 0$

6.3-9. Show that the Runge-Kutta method

$$k_1 = f(t_{i-1}, y_{i-1}) \qquad k_2 = f(t_{i-1} + \tfrac{1}{2}h, y_{i-1} + \tfrac{1}{2}k_1)$$

$$k_3 = f(t_{i-1} + h, y_{i-1} + 2k_2 - k_1) \tag{12}$$

$$y_i = y_{i-1} + \frac{h}{6}(k_1 + 4k_2 + k_3)$$

is locally of fourth order (and hence is a third order method).

*6.4 MULTISTEP PROCESSES: PREDICTOR-CORRECTOR METHODS

As in the previous section f is assumed to be a smooth function on $[a, b) \times (-\infty, \infty)$, and the initial value problem

$$y' = f(t, y) \qquad y(a) = y_0 \tag{1}$$

is considered. Also, y is assumed to be a solution to (1) on the interval $[a, c] \subset [a, b)$, and the purpose of this section is to indicate how multistep methods can be employed to approximate y. Multistep methods are discrete methods, so using the notations of the preceding two sections, it is assumed that $h > 0$ is the stepsize, that $t_0 = a$, and that $t_i = t_{i-1} + h = t_0 + ih$ are the mesh points in $[a, c]$. Also, y_i denotes the approximation of $y(t_i)$ for each $i = 0, \ldots, n$. Multistep methods are characterized by the fact that the computation of y_i depends not only on y_{i-1} but also on the approximations y_j, with $i - k \leq j \leq i - 1$, where $k \geq 1$ is a prescribed positive integer. In this case the method is called a *k-step method.* If $k = 1$, this terminology agrees with the one-step methods studied in the preceding two sections.

Now suppose that k is a positive integer and that $\alpha_1, \alpha_2, \ldots, \alpha_k$ and $\beta_0, \beta_1, \ldots, \beta_k$ are given numbers. We consider k-step methods of the general form

$$\begin{aligned} y_i = \alpha_1 y_{i-1} + \alpha_2 y_{i-2} + \cdots + \alpha_k y_{i-k} & \\ + h[\beta_0 y_i' + \beta_1 y_{i-1}' + \cdots + \beta_k y_{i-k}'] & \qquad \text{for } i = k, k+1, \ldots, n \end{aligned} \tag{2}$$

where $y_j' \equiv f(t_j, y_j)$ whenever $0 \leq j \leq n$. If $\beta_0 = 0$, the procedure (2) is called *explicit* (since y_i' does not appear on the right-hand side), and if $\beta_0 \neq 0$, it is called *implicit.* Although it will not be established, we point out that when $\beta_0 \neq 0$, equation (2) always has a unique solution y_i for given $y_{i-1}, \ldots, y_{i-k}$ if $h > 0$ is sufficiently small. It is important to note that in addition to the initial value y_0, initial approximations also must be obtained for $y_1, y_2, \ldots, y_{k-1}$ in order to apply the iterative scheme (2). Therefore, the *starting values* $y_0, y_1, \ldots, y_{k-1}$ must be computed by another process (Taylor's expansion or a Runge-Kutta method, for example). This starting process is a disadvantage of multistep methods. The principal advantage is that normally one can obtain the same order or accuracy as obtained by one-step methods, with fewer evaluations of f at each step. Since computations of f are usually the most time-consuming process in discrete methods, this is an important advantage.

The iterative method (2) is said to be *exact for the function y* if

$$y(t_i) = \alpha_1 y(t_{i-1}) + \cdots + \alpha_k y(t_{i-k}) + h[\beta_0 y'(t_i) + \cdots + \beta_k y'(t_{i-k})]$$

for all $i = k, k + 1, \ldots, n$. If m is a positive integer, the iterative process (2) is said to be of *order m* if it is exact for the functions $y(t) \equiv t^j, j = 0, 1, \ldots, m$, and not exact for $y(t) \equiv t^{m+1}$. Note that if (2) is of order m, it is exact for all polynomials of degree less than or equal to m.

The principal multistep procedures that are discussed in this section are known as *predictor-corrector methods.* This type of process employs a pair of iterative methods of the form (2), one of the pair being explicit (that is, $\beta_0 = 0$) and the other implicit (that is, $\beta_0 \neq 0$). Normally, the pairs are assumed to be of the same order. The explicit scheme is known as the *predictor*, and in going from y_{i-1} to y_i, it is used to give a first estimation (or "prediction") $\tilde{y}_i$ of y_i from the preceding values $y_{i-1}, \ldots, y_{i-k}$. The predicted value $\tilde{y}_i$ is then substituted into the right-hand side of the implicit scheme (which is called the *corrector*) in order to improve (or "correct") the initial estimation $\tilde{y}_i$. Once the corrected value is obtained, it can be used as y_i or it may be substituted back into the right-hand side of the corrector in order to obtain a still better approximation of y_i. This iteration process can be continued until one is convinced that the value of y_i satisfies the corrector equation within a prescribed tolerance. It is important to note that, in general, if the k-step method (2) is implicit, then one may not be able to obtain y_i exactly from $y_{i-1}, \ldots, y_{i-k}$. This is why one uses the predictor as an initial guess and then iterates the corrector to improve the original approximation.

As a first example consider the explicit second order scheme

$$y_i = y_{i-2} + 2hy'_{i-1} \qquad \text{for } i = 2, 3, \ldots, n \tag{3}$$

which is known as *Nystron's method.* Consider also the following implicit method, which is known as the *trapezoidal rule:*

$$y_i = y_{i-1} + \tfrac{1}{2}h[y'_{i-1} + y'_i] \qquad i = 2, 3, \ldots, n \tag{4}$$

In fact, the method (4) is actually a one-step implicit scheme of order 2. We combine Nystron's method with the trapezoidal rule to form a predictor-corrector process, and we indicate the procedure as follows:

$$\begin{aligned} &\tilde{y}_i = y_{i-2} + 2hy'_{i-1} \\ &y'_i = f(t_i, \tilde{y}_i) \qquad\qquad i = 2, 3, \ldots, n \\ &y_i = y_{i-l} + \tfrac{1}{2}h[y'_{i-l} + y'_i] \end{aligned} \tag{5}$$

The notation used for predictor-corrector methods can be confusing. The corrector is to be iterated and the value $y'_i = f(t_i, \tilde{y}_i)$ is only the starting value for the iteration. Therefore, once the predicted value $\tilde{y}_i$ is obtained, the procedure is to set $y_{i,0} = \tilde{y}_i$ and then use the iterative scheme

$$\begin{aligned} &y'_{i,j} = f(t_i, y_{i,j-1}) \\ &y_{i,j} = y_{i-1} + \tfrac{1}{2}h[y'_{i-1} + y'_{i,j}] \qquad j = 1, 2, \ldots \end{aligned} \tag{6}$$

in order to compute y_i. Once the difference $y_{i,j} - y_{i,j-1}$ is small relative to the allowable error, the iterations can be stopped and y_i can be set equal to $y_{i,j}$. However, the principal advantage of predictor-corrector methods is in the time saved by not having to make a large number of evaluations of f at each step. Therefore, if the convergence of the corrector iterations is too slow, alternative procedures should be tried (i.e., halving the stepsize h and restarting the method or changing to another method). The computational flow of predictor-corrector methods is illustrated by applying method (5) to a simple example.

EXAMPLE 6.4-1

Consider the equation $y' = -2ty^2$, $y(0) = 1$, whose exact solution is $y(t) = (1 + t^2)^{-1}$ for all $t \geq 0$. This solution is approximated on [0, 0.4] by taking $h = 0.1$ and using method (5). We have $y_0 = 1$, and we use the Runge-Kutta–type method of order 2, or the improved Euler method [see (7) in the preceding section] to obtain

$$y_1 = y_0 + \left[\frac{f(0, y_0) + f(h, y_0 + hf(0, y_0))}{2}\right]$$

$$= 1 + 0.1\left[\frac{-0 - 0.1(1 + 0)^2}{2}\right] = 0.995$$

Therefore $y_0 = 1$ and $y_1 = 0.995$ are the starting values. Applying method (5) we now obtain the following approximations (the corrector is iterated twice at each step):

$$\tilde{y}_2 = 1 + 2(0.1)[-2(0.2)(0.995)^2] = 0.9208$$

$$y'_{2,0} = -2(0.2)(0.9208)^2 = -0.3391$$

$$y_{2,1} = 0.995 + 0.05[-0.1980 - 0.3391] = 0.9681$$

$$y'_{2,1} = -2(0.2)(0.9681)^2 = -0.3749$$

$$y_{2,2} = 0.995 + 0.05[-0.1980 - 0.3749] = 0.9664$$

Therefore, take $y_2 = 0.9664$ [note that $y(0.2) = 0.9615$].

$$\tilde{y}_3 = 0.995 + 2(0.1)[-2(0.3)(0.9664)^2] = 0.8829$$

$$y'_{3,0} = -2(0.3)(0.8829)^2 = -0.4677$$

$$y_{3,1} = 0.9664 + 0.05[-0.3736 - 0.4677] = 0.9243$$

$$y'_{3,1} = -2(0.3)(0.9243)^2 = -0.5126$$

$$y_{3,2} = 0.9664 + 0.05[-0.3736 - 0.5126] = 0.9221$$

Therefore, take $y_3 = 0.9221$ [note that $y(0.3) = 0.9175$].

$$\tilde{y}_4 = 0.9664 + 2(0.1)(-2)(0.3)(0.9221)^2 = 0.8644$$

$$y'_{4,0} = -2(0.4)(0.8644)^2 = -0.5977$$

$$y_{4,1} = 0.9221 + 0.05[-0.5102 - 0.5977] = 0.8667$$

$$y'_{4,1} = -2(0.4)(0.8667)^2 = -0.6009$$

$$y_{4,2} = 0.9221 + 0.05[-0.5102 - 0.6009] = 0.8665$$

Therefore, take $y_4 = 0.8665$ [note that $y(0.4) = 0.8621$].

There is normally no advantage to using predictor-corrector methods of the second order since at least two evaluations of f are needed at each step, and both second order Runge-Kutta–type methods and the second order Taylor's expansion method require only two evaluations of f per step. In general, predictor-corrector methods can require fewer evaluations of f than one-step methods whenever the order is 4 or more. There are several well-known and frequently used fourth order predictor-corrector methods, and for many basic problems these methods provide suitable accuracy in a reasonable amount of computing time. We indicate here three of the most popular fourth order predictor-corrector methods and then comment on how to use these procedures.

The fourth order *Adams-Moulton predictor-corrector* method for equation (1) has the form

$$\tilde{y}_i = y_{i-1} + \frac{h}{24}[55y'_{i-1} - 59y'_{i-2} + 37y'_{i-3} - 9y'_{i-4}]$$

$$y'_i = f(t_i, \tilde{y}_i) \tag{7}$$

$$y_i = y_{i-1} + \frac{h}{24}[9y'_i + 19y'_{i-1} - 5y'_{i-2} + y'_{i-3}]$$

The predictor in (7) is said to be of the *Adams-Bashforth* type. The fourth order predictor-corrector method

$$\tilde{y}_i = y_{i-4} + \frac{4h}{3}[2y'_{i-1} - y'_{i-2} + 2y'_{i-3}]$$

$$y'_i = f(t_i, \tilde{y}_i) \tag{8}$$

$$y_i = y_{i-2} + \frac{h}{3}[y'_i + 4y'_{i-1} + y'_{i-2}]$$

is known as *Milne's method.* Note that the corrector is actually Simpson's rule for evaluating an integral. By changing the corrector in (8) we obtain *Hamming's method*, which is also fourth order:

$$\tilde{y}_i = y_{i-4} + \frac{4h}{3}[2y'_{i-1} - y'_{i-2} + 2y'_{i-3}]$$

$$y'_i = f(t_i, \tilde{y}_i) \tag{9}$$

$$y_i = \tfrac{1}{8}[9y_{i-1} - y_{i-3}] + \frac{3h}{8}[y'_i + 2y'_{i-1} - y'_{i-2}]$$

There are two principal properties to take into account when selecting a predictor-corrector method of a certain order: the *truncation error* estimation and the *regions of stability and relative stability.* Overall, the Adams-Moulton method (7) (as well as higher-order Adams-Moulton–type methods) seems to exhibit the best features and is recommended. Milne's method (8) is fairly accurate but has some instability properties that can cause small errors introduced early that become relatively large errors over a large number of steps. Although Hamming's method (9) has a reasonable region of relative stability, it is not as accurate as the other methods. It is, however, a good method to use if a large number of steps are required and if $f_y(t, y)$ is negative.

An analysis of truncation error and region of relative stability does not fit the aims of this book, and the interested student should consult texts on numerical methods for ordinary differential equations for further study. We do, however, indicate the flow of computations and analysis of implementing predictor-corrector methods. Suppose that we wish to approximate the solution to (1) over the interval $[a, c]$ and that we have decided to use a predictor-corrector k-step method of order n (note that the predictor and the corrector should be of the same order). To accompany the nth order predictor-corrector method, we choose a Runge-Kutta method, also of order n, as a starting method. An appropriate initial stepsize h must also be chosen. Normally, h is initially selected in accordance with the magnitude of accuracy that is needed. The process then flows as follows:

1. Compute the starting values $y_0, y_1, \ldots, y_{k-1}$ with the Runge-Kutta method and set $i = k$.

2. Compute the predicted value $\tilde{y}_i$ and set $y_{i,0} = \tilde{y}_i$.

3. Beginning with $j = 1$, iterate by computing $y'_{i,j} = f(t_i, y_{i,j-1})$ and then substituting this value into corrector to obtain $y_{i,j}$. After each iteration compare $|y_{i,j} - y_{i,j-1}|$ with an accuracy tolerance ε: if $|y_{i,j} - y_{i,j-1}| > \varepsilon$, replace j by $j + 1$ and iterate again; if $|y_{i,j} - y_{i,j-1}| \leq \varepsilon$, set $y_i = y_{i,j}$, replace i by $i + 1$, t_i by $t_i + h$; and if $t_i + h < c$, go back to step 2.

Another advantage of predictor-corrector methods is that there is a "built-in" way of estimating at least the order of magnitude of the error at each step. If the predicted value $\tilde{y}_i$ is considerably closer to the first corrected value $y_{i,1}$ than the needed accuracy, then the stepsize h should be doubled (and hence the number of steps needed to cover the interval $[a, c]$ is reduced). Also, if too many correction iterations are needed to be within the desired accuracy tolerance (see step 3 above), then the stepsize h should be halved. Normally, one should not have to iterate the corrector more than two or three times at each step.

Finally, it is important to note that predictor-corrector methods apply equally well to systems of equations, and the formulas are completely analogous. For example, the Adams-Moulton fourth order method for the two-dimensional system

$$\begin{aligned} x' &= f(x, y) \qquad x(0) = x_0 \\ y' &= g(x, y) \qquad y(0) = y_0 \end{aligned} \tag{10}$$

has the form

$$\tilde{x}_i = x_{i-1} + \frac{h}{24}[55x'_{i-1} - 59x'_{i-2} + 37x'_{i-3} - 9x'_{i-4}]$$

$$\tilde{y}_i = y_{i-1} + \frac{h}{24}[55y'_{i-1} - 59y'_{i-2} + 37y'_{i-3} - 9y'_{i-4}]$$

$$(11) \qquad x'_i = f(\tilde{x}_i, \tilde{y}_i) \qquad y'_i = g(\tilde{x}_i, \tilde{y}_i)$$

$$x_i = x_{i-1} + \frac{h}{24}[9x'_i + 19x'_{i-1} - 5x'_{i-2} + x'_{i-3}]$$

$$y_i = y_{i-1} + \frac{h}{24}[9y'_i + 19y'_{i-1} - 5y'_{i-2} + y'_{i-3}]$$

where the starting values (x_1, y_1), (x_2, y_2), and (x_3, y_3) must be computed with a one-step method.

PROBLEMS

6.4-1. Take $h = 0.1$ and use the improved Euler method [see (7) in the preceding section] to approximate the solutions to each of the following problems at $t = 0.1, 0.2$, and 0.3. Then, using the approximation at $t = 0$ and 0.1 as starting values, approximate the solutions at $t = 0.2$ and 0.3, using the Nystron-trapezoidal predictor-corrector method [see equation (4)]. Also, compute the exact solution and compare.

(a) $y' = y \quad y(0) = 1$ **(c)** $y' = -y \quad y(0) = 1$

(b) $y' = 2ty^2 \quad y(0) = 1$ **(d)** $y' = 1 + y^2 \quad y(0) = 0$

6.4-2. Consider the two-step predictor-corrector method

$$\tilde{y}_i = y_{i-1} + \frac{h}{2}[3y'_{i-1} - y_{i-2}]$$

$$y'_i = f(t_i, \tilde{y}_i)$$

$$y_i = y_{i-1} + \frac{h}{2}[y_{i-1} + y'_i]$$

Show that the predictor is of second order (note that the corrector is the trapezoidal rule). Use this method in place of the Nystron-trapezoidal method in Problem 6.4-1 and compare the results.

6.4-3. Determine all two-step, third order explicit methods of the form (2).

6.4-4. Determine all two-step, second order implicit methods of the form (2).

6.4-5. Use the classical Runge-Kutta fourth order method, with $h = 0.1$, to obtain three starting values: y_1, y_2, y_3. Then apply the fourth order Adams-Moulton method (7), Milne's method (8), and Hamming's method (9) to obtain y_4 and y_5. Use six significant figures in the computation.

(a) $y' = -4y \qquad y(0) = 1$

(b) $y' = t - y^2 \qquad y(0) = 1$

(c) $y' = y - y^2 \qquad y(0) = \frac{1}{2}$

Remarks A study of the numerical solution to ordinary differential equations at an intermediate level can be found in Birkhoff and Rota [1] and Conte and de Boor [5]. See also the book by McCormick and Salvadori [11], where computer programs for numerical methods are developed.

7 Higher-Order Linear Equations

Many of the ideas and techniques for second order linear equations extend naturally to higher-order equations, and it is the purpose of this chapter to indicate some of these extensions. The case where the equation has constant coefficients is the most important and is discussed in Section 7.2. Techniques for nonhomogeneous equations are indicated in Section 7.3, and Section 7.4 indicates results for systems of two equations.

7.1 FUNDAMENTAL CONCEPTS

Throughout this section n denotes a positive integer and I denotes an interval. If y is an n-times differentiable function on I, we sometimes use the notation

$$y^{(0)} = y, \; y^{(1)} = \frac{dy}{dt}, \; y^{(2)} = \frac{d^2y}{dt^2}, \; \ldots, \; y^{(n)} = \frac{d^ny}{dt^n}$$

Normally the prime notation is still used for derivatives of order 3 or less. There is a close analogy between the concepts discussed here and those in Chapter 2 concerning second order linear equations (which corresponds to the case $n = 2$). Suppose that f, $a_0, a_1, \ldots, a_n$ are continuous functions on the interval I and that $a_n(t) \neq 0$ for all $t \in I$. The differential equations

$$a_n(t)y^{(n)} + a_{n-1}(t)y^{(n-1)} + \cdots + a_1(t)y' + a_0(t)y = 0 \tag{1}$$

and

$$a_n(t)y^{(n)} + a_{n-1}(t)y^{(n-1)} + \cdots + a_1(t)y' + a_0(t)y = f(t) \tag{2}$$

are called *nth order linear differential equations.* Equation (1) is called *homogeneous* and equation (2) is called *nonhomogeneous* (or *inhomogeneous*).

An n-times differentiable function y on I is said to be a *solution to (2) on I* if

$$a_n(t)y^{(n)}(t) + \cdots + a_1(t)y'(t) + a_0(t)y(t) = f(t)$$

for all $t \in I$. Of course, y is a solution to (1) on I if it is a solution to (2) with $f(t) \equiv 0$ on I. The purpose of this chapter is to indicate that the basic results and ideas for second order linear equations discussed in Chapter 2 are also valid for nth order linear systems. As in the second order case it is necessary to consider initial value problems associated with (1) and (2). So suppose that t_0 is a given member of I and that $x_0, x_1, \ldots, x_{n-1}$ are given constants. We also consider the *nth order linear initial value problems*

$$\begin{aligned} &a_n(t)y^{(n)} + \cdots + a_1(t)y' + a_0(t)y = 0 \\ &y(t_0) = x_0, \; y'(t_0) = x_1, \ldots, y^{(n-1)}(t_0) = x_{n-1} \end{aligned} \tag{3}$$

and

$$\begin{aligned} &a_n(t)y^{(n)} + \cdots + a_1(t)y' + a_0(t)y = f(t) \\ &y(t_0) = x_0, \; y'(t_0) = x_1, \ldots, y^{(n-1)}(t_0) = x_{n-1} \end{aligned} \tag{4}$$

We state, without proof, a result on the existence and uniqueness of solutions to (4) [and hence (3)].

■ **Theorem 7.1-1**

For each given set $x_0, x_1, \ldots, x_{n-1}$ of initial data, there exists exactly one solution to the initial value problem (4).

As a simple illustration, consider the third order equation

$$y''' = 0 \qquad y(0) = 1 \qquad y'(0) = 2 \qquad y''(0) = 3 \tag{5}$$

The only functions whose third derivative is identically zero are polynomials whose degree is 2 or less. Hence any solution y to the differential equation in (5) is of the form

$$y(t) = c_0 + c_1 t + c_2 t^2$$

where c_0, c_1, and c_2 are constants. Since $y(0) = c_0$, $y'(0) = c_1$, and $y''(0) = 2c_2$, the initial conditions in (5) imply that

$$y(t) = 1 + 2t + \tfrac{3}{2}t^2$$

is a solution to (5), and Theorem 7.1-1 asserts that this is in fact the only solution to (5).

Now suppose that $y_1, y_2, \ldots, y_m$, m a positive integer, are functions on the interval I and that $c_1, c_2, \ldots, c_m$ are constants. The function y defined on I by

$$y(t) = c_1 y_1(t) + \cdots + c_m y_m(t)$$

is said to be a *linear combination of* $y_1, \ldots, y_m$. It follows readily from the properties of the derivative that the following lemma is valid:

□ **Lemma 7.1-1**

If $y_1, \ldots, y_m$ are solutions to the homogeneous equation (1), then any linear combination

$$y = c_1 y_1 + \cdots + c_m y_m$$

is also a solution to (1). Moreover, if y_p is a solution to the nonhomogeneous equation (2) and y_H is a solution to the homogeneous equation (1), then

$$y = y_H + y_p$$

is a solution to (2).

The details for the proof of this lemma can be readily checked. As an illustration, consider the simple third order nonhomogeneous equation

$$y''' = e^{2t} - t \tag{6}$$

From equation (5) we know that every solution y_H to the corresponding homogeneous equation is of the form

$$y_H = c_0 + c_1 t + c_2 t^2$$

Antidifferentiating repeatedly each side of equation (6) gives

$$y'' = \tfrac{1}{2}e^{2t} - \tfrac{1}{2}t^2 \qquad y' = \tfrac{1}{4}e^{2t} - \tfrac{1}{6}t^3 \qquad y = \tfrac{1}{8}e^{2t} - \tfrac{1}{24}t^4$$

and hence $y_p = \frac{1}{8}e^{2t} - \frac{1}{24}t^4$ is a particular solution to (6). According to the second part of Lemma 7.1-1,

$$y = c_0 + c_1 t + c_2 t^2 + \tfrac{1}{8}e^{2t} - \tfrac{1}{24}t^4$$

is a solution to (6) for any constants c_0, c_1, and c_2.

7.1*a* Linear Independence

The functions $y_1, \ldots, y_m$ are said to be *linearly dependent on I* if there are constants $c_1, \ldots, c_m$, not all zero, such that

$$c_1 y_1(t) + \cdots + c_m y_m(t) \equiv 0 \qquad \text{on } I \tag{7}$$

If $y_1, \ldots, y_m$ are not linearly dependent on I, they are said to be *linearly independent on I* [that is, (7) holds only in case all of the constants $c_1, \ldots, c_m$ are zero]. Suppose that the m functions $y_1, \ldots, y_m$ are $(m-1)$-times differentiable and that $c_1, \ldots, c_m$ are constants such that (7) holds. Differentiating each side of (7) $m - 1$ times, it follows that $c_1, \ldots, c_m$ must satisfy the linear system

$$
\begin{aligned}
&c_1 y_1(t) + \cdots + c_m y_m(t) = 0 \\
&c_1 y_1'(t) + \cdots + c_m y_m'(t) = 0 \\
&\vdots \\
&c_1 y^{(m-1)}(t) + \cdots + c_m y^{(m-1)}(t) = 0
\end{aligned} \tag{8}
$$

Analogous to the case of two functions, the *wronskian* of the m functions $y_1, \ldots, y_m$ is defined by

$$
W[y_1, \ldots, y_m; t] = \det \begin{pmatrix} y_1(t) & \cdots & y_m(t) \\ y_1'(t) & \cdots & y_m'(t) \\ \cdots & \cdots & \cdots \\ y_1^{(m-1)}(t) & \cdots & y_m^{(m-1)}(t) \end{pmatrix} \tag{9}
$$

Note that if $W[y_1, \ldots, y_m; t] \neq 0$ for some $t \in I$, then equation (8) has only the trivial solution $c_1 = c_2 = \cdots = c_m = 0$, and hence $y_1, \ldots, y_m$ are linearly independent. As indicated in Section 2.2, it may be possible for $y_1, \ldots, y_m$ to be linearly independent on I and also $W[y_1, \ldots, y_m; t] \equiv 0$ on I (see Problem 2.2-6). However, using the uniqueness of solutions to (3) [and (4)] asserted by Theorem 7.1-1, we can obtain the following important result relating the wronskian, linear independence, and general solution to the homogeneous equation (1):

Theorem 7.1-2

Suppose that $y_1, \ldots, y_n$ are n solutions to the homogeneous equation (1) on I. Then any one of the following statements implies the other two:

(i) Every solution y to (1) is a linear combination of $y_1, \ldots, y_n$ on I (that is, $y = c_1 y_1 + \cdots + c_n y_n$).

(ii) $y_1, \ldots, y_n$ are linearly independent on I.

(iii) The wronskian $W[y_1, \ldots, y_n; t_0] \neq 0$ for some $t_0 \in I$, and hence for all $t_0 \in I$.

The ideas behind the proof of Theorem 7.1-2 are exactly the same as those in the proof of Theorem 2.2-2. Note that, given the initial conditions $x_0, \ldots,$

x_{n-1} in the initial value problem (3), the linear combination $y = c_1y_1 + \cdots + c_n y_n$ is a solution to (3) only in case

$$
\begin{aligned}
&c_1y_1(t_0) + \cdots + c_n y_n(t_0) = x \\
&c_1y_1'(t_0) + \cdots + c_n y_n'(t_0) = x_1 \\
&\vdots \\
&c_1y_1^{(n-1)}(t_0) + \cdots + c_n y_n^{(n-1)}(t_0) = x_{n-1}
\end{aligned} \tag{10}
$$

Thus if $W[y_1, \ldots, y_n; t_0] \neq 0$, equation (10) has a unique solution, and once n linearly independent solutions to (1) are found, the *general solution to (1)* [i.e., the class of all solutions to (1)] is the set of all linear combinations of the n solutions, and the solution to the initial value problem (3) can be determined by solving the system (10) for $c_1, \ldots, c_n$.

Once the general solution to (1) and one particular solution to (2) are known, then all solutions to the nonhomogeneous equation (2) are known.

Theorem 7.1-3

Suppose that $y_1, \ldots, y_n$ are n linearly independent solutions to the homogeneous equation (1) and that y_p is a solution to the nonhomogeneous equation (2). Then every solution y of (2) has the form

$$y = c_1y_1 + \cdots + c_n y_n + y_p \tag{11}$$

where $c_1, \ldots, c_n$ are constants.

This theorem follows easily from Lemma 7.1-1 and Theorem 7.2-1 (see also Theorem 2.2-3). As in the homogeneous case, the family of all solutions (11) to equation (2) is called the *general solution of (2)*.

EXAMPLE 7.1-1

Consider the initial value problem

$$y''' + 3y'' = 1 \qquad y(0) = 0 \qquad y'(0) = 1 \qquad y''(0) = 1 \tag{12}$$

If $w = y''$ then $w' = y'''$, and the equation in (12) can be written

$$w' + 3w = 1$$

This equation is a first order linear equation (see Section 1.2), and multiplying by the integrating factor e^{3t} we have that $(e^{3t}w)' = e^{3t}$, and hence $e^{3t}w(t) = e^{3t}/3 + \bar{c}_1$. Therefore, $w(t) = \bar{c}_1e^{-3t} + \frac{1}{3}$, and since $y'' = w$ it follows that

$$y' = \frac{-\bar{c}_1e^{-3t}}{3} + \frac{t}{3} + c_2$$

and that

$$y(t) = \frac{\bar{c}_1 e^{-3t}}{9} + \frac{t^2}{6} + c_2 t + c_3$$

Therefore, replacing the constant $\bar{c}_1/9$ by c_1, it follows that

$$y(t) = c_1 e^{-3t} + c_2 t + c_3 + \frac{t^2}{6}$$

is a solution to the differential equation in (12) for any constants c_1, c_2, and c_3. In order to satisfy the initial conditions, the equations

$$\begin{aligned} y(0) &= c_1 + c_3 = 0 \\ y'(0) &= -3c_1 + c_2 = 1 \\ y''(0) &= 9c_1 + \tfrac{1}{3} = 1 \end{aligned}$$

must be satisfied. Therefore, $c_1 = \frac{2}{27}$, $c_2 = 1 + 3(\frac{2}{27}) = \frac{33}{27}$, $c_3 = -\frac{2}{27}$, and

$$y = \frac{2e^{-3t}}{27} - \frac{2}{27} + \frac{33t}{27} + \frac{t^2}{6}$$

is the solution to (12).

EXAMPLE 7.1-2

Consider the initial value problem

$$y^{(4)} - y = 0 \qquad y(0) = 4 \qquad y'(0) = -4 \qquad y''(0) = 2 \qquad y'''(0) = 2$$

The student should verify that e^t, e^{-t}, $\sin t$, and $\cos t$ are linearly independent solutions to the corresponding differential equation, and hence

$$y(t) = c_1 e^t + c_2 e^{-t} + c_3 \sin t + c_4 \cos t$$

is the general solution. Solving for the initial conditions

$$\begin{aligned} y(0) &= c_1 + c_2 + c_4 = 4 \\ y'(0) &= c_1 - c_2 + c_3 = -4 \\ y''(0) &= c_1 + c_2 - c_4 = 2 \\ y'''(0) &= c_1 - c_2 - c_3 = 2 \end{aligned}$$

it follows that $c_1 = 1$, $c_2 = 2$, $c_3 = -3$, and $c_4 = 1$. Hence

$$y = e^t + 2e^{-t} - 3 \sin t + \cos t$$

is the solution to the given initial value problem.

PROBLEMS

7.1-1. For each of the following problems, show that the given class of functions is the general solution to the corresponding differential equation and determine values for the constants so that the initial conditions are satisfied.

(a) $y''' - y'' - 2y' = 0$, $y(0) = 2$, $y'(0) = -3$, $y''(0) = -7$
$\{y = c_1 + c_2 e^{-t} + c_3 e^{2t}\}$

(b) $y''' + 5y'' + 9y' + 5y = 0$, $y(0) = 2$, $y'(0) = -2$, $y''(0) = -1$
$\{y = c_1 e^{-t} + c_2 e^{-2t} \cos t + c_3 e^{-2t} \sin t\}$

(c) $y^{(4)} + 8y'' + 16y = 0$, $y(0) = 2$, $y'(0) = 0$, $y''(0) = 0$, $y'''(0) = 0$
$\{y = c_1 \cos 2t + c_2 \sin 2t + c_3 t \cos 2t + c_4 t \sin 2t\}$

7.1-2. Determine a solution to each of the following initial value problems:

(a) $y''' = \sin 2t \qquad y(0) = 1 \qquad y'(0) = 0 \qquad y''(0) = 3$

(b) $y^{(4)} - y'' = 0 \qquad y(0) = 1 \qquad y'(0) = 0 \qquad y''(0) = 0 \qquad y'''(0) = -1$
(*Hint:* Try the substitution $w = y''$ to reduce the equation to second order in w.)

(c) $y''' + 2y'' + y' = t \qquad y(0) = 1 \qquad y'(0) = 1 \qquad y''(0) = 0$

7.1-3. Suppose that $\lambda_1 < \lambda_2 < \lambda_3$. Show that $e^{\lambda_1 t}$, $e^{\lambda_2 t}$, and $e^{\lambda_3 t}$ are linearly independent on any interval.

7.1-4. Suppose that λ_0 is a real number. Show that $e^{\lambda_0 t}$, $te^{\lambda_0 t}$, and $t^2 e^{\lambda_0 t}$ are linearly independent on any interval.

7.1-5. Suppose that $\beta_3 > \beta_2 > \beta_1 > 0$. Show that $\sin \beta_1 t$, $\sin \beta_2 t$, and $\sin \beta_3 t$ are linearly independent on any interval.

***7.1-6.** Suppose that n is a positive integer and that $p_0, p_1, \ldots, p_{n+1}$ are polynomials each having degree no larger than n.

(a) Show that each p_i is a solution to $y^{(n+2)} = 0$.

(b) Show that $W[p_0, p_1, \ldots, p_{n+1}; t] \equiv 0$ for all t.

(c) Deduce from parts *a* and *b* and Theorem 7.1-2 that any $n + 2$ polynomials of degree no larger than n must be linearly dependent on any interval.

7.1-7. Suppose that y_1 and y_2 are solutions to the nonhomogeneous equation (2), with f replaced by f_1 and f_2, respectively. Show that $y_1 + y_2$ is a solution to (2), with f replaced by $f_1 + f_2$.

7.2 HOMOGENEOUS EQUATIONS WITH CONSTANT COEFFICIENTS

As is the case with second order equations, solutions to nth order homogeneous linear equations can be reduced to determining the roots of a polynomial equation of degree n. Assume that $a_0, a_1, \ldots, a_n$ are real constants,

$a_n \neq 0$, and consider the equation

(1) $$a_n y^{(n)} + a_{n-1} y^{(n-1)} + \cdots + a_1 y' + a_0 y = 0$$

Assuming that $y(t) = e^{\lambda t}$ and noting that

$$y'(t) = \lambda e^{\lambda t}, \; y''(t) = \lambda^2 e^{\lambda t}, \ldots, y^{(n)}(t) = \lambda^n e^{\lambda t}$$

leads to the formula

$$a_n y^{(n)}(t) + \cdots + a_1 y'(t) + a_0 y(t) = (a_n \lambda^n + \cdots + a_1 \lambda + a_0)e^{\lambda t} = 0$$

The polynomial equation

(2) $$a_n \lambda^n + a_{n-1} \lambda^{n-1} + \cdots + a_1 \lambda + a_0 = 0$$

is called the *auxiliary equation* (*or characteristic equation*) of (1), and the roots of (2) are called the *auxiliary* (*or characteristic*) *roots.* Knowing the auxiliary roots and their multiplicities allows one to write explicitly the general solution to (1).

Before giving a precise formula for relating the auxiliary roots to the general solution of (1), we look at a couple of illustrations. Consider the third order equation

(3) $$y''' - 2y'' - 3y' = 0$$

According to (2) the auxiliary equation for (3) is

$$\lambda^3 - 2\lambda^2 - 3\lambda = 0$$

Since

$$\lambda^3 - 2\lambda^2 - 3\lambda = (\lambda^2 - 2\lambda - 3)\lambda = (\lambda - 3)(\lambda + 1)\lambda$$

it follows that $\lambda = 3, -1, 0$ are auxiliary roots for equation (3), and hence

$$y_1(t) = e^{3t} \qquad y_2(t) = e^{-t} \qquad y_3(t) = 1$$

are each solutions to (3). Since these three functions are linearly independent (note, for example, that the wronskian is nonzero), we have from Theorem 7.1-2 that

$$y = c_1 e^{3t} + c_2 e^{-t} + c_3$$

is the general solution to (3).

As a second illustration consider the fourth order equation

(4) $$y^{(4)} + 3y'' - 4y = 0$$

The auxiliary equation is

(5) $$\lambda^4 + 3\lambda^2 - 4 = 0$$

Even though the auxiliary equation is fourth order, it is in "quadratic form" (see Problem 7.2-2) and the roots can easily be obtained. By setting $\gamma = \lambda^2$ equation (5) becomes $\gamma^2 + 3\gamma - 4 = 0$. Since the left side of this equation factors as $(\gamma + 4)(\gamma - 1)$, the auxiliary equation (5) can be written

$$(\lambda^2 + 4)(\lambda^2 - 1) = 0$$

Setting the factor $\lambda^2 + 4$ equal to 0, we find that $\lambda = \pm 2i$ are auxiliary roots, and setting the factor $\lambda^2 - 1$ equal to 0, we find that $\lambda = \pm 1$ are auxiliary roots. Following the procedure for obtaining solutions to the differential equation from the auxiliary roots in the second order case (see Section 2.3), one can show that the four auxiliary roots ± 1, $\pm 2i$ indicate that

$$y_1 = e^t \qquad y_2 = e^{-t} \qquad y_3 = \cos 2t \qquad y_4 = \sin 2t$$

are each solutions to equation (4) (the student should verify this). Since these functions are linearly independent as well,

$$y = c_1 e^t + c_2 e^{-t} + c_3 \cos 2t + c_4 \sin 2t$$

is the general solution of (4) by Theorem 7.1-2.

As the above illustrations indicate, the pattern for determining the solutions to (1) from the auxiliary roots is a natural extension of the second order case, and we state (without proof) the fundamental relationships. Recall that a root λ_0 of (2) is of multiplicity p (p a positive integer) if $(\lambda - \lambda_0)^p$ is a factor of the auxiliary polynomial and $(\lambda - \lambda_0)^{p+1}$ is not a factor. Furthermore, if $\lambda_1, \ldots, \lambda_m$ are the distinct roots of (2), each having multiplicity $p_1, \ldots, p_m$, respectively, then $p_1 + \cdots + p_m = n$ (i.e., the sum of the multiplicities of the distinct auxiliary roots equals the order of the differential equation).

■ **Theorem 7.2-1**

Suppose that $\lambda = \lambda_0$ is a root of the auxiliary equation (2) and that λ_0 has multiplicity $m \geq 1$.

(i) If λ_0 is real then

$$y_1(t) = e^{\lambda_0 t},\ y_2(t) = te^{\lambda_0 t}, \ldots, y_m(t) = t^{m-1}e^{\lambda_0 t}$$

are m linearly independent solutions.

(ii) If $\lambda_0 = \alpha + i\beta$, where α and β are real, $\beta \neq 0$, and $i^2 = -1$, then $\alpha - i\beta$ is also a root and the functions

$$y_1(t) = e^{\alpha t}\cos \beta t,\ y_2(t) = te^{\alpha t}\cos \beta t, \ldots, y_m(t) = t^{m-1}e^{\alpha t}\cos \beta t$$

and

$$\bar{y}_1(t) = e^{\alpha t}\sin \beta t,\ \bar{y}_2(t) = te^{\alpha t}\sin \beta t, \ldots, \bar{y}_m(t) = t^{m-1}e^{\alpha t}\sin \beta t$$

are $2m$ linearly independent solutions to (1).

From Theorem 7.2-1 we can obtain n linearly independent solutions to (1) [and hence the general solution to (1)]. Note that if (2) has *n distinct real* roots $\lambda_1, \lambda_2, \ldots, \lambda_n$, then

$$y(t) = c_1 e^{\lambda_1 t} + c_2 e^{\lambda_2 t} + \cdots + c_n e^{\lambda_n t}$$

is the general solution to (1). The following examples illustrate further the relationship between the auxiliary roots and the general solution to (1).

EXAMPLE 7.2-1

Suppose that the auxiliary roots to a certain eighth order linear equation with constant coefficients are known to be 2, 2, 2, 3, $-1 \pm 2i$, $-1 \pm 2i$. The general solution to this equation is

$$y(t) = c_1 e^{2t} + c_2 te^{2t} + c_3 t^2 e^{2t} + c_4 e^{3t} + c_5 e^{-t} \sin 2t + c_6 e^{-t} \cos 2t + c_7 te^{-t} \sin 2t + c_8 te^{-t} \cos 2t$$

where $c_1, \ldots, c_8$ are arbitrary constants.

EXAMPLE 7.2-2

Consider the initial value problem

$$y''' - 3y' - 2y = 0 \qquad y(0) = 4 \qquad y'(0) = -2 \qquad y''(0) = 9$$

The auxiliary equation is $\lambda^3 - 3\lambda - 2 = 0$. Since this polynomial has integer coefficients, any rational root must be a factor of the constant term 2 divided by a factor of the leading coefficient 1 (see Problem 7.2-3). Thus, ± 1 and ± 2 are the only possible rational roots. It is certainly easy to check by direct substitution that $\lambda = -1$ is one root. Since $\lambda + 1$ must be a factor of the auxiliary polynomial, we may divide it by $\lambda + 1$ to obtain

$$\lambda^3 - 3\lambda - 3 = (\lambda + 1)(\lambda^2 - \lambda - 2)$$

Since $\lambda^2 - \lambda - 2 = (\lambda - 2)(\lambda + 1)$, we have further that

$$\lambda^3 - 3\lambda - 2 = (\lambda + 1)^2(\lambda - 2)$$

Hence $-1, -1, 2$ are the auxiliary roots (that is, $\lambda = -1$ has multiplicity 2 and $\lambda = 2$ has multiplicity 1), and it follows that

$$y = c_1 e^{2t} + c_2 e^{-t} + c_3 te^{-t}$$

is the general solution. Since

$$y'(t) = 2c_1 e^{2t} - c_2 e^{-t} + c_3 e^{-t} - c_3 te^{-t}$$

and

$$y''(t) = 4c_1 e^{2t} + c_2 e^{-t} - 2c_3 e^{-t} + c_3 te^{-t}$$

the system

$$\begin{aligned} y(0) &= c_1 + c_2 &&= 4 \\ y'(0) &= 2c_1 - c_2 + c_3 &&= -2 \\ y''(0) &= 4c_1 + c_2 - 2c_3 &&= 9 \end{aligned}$$

must be satisfied in order for y to satisfy the initial conditions. Solving for c_1, c_2, and c_3 we find that $c_1 = 1$, $c_2 = 3$, and $c_3 = -1$, and so

$$y(t) = e^{2t} + 3e^{-t} - te^{-t}$$

is the solution to the given initial value problem.

PROBLEMS

7.2-1. Each of the following equations is an auxiliary equation for a certain linear differential equation with constant coefficients. Determine the auxiliary roots (with their multiplicities) and write down the general solution of the corresponding differential equation.

(a) $(\lambda^2 + \lambda + 1)^3 = 0$ **(c)** $(\lambda^3 - 1)^2(\lambda^2 - 1) = 0$

(b) $(\lambda^2 + 9)^2(\lambda - 2)^3 = 0$ **(d)** $(\lambda - 1)(\lambda - 2)^2(\lambda - 3)^3 = 0$

7.2-2. A polynomial equation of the form

$$(*) \quad a\lambda^{2m} + b\lambda^m + c = 0$$

where m is a positive integer and $a \neq 0$, is said to have a quadratic form. Show that the roots λ of $(*)$ satisfy

$$(**) \quad \lambda^m = \frac{-b \pm \sqrt{b^2 - 4ac}}{2a}$$

and hence the roots of $(*)$ can be obtained by computing the mth roots of the right-hand side of $(**)$. Use this procedure to obtain the general solutions to the following differential equations:

(a) $y^{(4)} - 6y'' + 8y = 0$

(b) $y^{(6)} - 2y''' + y = 0$

(c) $y^{(4)} - 4y = 0$

(d) $y^{(6)} + 7y''' - 8y = 0$

(e) $y^{(4)} + 16y = 0$ {*Hint:* If $\beta > 0$ then $[(1 \pm i)\sqrt{\beta/2}]^2 = \pm i\beta$.}

7.2-3. If the auxiliary polynomial (2) has integer coefficients $a_n, a_{n-1}, \ldots, a_1, a_0$, then any rational root λ_0 of (2) must be a factor of a_0 divided by a factor of a_n. For example, the possible rational roots of

$$2\lambda^3 + \lambda^2 - 13\lambda + 6 = 0$$

are ± 1, ± 2, ± 3, ± 6, $\pm\frac{1}{2}$, and $\pm\frac{3}{2}$ (the factors of 6 are 1, 2, 3, 6 and those of 2 are 1, 2). The student should verify that the roots are in fact $\lambda = 2, -3, \frac{1}{2}$. The possible rational roots of

$$\lambda^3 - 2\lambda - 4 = 0$$

are $\lambda = \pm 1$, ± 2, and ± 4. In this instance the only rational root is $\lambda = 2$. Since this implies that $\lambda - 2$ is a factor, we may divide to obtain

$$\lambda^3 - 2\lambda - 4 = (\lambda - 2)(\lambda^2 + 2\lambda + 2)$$

and the remaining two roots can be obtained from the quadratic formula. Use this procedure for obtaining the general solution to the following differential equations.

(a) $y''' + 2y'' + 3y' + 2y = 0$

(b) $y^{(4)} - 3y''' + 3y'' - y' = 0$

(c) $2y''' + y'' + y' - y = 0$

(d) $2y^{(4)} + 7y''' - y'' - 14y' - 6y = 0$

7.2-4. Determine the general solution to each of the following homogeneous linear differential equations.

(a) $y^{(4)} - 3y'' + 2y = 0$

(b) $y''' + 8y = 0$

(c) $y''' - 2y' - 4y = 0$

(d) $3y''' - y'' - y' - y = 0$

(e) $y^{(6)} - 16y''' + 64y = 0$

(f) $y^{(4)} - y'' + 6y = 0$

(g) $y''' - 6y'' + 11y' - 6y = 0$

(h) $y^{(4)} + 4y'' + 4y = 0$

7.2-5. Determine a solution to each of the following linear initial value problems.

(a) $y''' = 0 \quad y(0) = 1 \quad y'(0) = -2 \quad y''(0) = 2$

(b) $y''' - y'' + 4y' - 4y = 0 \quad y(0) = 1 \quad y'(0) = 0 \quad y''(0) = -1$

(c) $y''' - y'' - y' + y = 0 \quad y(0) = 0 \quad y'(0) = 0 \quad y''(0) = 1$

(d) $y^{(4)} - y = 0 \quad y(0) = 0 \quad y'(0) = 1 \quad y''(0) = -2 \quad y'''(0) = 0$

7.3 NONHOMOGENEOUS EQUATIONS

In this section it is shown that the methods of undetermined coefficients and variation of parameters for second order equations have natural extensions to nth order systems. These techniques are only indicated here since the procedures are completely analogous to the second order case. The student should review the procedures for solving second order nonhomogeneous equations given in Sections 2.4 and 2.5.

7.3a Undetermined Coefficients

Suppose first that $a_0, a_1, \ldots, a_n$ are constants with $a_n \neq 0$, that I is an interval, and that f is a function on I that has a finite family of derivatives. Consider the homogeneous equation

$$a_n y^{(n)} + \cdots + a_1 y' + a_0 y = 0 \tag{1}$$

and the corresponding nonhomogeneous equation

$$a_n y^{(n)} + \cdots + a_1 y' + a_0 y = f(t) \tag{2}$$

Whenever the coefficients $a_0, \ldots, a_n$ are constants and the nonhomogeneous term f has a finite family of derivatives, then the method of undetermined coefficients applies in essentially the same manner as in the second order case.

As a specific illustration, consider the equation

$$y''' + 3y'' - y' - 3y = e^{2t} \tag{3}$$

Since the auxiliary polynomial to the corresponding homogeneous equation has the form

$$\lambda^3 + 3\lambda^2 - \lambda - 3 = (\lambda - 1)(\lambda + 1)(\lambda + 3)$$

it follows that $\lambda = 1, -1, -3$ are the auxiliary roots and

$$y_H = c_1 e^t + c_2 e^{-t} + c_3 e^{-3t}$$

is the general solution to the corresponding homogeneous problem. Since the differential family of e^{2t} is $\{e^{2t}\}$ and e^{2t} is not a solution to the homogeneous problem, it must be the case that $y_p = d_1 e^{2t}$ is a particular solution to the nonhomogeneous equation (3) for some constant d_1.

Substituting $y_p' = 2d_1 e^{2t}$, $y_p'' = 4d_1 e^{2t}$, and $y_p''' = 8d_1 e^{2t}$ into (3), we have

$$8d_1 e^{2t} + 12d_1 e^{2t} - 2d_1 e^{2t} - 3d_1 e^{2t} = 15d_1 e^{2t} = e^{2t}$$

Therefore $d_1 = \frac{1}{15}$ and it follows that $y_p = e^{2t}/15$ is a particular solution. Thus

$$y = c_1 e^t + c_2 e^{-t} + c_3 e^{-3t} + \frac{e^{2t}}{15}$$

is the general solution to (3).

Similar procedures apply for other types of nonhomogeneous terms with finite differential families. Consider, for example, the equation

$$y''' + 3y'' - y' - 3y = 4 \cos t - 8 \sin t + 1 - 6t \tag{3'}$$

In this case one should assume that a particular solution y_p has the form

$$y_p = d_1 \cos t + d_2 \sin t + d_3 t + d_4$$

where d_1, d_2, d_3, and d_4 are constants. Taking y_p', y_p'', and y_p''' and substituting into (3′), the student should verify that

$$y_p = -\cos t + \sin t + 2t - 1$$

is a particular solution to (3′). If some member of the differential family is a solution to the corresponding homogeneous equation, then the differential family must be altered by multiplying each member by an appropriate power of t. For example, the form of a particular solution y_p to the equation

$$y''' + 3y'' - y' - 3y = e^t \tag{3''}$$

should be $y_p = d_1 te^t$. That is, since e^t is a solution to the corresponding homogeneous equation (and te^t is not), then one should assume that $y_p = d_1 te^t$ (as opposed to $y_p = d_1 e^t$). The student should establish that $y_p = te^t/8$ is a particular solution to (3″).

If it is assumed that the nonhomogeneous term f has a particular form, then one can give a precise formula of the procedure to determine a particular solution to (2). Therefore, suppose that the function f can be written in the form

$$f(t) = P(t)e^{\alpha t} \cos \beta t + Q(t)e^{\alpha t} \sin \beta t \tag{4}$$

where α and β are real constants and P and Q are polynomials with degree no larger than k and at least one having degree exactly k. Recall that a differential family of the function f in (4) is

$$\{t^j e^{\alpha t} \cos \beta t, t^j e^{\alpha t} \sin \beta t : j = 0, 1, \ldots, k\} \tag{5}$$

Theorem 7.3-1

Suppose that f has the form (4), where α and β are real constants and P and Q are polynomials with the maximum degrees of P and Q being k. Let the integer v be the least nonnegative integer such that

$$y(t) \equiv t^v e^{\alpha t} \cos \beta t \qquad [\text{or } y(t) = t^v e^{\alpha t} \sin \beta t]$$

is not a solution to the homogeneous problem (2). Then the nonhomogeneous equation (1) has a solution y_p of the form

$$y_p(t) = (d_1 t^v + \cdots + d_k t^{v+k-1})e^{\alpha t} \cos \beta t + (\hat{d}_1 t^v + \cdots + \hat{d}_k t^{v+k-1})e^{\alpha t} \sin \beta t \tag{6}$$

where $d_1, \ldots, d_k$ and $\hat{d}_1, \ldots, \hat{d}_k$ are constants. Also, at least one of d_k, $\hat{d}_k$ is nonzero.

This theorem is analogous to Theorem 2.4-1, which applies to the second order case.

▶ **EXAMPLE 7.3-1**

Consider the equation $y''' - y = 6e^t + 2t - 3$. The auxiliary equation associated with the corresponding homogeneous equation is

$$\lambda^3 - 1 = (\lambda - 1)(\lambda^2 + \lambda + 1) = 0$$

and hence $\lambda = 1, -\frac{1}{2} \pm i\sqrt{3}/2$ are the auxiliary roots. Thus,

$$y_H(t) = c_1 e^t + c_2 e^{-t/2} \cos \frac{\sqrt{3}\,t}{2} + c_3 e^{-t/2} \sin \frac{\sqrt{3}\,t}{2}$$

is the general solution to the corresponding homogeneous equation. Separating the nonhomogeneous term into the two parts $6e^t$ and $2t - 3$ leads to the differential families $\{e^t\}$ and $\{1, t\}$. Since e^t is a solution to the corresponding homogeneous equation and te^t is not and since no linear combination of 1 and t is a solution, we can apply Theorem 7.3-1 (and the principle of superposition—see Problem 7.1-7) to assert that the given equation has a particular solution y_p of the form

$$y_p(t) = d_1 te^t + d_2 + d_3 t$$

where d_1, d_2, and d_3 are constants. Determining y_p''' and substituting into the equation leads to system

$$3d_1 e^t - d_2 - d_3 t = 6e^t + 2t - 3$$

and it follows that $d_1 = 2$, $d_2 = -2$, and $d_3 = 3$. Thus

$$y_p(t) = 2te^t - 2t + 3$$

is a particular solution.

*7.3*b* Variation of Parameters

The method of variation of parameters developed in Section 2.5 also can be extended to nth order equations. Assume that I is an interval and that a_0, a_1, ..., a_n and f are continuous functions on I such that $a_n(t) \neq 0$ for all $t \in I$. Consider the nonhomogeneous system

$$a_n(t)y^{(n)} + \cdots + a_1(t)y' + a_0(t)y = f(t) \tag{7}$$

along with the corresponding homogeneous system

$$a_n(t)y^{(n)} + \cdots + a_1(t)y' + a_0(t)y = 0 \tag{8}$$

Assume that $y_1, \ldots, y_n$ are n linear independent solutions to (8); that is, the general solution of (8) is

$$y = c_1 y_1 + \cdots + c_n y_n$$

where $c_1, \ldots, c_n$ are constants. The variation of constants formula has the following form for general nth order systems.

Theorem 7.3-2

Suppose that $y_1, y_2, \ldots, y_n$ are n linearly independent solutions to the homogeneous equation (8). Then the nonhomogeneous equation (7) has a particular solution y_p on I of the form

$$y_p(t) = A_1(t)y_1(t) + A_2(t)y_2(t) + \cdots + A_n(t)y_n(t) \tag{9}$$

where $A_1, A_2, \ldots, A_n$ are any continuously differentiable functions on I whose derivatives satisfy the algebraic system

$$\begin{aligned} &A_1'(t)y_1(t) + A_2'(t)y_2(t) + \cdots + A_n'(t)y_n(t) = 0 \\ &A_1'(t)y_1'(t) + A_2'(t)y_2'(t) + \cdots + A_n'(t)y_n(t) = 0 \\ &\vdots \\ &A_1'(t)y_1^{(n-2)}(t) + A_2'(t)y_2^{(n-2)}(t) + \cdots + A_n'(t)y_n^{(n-2)}(t) = 0 \\ &A_1'(t)y_1^{(n-1)}(t) + A_2'(t)y_2^{(n-1)}(t) + \cdots + A_n'(t)y_n^{(n-1)}(t) = \frac{f(t)}{a_n(t)} \end{aligned} \tag{10}$$

for all $t \in I$.

In order to see how the system (10) arises, assume that y_p has the form (9) and compute $y_p', y_p'', \ldots, y_p^{(n)}$, using the product rule. However, after taking each derivative, set equal to zero the sum of the terms that have some $A_i'(t)$ as a factor before taking the next derivative. This gives the first $n - 1$ equations in (10), and the final equation in (10) is obtained by substituting $y_p, y_p', \ldots, y_p^{(n)}$ into equation (7) [and using the fact that the y_i are solutions to the homogeneous equation (8)]. Since the determinant of the coefficients in (10) is the wronskian of the linear independent solutions $y_1, \ldots, y_n$ at time t, we have from Theorem 7.1-2 that this determinant is nonzero everywhere on I, and hence (10) has a unique solution for each t in I. The fact that the determinant of the coefficients in (10) is the wronskian also helps in remembering the form of the system (10) without having to rederive it.

EXAMPLE 7.3-3

Consider the system

$$y''' - 3y'' + 2y' = \frac{e^{2t}}{1 + e^t}$$

The auxiliary equation for the corresponding homogeneous system is

$$\lambda^3 - 3\lambda^2 + 2\lambda = \lambda(\lambda - 2)(\lambda - 1) = 0$$

and hence

$$y_p = A_1(t) + A_2(t)e^t + A_3(t)e^{2t}$$

is the form of a solution. Equation (10) takes the form

$$\begin{aligned} A_1' + A_2' e^t + A_3' e^{2t} &= 0 \\ A_2' e^t + A_3' 2e^{2t} &= 0 \\ A_2' e^t + A_3' 4e^{2t} &= \frac{e^{2t}}{1 + e^t} \end{aligned} \tag{11}$$

where the time variable t is suppressed in the A_i'. We will employ Cramer's method for solving this system. Since

$$\det \begin{pmatrix} 1 & e^t & e^{2t} \\ 0 & e^t & 2e^{2t} \\ 0 & e^t & 4e^{2t} \end{pmatrix} = 1(4e^{3t} - 2e^{3t}) = 2e^{3t}$$

we obtain

$$\begin{aligned} A_1'(t) &= \frac{1}{2} e^{-3t} \det \begin{pmatrix} 0 & e^t & e^{2t} \\ 0 & e^t & 2e^{2t} \\ \dfrac{e^{2t}}{1 + e^t} & e^t & 4e^{2t} \end{pmatrix} \\ &= \frac{1}{2} e^{-3t} \frac{e^{2t}}{1 + e^t} (2e^{3t} - e^{3t}) \\ &= \frac{1}{2} \frac{e^{2t}}{1 + e^t} \end{aligned}$$

and

$$\begin{aligned} A_2'(t) &= \frac{1}{2} e^{-3t} \det \begin{pmatrix} 1 & 0 & e^{2t} \\ 0 & 0 & 2e^{2t} \\ 0 & \dfrac{e^{2t}}{1 + e^t} & 4e^{2t} \end{pmatrix} \\ &= \frac{1}{2} e^{-3t} \left(-\frac{2e^{4t}}{1 + e^t} \right) = \frac{-e^t}{1 + e^t} \end{aligned}$$

and

$$A_3'(t) = \frac{1}{2} e^{-3t} \det \begin{pmatrix} 1 & e^t & 0 \\ 0 & e^t & 0 \\ 0 & e^t & \dfrac{e^{2t}}{1+e^t} \end{pmatrix}$$

$$= \frac{1}{2} e^{-3t} \frac{e^{3t}}{1+e^t} = \frac{1}{2} \frac{1}{1+e^t}$$

Therefore, using the substitution $u = 1 + e^t$,

$$A_1 = \frac{1}{2} \int \frac{e^t}{1+e^t} e^t\, dt = \frac{1}{2} \int \frac{u-1}{u}\, du$$

$$= \tfrac{1}{2}(1+e^t) - \tfrac{1}{2} \ln (1+e^t)$$

and

$$A_2 = \int \frac{-1}{1+e^t} e^t\, dt = -\ln (1+e^t)$$

Also,

$$A_3 = \frac{1}{2} \int \frac{1}{1+e^t}\, dt = \frac{1}{2} \int \frac{1}{e^{-t}+1} e^{-t}\, dt = \frac{-1}{2} \ln (e^{-t}+1)$$

and it follows that

$$y_p(t) = \tfrac{1}{2}(1+e^t) - \tfrac{1}{2} \ln (1+e^t) - e^t \ln (1+e^t) - \tfrac{1}{2} e^{2t} \ln (e^{-t}+1)$$

is a particular solution to the given equation. Including the first term into the homogeneous solution, it follows that

$$y(t) = c_1 + c_2 e^t + c_3 e^{2t} - (\tfrac{1}{2} + e^t) \ln (1+e^t) - \tfrac{1}{2} e^{2t} \ln (e^{-t}+1)$$

is the general solution to the given nonhomogeneous system.

PROBLEMS

7.3-1. Determine the general solution to each of the following equations.

(a) $y''' - y = e^{2t}$

(b) $y''' - y'' + 2y = 3t^2$

(c) $y''' + 6y'' + 11y' + 6y = e^t + t$

(d) $y^{(4)} - y = e^{-t} + t$

(e) $y''' - 3y'' + 3y' - y = e^t$

7.3-2. Determine a particular solution by variation of parameters.

(a) $y''' - 3y'' + 2y' = \dfrac{e^{3t}}{1 + e^t}$

(b) $y''' + y' = \sec t \qquad -\dfrac{\pi}{2} < t < \dfrac{\pi}{2}$

(c) $y''' - 6y'' + 11y' - 6y = \dfrac{e^t}{1 + e^{-t}}$

(d) $y''' - 3y'' + 3y' - y = \dfrac{e^t}{t} \qquad t > 0$

7.3-3. In parts *a* to *d* use the method of undetermined coefficients to determine a form of the particular solution. You need not evaluate the coefficients.

(a) $y^{(4)} + 2y'' + y = 4t^2 - 3t \sin t$

(b) $y^{(4)} - y'' = t + e^t$

(c) $y''' + 3y'' + 4y' + 2y = 2e^{-t} \cos t + \sin t$

(d) $y^{(4)} - y = 2 \sin t \cos 2t + te^t$

***7.3-4.** Suppose that a_0, a_1, a_2, a_3, A, and α are constants, with $a_3 \neq 0$, and consider the nonhomogeneous equation

$$(*) \qquad a_3 y''' + a_2 y'' + a_1 y' + a_0 y = Ae^{\alpha t}$$

Let $r(\lambda) = a_3 \lambda^3 + a_2 \lambda^2 + a_1 \lambda + a_0$ be the auxiliary polynomial by (*). Show that a particular solution y_p to (*) has the following form:

(a) If $r(\alpha) \neq 0$ then $y_p(t) \equiv Ae^{\alpha t}/r(\alpha)$.

(b) If $r(\alpha) = 0$ and $r'(\alpha) \neq 0$ then $y_p(t) \equiv Ate^{\alpha t}/r'(\alpha)$. Under what circumstances does (*) have a solution of the form $y_p(t) \equiv d_1 t^2 e^{\alpha t}$, where d_1 is a constant? What about $y_p(t) \equiv d_1 t^3 e^{\alpha t}$? (See also Problem 2.4-6.)

7.4 WEAKLY COUPLED EQUATIONS

In this section we consider a system of two simultaneous linear differential equations with constant coefficients. These systems are assumed to be coupled only through the functions and not through any derivatives of the functions (this is the same assumption made for the system of two first order equations in Section 5.2). This type of system can be reduced immediately to a linear equation of higher order but of one unknown, and so previous methods can be applied. The two unknown functions (dependent variables) are denoted by x and y and the independent variable by t.

Suppose that n and m are positive integers and that a_1, a_2, b_1, b_2 are constants. Also, let $\alpha_1, \alpha_2, \ldots, \alpha_n$ and $\beta_1, \beta_2, \ldots, \beta_m$ be constants, with $\alpha_n \neq 0$ and $\beta_m \neq 0$. Consider the homogeneous system

$$\begin{aligned} \alpha_n x^{(n)} + \cdots + \alpha_1 x' &= a_1 x + b_1 y \\ \beta_m y^{(m)} + \cdots + \beta_1 y' &= a_2 x + b_2 y \end{aligned} \tag{1}$$

as well as the nonhomogeneous system

$$\begin{aligned} \alpha_n x^{(n)} + \cdots + \alpha_1 x' &= a_1 x + b_1 y + f(t) \\ \beta_m y^{(m)} + \cdots + \beta_1 y' &= a_2 x + b_2 y + g(t) \end{aligned} \tag{2}$$

where f and g are continuous functions on an interval I. It is also supposed that f has m continuous derivatives and g has n continuous derivatives. A pair of functions x, y is said to be a solution to (2) [or (1)] on I if x has n derivatives, y has m derivatives, and equation (2) [or (1)] holds for all t in I.

We show that solutions to (2) [and hence (1), since (1) is a special case of (2)] are also solutions to an $(n + m)$th order nonhomogeneous equation of one unknown, with constant coefficients. Note that if both b_1 and a_2 are zero, then (2) is actually uncoupled, so the first equation can be solved directly for x and the second for y. Thus we assume at least one of b_1 and a_2 is nonzero, and for definiteness it is always assumed $b_1 \neq 0$. A procedure for determining x and y is as follows:

(a) Use the first equation in (2) to solve for y in terms of $x, x', \ldots, x^{(n)}$ and f:

$$y = \frac{a_n x^{(n)} + \cdots + \alpha_1 x' - a_1 x - f}{b_1}$$

(b) Compute $y', y'', \ldots, y^{(m)}$ from the expression for y in part a.

(c) Substitute these expressions for y and its derivatives into the second equation in (2) in order to obtain an $(n + m)$th order nonhomogeneous equation with constant coefficients in the variable x.

(d) Obtain the general solution of the x equation determined in part c.

(e) Use part a and the expression for x in part d to obtain y.

The techniques indicated by parts a to e are straightforward and lead to the determination of all the solutions to (2)—that is, the general solution of (2). Note that there are $m + n$ arbitrary constants involved in the general solution. Consider, for example, the system

$$\begin{aligned} x'' &= y \\ y'' &= 3y - 2x \end{aligned}$$

Since $y = x''$ and hence $y'' = x^{(4)}$, by substituting into the second equation we have

$$x^{(4)} = 3x'' - 2x$$

or

$$x^{(4)} - 3x'' + 2x = 0$$

The auxiliary equation for the linear differential equation is

$$\lambda^4 - 3\lambda^2 + 2 = (\lambda^2 - 2)(\lambda^2 - 1) = 0$$

and it follows that $\lambda = \pm 1, \pm\sqrt{2}$ are the auxiliary roots. Therefore, since $y = x''$, the general solution to this system is

$$x = c_1 e^t + c_2 e^{-t} + c_3 e^{\sqrt{2}t} + c_4 e^{-\sqrt{2}t}$$
$$y = c_1 e^t + c_2 e^{-t} + 2c_3 e^{\sqrt{2}t} + 2c_4 e^{-\sqrt{2}t}$$

The fact that these are indeed solutions is easy to verify.

There is also an initial value problem associated with equation (2). Suppose that $z_0, \ldots, z_{n-1}$ and $w_0, \ldots, w_{m-1}$ are given numbers and $t_0 \in I$. The initial value problem associated with (2) has the form

$$\begin{aligned} &\alpha_n x^{(n)} + \cdots + \alpha_1 x' = a_1 x + b_1 y + f(t) \\ &\beta_m y^{(m)} + \cdots + \beta_1 y' = a_2 x + b_2 y + g(t) \\ &x(t_0) = z_0, x'(t_0) = z_1, \ldots, x^{(n-1)}(t_0) = z_{n-1} \\ &y(t_0) = w_0, y'(t_0) = w_1, \ldots, y^{(m-1)}(t_0) = w_{m-1} \end{aligned} \tag{3}$$

In order to find a solution to the initial value problem (3), first obtain the general solution of (2) and then use the $m + n$ constants in the general solution to select the solution that satisfies the given initial conditions.

The procedure for solving equations (2) and (3) is illustrated further with the following two examples.

► **EXAMPLE 7.4-1**

Consider the system

$$\begin{aligned} x' &= 2x + y \\ y'' + y' &= -5x - 3y \end{aligned} \tag{4}$$

From the first equation we obtain

$$y = x' - 2x \qquad y' = x'' - 2x' \qquad \text{and} \qquad y'' = x''' - 2x''$$

Substituting into the second equation in (4), we obtain

$$x''' - 2x'' + x'' - 2x' = -5x - 3x' + 6x$$

and hence

$$x''' - x'' + x' - x = 0$$

The auxiliary equation is $\lambda^3 - \lambda^2 + \lambda - 1 = (\lambda - 1)(\lambda^2 + 1) = 0$, and it follows that

$$x(t) = c_1 e^t + c_2 \cos t + c_3 \sin t$$

where c_1, c_2, and c_3 are constants. Using $y = x' - 2x$,

$$y(t) = -c_1 e^t + (c_3 - 2c_2) \cos t - (c_2 + 2c_3) \sin t$$

and we have the general solution to (4).

▶ **EXAMPLE 7.4-2**

Consider the initial value system

$$\begin{aligned} &x'' = x + 2y + 2e^{-t} \\ (5)\quad &y'' = 3x + 2y - 1 \\ &x(0) = \tfrac{1}{6} \qquad x'(0) = -\tfrac{1}{6} \qquad y(0) = -\tfrac{1}{2} \qquad y'(0) = 1 \end{aligned}$$

Using the first equation we have

$$y = \tfrac{1}{2}(x'' - x) - e^{-t} \qquad y' = \tfrac{1}{2}(x''' - x') + e^{-t}$$
$$y'' = \tfrac{1}{2}(x^{(4)} - x'') - e^{-t}$$

and substituting in the second equation,

$$\tfrac{1}{2}(x^{(4)} - x'') - e^{-t} = 3x + (x'' - x) - 2e^{-t} - 1$$

and hence

$$(6)\qquad x^{(4)} - 3x'' - 4x = -2e^{-t} - 2$$

The auxiliary equation of the corresponding homogeneous equation is

$$\lambda^4 - 3\lambda^2 - 4 = (\lambda^2 - 4)(\lambda^2 + 1) = 0$$

and it follows that

$$x_H(t) = c_1 \cosh 2t + c_2 \sinh 2t + c_3 \cos t + c_4 \sin t$$

is the general solution to the homogeneous equation. Using the method of undetermined coefficients, a particular solution to (6) has the form $x_p = d_1 e^{-t} + d_2$, and it easily follows that $d_1 = \frac{1}{3}$ and $d_2 = \frac{1}{2}$. Using the fact that $y = \frac{1}{2}(x'' - x) - e^{-t}$, one can show that the general solution to the differential equation in (5) is

$$x(t) = c_1 \cosh 2t + c_2 \sinh 2t + c_3 \cos t + c_4 \sin t + \tfrac{1}{3}e^{-t} + \tfrac{1}{2}$$
$$y(t) = 2c_1 \cosh 2t + 2c_2 \sinh 2t - c_3 \cos t - c_4 \sin t - e^{-t} - \tfrac{1}{4}$$

From the initial conditions in (5) we obtain the equations

$$\begin{aligned} c_1 \quad + c_3 \quad &= -\tfrac{2}{3} \\ 2c_2 \quad + c_4 &= \tfrac{1}{6} \\ 2c_1 \quad + c_3 \quad &= \tfrac{3}{4} \\ 4c_2 \quad - c_4 &= 0 \end{aligned}$$

Solving this system we get $c_1 = \frac{1}{36}$, $c_2 = \frac{1}{36}$, $c_3 = -\frac{25}{36}$, and $c_4 = \frac{4}{36}$. Therefore

$$x(t) = \tfrac{1}{36}\cosh 2t + \tfrac{1}{36}\sinh 2t - \tfrac{25}{36}\cos t + \tfrac{4}{36}\sin t + \tfrac{1}{3}e^{-t} + \tfrac{1}{2}$$

$$y(t) = \tfrac{2}{36}\cosh 2t + \tfrac{2}{36}\sinh 2t + \tfrac{25}{36}\cos t - \tfrac{4}{36}\sin t - e^{-t} - \tfrac{1}{4}$$

is the solution to the initial value problem (5).

7.4a Mass-Spring Systems

The motion of two masses connected by linear springs can be described by a weakly coupled pair of second order linear equations. The basic ideas in deriving the equations of motion for linear springs can be found in Section 2.6. Here we consider two masses m_1 and m_2, the first mass suspended by a linear spring with spring constant $k_1 > 0$, and the second mass connected to a linear spring with spring constant $k_2 > 0$ that is suspended from the first mass (see Figure 7.1a). The equilibrium positions of the masses are indicated in

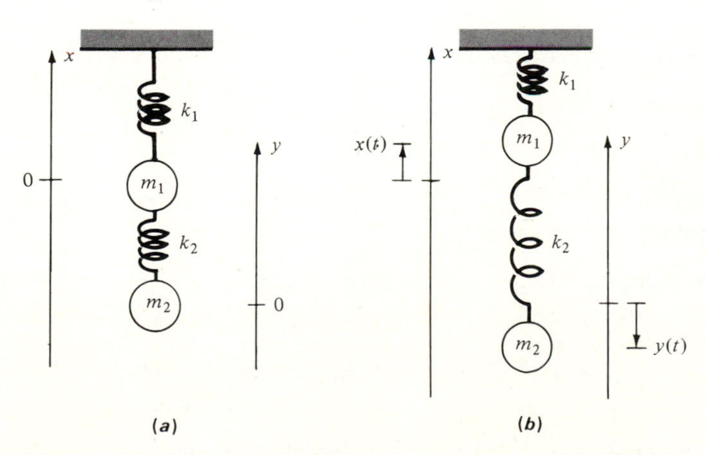

Figure 7.1 (a) Coupled mass-spring system at equilibrium. (b) Typical position of vibrating coupled mass-spring system.

Figure 7.1*a*, and typical positions of the masses at some time $t \geq 0$ are indicated in Figure 7.1*b*, where $x(t)$ is the position of m_1 relative to its equilibrium at time $t \geq 0$ and $y(t)$ is the positive of m_2 relative to its equilibrium at time $t \geq 0$.

The equations of motion are obtained from Newton's second law and Hooke's law for the force induced by a linear spring. The force from the springs acting on the mass m_2 depends only on how much the spring k_2 is stretched or compressed (see Figure 7.1*b*). Since this stretching is measured by the difference $y(t) - x(t)$ for each $t \geq 0$, it follows that

$$\text{Force on } m_2 \text{ at time } t \text{ is } -k_2[y(t) - x(t)] \tag{7}$$

The force from the springs acting on the mass m_1 depends on the stretching of both springs (see Figure 7.1*b*). Since the stretching of spring k_1 is measured by $x(t)$ and k_2 by $y(t) - x(t)$, it follows that

$$\text{Force on } m_1 \text{ at time } t \text{ is } -k_1 x(t) + k_2[y(t) - x(t)] \tag{8}$$

Since the total forces on mass m_1 and mass m_2 are $m_1 x''(t)$ and $m_2 y''(t)$ by Newton's second law, it follows directly from (7) and (8) that the equation of motion of the mass-spring system indicated in Figure 7.1 is the system

$$\begin{aligned} &m_1 x'' = -(k_1 + k_2)x + k_2 y \\ &m_2 y'' = k_2 x - k_2 y \\ &x(0) = x_0 \qquad x'(0) = z_0 \qquad y(0) = y_0 \qquad y'(0) = w_0 \end{aligned} \tag{9}$$

where x_0 and z_0 (or y_0 and w_0) are the initial position and velocity of the mass m_1 (or m_2). From the first equation in (9),

$$y = \frac{m_1 x'' + (k_1 + k_2)x}{k_2} \tag{10}$$

and computing y'', substituting into the second equation in (9), and simplifying, we obtain

$$m_1 m_2 x^{(4)} + [m_2(k_1 + k_2) + k_2 m_1]x'' + k_1 k_2 x = 0 \tag{11}$$

The auxiliary equation for (11) is

$$m_1 m_2 \lambda^4 + [m_2(k_1 + k_2) + k_2 m_1]\lambda^2 + k_1 k_2 = 0$$

Although this is fourth order in λ, it is quadratic in λ^2. Using the quadratic formula and noting that the discriminant is

$$\begin{aligned} [m_2(k_1 + k_2) + k_2 m_1]^2 - 4m_1 m_2 k_1 k_2 &= m_2^2 k_1^2 + 2m_2^2 k_1 k_2 + m_2^2 k_2^2 \\ &\quad + 2m_1 m_2 k_1 k_2 + 2m_1 m_2 k_2^2 \\ &\quad + k_2^2 m_1^2 - 4m_1 m_2 k_1 k_2 \\ &= (m_2 k_1 - m_1 k_2)^2 + m_2^2 k_2^2 \\ &\quad + 2m_1 m_2 k_2^2 + 2m_2^2 k_1 k_2 \\ &> 0 \end{aligned}$$

it follows that $\lambda^2 = -\omega_1^2, -\omega_2^2$, where ω_1 and ω_2 are real, positive, and distinct. Therefore $\lambda = \pm\omega_1 i$ and $\lambda = \pm\omega_2 i$ are the auxiliary roots of (11), and the general solution to (11) has the form

$$x(t) = c_1 \cos \omega_1 t + c_2 \sin \omega_1 t + c_3 \cos \omega_2 t + c_4 \sin \omega_2 t$$

From (10) it follows that y has a similar form, and hence each solution to (9) is bounded on $[0, \infty)$. The frequencies $\omega_1/2\pi$ and $\omega_2/2\pi$ are called the *natural frequencies* associated with (9).

As a particular case, suppose that $m_1 = m_2 = m$ and $k_1 = k_2 = k$. The auxiliary equation for (11) is then

$$m^2\lambda^4 + 3mk\lambda^2 + k^2 = 0$$

and it follows that

$$\lambda^2 = \frac{-3mk \pm \sqrt{9m^2k^2 - 4m^2k^2}}{2m^2} = \frac{(-3 \pm \sqrt{5})k}{2m}$$

Therefore, the natural frequencies are

$$\frac{\sqrt{3 - \sqrt{5}}\sqrt{k/2m}}{2\pi} \quad \text{and} \quad \frac{\sqrt{3 + \sqrt{5}}\sqrt{k/2m}}{2\pi}$$

in this special case.

PROBLEMS

7.4-1. Determine the general solution for each of the following linear systems.

(a) $x'' = y$, $y' = -x$

(b) $x' = -x + y$, $y'' + y' = 8x + 10y$

(c) $x'' = y + t^2$, $y'' = -x + 2$

(d) $x'' = 2y$, $y' = 4x + e^{2t}$

7.4-2. Determine a solution to the following initial value problems:

(a) $x'' + x' = -3x - 5y$, $y' = x + 2y$; $x(0) = 1$, $x'(0) = 2$, $y(0) = 4$

(b) $x'' = 4y + t$, $y'' = x + e^t$; $x(0) = x'(0) = 0$, $y(0) = y'(0) = 0$

7.4-3. Consider the mass-spring system

$$(*) \qquad \begin{aligned} 2x'' &= -6x + 2y \\ y'' &= 2x - 2y \end{aligned}$$

[equation (9), with $m_1 = 2$, $m_2 = 1$, $k_1 = 4$, and $k_2 = 2$].

(a) Determine the general solution to (*) and the natural frequencies.

(b) Describe the vibrations when both masses oscillate at each natural frequency.

(c) Determine the solution to (*) that initially satisfies

$$x(0) = 0 \qquad x'(0) = -2 \qquad y(0) = 1 \qquad y'(0) = 2$$

7.4-4. Determine an expression for the natural frequencies ω_1 and ω_2 for the system (9) in the following cases:

(a) $m_1 = m_2 = 4$ **(b)** $k_1 = k_2 = 2$

7.4-5. Using ideas analogous to those used in the derivation of equation (9), show that the equation of motion for the mass-spring system in Figure 7.2 is

$$m_1 x'' = -(k_1 + k_2)x + k_2 y$$
$$m_2 y'' = k_2 x - (k_1 + k_2)y$$

(a) Determine the natural frequencies of this system when $m_1 = m_2 = m$ and $k_1 = k_2 = k$

(b) Determine the general solution and the natural frequencies when $m_1 = m_2 = 1$, $k_1 = 2$, and $k_2 = 2$.

7.4-6. Suppose that a cannon is located at the origin of the xy coordinate system indicated in Figure 7.3. It is assumed that the cannon makes an angle α, $0 < \alpha < \pi/2$, with the x axis and that it fires the projectile with initial velocity v_0. So long as the projectile is above the x axis its position is denoted $(x(t), y(t))$. Neglecting any air resistance, observe that there are no forces acting on the projectile in the x direction and only a gravitational force in the y direction. Assuming that the gravitational attraction is constant (and letting g denote the gravitational constant), it can be routinely shown that the equations of motion for x and y are

$$(*) \qquad \begin{array}{lll} x'' = 0 & x(0) = 0 & x'(0) = v_0 \cos \alpha \\ y'' = -g & y(0) = 0 & y'(0) = v_0 \sin \alpha \end{array}$$

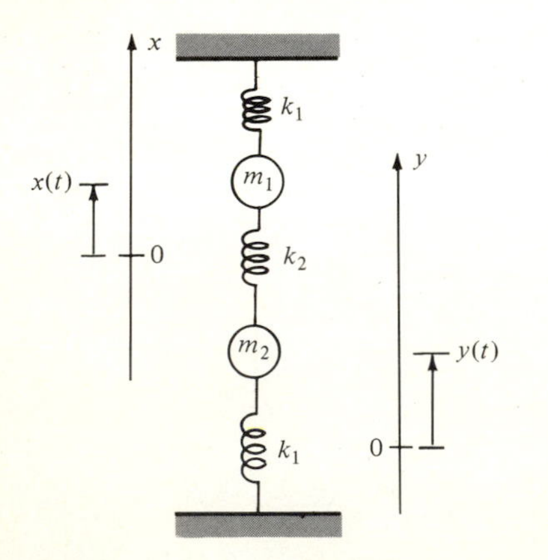

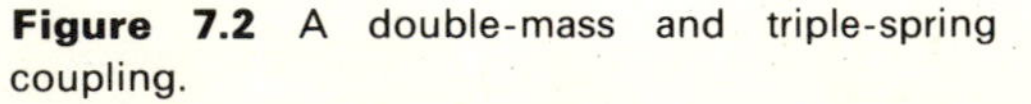
Figure 7.2 A double-mass and triple-spring coupling.

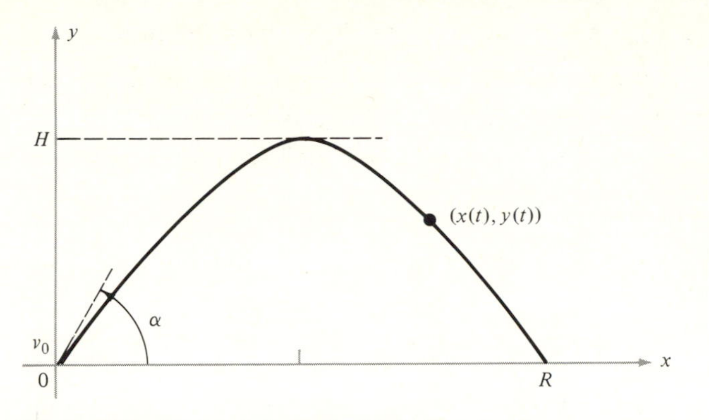

Figure 7.3 Flight of the projectile in Exercise 6.

(a) Show that the solution to (*) is

$$x(t) = tv_0 \cos \alpha \qquad \text{and} \qquad y(t) = tv_0 \sin \alpha - \frac{gt^2}{2}$$

(b) Eliminate the parameter t and determine what type of curve is described by the motion of the projectile.

(c) What is the time of flight of the projectile?

(d) Determine the range R of the projectile [note that R is the positive value of $x(t)$ when $y(t) = 0$]. What value or values of α give the maximum possible range R_{max}?

(e) For what value or values of α is the range equal to one-half the maximum range R_{max}?

(f) What is the maximum height of the projectile?

Remarks Additional examples of linear differential equations of higher order can be found in Hildebrand [6], Kreyszig [10], and Wylie and Barrett [15].

8 Matrix Methods for Linear Systems

In this chapter basic techniques for solving systems of linear differential equations are developed. These techniques are based on the concepts and properties of vectors and matrices and are applied to first order homogeneous linear systems with constant coefficients. The notations for elementary operations for vectors and matrices are discussed in the first section, and the direct relationship of these ideas to linear algebraic systems is indicated in the second section. Section 8.3 discusses the basic algebraic properties of solutions to homogeneous linear differential systems. Section 8.4 introduces the concepts of eigenvalues and eigenvectors for matrices, which are crucial to the study and computation of solutions to linear systems having constant coefficients; the basic techniques are developed in this section. The matrix exponential solution is introduced in Section 8.5, and several special methods for its computation are studied in Section 8.6. In Section 8.7 the important variation of constants formula for nonhomogeneous systems is derived, and some special techniques analogous to the method of undetermined coefficients are indicated for the computation of solutions to systems with particular types of nonhomogeneous terms.

8.1 NOTATIONS AND TERMINOLOGY

A method for solving systems of *two* first order linear differential equations with constant coefficients is studied in Section 5.3. In this chapter general methods are developed for solving systems of *two or more* first order linear differential equations with constant coefficients. Suppose that n is a positive integer and that a_{ij} is a real constant for each pair of integers i and j, with $1 \leq i, j \leq n$. The type of equations under consideration in this chapter have the form

$$
\begin{aligned}
y_1' &= a_{11}y_1 + a_{12}y_2 + \cdots + a_{1n}y_n + f_1(t) \\
y_2' &= a_{21}y_1 + a_{22}y_2 + \cdots + a_{2n}y_n + f_2(t) \\
&\vdots \\
y_n' &= a_{n1}y_1 + a_{n2}y_2 + \cdots + a_{nn}y_n + f_n(t)
\end{aligned} \tag{1}
$$

where f_i is a continuous function for each $i = 1, \ldots, n$. If $n = 1$ equation (1) is of the type studied in Section 1.2, and if $n = 2$ then (1) has the same form as equation (2) in Section 5.2. This equation is called a *first order, n-dimensional linear equation with constant coefficients*, and the functions f_i are called *nonhomogeneous* (*or inhomogeneous*) *terms.* In particular, if every f_i is identically zero then (1) is said to be *homogeneous.* Since the method for solving two-dimensional systems indicated in Section 5.3 is not immediately adaptable to general n-dimensional systems, the approaches developed here are somewhat different. The principal techniques and methods are based on the elementary properties of matrices and vectors.

8.1*a* Column Vectors

The notion of solution extends naturally from the two-dimensional case (see Section 5.2). A sequence of n continuously differentiable functions $y_1, y_2, \ldots, y_n$ is said to be a *solution to (1)* if

$$y_i'(t) = a_{i1}y_1(t) + a_{i2}y_2(t) + \cdots + a_{in}y_n(t) + f_i(t)$$

for every number t and every $i \in \{1, \ldots, n\}$. It is important to realize that the order in which the functions y_i are written is crucial in the definition of solution (e.g., if $n = 2$, then y_1, y_2 may be a solution and y_2, y_1 may not be a solution). To accommodate this concept, if $\xi_1, \xi_2, \ldots, \xi_n$ are *numbers* we say that

$$\boldsymbol{\xi} = \begin{pmatrix} \xi_1 \\ \xi_2 \\ \vdots \\ \xi_n \end{pmatrix} \tag{2}$$

is an *n-dimensional column vector.* In most cases when the dimension is known we say simply that $\boldsymbol{\xi}$ is a vector. The numbers $\xi_1, \xi_2, \ldots, \xi_n$ are called the *components* of the vector $\boldsymbol{\xi}$ and, in particular, ξ_i is the ith component. If $\eta_1, \eta_2, \ldots, \eta_n$ are numbers and $\boldsymbol{\eta}$ is the vector with η_i as its ith component, then $\boldsymbol{\xi} = \boldsymbol{\eta}$ if and only if $\xi_1 = \eta_1$, $\xi_2 = \eta_2, \ldots,$ and $\xi_n = \eta_n$.

Moreover, if k is a number then addition and scalar multiplication of vectors is defined as follows:

$$\begin{pmatrix} \xi_1 \\ \xi_2 \\ \vdots \\ \xi_n \end{pmatrix} + \begin{pmatrix} \eta_1 \\ \eta_2 \\ \vdots \\ \eta_n \end{pmatrix} = \begin{pmatrix} \xi_1 + \eta_1 \\ \xi_2 + \eta_2 \\ \vdots \\ \xi_n + \eta_n \end{pmatrix} \quad \text{and} \quad k\begin{pmatrix} \xi_1 \\ \xi_2 \\ \vdots \\ \xi_n \end{pmatrix} = \begin{pmatrix} k\xi_1 \\ k\xi_2 \\ \vdots \\ k\xi_n \end{pmatrix}$$

The vector that has zero in each component is called the *zero vector* and is denoted $\mathbf{0}$. Defining subtraction by $\boldsymbol{\xi} - \boldsymbol{\eta} = \boldsymbol{\xi} + (-1)\boldsymbol{\eta}$, we have that $\boldsymbol{\xi} - \boldsymbol{\xi} = \mathbf{0}$.

The concept of a *vector-valued function* is also important. Suppose that J is an interval and that $y_1, y_2, \ldots, y_n$ are each *functions* from J into the real numbers. The function $\mathbf{y}$ defined on J by

$$\mathbf{y}(t) = \begin{pmatrix} y_1(t) \\ y_2(t) \\ \vdots \\ y_n(t) \end{pmatrix} \tag{3}$$

is called an *n-dimensional vector-valued function* on J, and $y_1, y_2, \ldots, y_n$ are called the *components* of $\mathbf{y}$. If $\mathbf{x}$ is also a vector-valued function on J with components $x_1, x_2, \ldots, x_n$, then we define the functions $\mathbf{x} + \mathbf{y}$ and $k\mathbf{y}$ on J by

$$(\mathbf{x} + \mathbf{y})(t) = \mathbf{x}(t) + \mathbf{y}(t) \qquad \text{and} \qquad (k\mathbf{y})(t) = k\mathbf{y}(t)$$

Moreover, the derivatives and integrals of vector-valued functions are also defined componentwise:

$$\mathbf{y}'(t) = \begin{pmatrix} y_1'(t) \\ y_2'(t) \\ \vdots \\ y_n'(t) \end{pmatrix} \qquad \text{and} \qquad \int_\alpha^\beta \mathbf{y}(t)\,dt = \begin{pmatrix} \int_\alpha^\beta y_1(t)\,dt \\ \int_\alpha^\beta y_2(t)\,dt \\ \vdots \\ \int_\alpha^\beta y_n(t)\,dt \end{pmatrix}$$

Many of the usual properties of derivatives and integrals are easily seen to remain true in this context [for example, $(\mathbf{x} + \mathbf{y})'(t) = \mathbf{x}'(t) + \mathbf{y}'(t)$ and $(k\mathbf{x})'(t) = k\mathbf{x}'(t)$]. The function $\mathbf{y}$ is said to be continuous or differentiable only in case each of its components is continuous or differentiable, respectively.

Observe that the sequence of differentiable functions $y_1, \ldots, y_n$ is a solution to equation (1) if and only if

$$\begin{pmatrix} y_1'(t) \\ \vdots \\ y_n'(t) \end{pmatrix} = \begin{pmatrix} a_{11}y_1(t) + \cdots + a_{n1}y_n(t) \\ \vdots \\ a_{n1}y_1(t) + \cdots + a_{nn}y_n(t) \end{pmatrix} + \begin{pmatrix} f_1(t) \\ \vdots \\ f_n(t) \end{pmatrix} \tag{4}$$

for all t in $(-\infty, \infty)$.

8.1*b* Matrix Operations

The array of numbers

$$\mathbf{A} = \begin{pmatrix} a_{11} & a_{12} & \cdots & a_{1n} \\ a_{21} & a_{22} & \cdots & a_{2n} \\ \cdots & \cdots & \cdots & \cdots \\ a_{n1} & a_{n2} & \cdots & a_{nn} \end{pmatrix} \tag{5}$$

is called a *square matrix* and the a_{ij}'s are called the *components of* **A**. Since there are n rows and n columns in the matrix **A**, we sometimes say that **A** is an $n \times n$ *matrix*. Addition and scalar multiplication is also defined for matrices. If **B** is the $n \times n$ matrix with b_{ij} as its ijth component, then

$$\mathbf{A} + \mathbf{B} = \begin{pmatrix} a_{11} + b_{11} & \cdots & a_{1n} + b_{1n} \\ a_{21} + b_{21} & \cdots & a_{2n} + b_{2n} \\ \cdots & \cdots & \cdots \\ a_{n1} + b_{n1} & \cdots & a_{nn} + b_{nn} \end{pmatrix} \quad \text{and} \quad k\mathbf{A} = \begin{pmatrix} ka_{11} & \cdots & ka_{1n} \\ ka_{21} & \cdots & ka_{2n} \\ \cdots & \cdots & \cdots \\ ka_{n1} & \cdots & ka_{nn} \end{pmatrix}$$

for each scalar k. The matrix with all zero components is called the *zero matrix* and denoted **0**. Note that $\mathbf{A} + \mathbf{0} = \mathbf{0} + \mathbf{A} = \mathbf{A}$ for every matrix **A**, and if $\mathbf{A} - \mathbf{B} \equiv \mathbf{A} + (-1)\mathbf{B}$, then $\mathbf{A} - \mathbf{A} = \mathbf{0}$. A second important concept is the multiplication of an $n \times n$ matrix **A** with an n-dimensional column vector $\boldsymbol{\xi}$. If **A** is as in (5) and $\boldsymbol{\xi}$ is as in (2), then

$$\text{(6)} \qquad \mathbf{A}\boldsymbol{\xi} = \begin{pmatrix} a_{11}\xi_1 + a_{12}\xi_2 + \cdots + a_{1n}\xi_n \\ a_{21}\xi_1 + a_{22}\xi_2 + \cdots + a_{2n}\xi_n \\ \cdots \\ a_{n1}\xi_1 + a_{n2}\xi_2 + \cdots + a_{nn}\xi_n \end{pmatrix}$$

Therefore, $\mathbf{A}\boldsymbol{\xi}$ is the column vector $\boldsymbol{\eta}$ whose ith component η_i is

$$\eta_i = a_{i1}\xi_1 + a_{i2}\xi_2 + \cdots + a_{in}\xi_n$$

It is easy to check that the following formulas are valid:

$$\mathbf{A}(k_1\boldsymbol{\xi} + k_2\boldsymbol{\eta}) = k_1\mathbf{A}\boldsymbol{\xi} + k_2\mathbf{A}\boldsymbol{\eta} \qquad \text{and} \qquad (k_1\mathbf{A} + k_2\mathbf{B})\boldsymbol{\xi} = k_1\mathbf{A}\boldsymbol{\xi} + k_2\mathbf{B}\boldsymbol{\xi}$$

Note also that $\mathbf{A0} = \mathbf{0}$ and $\mathbf{0x} = \mathbf{0}$.

There is also a very important multiplicative operation for two $n \times n$ matrices **A** and **B**. The motivation for the definition of this matrix multiplication is somewhat subtle, and the definition is not what one might initially expect. If $\mathbf{A} = (a_{ij})$ and $\mathbf{B} = (b_{ij})$, where i and j run from 1 to n, then

$$\text{(7)} \qquad \mathbf{A} \cdot \mathbf{B} = \mathbf{C} \qquad \text{where the } ij\text{th component } c_{ij} \text{ of } \mathbf{C} \text{ has the form}$$

$$c_{ij} = a_{i1}b_{1j} + a_{i2}b_{2j} + \cdots + a_{in}b_{nj}$$

In the 2×2 case, for example, the product $\mathbf{A} \cdot \mathbf{B}$ can be written

$$\begin{pmatrix} a_{11} & a_{12} \\ a_{21} & a_{22} \end{pmatrix} \cdot \begin{pmatrix} b_{12} & b_{12} \\ b_{21} & b_{22} \end{pmatrix} = \begin{pmatrix} a_{11}b_{12} + a_{12}b_{21} & a_{11}b_{12} + a_{12}b_{22} \\ a_{21}b_{12} + a_{22}b_{21} & a_{21}b_{12} + a_{22}b_{22} \end{pmatrix}$$

A particular illustration in the 3×3 case is

$$\begin{pmatrix} 1 & 2 & -1 \\ 0 & 3 & 1 \\ -2 & 1 & 0 \end{pmatrix} \begin{pmatrix} 2 & 1 & 0 \\ -1 & 0 & 2 \\ 2 & 0 & 4 \end{pmatrix}$$

$$= \begin{pmatrix} 2-2-2 & 1+0+0 & 0+4-4 \\ 0-3+2 & 0+0+0 & 0+6+4 \\ -4-1+0 & -2+0+0 & 0+2+0 \end{pmatrix} = \begin{pmatrix} -2 & 1 & 0 \\ -1 & 0 & 10 \\ -5 & -2 & 2 \end{pmatrix}$$

Note also that, reversing the order of multiplication,

$$\begin{pmatrix} 2 & 1 & 0 \\ -1 & 0 & 2 \\ 2 & 0 & 4 \end{pmatrix}\begin{pmatrix} 1 & 2 & -1 \\ 0 & 3 & 1 \\ -2 & 1 & 0 \end{pmatrix}$$

$$= \begin{pmatrix} 2+0+0 & 4+3+0 & -2+1+0 \\ -1+0-4 & -2+0+2 & 1+0+0 \\ 2+0-8 & 4+0+4 & -2+0+0 \end{pmatrix} = \begin{pmatrix} 2 & 7 & -1 \\ -5 & 0 & 1 \\ -6 & 8 & -2 \end{pmatrix}$$

Comparing the preceding two computations shows that the *commutative law* $\mathbf{AB} = \mathbf{BA}$ *is not valid* in general. It is easy to check that $\mathbf{0A} = \mathbf{A0} = \mathbf{0}$. However, it is possible that $\mathbf{A} \cdot \mathbf{B} = \mathbf{0}$ with neither $\mathbf{A}$ nor $\mathbf{B}$ being the zero matrix. For example,

$$\begin{pmatrix} 3 & 0 \\ 0 & 0 \end{pmatrix}\begin{pmatrix} 0 & 0 \\ 1 & 2 \end{pmatrix} = \begin{pmatrix} 0 & 0 \\ 0 & 0 \end{pmatrix}$$

The elements a_{ii} of the matrix $\mathbf{A} = (a_{ij})$ are called the *diagonal* components of $\mathbf{A}$ and the elements a_{ij} with $i \neq j$ are called the off-diagonal components of $\mathbf{A}$. The $n \times n$ matrix with all zero off-diagonal components and all one-diagonal components is called the $n \times n$ *identity matrix* and is denoted by $\mathbf{I}$ (or $\mathbf{I}_n$ if needed for clarity). Specifically, $\mathbf{I} = (\delta_{ij})$, where $\delta_{ij} = \mathbf{0}$ if $i \neq j$ and $\delta_{ii} = 1$. As an illustration, the 2×2, 3×3, and 4×4 identity matrices have the following forms:

$$\begin{pmatrix} 1 & 0 \\ 0 & 1 \end{pmatrix} \qquad \begin{pmatrix} 1 & 0 & 0 \\ 0 & 1 & 0 \\ 0 & 0 & 1 \end{pmatrix} \qquad \text{and} \qquad \begin{pmatrix} 1 & 0 & 0 & 0 \\ 0 & 1 & 0 & 0 \\ 0 & 0 & 1 & 0 \\ 0 & 0 & 0 & 1 \end{pmatrix}$$

The identity matrix $\mathbf{I}$ has the following important property. If $\mathbf{A}$ is *any* $n \times n$ matrix and $\mathbf{I}$ is the $n \times n$ identity matrix, then

$$\mathbf{A} \cdot \mathbf{I} = \mathbf{I} \cdot \mathbf{A} = \mathbf{A} \tag{8}$$

Formula (8) can be routinely checked directly from the definition.

It is convenient in many cases to associate with each $n \times n$ matrix $\mathbf{A} = (a_{ij})$ the n vectors determined by the columns of $\mathbf{A}$. Defining

$$\mathbf{a}_1 = \begin{pmatrix} a_{11} \\ a_{21} \\ \vdots \\ a_{n1} \end{pmatrix}, \mathbf{a}_2 = \begin{pmatrix} a_{12} \\ a_{22} \\ \vdots \\ a_{n2} \end{pmatrix}, \ldots, \mathbf{a}_n = \begin{pmatrix} a_{1n} \\ a_{2n} \\ \vdots \\ a_{nn} \end{pmatrix} \tag{9}$$

we use the notation

$$\mathbf{A} = (\mathbf{a}_1, \mathbf{a}_2, \ldots, \mathbf{a}_n) \tag{10}$$

Therefore, given any matrix $\mathbf{A}$ we can associate the n vectors determined by the columns of $\mathbf{A}$, and conversely, given n vectors there is a matrix whose columns are determined by these vectors. Note in particular that if $\mathbf{B} = (\mathbf{b}_1, \mathbf{b}_2, \ldots, \mathbf{b}_n)$, then

(11) $$\mathbf{A} \cdot \mathbf{B} = (\mathbf{A}\mathbf{b}_1, \mathbf{A}\mathbf{b}_2, \ldots, \mathbf{A}\mathbf{b}_n)$$

That is, each column vector of the product of $\mathbf{A} \cdot \mathbf{B}$ is determined by the multiplication of $\mathbf{A}$ with the corresponding column of the matrix $\mathbf{B}$.

The following example illustrates routine vector and matrix calculations.

▶ **EXAMPLE 8.1-1**

Consider the following operations:

$$2\begin{pmatrix}-2\\3\end{pmatrix} - 7\begin{pmatrix}4\\1\end{pmatrix} = \begin{pmatrix}-4\\6\end{pmatrix} + \begin{pmatrix}-28\\-7\end{pmatrix} = \begin{pmatrix}-32\\-1\end{pmatrix}$$

$$\begin{pmatrix}3 & -1\\1 & 6\end{pmatrix}\begin{pmatrix}e^t\\2e^t\end{pmatrix} = \begin{pmatrix}3e^t - 2e^t\\e^t + 12e^t\end{pmatrix} = \begin{pmatrix}e^t\\13e^t\end{pmatrix}$$

$$\begin{pmatrix}1 & 2 & 3\\0 & 2 & -1\\4 & 0 & 7\end{pmatrix}\begin{pmatrix}2\\-3\\2\end{pmatrix} = \begin{pmatrix}2-6+6\\0-6-2\\8+0+14\end{pmatrix} = \begin{pmatrix}2\\-8\\22\end{pmatrix}$$

and

$$\begin{pmatrix}1 & 1 & 1\\-1 & 0 & 2\\3 & 1 & 0\end{pmatrix}\begin{pmatrix}e^t & 0 & 1\\0 & \sin t & 2\\2 & \cos t & 0\end{pmatrix} = \begin{pmatrix}e^t+2 & \sin t + \cos t & 3\\-e^t+4 & 2\cos t & -1\\3e^t & \sin t & 5\end{pmatrix}$$

8.1c Matrix Notation for Differential Systems

If equation (4) is compared with equation (6), it is easy to see that (4) can be rewritten in the form

$$\begin{pmatrix}y_1'(t)\\\vdots\\y_n'(t)\end{pmatrix} = \begin{pmatrix}a_{11} & \cdots & a_{1n}\\a_{21} & \cdots & a_{2n}\\a_{n1} & \cdots & a_{nn}\end{pmatrix}\begin{pmatrix}y_1(t)\\\vdots\\y_n(t)\end{pmatrix} + \begin{pmatrix}f_1(t)\\\vdots\\f_n(t)\end{pmatrix}$$

Therefore, if $\mathbf{f}$ is the vector-valued function

$$\mathbf{f}(t) = \begin{pmatrix}f_1(t)\\f_2(t)\\\vdots\\f_n(t)\end{pmatrix}$$

and $\mathbf{A}$ is defined by (5), then equation (1) can be written in the compact form

(12) $$\mathbf{y}' = \mathbf{A}\mathbf{y} + \mathbf{f}(t)$$

Not only are properties of matrices important in the study of systems of linear differential equations, but the convenience of notation is also a distinct asset. Note that if the vector-valued function $\mathbf{y}$ is defined by (3), then $y_1, y_2, \ldots, y_n$ is a solution to (1) only in case $\mathbf{y}'(t) = \mathbf{A}\mathbf{y}(t) + \mathbf{f}(t)$ for all t [that is, $\mathbf{y}$ is a solution to (12)]. Therefore, solutions to (1) are usually considered to be column vector–valued functions.

There is, of course, an initial value problem associated with the differential equation (8). If t_0 is a given time (usually taken to be 0) and if $\boldsymbol{\eta}$ is a given vector, then the system

$$\mathbf{y}' = \mathbf{A}\mathbf{y} + \mathbf{f}(t) \qquad \mathbf{y}(t_0) = \boldsymbol{\eta} \tag{13}$$

is a *first order, n-dimensional linear initial value problem with constant coefficients.* Observe that (13) can be written

$$\begin{aligned} y_1' &= a_{11}y_1 + \cdots + a_{1n}y_n + f_1(t) \quad & y_1(t_0) &= \eta_1 \\ y_2' &= a_{21}y_1 + \cdots + a_{2n}y_n + f_2(t) \quad & y_2(t_0) &= \eta_2 \\ &\vdots \\ y_n' &= a_{n1}y_1 + \cdots + a_{nn}y_n + f_n(t) \quad & y_n(t_0) &= \eta_n \end{aligned} \tag{13'}$$

A vector-valued function $\mathbf{y}$ is a solution to (13) only in case it is a solution to (12) and has the value $\boldsymbol{\eta}$ when $t = t_0$.

As an illustration, consider the two-dimensional system

$$\begin{aligned} y_1' &= y_1 + 2y_2 - 5t \\ y_2' &= 6y_1 - 3y_2 - 19 \end{aligned} \tag{14}$$

Using the matrix and vector notation this equation can be written

$$\begin{pmatrix} y_1 \\ y_2 \end{pmatrix}' = \begin{pmatrix} 1 & 2 \\ 6 & -3 \end{pmatrix}\begin{pmatrix} y_1 \\ y_2 \end{pmatrix} - \begin{pmatrix} 5t \\ 19 \end{pmatrix} \tag{14'}$$

From the method of solving two-dimensional systems in Section 5.3, it follows that if c_1 and c_2 are constants and

$$\begin{aligned} y_1(t) &= c_1e^{3t} + c_2e^{-5t} + t + 3 \\ y_2(t) &= c_1e^{3t} - 3c_2e^{-5t} + 2t - 1 \end{aligned}$$

then y_1, y_2 is a solution to (14). Using vector notation,

$$\begin{pmatrix} y_1(t) \\ y_2(t) \end{pmatrix} = \begin{pmatrix} c_1e^{3t} \\ c_1e^{3t} \end{pmatrix} + \begin{pmatrix} c_2e^{-5t} \\ -3c_2e^{-5t} \end{pmatrix} + \begin{pmatrix} t+3 \\ 2t-1 \end{pmatrix}$$

and hence

$$\begin{pmatrix} y_1(t) \\ y_2(t) \end{pmatrix} = c_1e^{3t}\begin{pmatrix} 1 \\ 1 \end{pmatrix} + c_2e^{-5t}\begin{pmatrix} 1 \\ -3 \end{pmatrix} + \begin{pmatrix} t+3 \\ 2t-1 \end{pmatrix}$$

is a solution to (14') for all constants c_1 and c_2.

There is a fundamental result regarding the existence and uniqueness of a solution to the initial value problem (13); this result is stated here. The proof of this assertion is beyond the scope of this text and is omitted.

Theorem 8.1-1

Suppose that $t_0 \in (-\infty, \infty)$ is given. Then for each given vector $\boldsymbol{\eta}$ there exists precisely one solution $\mathbf{y}$ to the initial value problem (13) on $(-\infty, \infty)$.

This theorem plays essentially the same role for systems as Theorem 2.2-1 played for second order linear equations and that Theorem 7.1-1 played for nth order linear systems. In fact it is important to realize that nth order linear differential equations are actually equivalent to a system of n first order linear equations with constant coefficients. To see that this is so, let α_0, α_1, $\alpha_2, \ldots, \alpha_{n-1}$ be constants, let g be a continuous real-valued function on $(-\infty, \infty)$, and consider the nth order linear initial value problem

$$(15)\qquad \begin{aligned} &y^{(n)} + \alpha_{n-1}y^{(n-1)} + \cdots + \alpha_1 y' + \alpha_0 y = g(t) \\ &y(t_0) = \eta_1,\ y'(t_0) = \eta_2, \ldots, y^{(n-1)}(t_0) = \eta_n \end{aligned}$$

[see equation (4) in Section 6.1]. Now define the functions $y_1, y_2, \ldots, y_n$ by

$$(16)\qquad y_1 = y,\ y_2 = y',\ y_3 = y'', \ldots, y_{n-1} = y^{(n-2)},\ y_n = y^{(n-1)}$$

and then note that

$$y_1' = y' = y_2,\ y_2' = y'' = y_3, \ldots, y_{n-1}' = y^{(n-1)} = y_n$$

Moreover, it follows directly from (15) that

$$\begin{aligned} y_n' = y^{(n)} &= -[\alpha_0 y + \alpha_1 y' + \cdots + \alpha_{n-1}y^{(n-1)}] + g(t) \\ &= -[\alpha_0 y_1 + \alpha_1 y_2 + \cdots + \alpha_{n-1}y_n] + g(t) \end{aligned}$$

Therefore, if y is a solution to the differential equation in (15) and $y_1, \ldots, y_n$ are defined by (16), then

$$(17)\qquad \begin{pmatrix} y_1 \\ y_2 \\ \cdot \\ y_{n-1} \\ y_n \end{pmatrix}' = \begin{pmatrix} 0 & 1 & 0 & \cdots & 0 \\ 0 & 0 & 1 & \cdots & 0 \\ \cdot & \cdot & \cdot & \cdot & \cdot \\ 0 & 0 & 0 & \cdots & 1 \\ -\alpha_0 & -\alpha_1 & -\alpha_2 & \cdots & -\alpha_{n-1} \end{pmatrix} \begin{pmatrix} y_1 \\ y_2 \\ \cdot \\ y_{n-1} \\ y_n \end{pmatrix} + \begin{pmatrix} 0 \\ 0 \\ \cdot \\ 0 \\ g(t) \end{pmatrix}$$

Also, the initial condition in (15) can be written

$$(18)\qquad \begin{pmatrix} y_1(t_0) \\ y_2(t_0) \\ \vdots \\ y_n(t_0) \end{pmatrix} = \begin{pmatrix} \eta_1 \\ \eta_2 \\ \vdots \\ \eta_n \end{pmatrix}$$

It is easily seen that (17) and (18) have exactly the form of the initial value problem (13), with

$$\mathbf{A} = \begin{pmatrix} 0 & 1 & 0 & \cdots & 0 \\ 0 & 0 & 1 & \cdots & 0 \\ \cdots & \cdots & \cdots & \cdots & \cdots \\ 0 & 0 & 0 & \cdots & 1 \\ -\alpha_0 & -\alpha_1 & -\alpha_2 & \cdots & -\alpha_{n-1} \end{pmatrix} \quad \text{and} \quad \mathbf{f}(t) = \begin{pmatrix} 0 \\ 0 \\ \vdots \\ 0 \\ g(t) \end{pmatrix}$$

If $\mathbf{y}(t)$ is a solution to (17) and (18) and $y(t) \equiv y_1(t)$ for all t, then necessarily $y' = y_2, y'' = y_3, \ldots, y^{(n-1)} = y_n$, and y is the solution to (15).

▶ **EXAMPLE 8.1-2**

Consider the fourth order equation

$$\begin{gathered} y^{(4)} - 2y''' + y' - 6y = \sin 2t \\ y(0) = 1 \qquad y'(0) = 2 \qquad y''(0) = -2 \qquad y'''(0) = 0 \end{gathered} \tag{19}$$

Setting $y_1 = y$, $y_2 = y'$, $y_3 = y''$, $y_4 = y'''$ the first order system

$$\begin{gathered} \begin{pmatrix} y_1 \\ y_2 \\ y_3 \\ y_4 \end{pmatrix}' = \begin{pmatrix} 0 & 1 & 0 & 0 \\ 0 & 0 & 1 & 0 \\ 0 & 0 & 0 & 1 \\ 6 & -1 & 0 & 2 \end{pmatrix} \begin{pmatrix} y_1 \\ y_2 \\ y_3 \\ y_4 \end{pmatrix} + \begin{pmatrix} 0 \\ 0 \\ 0 \\ \sin 2t \end{pmatrix} \\ \begin{pmatrix} y_1(0) \\ y_2(0) \\ y_3(0) \\ y_4(0) \end{pmatrix} = \begin{pmatrix} 1 \\ 2 \\ -2 \\ 0 \end{pmatrix} \end{gathered} \tag{20}$$

is obtained.

PROBLEMS

8.1-1. Perform each of the indicated matrix and vector operations:

(a) $2\begin{pmatrix} 1 \\ -5 \end{pmatrix} - 6\begin{pmatrix} 3 \\ -1 \end{pmatrix}$

(b) $3\begin{pmatrix} 1 \\ 0 \\ -2 \end{pmatrix} + \begin{pmatrix} 1 & 1 & 1 \\ 2 & 0 & -1 \\ 0 & 2 & 3 \end{pmatrix}\begin{pmatrix} -3 \\ 1 \\ 2 \end{pmatrix}$

(c) $\begin{pmatrix} 2 & -1 & 1 \\ 0 & 4 & 0 \\ 3 & -2 & 1 \end{pmatrix}\left(3\begin{pmatrix} 2 \\ 2 \\ 0 \end{pmatrix} - 2\begin{pmatrix} 0 \\ 1 \\ 3 \end{pmatrix}\right)$

(d) $\begin{pmatrix} 1 & -1 & 1 \\ 4 & 0 & 0 \\ -3 & 2 & -2 \end{pmatrix}\begin{pmatrix} 0 \\ 2 \\ 2 \end{pmatrix}$

(e) $\begin{pmatrix} 2 & 2 & 0 & 0 \\ 0 & 1 & 1 & 3 \\ 4 & 0 & -1 & 2 \\ -1 & -3 & 2 & 0 \end{pmatrix}\begin{pmatrix} 1 \\ -1 \\ 2 \\ -2 \end{pmatrix}$

(f) $\begin{pmatrix} 1 & 2 & 3 \\ 2 & 3 & 1 \\ 3 & 1 & 2 \end{pmatrix}\begin{pmatrix} 1 & 2 & 0 \\ 1 & 0 & -2 \\ 1 & 2 & 3 \end{pmatrix}$

(g) $\begin{pmatrix} 1 & 0 & 2 & 3 \\ 1 & -1 & 1 & 0 \\ 2 & 0 & 1 & 1 \\ -2 & 2 & 0 & 0 \end{pmatrix}\begin{pmatrix} 4 & 3 & 2 & 1 \\ 0 & 1 & 0 & 2 \\ 1 & 0 & -1 & 0 \\ 0 & -1 & 2 & 1 \end{pmatrix}$

8.1-2. Consider the following matrices and vectors:

$$\mathbf{A} = \begin{pmatrix} 1 & -1 & 0 \\ 2 & 0 & -1 \\ -1 & 1 & 1 \end{pmatrix} \quad \mathbf{B} = \begin{pmatrix} 3 & 2 & 1 \\ 1 & 3 & 2 \\ 2 & 1 & 3 \end{pmatrix} \quad \boldsymbol{\eta} = \begin{pmatrix} 1 \\ -1 \\ 1 \end{pmatrix} \quad \boldsymbol{\xi} = \begin{pmatrix} 0 \\ 1 \\ 2 \end{pmatrix}$$

Perform each of the following indicated operations:

(a) $\mathbf{A}(2\boldsymbol{\eta} - \boldsymbol{\xi})$ **(c)** $\mathbf{A}^2\ (\equiv \mathbf{A} \cdot \mathbf{A})$

(b) $2\mathbf{A} - 3\mathbf{B}$ **(d)** $2\mathbf{A}\boldsymbol{\xi} + 3\mathbf{B}\boldsymbol{\eta}$

***8.1-3.** Establish the associative law for matrix multiplication: $(\mathbf{AB})\mathbf{C} = \mathbf{A}(\mathbf{BC})$ for all $n \times n$ matrices **A**, **B**, and **C**.

8.1-4. Write each of the following linear differential systems, using vector and matrix notation:

(a)
$$y_1' = 6y_1 - 2y_2 + te^t$$
$$y_2' = y_2 - 7e^{2t}$$

(b)
$$y_1' = y_2 - y_3 + \sin 2t$$
$$y_2' = 2y_2 + y_3 + 2e^{-t}$$
$$y_3' = -y_1 + y_2 - y_3 - \cos 2t$$

(c)
$$y_1' = 3y_3$$
$$y_2' = -4y_1 + y_2$$
$$y_3' = 6y_1$$

(d)
$$y_1' = y_1 + y_2 + y_3 + y_4 + e^t$$
$$y_2' = y_1 + y_2 + y_3 + y_4 - e^{-t}$$
$$y_3' = y_3 - y_4 + e^t$$
$$y_4' = y_1 - y_3 - e^{-t}$$

8.1-5. Consider the homogeneous and nonhomogeneous equations:

$$\text{(H)} \qquad \begin{pmatrix} y_1 \\ y_2 \end{pmatrix}' = \begin{pmatrix} 0 & -1 \\ 6 & 5 \end{pmatrix} \begin{pmatrix} y_1 \\ y_2 \end{pmatrix} \qquad \text{and}$$

$$\text{(NH)} \qquad \begin{pmatrix} y_1 \\ y_2 \end{pmatrix}' = \begin{pmatrix} 0 & -1 \\ 6 & 5 \end{pmatrix} \begin{pmatrix} y_1 \\ y_2 \end{pmatrix} + \begin{pmatrix} e^t \\ 2e^t \end{pmatrix}$$

(a) Show that

$$\mathbf{y}(t) = e^{2t} \begin{pmatrix} 1 \\ -2 \end{pmatrix}$$

is a solution to (H).

(b) Determine, if possible, constants d_1 and d_2 such that

$$\mathbf{y}(t) = e^t \begin{pmatrix} d_1 \\ d_2 \end{pmatrix}$$

is a solution to (NH).

(c) Determine, if possible, a number c such that

$$\mathbf{y}(t) = e^{3t} \begin{pmatrix} c \\ 1 \end{pmatrix}$$

is a solution to (H).

8.1-6. Write each of the following higher-order linear equations as an equivalent first order linear system:

(a) $y''' - y'' + y' - y = e^{2t}$

(b) $y^{(4)} - 2y'' + y = e^{-t} + 1$

(c) $y''' - 8y = 0$

8.2 LINEAR COMBINATIONS AND ALGEBRAIC SYSTEMS

Linear combinations of vectors and systems of linear algebraic equations play important roles in the study of both the properties and the computation of solutions to n-dimensional linear differential equations. Throughout this section n is a positive integer, and, unless stated otherwise, all column vectors are n-dimensional and all matrices are $n \times n$. Also, the notations follow those of the preceding section. It is the purpose of this section to indicate some basic results associated with linear systems of algebraic equations. Their application to linear differential equations begins in the next section.

Suppose that m is a positive integer and that $\xi_1, \xi_2, \ldots, \xi_m$ are the vectors

$$\xi_1 = \begin{pmatrix} \xi_{11} \\ \xi_{21} \\ \vdots \\ \xi_{n1} \end{pmatrix}, \; \xi_2 = \begin{pmatrix} \xi_{12} \\ \xi_{22} \\ \vdots \\ \xi_{n2} \end{pmatrix}, \; \ldots, \; \xi_m = \begin{pmatrix} \xi_{1m} \\ \xi_{2m} \\ \vdots \\ \xi_{nm} \end{pmatrix}$$

It is important to realize that $\boldsymbol{\xi}_j$ is the jth vector and that ξ_{ij} is the ith component of the jth vector. A vector $\boldsymbol{\eta}$ is said to be a *linear combination of* $\boldsymbol{\xi}_1, \ldots, \boldsymbol{\xi}_m$ if there are constants $c_1, c_2, \ldots, c_m$ such that

$$\boldsymbol{\eta} = c_1 \boldsymbol{\xi}_1 + c_2 \boldsymbol{\xi}_2 + \cdots + c_m \boldsymbol{\xi}_m \tag{1}$$

Notice that (1) holds only in case the linear system

$$\begin{aligned} \xi_{11} c_1 + \xi_{12} c_2 + \cdots + \xi_{1m} c_m &= \eta_1 \\ \xi_{21} c_1 + \xi_{22} c_2 + \cdots + \xi_{2m} c_m &= \eta_2 \\ \cdots\cdots\cdots\cdots\cdots\cdots\cdots\cdots& \\ \xi_{n1} c_1 + \xi_{n2} c_2 + \cdots + c_{nm} c_m &= \eta_n \end{aligned} \tag{2}$$

has a solution $c_1, c_2, \ldots, c_n$.

EXAMPLE 8.2-1

Define the three-dimensional vectors $\boldsymbol{\xi}_1$ and $\boldsymbol{\xi}_2$ by

$$\boldsymbol{\xi}_1 = \begin{pmatrix} 1 \\ 1 \\ -1 \end{pmatrix} \quad \text{and} \quad \boldsymbol{\xi}_2 = \begin{pmatrix} -2 \\ 0 \\ 2 \end{pmatrix}$$

Since

$$\begin{pmatrix} -4 \\ 2 \\ 4 \end{pmatrix} = 2 \begin{pmatrix} 1 \\ 1 \\ -1 \end{pmatrix} + 3 \begin{pmatrix} -2 \\ 0 \\ 2 \end{pmatrix}$$

it follows that the vector

$$\begin{pmatrix} -4 \\ 2 \\ 4 \end{pmatrix}$$

is a linear combination of $\boldsymbol{\xi}_1$ and $\boldsymbol{\xi}_2$.

One very important concept deals with the question of when the zero vector $\mathbf{0}$ is a *nontrivial* linear combination of $\boldsymbol{\xi}_1, \boldsymbol{\xi}_2, \ldots, \boldsymbol{\xi}_m$. The vectors $\boldsymbol{\xi}_1, \boldsymbol{\xi}_2, \ldots, \boldsymbol{\xi}_m$ are said to be *linearly dependent* if there exist constants $c_1, c_2, \ldots, c_m$, *not all zero*, such that

$$c_1 \boldsymbol{\xi}_1 + c_2 \boldsymbol{\xi}_2 + \cdots + c_m \boldsymbol{\xi}_m = \mathbf{0} \tag{3}$$

In other words, $\mathbf{0}$ is a *nontrivial linear combination* of $\boldsymbol{\xi}_1, \ldots, \boldsymbol{\xi}_m$ [if $c_2 = c_2 = \cdots = c_m = 0$, then (3) is certainly satisfied—therefore it is required that at least one of the c_i's is not zero for (3) to be nontrivial]. If the vectors $\boldsymbol{\xi}_1, \boldsymbol{\xi}_2, \ldots, \boldsymbol{\xi}_m$

are not linearly dependent, then they are said to be *linearly independent*. Therefore, if (3) is valid only when all of the c_i's are zero, then $\xi_1, \ldots, \xi_m$ are linearly independent.

EXAMPLE 8.2-2

Consider the three vectors

$$\xi_1 = \begin{pmatrix} 1 \\ 0 \\ 1 \end{pmatrix} \qquad \xi_2 = \begin{pmatrix} 2 \\ -1 \\ 2 \end{pmatrix} \qquad \text{and} \qquad \xi_3 = \begin{pmatrix} 0 \\ 2 \\ 0 \end{pmatrix}$$

The equation $c_1 \xi_1 + c_2 \xi_2 + c_3 \xi_3 = \mathbf{0}$ can be written

$$\begin{aligned} c_1 + 2c_2 \qquad &= 0 \\ - c_2 + 2c_3 &= 0 \\ c_1 + 2c_2 \qquad &= 0 \end{aligned} \tag{4}$$

Subtracting each side of the first equation from the third, the system

$$\begin{aligned} c_1 + 2c_2 \qquad &= 0 \\ - c_2 + 2c_3 &= 0 \\ 0 &= 0 \end{aligned}$$

is obtained. If k is a nonzero constant and $c_3 = k$, then $c_2 = 2k$ and $c_1 = -4k$ is a solution; and it follows that $-4k\xi_1 + 2k\xi_2 + k\xi_3 = \mathbf{0}$ for all $k \neq 0$. Therefore, ξ_1, ξ_2, and ξ_3 are linearly dependent.

8.2*a* Algebraic Systems of Linear Equations

Determining the linear independence or dependence of vectors and determining if a certain vector can be written as a linear combination of given vectors is equivalent to solving a system of linear equations [see equation (2)]. The most important case for our needs is when the number of equations and the number of unknowns are the same [for example, $m = n$ in (2)]. Therefore, define

$$\mathbf{A} = \begin{pmatrix} a_{11} & a_{12} & \cdots & a_{1n} \\ a_{21} & a_{22} & \cdots & a_{2n} \\ \cdots & \cdots & \cdots & \cdots \\ a_{n1} & a_{n2} & \cdots & a_{nn} \end{pmatrix} \qquad \mathbf{b} = \begin{pmatrix} b_1 \\ b_2 \\ \vdots \\ b_n \end{pmatrix} \qquad \text{and} \qquad \mathbf{x} = \begin{pmatrix} x_1 \\ x_2 \\ \vdots \\ x_n \end{pmatrix}$$

The problem under consideration is to determine the set of all vectors $\mathbf{x}$ such that, for given $\mathbf{A}$ and $\mathbf{b}$,

$$\mathbf{A}\mathbf{x} = \mathbf{b} \tag{5}$$

It is easy to check that (5) can be written in the form

$$
\begin{aligned}
a_{11}x_1 + a_{12}x_2 + \cdots + a_{1n}x_n &= b_1 \\
a_{21}x_1 + a_{22}x_2 + \cdots + a_{2n}x_n &= b_2 \\
\cdots\cdots\cdots\cdots\cdots\cdots\cdots\cdots & \\
a_{n1}x_1 + a_{n2}x_2 + \cdots + a_{nn}x_n &= b_n
\end{aligned}
\tag{6}
$$

For systems that are reasonably simple, there are several elementary procedures for computing the solution set to (6), and we indicate one such procedure here.

First we note several elementary operations that can be applied to the equations in (6) while still maintaining an equivalent system (that is, a system with precisely the same solutions). The following three operations do not change the solution set:

(7)
- **(a)** Interchanging any pair of equations in (6)
- **(b)** Multiplying each side of an equation in (6) by a nonzero number
- **(c)** Adding each side of one of the equations in (6) to the corresponding side of another equation in (6)

In particular we want to use these operations to reduce the system (6) to an equivalent system where the first $i - 1$ coefficients in the ith equation are zero for $i = 2, 3, \ldots, n$. This new system has the form

$$
\begin{aligned}
\alpha_{11}x_1 + \alpha_{12}x_2 + \cdots + \alpha_{1n-1}x_{n-1} + \alpha_{1n}x_n &= \beta_1 \\
\alpha_{22}x_2 + \cdots + \alpha_{2n-1}x_{n-1} + \alpha_{2n}x_n &= \beta_2 \\
&\vdots \\
\alpha_{n-1n-1}x_{n-1} + \alpha_{n-1n}x_n &= \beta_{n-1} \\
\alpha_{nn}x_n &= \beta_n
\end{aligned}
\tag{8}
$$

A system having the form (8) is called *upper triangular*. It is routine to determine the solution set for an upper triangular system. For example if $\alpha_{nn} = 0$ and $\beta_n \neq 0$, there are *no solutions*, and if each of the diagonal terms α_{ii} are nonzero, there is a unique solution which can be computed as follows: solve for x_n in the nth equation; substitute for x_n in the $(n - 1)$st equation and for x_{n-1}; substitute for x_n and x_{n-1} in the $(n - 2)$nd equation and solve for x_{n-2}; and continue until x_1 is computed. The procedure is known as the *Gauss reduction method*. The following example illustrates this procedure.

► **EXAMPLE 8.2-3**

Consider the system

$$
\begin{pmatrix} 1 & 1 & 0 \\ 0 & 1 & 1 \\ 2 & 1 & -1 \end{pmatrix}
\begin{pmatrix} x_1 \\ x_2 \\ x_3 \end{pmatrix}
=
\begin{pmatrix} 3 \\ 2 \\ \alpha \end{pmatrix}
\tag{9}
$$

where α is a real constant. The equivalent simultaneous system is

$$\begin{aligned} x_1 + x_2 \qquad &= 3 \\ x_2 + x_3 &= 2 \\ 2x_1 + x_2 - x_3 &= \alpha \end{aligned} \tag{10}$$

Multiplying each side of the first equation in (10) by -2 and adding to the third equation, and then adding the second equation to the resulting third equation, yields the system

$$\begin{aligned} x_1 + x_2 \qquad &= 3 \\ x_2 + x_3 &= 2 \\ 0 &= \alpha - 4 \end{aligned} \tag{11}$$

Since the last equation is satisfied only if $\alpha = 4$, it is easy to see that (9) has a solution if and only if $\alpha = 4$. Moreover, if $\alpha = 4$ then (9) has an *infinite number of solutions.* Letting $\alpha = 4$, letting k be any constant, and setting $x_3 = k$, it follows from the second equation in (11) that $x_2 = 2 - k$, and then from the first equation that $x_1 = 3 - x_2 = 1 + k$. Therefore, if $\alpha = 4$ then $x_1 = 1 + k$, $x_2 = 2 - k$, and $x_3 = k$ is a solution to (10) for every constant k. Notice then that

$$\mathbf{x} = \begin{pmatrix} 1+k \\ 2-k \\ k \end{pmatrix} = k \begin{pmatrix} 1 \\ -1 \\ 1 \end{pmatrix} + \begin{pmatrix} 1 \\ 2 \\ 0 \end{pmatrix} \tag{12}$$

is a solution to (10) for every k. If $\alpha \neq 4$ then (10) has *no solutions.*

8.2*b* Determinants

Determinants also play an important role in the study of linear systems of differential equations; some of the most basic ideas are indicated here. Assume that $\mathbf{A} = (a_{ij})$ is an $n \times n$ matrix and use the notation det $\mathbf{A}$ to denote the *determinant of* $\mathbf{A}$. Recall that

$$\begin{cases} \det (a_{11}) \equiv a_{11} \\ \det \begin{pmatrix} a_{11} & a_{12} \\ a_{21} & a_{22} \end{pmatrix} \equiv a_{11}a_{22} - a_{21}a_{12} \\ \text{and} \\ \det \begin{pmatrix} a_{11} & a_{12} & a_{13} \\ a_{21} & a_{22} & a_{23} \\ a_{31} & a_{32} & a_{33} \end{pmatrix} \equiv a_{11}a_{22}a_{33} + a_{21}a_{32}a_{13} + a_{31}a_{23}a_{12} \\ \qquad\qquad - a_{31}a_{22}a_{13} - a_{32}a_{23}a_{11} - a_{33}a_{12}a_{21} \end{cases} \tag{13}$$

The definition of the determinant for a general $n \times n$ matrix can be given inductively on n. Suppose that $n > 1$, that the determinant is defined for all

$(n-1) \times (n-1)$ matrices, and that $\mathbf{A}$ is $n \times n$. For each $i, j \in \{1, \ldots, n\}$ let M_{ij} be the determinant of the $(n-1) \times (n-1)$ matrix obtained from $\mathbf{A}$ by deleting the ith row and the jth column from $\mathbf{A}$. This number $\mathbf{M}_{ij}$ is called the *minor* of a_{ij}. If k is any member of $\{1, \ldots, n\}$, then we define

$$\det \mathbf{A} \equiv (-1)^{k+1} a_{k1} \mathbf{M}_{k1} + (-1)^{k+2} a_{k2} \mathbf{M}_{k2} + \cdots + (-1)^{k+n} a_{kn} \mathbf{M}_{kn} \tag{14}$$

It is the case that (14) is *independent of the row* k that is selected. In fact, instead of expanding across the kth row we can expand down the kth column:

$$\det \mathbf{A} = (-1)^{1+k} a_{1k} \mathbf{M}_{1k} + (-1)^{2+k} a_{1k} \mathbf{M}_{2k} + \cdots + (-1)^{n+k} a_{nk} \mathbf{M}_{nk} \tag{14'}$$

Therefore, once the determinant is defined for $(n-1) \times (n-1)$ matrices it can be extended to $n \times n$ matrices [and hence then to $(n+1) \times (n+1)$ matrices, etc.].

Since $\det (a_{11}) = a_{11}$ we have from (14) with $k = 1$ and $n = 2$ that

$$\begin{aligned} \det \begin{pmatrix} a_{11} & a_{12} \\ a_{21} & a_{22} \end{pmatrix} &= (-1)^{1+1} a_{11} \det (a_{22}) + (-1)^{1+2} a_{12} \det (a_{21}) \\ &= a_{11} a_{22} - a_{12} a_{21} \end{aligned}$$

which agrees with the computation given in (13). There are several properties of determinants that are helpful in computations, and they are listed here for convenience:

(15)

- **(a)** If any two rows of $\mathbf{A}$ (or any two columns) are interchanged, the determinant of the resulting matrix is the negative of the determinant of $\mathbf{A}$.
- **(b)** If each element in any row (or column) of $\mathbf{A}$ is multiplied by a number c, then the determinant of the resulting matrix is c times the determinant of $\mathbf{A}$.
- **(c)** If the elements of one row (or column) of $\mathbf{A}$ are altered by adding some scalar multiple of another row (or column), then the determinant of the resulting matrix is the same as that of $\mathbf{A}$.

As an illustration the following determinant is computed.

EXAMPLE 8.2-4

Consider the 4×4 matrix

$$\mathbf{A} = \begin{pmatrix} 1 & 2 & 1 & 2 \\ 2 & 1 & 0 & -1 \\ 1 & 0 & 2 & 3 \\ 0 & 2 & -1 & -2 \end{pmatrix}$$

Multiplying the first row of $\mathbf{A}$ by -2 and adding it to the second row, and then multiplying the first row by -1 and adding to the third row, we obtain from part c of (15) that

$$\det \mathbf{A} = \det \begin{pmatrix} 1 & 2 & 1 & 2 \\ 0 & -3 & -2 & -5 \\ 1 & 0 & 2 & 3 \\ 0 & 2 & -1 & -2 \end{pmatrix} = \det \begin{pmatrix} 1 & 2 & 1 & 2 \\ 0 & -3 & -2 & -5 \\ 0 & -2 & 1 & 1 \\ 0 & 2 & -1 & -2 \end{pmatrix}$$

Expanding by minors down the first column [see (14′) with $n = 4$ and $k = 1$] implies that

$$\det \mathbf{A} = (-1)^{1+1} \cdot 1 \cdot \det \begin{pmatrix} -3 & -2 & -5 \\ -2 & 1 & 1 \\ 2 & -1 & -2 \end{pmatrix} + 0 + 0 + 0$$

Using the formula for computing the determinant of a 3×3 matrix, it follows that

$$\det \mathbf{A} = \det \begin{pmatrix} -3 & -2 & -5 \\ -2 & 1 & 1 \\ 2 & -1 & -2 \end{pmatrix} = 6 - 10 - 4 + 10 - 3 + 8 = 7$$

8.2c Fundamental Theorem of Linear Algebra and Matrix Inversion

Our main result in this section is to connect the concepts of linearly independent vector, the solvability of the linear system (5), and the determinant. $\mathbf{A}$ is assumed to be the $n \times n$ matrix

$$\text{(16)} \qquad \mathbf{A} = \begin{pmatrix} a_{11} & a_{11} & \cdots & a_{1n} \\ a_{21} & a_{22} & \cdots & a_{2n} \\ \cdots & \cdots & \cdots & \cdots \\ a_{n1} & a_{n2} & \cdots & a_{nn} \end{pmatrix}$$

and for each $k \in \{1, \ldots, n\}$ we let $\mathbf{a}_k$ denote the column vector determined by the kth column of $\mathbf{A}$:

$$\text{(17)} \qquad \mathbf{a}_1 = \begin{pmatrix} a_{11} \\ a_{21} \\ \vdots \\ a_{n1} \end{pmatrix}, \mathbf{a}_2 = \begin{pmatrix} a_{12} \\ a_{22} \\ \vdots \\ a_{n2} \end{pmatrix}, \ldots, \mathbf{a}_n = \begin{pmatrix} a_{1n} \\ a_{2n} \\ \vdots \\ a_{nn} \end{pmatrix}$$

Sometimes we write $\mathbf{A} = (\mathbf{a}_1, \mathbf{a}_2, \ldots, \mathbf{a}_n)$ for convenience. See (10) in Section 8.1. Now let

$$\text{(18)} \qquad \mathbf{b} = \begin{pmatrix} b_1 \\ b_2 \\ \vdots \\ b_n \end{pmatrix}$$

and consider the solvability of the system $\mathbf{Ax} = \mathbf{b}$. The following theorem is the *fundamental theorem of linear algebra.*

Theorem 8.2-1

Suppose that $\mathbf{A}$, $\mathbf{b}$, $\mathbf{a}_1, \mathbf{a}_2, \ldots,$ and $\mathbf{a}_n$ are as in the preceding paragraph. Then any one of the following statements implies the other three.

(a) $\det \mathbf{A} \neq 0$.

(b) The column vectors $\mathbf{a}_1, \mathbf{a}_2, \ldots, \mathbf{a}_n$ of $\mathbf{A}$ are linearly independent.

(c) The equation $\mathbf{Ax} = \mathbf{b}$ has a unique solution $\mathbf{x}$ for every $\mathbf{b}$.

(d) The equation $\mathbf{Ax} = \mathbf{0}$ has only the trivial solution $\mathbf{x} = \mathbf{0}$.

There are many important implications of Theorem 8.2-1. For example, if we are given n column vectors $\mathbf{a}_1, \mathbf{a}_2, \ldots, \mathbf{a}_n$ and form the matrix $\mathbf{A}$ by letting the kth column of $\mathbf{A}$ be the components of $\mathbf{a}_k$ [that is, $\mathbf{A} = (\mathbf{a}_1, \ldots, \mathbf{a}_n)$], then $\mathbf{a}_1, \ldots, \mathbf{a}_n$ are linearly independent if and only if $\det \mathbf{A} \neq 0$. Moreover, if $\mathbf{a}_1, \ldots, \mathbf{a}_n$ are linearly independent, then part c implies that *every* n-dimensional column vector is a linear combination of $\mathbf{a}_1, \ldots, \mathbf{a}_n$ (compare the equation $\mathbf{Ax} = \mathbf{b}$ with the system (2) where $n = m$). Since they are so crucial to the applications of matrix methods of linear differential equations, two of the implications of Theorem 8.2-1 are explicitly stated.

Theorem 8.2-2

Suppose that $\mathbf{A}$ is an $n \times n$ matrix. Then $\det \mathbf{A} = 0$ if and only if the vector equation $\mathbf{Ax} = \mathbf{0}$ has a nontrivial solution (i.e., there is an $\mathbf{x} \neq \mathbf{0}$ such that $\mathbf{Ax} = \mathbf{0}$).

Theorem 8.2-3

The column vectors $\mathbf{a}_1, \mathbf{a}_2, \ldots, \mathbf{a}_n$ are linearly independent if and only if for each column vector $\mathbf{b}$ there are unique constants $c_1, c_2, \ldots, c_n$ such that

$$\mathbf{b} = c_1 \mathbf{a}_1 + c_2 \mathbf{a}_1 + \cdots + c_n \mathbf{a}_n$$

Theorem 8.2-2 is just a restatement of the equivalence of part a with part d of Theorem 8.2-1, and an indication of how Theorem 8.2-3 follows from Theorem 8.2-1 is given in Problem 8.2-4.

EXAMPLE 8.2-5

Define the 3×3 matrix $\mathbf{A}$ by

$$\mathbf{A} = \begin{pmatrix} 1 & -1 & 3 \\ 3 & 0 & -1 \\ -1 & -2 & 7 \end{pmatrix}$$

Since

$$\det \mathbf{A} = 0 - 18 - 1 - 0 - 2 + 21 = 0$$

the system $\mathbf{Ax} = \mathbf{0}$ has a nontrivial solution by Theorem 8.2-2, so consider the system

$$\begin{aligned} x_1 - x_2 + 3x_3 &= 0 \\ 3x_1 \qquad - x_3 &= 0 \\ -x_1 - 2x_2 + 7x_3 &= 0 \end{aligned}$$

Multiplying each side of the first equation by -3 and adding to the second equation, and then adding the first equation to the third equation, implies that

$$\begin{aligned} x_1 - x_2 + 3x_3 &= 0 \\ 3x_2 - 10x_3 &= 0 \\ -3x_2 + 10x_3 &= 0 \end{aligned}$$

Adding each side of the second equation to the third, we obtain

$$\begin{aligned} x_1 - x_2 + 3x_3 &= 0 \\ 3x_2 - 10x_3 &= 0 \\ 0 &= 0 \end{aligned}$$

Therefore, if k is any constant we set $x_3 = k$ so that by the second equation $x_2 = 10k/3$, and then by the first equation $x_1 = k/3$. Therefore, the vector

$$\mathbf{x} = k \begin{pmatrix} \frac{1}{3} \\ \frac{10}{3} \\ 1 \end{pmatrix}$$

is a solution to $\mathbf{Ax} = \mathbf{0}$ for *every* constant k.

Again let $\mathbf{A}$ be an $n \times n$ matrix. Then $\mathbf{A}$ is said to be *invertible* if there is an $n \times n$ matrix $\mathbf{B}$ such that $\mathbf{AB} = \mathbf{BA} = \mathbf{I}$. Such a matrix $\mathbf{B}$ is called the *inverse of* $\mathbf{A}$ and is denoted $\mathbf{A}^{-1}$:

$$\mathbf{A} \cdot \mathbf{A}^{-1} = \mathbf{A}^{-1}\mathbf{A} = \mathbf{I}$$

Define the vectors

$$\mathbf{e}_1 = \begin{pmatrix} 1 \\ 0 \\ 0 \\ \vdots \\ 0 \end{pmatrix}, \mathbf{e}_2 = \begin{pmatrix} 0 \\ 1 \\ 0 \\ \vdots \\ 0 \end{pmatrix}, \mathbf{e}_3 = \begin{pmatrix} 0 \\ 0 \\ 1 \\ \vdots \\ 0 \end{pmatrix}, \ldots, \mathbf{e}_n = \begin{pmatrix} 0 \\ 0 \\ 0 \\ \vdots \\ 1 \end{pmatrix} \tag{19}$$

and note that $\mathbf{I} = (\mathbf{e}_1, \mathbf{e}_2, \ldots, \mathbf{e}_n)$. Therefore, if $\mathbf{A}^{-1}$ exists and $\boldsymbol{\xi}_1, \boldsymbol{\xi}_2, \ldots, \boldsymbol{\xi}_n$ are the columns of $\mathbf{A}^{-1}$ [that is, $\mathbf{A}^{-1} = (\boldsymbol{\xi}_1, \boldsymbol{\xi}_2, \ldots, \boldsymbol{\xi}_n)$], then

$$(\mathbf{e}_1, \mathbf{e}_2, \ldots, \mathbf{e}_n) = \mathbf{A}\mathbf{A}^{-1} = \mathbf{A}(\boldsymbol{\xi}_1, \boldsymbol{\xi}_2, \ldots, \boldsymbol{\xi}_n) = (\mathbf{A}\boldsymbol{\xi}_1, \mathbf{A}\boldsymbol{\xi}_2, \ldots, \mathbf{A}\boldsymbol{\xi}_n)$$

Hence the n systems of equation

$$\mathbf{A}\boldsymbol{\xi}_1 = \mathbf{e}_1, \mathbf{A}\boldsymbol{\xi}_2 = \mathbf{e}_2, \ldots, \mathbf{A}\boldsymbol{\xi}_n = \mathbf{e}_n \tag{20}$$

have solutions. Conversely, if each of the n equations in (20) has a solution, then $\mathbf{A}$ is invertible and $\mathbf{A}^{-1} = (\boldsymbol{\xi}_1, \boldsymbol{\xi}_2, \ldots, \boldsymbol{\xi}_n)$. Since the set of linear combinations of $\mathbf{e}_1, \mathbf{e}_2, \ldots, \mathbf{e}_n$ is all n vectors (in fact, each vector $\boldsymbol{\eta}$ can be written $\boldsymbol{\eta} = \eta_1 \mathbf{e}_1 + \eta_2 \mathbf{e}_2 + \cdots + \eta_n \mathbf{e}_n$), it follows that all of the systems in (20) have solutions if and only if the equation $\mathbf{A}\boldsymbol{\xi} = \mathbf{b}$ has a solution for every given vector $\mathbf{b}$. By the fundamental theorem of linear algebra (Theorem 8.2-1) it follows that $\mathbf{A}$ is invertible if and only if $\det \mathbf{A} \neq 0$. From the above discussion we have the following theorem.

Theorem 8.2-4

Suppose that $\mathbf{A}$ is an $n \times n$ matrix. Then $\mathbf{A}$ is invertible if and only if $\det \mathbf{A} \neq 0$. Moreover, if $\mathbf{A}$ is invertible and $k = 1, \ldots, n$, then the kth column of $\mathbf{A}^{-1}$ is the solution $\boldsymbol{\xi}_k$ of the equation $\mathbf{A}\boldsymbol{\xi}_k = \mathbf{e}_k$.

Theorem 8.2-4 provides a means of checking if a matrix has an inverse and also a method for computing the inverse if it exists. There is a second procedure to compute $\mathbf{A}^{-1}$ using determinants. For each $i, j \in \{1, \ldots, m\}$ let $\mathbf{M}_{ij}$ be the determinant of the $(n-1) \times (n-1)$ matrix obtained by deleting the ith row and the jth column from $\mathbf{A}$ ($\mathbf{M}_{ij}$ is the minor of a_{ij}). If $\det \mathbf{A} \neq 0$ then

$$\mathbf{A}^{-1} = \frac{1}{\det \mathbf{A}} \begin{pmatrix} \mathbf{M}_{11} & (-1)\mathbf{M}_{21} & \cdots & (-1)^{n+1}\mathbf{M}_{n1} \\ -\mathbf{M}_{12} & \mathbf{M}_{22} & \cdots & (-1)^{n+2}\mathbf{M}_{n2} \\ \cdots & \cdots & \cdots & \cdots \\ (-1)^{1+n}\mathbf{M}_{1n} & (-1)^{2+n}\mathbf{M}_{2n} & \cdots & \mathbf{M}_{nn} \end{pmatrix} \tag{21}$$

Using the definition of determinant [see (14) and (14′) in Section 8.2] one can indeed check that if $\mathbf{A}^{-1}$ is given by (21) then $\mathbf{A}\mathbf{A}^{-1} = \mathbf{A}^{-1}\mathbf{A} = \mathbf{I}$.

Note that if **A** is the 2×2 matrix

$$\begin{pmatrix} a_{11} & a_{12} \\ a_{21} & a_{22} \end{pmatrix}$$

then **A** is invertible if and only if $a_{11}a_{22} - a_{21}a_{12} \neq 0$, and in this case,

$$\begin{pmatrix} a_{11} & a_{12} \\ a_{21} & a_{22} \end{pmatrix}^{-1} = \frac{1}{a_{11}a_{22} - a_{21}a_{12}} \begin{pmatrix} a_{22} & -a_{12} \\ -a_{21} & a_{11} \end{pmatrix}$$

EXAMPLE 8.2-6

Using formula (21) shows that

$$\begin{pmatrix} 1 & 0 & -1 \\ 0 & 2 & 3 \\ 0 & 1 & 1 \end{pmatrix}^{-1} = \begin{pmatrix} 1 & 1 & -2 \\ 0 & -1 & 3 \\ 0 & 1 & -2 \end{pmatrix}$$

This can be easily verified by showing that the appropriate product is the identity matrix.

PROBLEMS

8.2-1. Determine the linear independence or dependence of each of the following sets of vectors. If the set of vectors is linearly dependent, then determine a nontrivial linear combination that sums up to the zero vector.

(a) $\begin{pmatrix} 1 \\ 2 \end{pmatrix} \quad \begin{pmatrix} 2 \\ 3 \end{pmatrix}$

(b) $\begin{pmatrix} 1 \\ -1 \end{pmatrix} \quad \begin{pmatrix} 1 \\ 2 \end{pmatrix} \quad \begin{pmatrix} 3 \\ 4 \end{pmatrix}$

(c) $\begin{pmatrix} 1 \\ 2 \\ 3 \end{pmatrix} \quad \begin{pmatrix} 2 \\ 3 \\ 1 \end{pmatrix} \quad \begin{pmatrix} 3 \\ 1 \\ 2 \end{pmatrix}$

(d) $\begin{pmatrix} 2 \\ 4 \\ 0 \end{pmatrix} \quad \begin{pmatrix} 3 \\ -1 \\ 7 \end{pmatrix} \quad \begin{pmatrix} 1 \\ 0 \\ 2 \end{pmatrix}$

(e) $\begin{pmatrix} 1 \\ 2 \\ 3 \end{pmatrix} \quad \begin{pmatrix} 1 \\ 1 \\ 2 \end{pmatrix} \quad \begin{pmatrix} 2 \\ 2 \\ 2 \end{pmatrix} \quad \begin{pmatrix} 3 \\ 2 \\ 1 \end{pmatrix}$

(f) $\begin{pmatrix} 1 \\ 1 \\ 0 \\ 0 \end{pmatrix} \quad \begin{pmatrix} 0 \\ 1 \\ 1 \\ 0 \end{pmatrix} \quad \begin{pmatrix} 0 \\ 0 \\ 1 \\ 1 \end{pmatrix} \quad \begin{pmatrix} 1 \\ 1 \\ 1 \\ 1 \end{pmatrix}$

8.2-2. Determine the set of *all* solutions for each of the following systems of equations.

(a)
$$\begin{aligned} x_1 \quad + x_3 &= -1 \\ 2x_2 + x_3 &= -2 \\ -x_1 \quad + 3x_3 &= 2 \end{aligned}$$

(b)
$$\begin{aligned} x_1 - 2x_2 + 3x_3 &= 0 \\ 4x_1 - x_2 + 5x_3 &= 0 \\ 2x_1 + 3x_2 - x_3 &= 0 \end{aligned}$$

(c)
$$\begin{aligned} 3x_1 + x_2 - x_3 &= 0 \\ 2x_1 + x_2 - 2x_3 &= 0 \\ x_2 - 4x_3 &= 0 \end{aligned}$$

(d)
$$\begin{aligned} x_1 + 4x_2 \quad + x_4 &= 0 \\ 2x_1 + 3x_2 - x_3 + 6x_4 &= 0 \\ 3x_1 + 2x_2 + 2x_3 + 4x_4 &= 1 \\ 4x_1 + x_2 + 3x_3 - 2x_4 &= -6 \end{aligned}$$

8.2-3. Compute the determinant of each of the following matrices **A**. Moreover, if det **A** = 0 determine the set of all vectors **x** such that **Ax** = **0**.

(a) $\mathbf{A} = \begin{pmatrix} 1 & 2 & 3 \\ 3 & 1 & 2 \\ 2 & 3 & 1 \end{pmatrix}$ **(c)** $\mathbf{A} = \begin{pmatrix} 1 & 1 & -2 \\ 0 & 2 & -2 \\ 3 & 1 & -4 \end{pmatrix}$

(b) $\mathbf{A} = \begin{pmatrix} 1 & 2 & 1 \\ 2 & 3 & 3 \\ -3 & -1 & 10 \end{pmatrix}$ **(d)** $\mathbf{A} = \begin{pmatrix} 1 & 1 & 1 \\ 2 & 2 & 2 \\ 3 & 3 & 3 \end{pmatrix}$

8.2-4. Suppose that the $n \times n$ matrix **A** is as in (16) and the vectors $\mathbf{a}_1, \mathbf{a}_2, \ldots, \mathbf{a}_n$ are as in (17).

(a) Show that $\mathbf{Ax} = x_1\mathbf{a}_1 + x_2\mathbf{a}_2 + \cdots + x_n\mathbf{a}_n$ for all column vectors **x**.

(b) Deduce from part *a* above and the equivalence of parts *b* and *c* of Theorem 8.2-1 that if $\mathbf{a}_1, \mathbf{a}_2, \ldots, \mathbf{a}_n$ are linearly independent, then *every* n-dimensional column vector **b** is a linear combination of $\mathbf{a}_1, \ldots, \mathbf{a}_n$.

8.2-5. Compute $\mathbf{A}^{-1}$ for each of the following matrices **A**.

(a) $\mathbf{A} = \begin{pmatrix} 1 & 2 \\ 3 & 4 \end{pmatrix}$

(b) $\mathbf{A} = \begin{pmatrix} 0 & -6 & 3 \\ -1 & -2 & 1 \\ 1 & 7 & -1 \end{pmatrix}$

(c) $\mathbf{A} = \begin{pmatrix} 0 & 1 & 0 \\ 0 & 0 & 1 \\ 1 & -1 & 1 \end{pmatrix}$

8.3 ALGEBRAIC PROPERTIES OF LINEAR DIFFERENTIAL SYSTEMS

There are many important algebraic concepts associated with the study of linear differential systems, and some of the most basic are discussed in this section. Throughout this section n is a positive integer and **A** is the $n \times n$

matrix

$$\mathbf{A} = \begin{pmatrix} a_{11} & a_{12} & \cdots & a_{1n} \\ a_{21} & a_{22} & \cdots & a_{2n} \\ \cdots & \cdots & \cdots & \cdots \\ a_{n1} & a_{n2} & \cdots & a_{nn} \end{pmatrix}$$

First we consider the homogeneous system

$$\mathbf{y}' = \mathbf{A}\mathbf{y} \tag{1}$$

where

$$\mathbf{y}(t) = \begin{pmatrix} y_1(t) \\ y_2(t) \\ \vdots \\ y_n(t) \end{pmatrix}$$

is a vector-valued function for t in $(-\infty, \infty)$. For notational convenience we let $\mathbf{y}_k$ denote a vector-valued function with components $y_{1k}, \ldots, y_{nk}$; that is

$$\mathbf{y}_k(t) = \begin{pmatrix} y_{1k}(t) \\ y_{2k}(t) \\ \vdots \\ y_{nk}(t) \end{pmatrix} \qquad \text{for } t \in (-\infty, \infty) \text{ and } k = 1, 2, \ldots$$

Many of the results indicated here have counterparts in the theory of second order linear equations (Chapter 2) as well as in the theory of two-dimensional linear systems (Chapter 5).

□ **Lemma 1**

Suppose that $\mathbf{y}_1$ and $\mathbf{y}_2$ are solutions to (1) and c_1 and c_2 are constants. Then $\mathbf{y} = c_1\mathbf{y}_1 + c_2\mathbf{y}_2$ is also a solution to (1): that is, any linear combination of solutions to (1) is also a solution to (1).

This lemma follows easily, using the linearity properties of matrices. For if $\mathbf{y} = c_1\mathbf{y} + c_2\mathbf{y}_2$ then

$$\begin{aligned} \mathbf{y}'(t) &= c_1\mathbf{y}_1'(t) + c_2 y_2'(t) = c_1\mathbf{A}\mathbf{y}_1(t) + c_2\mathbf{A}\mathbf{y}_2(t) \\ &= \mathbf{A}(c_1\mathbf{y}_1(t) + c_2\mathbf{y}_2(t)) = \mathbf{A}\mathbf{y}(t) \end{aligned}$$

and so $\mathbf{y}$ is a solution to (1) by definition.

As in the previous linear cases it is of interest to determine what is required to generate the class of all solutions to (1) [i.e., the *general solution* to (1)]. The concept of linear independent vectors plays a crucial role, and the main result of this section is given by the following theorem.

Theorem 8.3-1

Suppose that $\mathbf{y}_1, \mathbf{y}_2, \ldots, \mathbf{y}_n$ are solutions to the homogeneous system (1) and consider the class of all linear combinations of $\mathbf{y}_1, \ldots, \mathbf{y}_n$:

$$\mathbf{y} = c_1\mathbf{y}_1 + c_2\mathbf{y}_2 + \cdots + c_n\mathbf{y}_n \qquad c_1, c_2, \ldots, c_n \text{ constants} \tag{2}$$

Then the class (2) is the general solution to (1) if and only if the set of vectors $\mathbf{y}_1(t_0), \mathbf{y}_2(t_0), \ldots, \mathbf{y}_n(t_0)$ is linearly independent for some t_0.

The assertions in Theorem 8.3-1 follow from the existence and uniqueness result stated in Section 8.1 (see Theorem 8.1-1). For suppose that $t_0 \in (-\infty, \infty)$ and $\mathbf{y}_1(t_0), \ldots, \mathbf{y}_n(t_0)$ are linearly independent. Now let $\mathbf{y}$ be *any* solution to (1). It follows from Theorem 8.2-3 that there are constants $c_1, c_2, \ldots, c_n$ such that

$$\mathbf{y}(t_0) = c_1\mathbf{y}_1(t_0) + \cdots + c_n\mathbf{y}_n(t_0)$$

Let $\mathbf{x} = c_1\mathbf{y}_1 + \cdots + c_n\mathbf{y}_n$. Then $\mathbf{x}$ is a solution to (1) by Lemma 1 (why?), and since

$$\mathbf{x}(t_0) = c_1\mathbf{y}_1(t_0) + \cdots + c_n\mathbf{y}_n(t_0) = \mathbf{y}(t_0)$$

we have from the uniqueness assertion of Theorem 8.1-1 that $\mathbf{x}(t) \equiv \mathbf{y}(t)$ for *all* $t \in (-\infty, \infty)$. Therefore,

$$\mathbf{y}(t) \equiv c_1\mathbf{y}_1(t) + \cdots + c_n\mathbf{y}_n(t) \qquad \text{for all } t \in (-\infty, \infty)$$

and so $\mathbf{y}$ is a linear combination of $\mathbf{y}_1, \ldots, \mathbf{y}_n$. This shows that if $\mathbf{y}_1(t_0), \ldots, \mathbf{y}_n(t_0)$ are linearly independent for some $t_0 \in (-\infty, \infty)$, then (2) defines the general solution to (1).

Instead of the converse implication in Theorem 8.3-1, we establish a somewhat stronger result:

(3) If (2) defines the general solution, then the vectors $\mathbf{y}_1(t), \ldots, \mathbf{y}_n(t)$ are linearly independent for all $t \in (-\infty, \infty)$.

Assertion (3) also follows easily from Theorem 8.2-3. For let $t_0 \in (-\infty, \infty)$ and let $\boldsymbol{\eta}$ be a given vector. By the existence of a solution to (1) satisfying any initial condition at time t_0 (see Theorem 8.1-1) there is a solution $\mathbf{y}$ to (1) such that $\mathbf{y}(t_0) = \boldsymbol{\eta}$. Since (2) defines the general solution to (1), there are constants $c_1, \ldots, c_n$ such that $\mathbf{y} = c_1\mathbf{y}_1 + \cdots + c_n\mathbf{y}_n$, and hence

$$\mathbf{y}(t_0) = c_1\mathbf{y}_1(t_0) + \cdots + c_n\mathbf{y}_n(t_0) = \boldsymbol{\eta}$$

By Theorem 8.2-3 the vectors $\mathbf{y}_1(t_0), \ldots, \mathbf{y}_n(t_0)$ must be linearly independent. This establishes (3), and the proof of Theorem 8.3-1 is complete.

A convenient criterion for determining if a given set of solutions to (1) generate the general solution to (1) is the following:

□ **Lemma 2**

Suppose that $y_1, y_2, \ldots, y_n$ are solutions to (1) and that

$$\det(\mathbf{y}_1(t), \mathbf{y}_2(t), \ldots, \mathbf{y}_n(t)) \equiv \det \begin{pmatrix} y_{11}(t) & y_{12}(t) & \cdots & y_{1n}(t) \\ y_{21}(t) & y_{22}(t) & \cdots & y_{2n}(t) \\ \cdots & \cdots & \cdots & \cdots \\ y_{n1}(t) & y_{n2}(t) & \cdots & y_{nn}(t) \end{pmatrix} \neq 0$$

for some $t \in (-\infty, \infty)$. Then the family (2)

$$\mathbf{y} = c_1 \mathbf{y}_1 + \cdots + c_n \mathbf{y}_n \qquad \text{where } c_1, \ldots, c_n \text{ are constants}$$

is the general solution to (1).

This lemma is immediate from Theorem 8.3-1, and the fact that part *a* of Theorem 8.2-1 implies part *b* of Theorem 8.2-1.

Once the general solution (2) is obtained, one is assured that given any initial condition, say $\mathbf{y}(t_0) = \boldsymbol{\eta}$, unique constants $c_1, c_2, \ldots, c_n$ can be found so that the function $\mathbf{y}$ satisfies $\mathbf{y}(t_0) = \boldsymbol{\eta}$. In fact, the algebraic system

$$c_1 \mathbf{y}_1(t_0) + c_2 \mathbf{y}_2(t_0) + \cdots + c_n \mathbf{y}_n(t_0) = \boldsymbol{\eta}$$

has a unique solution $c_1, \ldots, c_n$ since the vectors $\mathbf{y}_1(t_0), \ldots, \mathbf{y}_n(t_0)$ are linearly independent (see part *c* of Theorem 8.2-1). Finally, note that solutions $\mathbf{y}_1, \mathbf{y}_2, \ldots, \mathbf{y}_k$ to (1) with $k < n$ cannot possibly generate the general solution to (1). This follows, for example, since the system

$$c_1 \mathbf{y}_1(t_0) + \cdots + c_k \mathbf{y}_k(t_0) = \boldsymbol{\eta}$$

cannot have a solution for all initial vectors $\boldsymbol{\eta}$.

As in the case of second order linear equations we can also use the concept of vector-valued *functions* $\mathbf{y}_1, \mathbf{y}_2, \ldots, \mathbf{y}_k$ being linearly independent or dependent [on the interval $(-\infty, \infty)$]. They are said to be *linearly independent* if the identity

$$c_1 \mathbf{y}_1(t) + c_2 \mathbf{y}_2(t) + \cdots + c_k \mathbf{y}_k(t) \equiv 0 \qquad \text{for all } t \in (-\infty, \infty)$$

implies that the constants $c_1, c_2, \ldots, c_k$ are all zero. Otherwise, they are *linearly dependent* [i.e., some nontrivial linear combination is identically zero on $(-\infty, \infty)$]. It is easy to see that if the vectors $\mathbf{y}_1(t_0), \mathbf{y}_2(t_0), \ldots, \mathbf{y}_k(t_0)$ are linearly independent for some $t_0 \in (-\infty, \infty)$, then the functions $\mathbf{y}_1, \mathbf{y}_2, \ldots, \mathbf{y}_k$ must also be linearly dependent. The converse is not true for general functions. However, if $\mathbf{y}_1, \ldots, \mathbf{y}_k$ are solutions to (1), then $\mathbf{y}_1, \ldots, \mathbf{y}_k$ are linearly independent functions if and only if $\mathbf{y}_1(t), \ldots, \mathbf{y}_k(t)$ are linearly independent for *all* $t \in (-\infty, \infty)$ [see (3)]. If the n solutions $\mathbf{y}_1, \mathbf{y}_2, \ldots, \mathbf{y}_n$ to (1) are linearly independent, then we say that they *generate the general solution to (1)* or are a *basis* for the general solution to (1).

EXAMPLE 8.3-1

Consider the 3×3 system

$$\mathbf{y}' = \begin{pmatrix} 3 & 0 & -2 \\ 1 & 2 & -2 \\ 1 & 3 & -3 \end{pmatrix} \mathbf{y} \tag{4}$$

along with the initial condition

$$\mathbf{y}(0) = \begin{pmatrix} 4 \\ 2 \\ 1 \end{pmatrix} \tag{5}$$

Consider also the functions

$$\mathbf{y}_1(t) = e^t \begin{pmatrix} 1 \\ 1 \\ 1 \end{pmatrix} \qquad \mathbf{y}_2(t) = e^{-t} \begin{pmatrix} 1 \\ 1 \\ 2 \end{pmatrix} \qquad \mathbf{y}_3(t) = e^{2t} \begin{pmatrix} 2 \\ 1 \\ 1 \end{pmatrix}$$

It is easy to check that $\mathbf{y}_1$, $\mathbf{y}_2$, and $\mathbf{y}_3$ are solutions to (4). For example, since

$$\begin{pmatrix} 3 & 0 & -2 \\ 1 & 2 & -2 \\ 1 & 3 & -3 \end{pmatrix} \left(e^{2t} \begin{pmatrix} 2 \\ 1 \\ 1 \end{pmatrix} \right) = e^{2t} \begin{pmatrix} 3 & 0 & -2 \\ 1 & 2 & -2 \\ 1 & 3 & -3 \end{pmatrix} \begin{pmatrix} 2 \\ 1 \\ 1 \end{pmatrix} = e^{2t} \begin{pmatrix} 4 \\ 2 \\ 2 \end{pmatrix}$$

and

$$\frac{d}{dt} \; e^{2t} \begin{pmatrix} 2 \\ 1 \\ 1 \end{pmatrix} = 2e^{2t} \begin{pmatrix} 2 \\ 1 \\ 1 \end{pmatrix} = e^{2t} \begin{pmatrix} 4 \\ 2 \\ 2 \end{pmatrix}$$

it follows that $\mathbf{y}_3$ is a solution to (4). Since

$$\det\,[\mathbf{y}_1(0), \mathbf{y}_2(0), \mathbf{y}_3(0)] = \det \begin{pmatrix} 1 & 1 & 2 \\ 1 & 1 & 1 \\ 1 & 2 & 1 \end{pmatrix} = 1 \neq 0$$

it follows from Lemma 2 that

$$\mathbf{y}(t) = c_1 e^t \begin{pmatrix} 1 \\ 1 \\ 1 \end{pmatrix} + c_2 e^{-t} \begin{pmatrix} 1 \\ 1 \\ 2 \end{pmatrix} + c_3 e^{2t} \begin{pmatrix} 2 \\ 1 \\ 1 \end{pmatrix} \tag{6}$$

where c_1, c_2, c_3 are constants, is the general solution to (4). In order to determine a solution to (4) that also satisfies the initial condition (5), we see from (6) that numbers c_1, c_2, and c_3 need to be computed so that

$$c_1 \begin{pmatrix} 1 \\ 1 \\ 1 \end{pmatrix} + c_2 \begin{pmatrix} 1 \\ 1 \\ 2 \end{pmatrix} + c_3 \begin{pmatrix} 2 \\ 1 \\ 1 \end{pmatrix} = \begin{pmatrix} 4 \\ 2 \\ 1 \end{pmatrix}$$

Equivalently,

$$c_1 + c_2 + 2c_3 = 4$$
$$c_1 + c_2 + c_3 = 2$$
$$c_1 + 2c_2 + c_3 = 1$$

Subtracting each side of the first equation from the second and from the third implies that

$$c_1 + c_2 + 2c_3 = 4$$
$$-c_3 = -2$$
$$c_2 - c_3 = -3$$

By the second equation we obtain $c_3 = 2$, and substituting this into the third we obtain $c_2 = -1$. Moreover, $c_1 = 1$ by substituting for c_2 and c_3 in the first equation. Therefore

$$\mathbf{y}(t) = e^t \begin{pmatrix} 1 \\ 1 \\ 1 \end{pmatrix} - e^{-t} \begin{pmatrix} 1 \\ 1 \\ 2 \end{pmatrix} + 2e^{2t} \begin{pmatrix} 2 \\ 1 \\ 1 \end{pmatrix}$$
$$= \begin{pmatrix} e^t - e^{-t} + 4e^{2t} \\ e^t - e^{-t} + 2e^{2t} \\ e^t - 2e^{-t} + 2e^{2t} \end{pmatrix}$$

is the solution to (4) that satisfies (5).

Now suppose that $\mathbf{f}$ is a vector-valued function on $(-\infty, \infty)$ and that

$$\mathbf{f}(t) = \begin{pmatrix} f_1(t) \\ f_2(t) \\ \vdots \\ f_n(t) \end{pmatrix}$$

Consider the nonhomogeneous system

(7) $$\mathbf{y}' = \mathbf{A}\mathbf{y} + \mathbf{f}(t)$$

whose corresponding homogeneous system is (1). We want to describe the general solution to (7), and the following lemma is a fundamental observation:

□ **Lemma 3**

Suppose that $\mathbf{y}_H$ is a solution to the homogeneous system (1) and $\mathbf{y}_p$ is a solution to the nonhomogeneous system (7). Then $\mathbf{y} = \mathbf{y}_H + \mathbf{y}_p$ is also a solution to the nonhomogeneous system (7).

The assertion of the lemma follows easily from the linearity properties of matrices. For since

$$\begin{aligned}\mathbf{y}'(t) &= \mathbf{y}_H'(t) + \mathbf{y}_p'(t) = \mathbf{A}\mathbf{y}_H(t) + \mathbf{A}\mathbf{y}_p(t) + \mathbf{f}(t)\\ &= \mathbf{A}[\mathbf{y}_H(t) + \mathbf{y}_p(t)] + \mathbf{f}(t) = \mathbf{A}\mathbf{y}(t) + \mathbf{f}(t)\end{aligned}$$

it follows that $\mathbf{y}$ is a solution to (7) by definition.

The fundamental result describing the general solution to (7) is the following theorem.

Theorem 8.3-2

Suppose that $\mathbf{y}_p$ is a solution to the nonhomogeneous system (7) and that $\mathbf{y}_1, \mathbf{y}_2, \ldots, \mathbf{y}_n$ are solutions to the homogeneous system (1) such that

$$\mathbf{y}_H \equiv c_1 \mathbf{y}_1 + \cdots + c_n \mathbf{y}_n \qquad \text{where } c_1, \ldots, c_n \text{ are constants}$$

is the general solution to (1). Then the family

$$\mathbf{y} \equiv c_1 \mathbf{y}_1 + \cdots + c_n \mathbf{y}_n + \mathbf{y}_p \qquad \text{where } c_1, \ldots, c_n \text{ are constants} \tag{8}$$

is the general solution to the nonhomogeneous system (7).

We know already from Lemma 3 (and Lemma 1) that each function $\mathbf{y}$ of the form (8) is a solution to (7). It remains to show that *every* solution to (7) is of the form (8). So suppose that $\mathbf{y}$ is a solution to (7) and define $\mathbf{x} = \mathbf{y} - \mathbf{y}_p$. Then

$$\begin{aligned}\mathbf{x}' &= \mathbf{y}' - \mathbf{y}_p' = \mathbf{A}\mathbf{y} + \mathbf{f}(t) - [\mathbf{A}\mathbf{y}_p + \mathbf{f}(t)]\\ &= \mathbf{A}(\mathbf{y} - \mathbf{y}_p) = \mathbf{A}\mathbf{x}\end{aligned}$$

and it follows that $\mathbf{y} - \mathbf{y}_p$ is a solution to (1). Since the general solution to (1) is of the form (2), we see that there are constants $c_1, \ldots, c_n$ such that

$$\mathbf{y} - \mathbf{y}_p = c_1 \mathbf{y}_1 + \cdots + c_n \mathbf{y}_n$$

Therefore, $\mathbf{y}$ is of the form (7) and the proof of Theorem 8.3-2 is complete.

Theorem 8.3-2 asserts that if the general solution to the homogeneous system (1) is known and one *particular solution* to the nonhomogeneous system (7) is known, then the general solution to (7) is also known. Of course Theorem 8.3-2 can be applied to initial value problems as well. Consider the initial value system

$$\mathbf{y}' = \mathbf{A}\mathbf{y} + \mathbf{f}(t) \qquad \mathbf{y}(t_0) = \boldsymbol{\eta} \tag{9}$$

If $\mathbf{y}_p, \mathbf{y}_1, \mathbf{y}_2, \ldots, \mathbf{y}_n$ are as in Theorem 8.3-2, then the solution to (9) can be determined by finding constants $c_1, c_2, \ldots, c_n$ such that

$$c_1 \mathbf{y}_1(t_0) + c_2 \mathbf{y}_2(t_0) + \cdots + c_n \mathbf{y}_n(t_0) + \mathbf{y}_p(t_0) = \boldsymbol{\eta}$$

Therefore, if the c_i's are solutions to the algebraic system

$$c_1 \mathbf{y}_1(t_0) + c_2 \mathbf{y}_2(t_0) + \cdots + c_n \mathbf{y}_n(t_0) = \boldsymbol{\eta} - \mathbf{y}_p(t_0)$$

then the solution (8) is the solution to the initial value problem (9).

PROBLEMS

8.3-1. In each of the following problems a homogeneous differential system is given, along with a number of solutions to this system. Verify that the given functions are indeed solutions and determine if the family of all linear combinations is the general solution. If the given solutions do not generate the general solution, classify all vectors $\boldsymbol{\eta}$ such that $\mathbf{y}(0) = \boldsymbol{\eta}$ for some solution $\mathbf{y}$ that is a linear combination of the given solutions.

(a) $\mathbf{y}' = \begin{pmatrix} 0 & 2 \\ -2 & 0 \end{pmatrix}\mathbf{y} \qquad \mathbf{y}_1 = \begin{pmatrix} \sin 2t \\ \cos 2t \end{pmatrix} \qquad \mathbf{y}_2 = \begin{pmatrix} -\cos 2t \\ \sin 2t \end{pmatrix}$

(b) $\mathbf{y}' = \begin{pmatrix} 1 & 2 \\ 6 & -3 \end{pmatrix}\mathbf{y} \qquad \mathbf{y}_1 = \begin{pmatrix} e^{3t} + e^{5t} \\ e^{3t} - 3e^{5t} \end{pmatrix} \qquad \mathbf{y}_2 = \begin{pmatrix} e^{3t} \\ e^{3t} \end{pmatrix}$

(c) $\mathbf{y}' = \begin{pmatrix} 0 & 1 & 0 \\ 0 & 0 & 1 \\ 2 & 1 & -2 \end{pmatrix}\mathbf{y} \qquad \mathbf{y}_1 = \begin{pmatrix} e^{t} \\ e^{t} \\ e^{t} \end{pmatrix} \qquad \mathbf{y}_2 = \begin{pmatrix} e^{-t} \\ -e^{-t} \\ e^{-t} \end{pmatrix} \qquad \mathbf{y}_3 = \begin{pmatrix} \cosh t \\ \sinh t \\ \cosh t \end{pmatrix}$

(d) $\mathbf{y}' = \begin{pmatrix} 0 & 1 & 0 \\ 0 & 0 & 1 \\ 1 & -1 & 1 \end{pmatrix}\mathbf{y} \quad \mathbf{y}_1 = \begin{pmatrix} e^{t} \\ e^{t} \\ e^{t} \end{pmatrix} \quad \mathbf{y}_2 = \begin{pmatrix} \cos t \\ -\sin t \\ -\cos t \end{pmatrix} \quad \mathbf{y}_3 = \begin{pmatrix} \sin t - \cos t \\ \cos t + \sin t \\ \cos t - \sin t \end{pmatrix}$

(e) $\mathbf{y}' = \begin{pmatrix} 0 & 1 & 0 \\ 0 & 0 & 1 \\ 0 & 1 & 0 \end{pmatrix}\mathbf{y} \qquad \mathbf{y}_1 = \begin{pmatrix} 1 \\ 0 \\ 0 \end{pmatrix} \qquad \mathbf{y}_2 = \begin{pmatrix} e^{-t} \\ -e^{-t} \\ e^{-t} \end{pmatrix} \qquad \mathbf{y}_3 = \begin{pmatrix} 1 - e^{-t} \\ e^{-t} \\ -e^{-t} \end{pmatrix}$

8.3-2. Show that the given family of functions is the general solution to the corresponding differential equation and then determine constants so that a solution to the given initial value problem is obtained.

(a) $\mathbf{y}' = \begin{pmatrix} 0 & 1 \\ 4 & 0 \end{pmatrix}\mathbf{y} - \begin{pmatrix} e^{t} + e^{2t} \\ 4e^{t} - 2e^{2t} \end{pmatrix} \qquad \mathbf{y}(0) = \begin{pmatrix} 0 \\ 11 \end{pmatrix}$

$$\mathbf{y} = c_1 \begin{pmatrix} e^{2t} \\ 2e^{2t} \end{pmatrix} + c_2 \begin{pmatrix} e^{-2t} \\ -2e^{-2t} \end{pmatrix} + \begin{pmatrix} e^{t} \\ e^{2t} \end{pmatrix}$$

(b) $\mathbf{y}' = \begin{pmatrix} 2 & 1 & -2 \\ 0 & 3 & -2 \\ 3 & 1 & -3 \end{pmatrix}\mathbf{y} + \begin{pmatrix} -2 \\ 0 \\ -4 \end{pmatrix}$

$$\mathbf{y} = c_1 \begin{pmatrix} e^{-t} \\ e^{-t} \\ 2e^{-t} \end{pmatrix} + c_2 \begin{pmatrix} e^{t} \\ e^{t} \\ e^{t} \end{pmatrix} + c_3 \begin{pmatrix} e^{2t} \\ 2e^{2t} \\ e^{2t} \end{pmatrix} + \begin{pmatrix} 1 \\ 2 \\ 3 \end{pmatrix}$$

8.3-3. Each of the following linear differential systems have lower triangular matrices. Observe that the solutions can be determined by solving the first equation for y_1, substituting for y_1 into the second equation and solving for y_2, substituting for y_1 and y_2 into the third equation and solving for y_3, etc. Determine the general solution for each of the following triangular systems:

(a) $\mathbf{y}' = \begin{pmatrix} 1 & 0 & 0 \\ 0 & -1 & 0 \\ 1 & 0 & 2 \end{pmatrix} \mathbf{y}$

(b) $\mathbf{y}' = \begin{pmatrix} 2 & 0 & 0 \\ 1 & 2 & 0 \\ 1 & 0 & 2 \end{pmatrix} \mathbf{y} + \begin{pmatrix} 0 \\ 0 \\ t \end{pmatrix}$

(c) $\mathbf{y}' = \begin{pmatrix} 1 & 0 & 0 \\ 1 & 1 & 0 \\ 1 & 1 & 1 \end{pmatrix} \mathbf{y}$

(d) $\mathbf{y}' = \begin{pmatrix} 0 & 0 & 0 & 0 \\ 0 & 2 & 0 & 0 \\ 1 & 0 & -2 & 0 \\ 0 & 1 & 0 & 2 \end{pmatrix} \mathbf{y} + \begin{pmatrix} 0 \\ 0 \\ 0 \\ 1 \end{pmatrix}$

8.4 SOLUTION COMPUTATION USING EIGENVALUES

The purpose of this section is to indicate a method of computing solutions to homogeneous systems with constant coefficients. This technique is the *eigenvalue-eigenvector method* and is an extension of the procedure using the auxiliary equation for second order equations with constant coefficients given in Chapter 2 (see also the higher-order equations treated in Chapter 7). Again $\mathbf{A}$ is the $n \times n$ matrix

$$\mathbf{A} = \begin{pmatrix} a_{11} & a_{12} & \cdots & a_{1n} \\ a_{21} & a_{22} & \cdots & a_{2n} \\ \cdots & \cdots & \cdots & \cdots \\ a_{n1} & a_{n2} & \cdots & a_{nn} \end{pmatrix}$$

and we consider the homogeneous system

$$\mathbf{y}' = \mathbf{A}\mathbf{y} \tag{1}$$

The problem is to determine n linearly independent solutions to (1) and hence obtain the general solution (see Theorem 8.3-1).

The procedure is to determine all solutions to (1) that are of the form

$$\mathbf{y}(t) = e^{\lambda t}\boldsymbol{\xi} \tag{2}$$

where λ is a number and $\boldsymbol{\xi}$ is a constant vector. Noting that $\mathbf{y}'(t) = \lambda e^{\lambda t}\boldsymbol{\xi}$ and $\mathbf{A}\mathbf{y}(t) = \mathbf{A}(e^{\lambda t}\boldsymbol{\xi}) = e^{\lambda t}\mathbf{A}\boldsymbol{\xi}$, it follows that $\mathbf{y}$ satisfies (1) only in case

$$\mathbf{y}'(t) - \mathbf{A}\mathbf{y}(t) = \lambda e^{\lambda t}\boldsymbol{\xi} - e^{\lambda t}\mathbf{A}\boldsymbol{\xi} = \mathbf{0}$$

Canceling the factor $e^{\lambda t}$ we see that the function $\mathbf{y}(t) = e^{\lambda t}\boldsymbol{\xi}$ is a solution to (1) if and only if

$$\mathbf{A}\boldsymbol{\xi} = \lambda\boldsymbol{\xi} \tag{3}$$

or equivalently,

$$(\mathbf{A} - \lambda\mathbf{I})\boldsymbol{\xi} = 0 \tag{3'}$$

where $\mathbf{I}$ is the $n \times n$ identity matrix [$\mathbf{I}$ is the matrix with 1 as each diagonal component and 0 for all other components—see (8) in Section 8.1]. If $\boldsymbol{\xi} = \mathbf{0}$ then (3) is satisfied for *every* number λ, and in this case $\mathbf{y}(t) \equiv \mathbf{0}$ for every λ. The interest is, of course, to obtain nontrivial solutions to (1), and hence we need to *determine those numbers λ such that (3) [or (3')] has a nontrivial solution $\boldsymbol{\xi}$.* A number λ such that (3) has a nonzero solution $\boldsymbol{\xi}$ is called an *eigenvalue* of $\mathbf{A}$, and any nonzero vector solution $\boldsymbol{\xi}$ to (3) is called an *eigenvector* of $\mathbf{A}$ corresponding to the eigenvalue λ. According to Theorem 8.2-2, equations (3) and (3') have nontrivial solutions $\boldsymbol{\xi}$ for a given λ if $\det(\mathbf{A} - \lambda\mathbf{I}) = 0$. Therefore, define

$$p_A(\lambda) \equiv \det(\mathbf{A} - \lambda\mathbf{I}) = \det\begin{pmatrix} a_{11} - \lambda & a_{12} & \cdots & a_{1n} \\ a_{21} & a_{22} - \lambda & \cdots & a_{2n} \\ \cdots & \cdots & \cdots & \cdots \\ a_{n1} & a_{n2} & \cdots & a_{nn} - \lambda \end{pmatrix} \tag{4}$$

The function $p_A(\lambda)$ is a polynomial in λ of degree n [in fact the leading term is $(-1)^n\lambda^n$] and is called the *characteristic polynomial* of $\mathbf{A}$. Therefore, the following important result is immediate.

□ **Lemma 1**

A number λ is an eigenvalue of $\mathbf{A}$ if and only if it is a zero of the characteristic polynomial p_A [that is, $p_A(\lambda) = 0$].

Since every polynomial has at least one root (which may be complex), we see that every matrix has at least one eigenvalue. In fact, since a polynomial has n roots, when counting the multiplicity of each distinct root, the following is true:

□ **Lemma 2**

Let $\lambda_1, \ldots, \lambda_k$ be the distinct eigenvalues of the $n \times n$ matrix $\mathbf{A}$, and for each $i \in \{1, \ldots, k\}$ let m_i be the multiplicity of the root $\lambda = \lambda_i$ of the equation $p_A(\lambda) = 0$. Then $m_1 + m_2 + \cdots + m_k = n$ [i.e., counting multiplicities, an $n \times n$ matrix has n eigenvalues].

As in the second order equations in Section 2.3, the analysis of the eigenvalue-eigenvector method divides naturally into three cases: real eigenvalues, complex conjugate eigenvalues, and multiple eigenvalues. We consider all three cases.

8.4a Real Eigenvalues

Note that if $\lambda = \lambda_0$ is a real eigenvalue of $\mathbf{A}$ then every nontrivial solution $\boldsymbol{\xi}$ to the system

$$
\begin{aligned}
&(a_{11} - \lambda_0)\xi_1 + a_{12}\xi_2 + \cdots + a_{1n}\xi_n = 0 \\
&a_{21}\xi_1 + (a_{22} - \lambda_0)\xi_2 + \cdots + a_{2n}\xi_n = 0 \\
&\vdots \\
&a_{n1}\xi_1 + a_{n2}\xi_2 + \cdots + (a_{nn} - \lambda_0)\xi_n = 0
\end{aligned} \tag{5}
$$

is an eigenvector corresponding to λ_0 [and hence $\mathbf{y}(t) = e^{\lambda_0 t}\boldsymbol{\xi}$ is a nontrivial solution to (1)]. For the simple types of systems considered here, the most effective way of determining nontrivial solutions to (5) (which are guaranteed to exist since the determinant of the system is zero by the definition of an eigenvalue) is by the Gauss reduction method [see (6), (7), and (8) as well as Example 8.2-7 in Section 8.2]. This is illustrated with two examples.

▶ **EXAMPLE 8.4-1**

Consider the 2×2 system

$$\mathbf{y}' = \begin{pmatrix} 1 & 2 \\ 2 & 1 \end{pmatrix} \mathbf{y} \tag{6}$$

In this case the characteristic polynomial p_A of the matrix in (6) has the form

$$p_A(\lambda) = \det \begin{pmatrix} 1 - \lambda & 2 \\ 2 & 1 - \lambda \end{pmatrix} = (1 - \lambda)^2 - 4 = \lambda^2 - 2\lambda - 3$$

Therefore, since the eigenvalues are solutions to

$$p_A(\lambda) = \lambda^2 - 2\lambda - 3 = (\lambda - 3)(\lambda + 1) = 0$$

we see that $\lambda = 3$ and $\lambda = -1$ are the eigenvalues of the matrix in (6). In order to compute an eigenvector corresponding to $\lambda = 3$, a nontrivial solution to

$$\begin{pmatrix} 1 & 2 \\ 2 & 1 \end{pmatrix} \begin{pmatrix} \xi_1 \\ \xi_2 \end{pmatrix} - 3 \begin{pmatrix} \xi_1 \\ \xi_2 \end{pmatrix} = \begin{pmatrix} 0 \\ 0 \end{pmatrix}$$

needs to be computed. Rewriting as an algebraic system and simplifying, we obtain

$$
\begin{aligned}
-2\xi_1 + 2\xi_2 &= 0 \\
2\xi_1 - 2\xi_2 &= 0
\end{aligned}
$$

Every solution to this system is of the form $\xi_1 = k$, $\xi_2 = k$, where k is an arbitrary constant. Taking $k = 1$ for definiteness, we have

$$\begin{pmatrix} 1 \\ 1 \end{pmatrix} \quad \text{is an eigenvector of} \quad \begin{pmatrix} 1 & 2 \\ 2 & 1 \end{pmatrix} \quad \text{corresponding to } \lambda = 3$$

Therefore,

$$\mathbf{y}_1(t) \equiv e^{3t}\begin{pmatrix} 1 \\ 1 \end{pmatrix}$$

is a solution to (6). Taking $\lambda = -1$ we obtain the system

$$\begin{pmatrix} 1 & 2 \\ 2 & 1 \end{pmatrix}\begin{pmatrix} \xi_1 \\ \xi_2 \end{pmatrix} - (-1)\begin{pmatrix} \xi_1 \\ \xi_2 \end{pmatrix} = \begin{pmatrix} 2\xi_1 + 2\xi_2 \\ 2\xi_1 + 2\xi_2 \end{pmatrix} = \begin{pmatrix} 0 \\ 0 \end{pmatrix}$$

As in the preceding case we find that $\xi_1 = 1$ and $\xi_2 = -1$ is a solution to this system. Therefore,

$$\begin{pmatrix} 1 \\ -1 \end{pmatrix} \quad \text{is an eigenvalue of} \quad \begin{pmatrix} 1 & 2 \\ 2 & 1 \end{pmatrix} \quad \text{corresponding to } \lambda = -1$$

and hence

$$\mathbf{y}_2(t) \equiv e^{-t}\begin{pmatrix} 1 \\ -1 \end{pmatrix}$$

is a solution to (6). Since

$$\det\,[\mathbf{y}_1(0)\mathbf{y}_2(0)] = \det\begin{pmatrix} 1 & 1 \\ 1 & -1 \end{pmatrix} = -2 \neq 0$$

we see that $\mathbf{y}_1$ and $\mathbf{y}_2$ are linearly independent, and hence

$$\mathbf{y}(t) = c_1 e^{3t}\begin{pmatrix} 1 \\ 1 \end{pmatrix} + c_2 e^{-t}\begin{pmatrix} 1 \\ -1 \end{pmatrix} = \begin{pmatrix} c_1 e^{3t} + c_2 e^{-t} \\ c_1 e^{3t} - c_2 e^{-t} \end{pmatrix}$$

where c_1 and c_2 are constants, is the general solution to (6) (see Lemma 2 in the preceding section).

▶ **EXAMPLE 8.4-2**

Consider the 3×3 system

$$\text{(7)} \qquad \mathbf{y}' = \begin{pmatrix} 0 & 1 & 0 \\ 0 & 0 & 1 \\ 1 & -1 & 1 \end{pmatrix}\mathbf{y}$$

The characteristic polynomial p_A of the matrix in (7) has the form

$$\begin{aligned} p_A(\lambda) &= \det \begin{pmatrix} -\lambda & 1 & 0 \\ 0 & -\lambda & 1 \\ 1 & -1 & 1-\lambda \end{pmatrix} \\ &= (-\lambda)(-\lambda)(1-\lambda) + 1 - (-\lambda)(-1)1 \\ &= -\lambda^3 + \lambda^2 - \lambda + 1 = (1-\lambda)(\lambda^2+1) \end{aligned}$$

Since the roots of $(1-\lambda)(\lambda^2+1) = 0$ are $\lambda = 1, \pm i$, the only real eigenvalue of the matrix in (7) is $\lambda = 1$. An eigenvector corresponding to $\lambda = 1$ is any nontrivial solution to the system

$$\begin{pmatrix} 0 & 1 & 0 \\ 0 & 0 & 1 \\ 1 & -1 & 1 \end{pmatrix} \begin{pmatrix} \xi_1 \\ \xi_2 \\ \xi_3 \end{pmatrix} - \begin{pmatrix} \xi_1 \\ \xi_2 \\ \xi_3 \end{pmatrix} = \begin{pmatrix} \xi_2 - \xi_1 \\ \xi_3 - \xi_2 \\ \xi_1 - \xi_2 \end{pmatrix} = \begin{pmatrix} 0 \\ 0 \\ 0 \end{pmatrix}$$

It is routine to check that $\xi_1 = \xi_2 = \xi_3 = 1$ is a solution and so

$$\begin{pmatrix} 1 \\ 1 \\ 1 \end{pmatrix} \quad \text{is an eigenvalue of} \quad \begin{pmatrix} 0 & 1 & 1 \\ 0 & 0 & 1 \\ 1 & -1 & 1 \end{pmatrix} \quad \text{corresponding to } \lambda = 1$$

Therefore,

$$\mathbf{y}_1(t) \equiv e^t \begin{pmatrix} 1 \\ 1 \\ 1 \end{pmatrix}$$

is a solution to (7). A procedure for using the complex roots $\pm i$ to compute two more solutions to $\mathbf{y}_2$ and $\mathbf{y}_3$ such that $\mathbf{y}_1, \mathbf{y}_2, \mathbf{y}_3$ are linearly independent is indicated in Example 8.4-7 in the next subsection.

A fundamental result of this section is the following theorem:

Theorem 8.4-1

Suppose that $\lambda_1, \lambda_2, \ldots, \lambda_k$, $k \le n$, are real and distinct eigenvalues of $\mathbf{A}$ and for each $i \in \{1, \ldots, k\}$ that $\boldsymbol{\xi}_i$ is an eigenvector corresponding to λ_i. Then the solutions

$$\mathbf{y}_1(t) \equiv e^{\lambda_1 t}\boldsymbol{\xi}_1, \mathbf{y}_2(t) \equiv e^{\lambda_2 t}\boldsymbol{\xi}_2, \ldots, \mathbf{y}_k(t) \equiv e^{\lambda_k t}\boldsymbol{\xi}_k$$

are linearly independent. In particular, if $k = n$ (and hence all n of the eigenvalues of $\mathbf{A}$ are real and distinct), then

$$\mathbf{y}(t) \equiv c_1 e^{\lambda_1 t}\boldsymbol{\xi}_1 + c_2 e^{\lambda_2 t}\boldsymbol{\xi}_2 + \cdots + c_n e^{\lambda_n t}\boldsymbol{\xi}_n$$

where $c_1, c_2, \ldots, c_n$ are constants, is the general solution to (1).

In order to establish the linear independence of the functions $\mathbf{y}_1, \ldots, \mathbf{y}_k$ it is sufficient to show that the vectors $\mathbf{y}_1(t), \ldots, \mathbf{y}_k(t)$ are linearly independent for some t. Taking $t = 0$ we have that $\mathbf{y}_1(0) = \xi_1, \ldots, \mathbf{y}_k(0) = \xi_k$. Thus if the eigenvectors $\xi_1, \xi_2, \ldots, \xi_k$ are linearly independent, so are the solutions $\mathbf{y}_1, \mathbf{y}_2, \ldots, \mathbf{y}_k$. Therefore, it suffices to show the following:

(8) If $\xi_1, \xi_2, \ldots, \xi_k$ are eigenvectors of $\mathbf{A}$ corresponding to distinct eigenvalues $\lambda_1, \lambda_2, \ldots, \lambda_k$, then the vectors $\xi_1, \xi_2, \ldots, \xi_k$ are linearly independent.

Assertion (8) is established using a contradiction argument. So suppose, for contradiction, that the vectors $\xi_1, \xi_2, \ldots, \xi_k$ are linearly dependent. By considering first ξ_1; then ξ_1, ξ_2; then ξ_1, ξ_2, ξ_3; etc. it follows that there is an integer $j < k$ such that $\xi_1, \xi_2, \ldots, \xi_j$ are linearly independent and

$$\xi_{j+1} = \alpha_1\xi_1 + \cdots + \alpha_j\xi_j \tag{9}$$

Applying the matrix $\mathbf{A}$ to each side of (9) and using the properties of linearity,

$$\mathbf{A}\xi_{j+1} = \alpha_1\mathbf{A}\xi_1 + \cdots + \alpha_j\mathbf{A}\xi_j$$

Therefore, by the definition of eigenvector,

$$\lambda_{j+1}\xi_{j+1} = \alpha_1\lambda_1\xi_1 + \cdots + \alpha_j\lambda_j\xi_j \tag{10}$$

Multiplying each side of (9) by λ_{j+1} and subtracting the resulting equation from (10), we obtain

$$\mathbf{0} = \alpha_1(\lambda_1 - \lambda_{j+1})\xi_1 + \cdots + \alpha_j(\lambda_j - \lambda_{j+1})\xi_j$$

But $\xi_1, \ldots, \xi_j$ are linearly independent by assumption, so

$$\alpha_1(\lambda_1 - \lambda_{j+1}) = \cdots = \alpha_j(\lambda_j - \lambda_{j+1}) = 0$$

Since $\lambda_i \neq \lambda_{j+1}$ for $i = 1, \ldots, j$ (i.e., the eigenvalues are distinct), it must be the case that $\alpha_1 = \cdots = \alpha_j = 0$. However, this implies from (9) that $\xi_{j+1} = \mathbf{0}$, which is impossible since eigenvectors are nonzero by definition. This contradiction shows that (8) is valid.

Since the eigenvalues of the 2×2 matrix in Example 8.4-1 are real and distinct, the two solutions obtained by the eigenvalue method must be linearly independent, and this was indeed shown to be the case. The next example is a three-dimensional system.

EXAMPLE 8.4-3

Consider the 3×3 matrix

$$\mathbf{y}' = \begin{pmatrix} 4 & 6 & 6 \\ 1 & 3 & 2 \\ -1 & -4 & -3 \end{pmatrix}\mathbf{y} \tag{11}$$

Since

$$P_A(\lambda) = \det \begin{pmatrix} 4-\lambda & 6 & 6 \\ 1 & 3-\lambda & 2 \\ -1 & -4 & -3-\lambda \end{pmatrix}$$

$$\begin{aligned} &= (4-\lambda)(3-\lambda)(-3-\lambda) + (-4)(6) + (-1)(2)(6) \\ &= (4-\lambda)(3-\lambda)(-3-\lambda) - 24 - 12 + (3-\lambda)6 + 8(4-\lambda) + (3+\lambda)6 \\ &= -\lambda^3 + \lambda + 4\lambda^2 - 4 = -(\lambda-1)(\lambda+1)(\lambda-4) \end{aligned}$$

Hence the eigenvalues of the 3×3 matrix $\mathbf{A}$ in (11) are $1, -1, 4$. Now we obtain the corresponding eigenvectors. For $\lambda = 1$,

$$(\mathbf{A} - \mathbf{I})\boldsymbol{\xi} = \begin{pmatrix} 3\xi_1 + 6\xi_2 + 6\xi_3 \\ \xi_1 + 2\xi_2 + 2\xi_3 \\ -\xi_1 - 4\xi_2 - 4\xi_3 \end{pmatrix} = \begin{pmatrix} 0 \\ 0 \\ 0 \end{pmatrix}$$

and hence the system

$$\begin{aligned} 3\xi_1 + 6\xi_2 + 6\xi_3 &= 0 \\ \xi_1 + 2\xi_2 + 2\xi_3 &= 0 \\ -\xi_1 - 4\xi_2 - 4\xi_3 &= 0 \end{aligned}$$

is obtained. Multiplying each side of the first equation by $-\frac{1}{3}$ and adding to the second, we have

$$\begin{aligned} 3\xi_1 + 6\xi_2 + 6\xi_3 &= 0 \\ 0 &= 0 \\ -\xi_1 - 4\xi_2 - 4\xi_3 &= 0 \end{aligned}$$

Setting $\xi_3 = k$ we obtain the 2×2 system

$$\begin{aligned} 3\xi_1 + 6\xi_2 &= -6k \\ -\xi_1 - 4\xi_2 &= 4k \end{aligned}$$

The solution is easily seen to be $\xi_1 = 0$, $\xi_2 = -k$. Setting $k = 1$ we have that

$$\begin{pmatrix} 0 \\ -1 \\ 1 \end{pmatrix} \text{ is an eigenvector of } \begin{pmatrix} 4 & 6 & 6 \\ 1 & 3 & 2 \\ -1 & -4 & -3 \end{pmatrix} \text{ corresponding to } \lambda = 1$$

In a similar manner it can be shown that

$$\begin{pmatrix} -6 \\ -2 \\ 7 \end{pmatrix} \text{ is an eigenvalue of } \begin{pmatrix} 4 & 6 & 6 \\ 1 & 3 & 2 \\ -1 & -4 & -3 \end{pmatrix} \text{ corresponding to } \lambda = -1$$

and that

$$\begin{pmatrix} 3 \\ 1 \\ -1 \end{pmatrix} \text{ is an eigenvector of } \begin{pmatrix} 4 & 6 & 6 \\ 1 & 3 & 2 \\ -1 & -4 & -3 \end{pmatrix} \text{ corresponding to } \lambda = 4$$

Applying Theorem 8.4-1,

$$\mathbf{y}(t) = c_1 e^t \begin{pmatrix} 0 \\ -1 \\ 1 \end{pmatrix} + c_2 e^{-t} \begin{pmatrix} -6 \\ -2 \\ 7 \end{pmatrix} + c_3 e^{4t} \begin{pmatrix} 3 \\ 1 \\ -1 \end{pmatrix}$$

is the general solution to (11).

8.4*b* Complex Eigenvalues

In this subsection we consider the case when λ is complex-valued. Since $\mathbf{A}$ has only real entries, the characteristic polynomial p_A has real coefficients, and hence complex roots occur in conjugate pairs. Therefore, it is assumed that α and β are real numbers with $\beta \neq 0$, and that $\lambda = \alpha \pm i\beta$ is a pair of conjugate eigenvalues of $\mathbf{A}$. Since λ is complex and $\mathbf{A}$ has only real components, it is clear that in order for equation (3) to possibly be satisfied, the eigenvector $\boldsymbol{\xi}$ must be allowed to have complex components. It is the purpose of this subsection to indicate the procedure for computing complex eigenvectors [i.e., solving equation (5) when λ_0 is complex] and also giving an appropriate interpretation of $e^{\lambda t}$ when λ is complex.

Recall that if α_1, α_2, β_1, and β_2 are real numbers and the imaginary number denoted i satisfies $i^2 = -1$, then the addition and multiplication of the complex numbers $\alpha_1 + i\beta$ and $\alpha_2 + i\beta_2$ can be computed as follows:

$$(\alpha_1 + i\beta_1) + (\alpha_2 + i\beta_2) = (\alpha_1 + \alpha_2) + i(\beta_1 + \beta_2)$$

and

$$\begin{aligned}(\alpha_1 + i\beta_1) \cdot (\alpha_2 + i\beta_2) &= \alpha_1(\alpha_2 + i\beta_2) + i\beta_1(\alpha_2 + i\beta_2) \\ &= \alpha_1\alpha_2 + i\alpha_1\beta_2 + \beta_1\alpha_2 i + i^2\beta_1\beta_2 \\ &= (\alpha_1\alpha_2 - \beta_1\beta_2) + i(\alpha_1\beta_2 + \alpha_2\beta_1)\end{aligned}$$

If $z = \alpha_1 + i\beta_1$ then α_1 is called the *real part* of z and β_1 the *imaginary part*, and the notations $\operatorname{Re} z = \alpha_1$ and $\operatorname{Im} z = \beta_1$ are frequently used. Moreover, the *conjugate of* z is $\alpha_1 - i\beta_1$ and the notation $\bar{z} = \alpha_1 - i\beta_1$ is used. The *absolute value* of z (or *modulus* of z) is denoted $|z|$ and defined by $|z| = \sqrt{\alpha_1^2 + \beta_1^2}$. Note that $|z|^2 = z \cdot \bar{z}$ and that z is zero only in case $|z| = 0$. If $z \neq 0$, then $1/z$ exists ($1/z$ is by definition the number w such that $wz = 1$), and we have that

$$\frac{1}{z} = \frac{\bar{z}}{|z|^2} = \frac{\alpha_1 - i\beta_1}{\alpha_1^2 + \beta_2^2}$$

The reciprocal $1/z$ is often denoted z^{-1}.

EXAMPLE 8.4-4

Determine all numbers z such that

$$(1 + i)z = 2 - 3i$$

Multiplying each side by

$$(1 + i)^{-1} = \frac{1 - i}{|1 + i|^2} = \frac{1}{2} - \frac{1}{2}i$$

implies that

$$z = (1 + i)^{-1}(2 - 3i) = (\tfrac{1}{2} - \tfrac{1}{2}i)(2 - 3i) = -\tfrac{1}{2} - \tfrac{5}{2}i$$

Vectors with complex components are also considered. Let $\xi_1, \ldots, \xi_n$ and $\eta_1, \ldots, \eta_n$ be real numbers and let $\mathbf{z}$ be the vector with jth component $\xi_j + i\eta_j$. Then

$$\mathbf{z} = \begin{pmatrix} \xi_1 + i\eta_1 \\ \xi_2 + i\eta_2 \\ \vdots \\ \xi_n + i\eta_n \end{pmatrix} = \begin{pmatrix} \xi_1 \\ \xi_2 \\ \vdots \\ \xi_n \end{pmatrix} + i \begin{pmatrix} \eta_1 \\ \eta_2 \\ \vdots \\ \eta_n \end{pmatrix} = \boldsymbol{\xi} + i\boldsymbol{\eta}$$

and so the vector $\boldsymbol{\xi}$ is called the *real part* of $\mathbf{z}$ and the vector $\boldsymbol{\eta}$ is called the *imaginary part* of $\mathbf{z}$. Notice that two complex vectors $\mathbf{z}$ and $\mathbf{w}$ are equal if and only if they have the same real part and the same imaginary part. These concepts can be applied in a straightforward manner to eigenvalues and eigenvectors.

EXAMPLE 8.4-5

Consider the 2×2 matrix

$$\mathbf{A} = \begin{pmatrix} 1 & 2 \\ -2 & 1 \end{pmatrix}$$

Then

$$p_A(\lambda) = \det \begin{pmatrix} 1 - \lambda & 2 \\ -2 & 1 - \lambda \end{pmatrix}$$

$$= (1 - \lambda)^2 + 4 = \lambda^2 - 2\lambda + 5$$

Since the roots of the quadratic equation $\lambda^2 - 2\lambda + 5 = 0$ are $\lambda = 1 \pm 2i$, we have that $\mathbf{A}$ has the complex conjugate eigenvalues $1 \pm 2i$. Using $\lambda = 1 + 2i$

we obtain the system

$$\begin{pmatrix} 1 & 2 \\ -2 & 1 \end{pmatrix}\begin{pmatrix} z_1 \\ z_2 \end{pmatrix} - (1 + 2i)\begin{pmatrix} z_1 \\ z_2 \end{pmatrix} = \begin{pmatrix} -2iz_1 - 2z_2 \\ -2z_1 - 2iz_2 \end{pmatrix} = \begin{pmatrix} 0 \\ 0 \end{pmatrix}$$

Therefore, we need to determine nontrivial solutions z_1 and z_2 to the system

$$-2iz_1 + 2z_2 = 0$$
$$-2z_1 - 2iz_2 = 0$$

Multiplying each side of the second equation by i shows that the second equation is a multiple of the first, and hence z_1 and z_2 must satisfy $-2iz_1 + 2z_2 = 0$. Therefore, if $z_2 = iz_1$ then $\mathbf{z}$ is an eigenvector. Taking $z_1 = 1$ and hence $z_2 = i$ we obtain

$$\begin{pmatrix} 1 \\ i \end{pmatrix} \quad \text{is an eigenvector of} \quad \begin{pmatrix} 1 & 2 \\ -2 & 1 \end{pmatrix} \quad \text{corresponding to } \lambda = 1 + 2i \tag{12}$$

In a similar manner it can be shown that

$$\begin{pmatrix} 1 \\ -i \end{pmatrix} \quad \text{is an eigenvector of} \quad \begin{pmatrix} 1 & 2 \\ -2 & 1 \end{pmatrix} \quad \text{corresponding to } \lambda = 1 - 2i \tag{13}$$

Notice that the eigenvector in (13) is the conjugate (componentwise) of the eigenvector in (12). This follows from the facts that the eigenvalues are complex conjugates and the matrix **A** is real and is true in general:

(14) If λ is a complex eigenvalue of the real matrix $\mathbf{A}$ and $\mathbf{z} = \boldsymbol{\xi} + i\boldsymbol{\eta}$ is a corresponding eigenvector, then $\mathbf{w} = \boldsymbol{\xi} - i\boldsymbol{\eta}$ is a corresponding eigenvector of the conjugate eigenvalue $\bar{\lambda}$.

To see that (14) is valid note that

$$\mathbf{A}(\boldsymbol{\xi} + i\boldsymbol{\eta}) = \lambda(\boldsymbol{\xi} + i\boldsymbol{\eta})$$

by definition. Using the fact that the conjugate of the product is the product of the conjugates, it follows that

$$\begin{aligned} \bar{\lambda}(\boldsymbol{\xi} - i\boldsymbol{\eta}) &= \overline{\lambda(\boldsymbol{\xi} + i\boldsymbol{\eta})} = \overline{\mathbf{A}(\boldsymbol{\xi} + i\boldsymbol{\eta})} \\ &= \overline{\mathbf{A}\boldsymbol{\xi} + i\mathbf{A}\boldsymbol{\eta}} = \mathbf{A}\boldsymbol{\xi} - i\mathbf{A}\boldsymbol{\eta} \\ &= \mathbf{A}(\boldsymbol{\xi} - i\boldsymbol{\eta}) \end{aligned}$$

and (14) is established.

Complex-valued functions of time t can be considered as functions into the plane, with the first component the real part and the second component the imaginary part. Moreover, if $\phi(t)$ is a complex-valued function with real part $\alpha(t)$ and imaginary part $\beta(t)$, we write $\phi(t) = \alpha(t) + i\beta(t)$ [instead of $\phi(t) = (\alpha(t), \beta(t))$]. Note that

$$\phi'(t) = \alpha'(t) + i\beta'(t) \qquad \text{and} \qquad \int_a^b \phi(t)\,dt = \int_a^b \alpha(t)\,dt + i\int_a^b \beta(t)\,dt \tag{15}$$

Our problem is to define $e^{\lambda t}$ ($= e^{\alpha t + i\beta t}$) so that the function $\mathbf{y}$ defined by (2) is a solution to (1). The crucial property that $e^{\lambda t}$ must have is that $d/dt\, e^{\lambda t} = \lambda e^{\lambda t}$. Using Euler's formula

$$e^{\lambda t} = e^{\alpha t + i\beta t} \equiv e^{\alpha t} \cos \beta t + i e^{\alpha t} \sin \beta t \tag{16}$$

we have, if $\beta = 0$, that this definition agrees with the real exponential function, and using (15) we have

$$\begin{aligned} \frac{d}{dt} e^{\alpha t + i\beta t} &= \frac{d}{dt}(e^{\alpha t} \cos \beta t) + i \frac{d}{dt}(e^{\alpha t} \sin \beta t) \\ &= \alpha e^{\alpha t} \cos \beta t - \beta e^{\alpha t} \sin \beta t + i(\alpha e^{\alpha t} \sin \beta t + \beta e^{\alpha t} \cos \beta t) \end{aligned}$$

Comparing this with

$$\begin{aligned} (\alpha + i\beta)e^{\alpha t + i\beta t} &= (\alpha + i\beta)(e^{\alpha t} \cos \beta t + i e^{\alpha t} \sin \beta t) \\ &= \alpha e^{\alpha t} \cos \beta t - \beta e^{\alpha t} \sin \beta t + i(\alpha e^{\alpha t} \sin \beta t + \beta e^{\alpha t} \cos \beta t) \end{aligned}$$

it follows that $d/dt\, e^{\lambda t} = \lambda e^{\lambda t}$. Therefore, the following result is valid:

(17) If $\lambda = \alpha + i\beta$ is an eigenvalue of $\mathbf{A}$ and $\mathbf{z} = \boldsymbol{\xi} + i\boldsymbol{\eta}$ is a corresponding eigenvector, then

$$\mathbf{y}(t) \equiv (\boldsymbol{\xi} + i\boldsymbol{\eta})e^{\alpha t + i\beta t}$$

is a complex-valued solution to (1).

What is interesting (and not quite so evident) is that we can obtain two linearly independent, *real vector-valued* solutions to (1) from the complex-valued solution in (17):

Theorem 8.4-2

Suppose that $\mathbf{A}$ has real components, that $\lambda = \alpha + i\beta$, $\beta \neq 0$, is an eigenvalue of $\mathbf{A}$, and that $\mathbf{z} = \boldsymbol{\xi} + i\boldsymbol{\eta}$ is a corresponding eigenvector. Define

$$\text{(18)} \quad \begin{cases} \mathbf{y}_1(t) \equiv \text{Re}\,[(\boldsymbol{\xi} + i\boldsymbol{\eta})e^{\alpha t + i\beta t}] \\ \text{and} \\ \mathbf{y}_2(t) \equiv \text{Im}\,[(\boldsymbol{\xi} + i\boldsymbol{\eta})e^{\alpha t + i\beta t}] \end{cases}$$

for all t. Then $\mathbf{y}_1$ and $\mathbf{y}_2$ are real vector-valued solutions to (1) and are linearly independent.

The solutions $\mathbf{y}_1$ and $\mathbf{y}_2$ defined by (18) can be written explicitly using (16):

$$\text{(18')} \quad \begin{cases} \mathbf{y}_1(t) = e^{\alpha t} (\cos \beta t)\boldsymbol{\xi} - e^{\alpha t} (\sin \beta t)\boldsymbol{\eta} \\ \text{and} \\ \mathbf{y}_2(t) = e^{\alpha t} (\sin \beta t)\boldsymbol{\xi} + e^{\alpha t} (\cos \beta t)\boldsymbol{\eta} \end{cases}$$

To see that $\mathbf{y}_1$ and $\mathbf{y}_2$ are solutions to (1), set $\mathbf{z}(t) = (\boldsymbol{\xi} + i\boldsymbol{\eta})e^{\alpha t + i\beta t}$ for all t and note that

$$\begin{aligned}\mathbf{y}_1'(t) + i\mathbf{y}_2'(t) &= \mathbf{z}'(t) = \mathbf{A}\mathbf{z}(t)\\ &= \mathbf{A}(\mathbf{y}_1(t) + i\mathbf{y}_2(t))\\ &= \mathbf{A}\mathbf{y}_1(t) + i\mathbf{A}\mathbf{y}_2(t)\end{aligned}$$

Equating the real and imaginary parts shows that both $\mathbf{y}_1$ and $\mathbf{y}_2$ are solutions. We omit the proof that $\mathbf{y}_1$ and $\mathbf{y}_2$ are linearly independent.

The procedure for computing $\mathbf{y}_1$ and $\mathbf{y}_2$ in simple cases is illustrated by two examples.

EXAMPLE 8.4-6

(See also Example 8.4-5.) Consider the 2×2 system

$$\mathbf{y}' = \begin{pmatrix} 1 & 2 \\ -2 & 1 \end{pmatrix}\mathbf{y} \tag{19}$$

As shown in Example 8.4-5, $\lambda = 1 \pm 2i$ are the eigenvalues of the matrix in (19), and according to (12) and (17),

$$\begin{aligned}\mathbf{y}(t) &= \begin{pmatrix} 1 \\ i \end{pmatrix} e^{t + i2t}\\ &= \left[\begin{pmatrix} 1 \\ 0 \end{pmatrix} + i\begin{pmatrix} 0 \\ 1 \end{pmatrix}\right](e^t \cos 2t + ie^t \sin 2t)\end{aligned}$$

is a complex-valued solution to (19). Upon multiplication,

$$\begin{aligned}\mathbf{y}(t) &= \left[\begin{pmatrix} 1 \\ 0 \end{pmatrix} e^t \cos 2t - \begin{pmatrix} 0 \\ 1 \end{pmatrix} e^t \sin 2t\right] + i\left[\begin{pmatrix} 0 \\ 1 \end{pmatrix} e^t \cos 2t + \begin{pmatrix} 1 \\ 0 \end{pmatrix} e^t \sin 2t\right]\\ &= \begin{pmatrix} e^t \cos 2t \\ -e^t \sin 2t \end{pmatrix} + i\begin{pmatrix} e^t \sin 2t \\ e^t \cos 2t \end{pmatrix}\end{aligned}$$

and it follows from Theorem 8.4-2 that

$$\mathbf{y}_1(t) \equiv \begin{pmatrix} e^t \cos 2t \\ -e^t \sin 2t \end{pmatrix} \quad \text{and} \quad \mathbf{y}_2(t) \equiv \begin{pmatrix} e^t \sin 2t \\ e^t \cos 2t \end{pmatrix}$$

are linearly independent, real-valued solutions to (19).

EXAMPLE 8.4-7

(See also Example 8.4-2.) Consider again the system (7):

$$\mathbf{y}' = \begin{pmatrix} 0 & 1 & 0 \\ 0 & 0 & 1 \\ 1 & -1 & 1 \end{pmatrix} \mathbf{y} \tag{7}$$

As shown in Example 8.4-2, $\lambda = 1, \pm i$ are the eigenvalues of the matrix in (7). In order to determine an eigenvalue corresponding to $\lambda = i$, consider the system

$$\begin{aligned} -iz_1 + z_2 \qquad\qquad &= 0 \\ -iz_2 + \qquad z_3 &= 0 \\ z_1 - z_2 + (1 - i)z_3 &= 0 \end{aligned}$$

Multiplying each side of the first equation by $-i$ and adding to the third implies that

$$\begin{aligned} -iz_1 + \qquad z_2 \qquad\qquad &= 0 \\ - \qquad iz_2 + \qquad z_3 &= 0 \\ -(1 + i)z_2 + (1 - i)z_3 &= 0 \end{aligned}$$

Multiplying each side of the second equation by $(1 + i)/(-i) = i - 1$ and adding to the third equation cancels the third equation, and the system

$$\begin{aligned} -iz_1 + z_2 \qquad &= 0 \\ -iz_2 + z_3 &= 0 \\ 0 &= 0 \end{aligned}$$

is obtained. Taking $z_3 = i$ implies that $z_2 = 1$ and $z_1 = -i$, and so

$$\begin{pmatrix} -i \\ 1 \\ i \end{pmatrix} \quad \text{is an eigenvector of} \quad \begin{pmatrix} 0 & 1 & 0 \\ 0 & 0 & 1 \\ 1 & -1 & 1 \end{pmatrix} \quad \text{corresponding to } \lambda = i$$

Therefore,

$$\begin{aligned} \mathbf{y}(t) &= \begin{pmatrix} -i \\ 1 \\ i \end{pmatrix} e^{it} = \left[\begin{pmatrix} 0 \\ 1 \\ 0 \end{pmatrix} + i \begin{pmatrix} -1 \\ 0 \\ 1 \end{pmatrix} \right] (\cos t + i \sin t) \\ &= \begin{pmatrix} \sin t \\ \cos t \\ -\sin t \end{pmatrix} + i \begin{pmatrix} -\cos t \\ \sin t \\ \cos t \end{pmatrix} \end{aligned}$$

is a complex-valued solution to (7), and hence

$$\mathbf{y}_1(t) = \begin{pmatrix} \sin t \\ \cos t \\ -\sin t \end{pmatrix} \quad \text{and} \quad \mathbf{y}_2(t) = \begin{pmatrix} -\cos t \\ \sin t \\ \cos t \end{pmatrix}$$

are linearly independent solutions to (7) by Theorem 8.4-2. Combining these two solutions with the one obtained in Example 8.4-2, it is easy to check that the family

$$\mathbf{y}(t) = c_1 \begin{pmatrix} e^t \\ e^t \\ e^t \end{pmatrix} + c_2 \begin{pmatrix} \sin t \\ \cos t \\ -\sin t \end{pmatrix} + c_3 \begin{pmatrix} -\cos t \\ \sin t \\ \cos t \end{pmatrix}$$

where c_1, c_2, and c_3 are constants, is the general solution to (7).

8.4c Multiple Eigenvalues

The case when **A** has an eigenvalue with multiplicity larger than 1 can be rather complicated, and so we indicate the main ideas involved in the computations only in some special cases. Recall that the multiplicity of an eigenvalue λ_0 of **A** is determined by the largest integer m_0 such that $(\lambda - \lambda_0)^{m_0}$ is a factor of the characteristic polynomial $p_A(\lambda)$ of **A**. Therefore, if

$$p_A(\lambda) = (\lambda - \lambda_0)^{m_0} q(\lambda) \qquad \text{for all } \lambda \tag{20}$$

where $q(\lambda)$ is a polynomial and $q(\lambda_0) \neq 0$, then λ_0 is an eigenvalue of **A** with multiplicity m_0. An eigenvalue of multiplicity m_0 must generate m_0 linearly independent solutions to the homogeneous system (1), and the concern here is to indicate the procedure for determining these solutions. Throughout this discussion it is assumed that λ_0 is an eigenvalue of **A** with multiplicity m_0.

The first problem is to determine all the eigenvectors corresponding to λ_0: that is, determine the solution set to

$$(\mathbf{A} - \lambda_0 \mathbf{I})\boldsymbol{\xi} = \mathbf{0} \tag{21}$$

The number of linearly independent eigenvectors satisfying (21) depends on **A** and can be *any* integer k in $\{1, \ldots, m_0\}$. If $\boldsymbol{\xi}_1, \ldots, \boldsymbol{\xi}_k$ are linearly independent vector solutions to (21), then

$$\mathbf{y}_1(t) \equiv e^{\lambda_0 t}\boldsymbol{\xi}_1, \ldots, \mathbf{y}_k(t) \equiv e^{\lambda_0 t}\boldsymbol{\xi}_k \tag{22}$$

are k linearly independent solutions to the differential equation (1), and

$$\mathbf{y}(t) \equiv e^{\lambda_0 t}(c_1 \boldsymbol{\xi}_1 + \cdots + c_k \boldsymbol{\xi}_k) \tag{23}$$

where $c_1, \ldots, c_k$ are constants, are all the solutions to (1) of the form $e^{\lambda_0 t}\boldsymbol{\xi}$ for some vector $\boldsymbol{\xi}$. Therefore, if $k = m_0$ the family (23) is *all* the solutions generated by the eigenvalue λ_0. This is the case in the following example.

▶ **EXAMPLE 8.4-8**

Consider the 3×3 system

$$\mathbf{y}' = \begin{pmatrix} 2 & -3 & 3 \\ 0 & 5 & -3 \\ 0 & 6 & -4 \end{pmatrix} \mathbf{y} \tag{24}$$

Then

$$p_A(\lambda) = \det \begin{pmatrix} 2-\lambda & -3 & 3 \\ 0 & 5-\lambda & -3 \\ 0 & 6 & -4-\lambda \end{pmatrix}$$

$$= (2-\lambda)[(5-\lambda)(-4-\lambda)+18]$$

$$= (2-\lambda)[\lambda^2-\lambda-2] = -(\lambda-2)^2(\lambda+1)$$

and hence the eigenvalues of the matrix in (24) are $\lambda = 2, 2, -1$. To find the eigenvectors corresponding to $\lambda = 2$, consider the system

$$\begin{pmatrix} 2 & -3 & 3 \\ 0 & 5 & -3 \\ 0 & 6 & -4 \end{pmatrix}\begin{pmatrix} \xi_1 \\ \xi_2 \\ \xi_3 \end{pmatrix} - 2\begin{pmatrix} \xi_1 \\ \xi_2 \\ \xi_3 \end{pmatrix} = \begin{pmatrix} -\xi_2 + 3\xi_3 \\ 3\xi_2 - 3\xi_3 \\ 6\xi_2 - 6\xi_3 \end{pmatrix} = \begin{pmatrix} 0 \\ 0 \\ 0 \end{pmatrix}$$

Therefore, the nontrivial solutions to the system

$$\begin{aligned} -3\xi_2 + 3\xi_3 &= 0 \\ 3\xi_2 - 3\xi_3 &= 0 \\ 6\xi_2 - 6\xi_3 &= 0 \end{aligned}$$

determine the eigenvectors corresponding to $\lambda = 2$. Since these equations are multiples of each other and ξ_1 does not appear, we can let h and k be arbitrary constants and take $\xi_1 = h$, $\xi_2 = k$, $\xi_3 = k$. Therefore

$$\begin{pmatrix} h \\ k \\ k \end{pmatrix}$$

is an eigenvector corresponding to 2 for every h and k. Taking first $h = 1$, $k = 0$ and then $h = 0$, $k = 1$, we see that

$$\mathbf{y}_1(t) = e^{2t}\begin{pmatrix} 1 \\ 0 \\ 0 \end{pmatrix} \qquad \text{and} \qquad \mathbf{y}_2(t) = e^{2t}\begin{pmatrix} 0 \\ 1 \\ 1 \end{pmatrix}$$

is a pair of linearly independent solutions to (1) generated by $\lambda = 2$ and its eigenvectors. There is a third independent solution generated by the eigenvalue $\lambda = -1$ and its eigenvector (see Problem 8.4-4*a*).

The difficulty with the multiple eigenvalue λ_0 occurs if there are not m_0 linearly independent eigenvectors corresponding to λ_0 [that is, $k < m_0$ in (22)]. A procedure analogous to the case when the auxiliary equation for a second order equation has a double root also applies for systems. In particular, if

$k < m_0$ in (22) then there is a solution to (1) of the form

$$\mathbf{y}(t) \equiv e^{\lambda_0 t}(\boldsymbol{\xi} + t\boldsymbol{\eta}) \tag{25}$$

Not only is there a solution of the form (25), but the vectors $\boldsymbol{\xi}$ and $\boldsymbol{\eta}$ can be computed by solving a system of equations involving $(\mathbf{A} - \lambda_0 \mathbf{I})^2$ [recall that $(\mathbf{A} - \lambda_0 \mathbf{I})^2 = (\mathbf{A} - \lambda_0 \mathbf{I})(\mathbf{A} - \lambda_0 \mathbf{I})$]. The basic result is the following theorem:

■ **Theorem 8.4-3**

Suppose that λ_0 is an eigenvalue of multiplicity $m_0 > 1$ and that the number k of linearly independent eigenvectors corresponding to λ_0 is less than m_0. Then there is at least one nontrivial vector $\boldsymbol{\xi}$ such that

$$(\mathbf{A} - \lambda_0 \mathbf{I})^2\boldsymbol{\xi} = \mathbf{0} \qquad \text{and} \qquad (\mathbf{A} - \lambda_0 \mathbf{I})\boldsymbol{\xi} \neq \mathbf{0} \tag{26}$$

Moreover, if $\boldsymbol{\xi}$ is any such vector satisfying (26), then

$$\mathbf{y}(t) \equiv e^{\lambda_0 t}[\boldsymbol{\xi} + t(\mathbf{A} - \lambda_0 \mathbf{I})\boldsymbol{\xi}] \tag{27}$$

is a solution to (1) that is linearly independent from every solution of the form (23).

We show only that if $\boldsymbol{\xi}$ satisfies (26) and $\mathbf{y}$ is defined by (27) then $\mathbf{y}$ is a solution to (1). This actually follows by direct computation. For

$$\begin{aligned}\mathbf{y}'(t) &= e^{\lambda_0 t}(\mathbf{A} - \lambda_0 \mathbf{I})\boldsymbol{\xi} + \lambda_0 e^{\lambda_0 t}[\boldsymbol{\xi} + t(\mathbf{A} - \lambda_0 \mathbf{I})\boldsymbol{\xi}] \\ &= e^{\lambda_0 t}(\mathbf{A} - \lambda_0 \mathbf{I})\boldsymbol{\xi} + \lambda_0 \mathbf{y}(t)\end{aligned}$$

and

$$\begin{aligned}\mathbf{A}\mathbf{y}(t) &= (\mathbf{A} - \lambda_0 \mathbf{I})\mathbf{y}(t) + \lambda_0 \mathbf{I}\mathbf{y}(t) \\ &= (\mathbf{A} - \lambda_0 \mathbf{I})\{e^{\lambda_0 t}[\boldsymbol{\xi} + t(\mathbf{A} - \lambda_0 \mathbf{I})\boldsymbol{\xi}]\} + \lambda_0 \mathbf{y}(t) \\ &= e^{\lambda_0 t}(\mathbf{A} - \lambda_0 \mathbf{I})\boldsymbol{\xi} + e^{\lambda_0 t}t(\mathbf{A} - \lambda_0 \mathbf{I})^2\boldsymbol{\xi} + \lambda_0 \mathbf{y}(t) \\ &= e^{\lambda_0 t}(\mathbf{A} - \lambda_0 \mathbf{I})\boldsymbol{\xi} + \lambda_0 \mathbf{y}(t)\end{aligned}$$

Comparing, we have $\mathbf{y}'(t) = \mathbf{A}\mathbf{y}(t)$ and so $\mathbf{y}$ is a solution to (1).

▶ **EXAMPLE 8.4-9**

Consider the 3×3 system

$$\mathbf{y}' = \begin{pmatrix} -6 & -7 & -13 \\ 5 & 6 & 9 \\ 2 & 2 & 5 \end{pmatrix}\mathbf{y} \tag{28}$$

Then

$$p_A(\lambda) = \det\begin{pmatrix} -6-\lambda & -7 & -13 \\ 5 & 6-\lambda & 9 \\ 2 & 2 & 5-\lambda \end{pmatrix}$$

$$\begin{aligned} &= (-6-\lambda)(6-\lambda)(5-\lambda) - 130 - 126 \\ &\quad + 35(5-\lambda) + 26(6-\lambda) + 18(6+\lambda) \\ &= -(\lambda^3 - 5\lambda^2 - 36\lambda + 180) + 183 - 43\lambda \\ &= -\lambda^3 + 5\lambda^2 - 7\lambda + 3 = -(\lambda-1)^2(\lambda-3) \end{aligned}$$

and so the eigenvalues of the matrix in (28) are $\lambda = 1, 1, 3$. To determine the eigenvalues corresponding to $\lambda = 1$, consider the system

$$\begin{aligned} -7\xi_1 - 7\xi_2 - 13\xi_3 &= 0 \\ 5\xi_1 + 5\xi_2 + 9\xi_3 &= 0 \\ 2\xi_1 + 2\xi_2 + 4\xi_3 &= 0 \end{aligned}$$

Multiplying each side of the first equation by $\frac{5}{7}$ and adding to the second, and then multiplying each side of the first equation by $\frac{2}{7}$ and adding to the third, the system

$$\begin{aligned} -7\xi_1 - 7\xi_2 - 13\xi_3 &= 0 \\ -\tfrac{2}{7}\xi_3 &= 0 \\ \tfrac{2}{7}\xi_3 &= 0 \end{aligned}$$

is obtained. Hence $\xi_3 = 0$ and $\xi_1 = -\xi_2$, and it follows that

$$\begin{pmatrix} 1 \\ -1 \\ 0 \end{pmatrix} \text{ is an eigenvector of } \begin{pmatrix} -6 & -7 & -13 \\ 5 & 6 & 9 \\ 2 & 2 & 5 \end{pmatrix} \text{ corresponding to } \lambda = 1 \tag{29}$$

and there is no eigenvector linearly independent from the one in (29) that corresponds to $\lambda = 1$. According to (26) in Theorem 8.4-3, the system

$$\begin{pmatrix} -7 & -7 & -13 \\ 5 & 5 & 9 \\ 2 & 2 & 3 \end{pmatrix}^2 \begin{pmatrix} \xi_1 \\ \xi_2 \\ \xi_3 \end{pmatrix} = \begin{pmatrix} 0 \\ 0 \\ 0 \end{pmatrix}$$

should be considered. Since

$$\begin{pmatrix} -7 & -7 & -13 \\ 5 & 5 & 9 \\ 2 & 2 & 4 \end{pmatrix}^2 = \begin{pmatrix} -12 & -12 & -24 \\ 8 & 8 & 16 \\ 4 & 4 & 8 \end{pmatrix}$$

the system

$$\begin{aligned} -12\xi_1 - 12\xi_2 - 24\xi_3 &= 0 \\ 8\xi_1 + 8\xi_2 + 16\xi_3 &= 0 \\ 4\xi_1 + 4\xi_2 + 8\xi_3 &= 0 \end{aligned}$$

is obtained. All three of these equations are equivalent to the single equation

$$\xi_1 + \xi_2 + 2\xi_3 = 0$$

This equation has two linearly independent vector solutions:

$$\begin{pmatrix} 1 \\ -1 \\ 0 \end{pmatrix} \quad \text{and} \quad \begin{pmatrix} 1 \\ 1 \\ -1 \end{pmatrix}$$

The first vector is an eigenvector corresponding to $\lambda = 1$, and according to (27) in Theorem 8.4-3, a solution to (28) is

$$\mathbf{y}(t) = e^t \left[\begin{pmatrix} 1 \\ 1 \\ -1 \end{pmatrix} + t \begin{pmatrix} -7 & -7 & -13 \\ 5 & 5 & 9 \\ 2 & 2 & 4 \end{pmatrix} \begin{pmatrix} 1 \\ 1 \\ -1 \end{pmatrix} \right]$$

$$= e^t \begin{pmatrix} 1-t \\ 1+t \\ -1 \end{pmatrix}$$

Therefore,

$$\mathbf{y}_1(t) = e^t \begin{pmatrix} 1 \\ -1 \\ 0 \end{pmatrix} \quad \text{and} \quad \mathbf{y}_2(t) = e^t \begin{pmatrix} 1-t \\ 1+t \\ -1 \end{pmatrix}$$

are two linearly independent solutions to (28) generated by the double eigenvalue $\lambda = 1$. A third independent solution is generated by the eigenvalue $\lambda = 3$ and its eigenvector (see Problem 8.4-4*b*).

If the eigenvalue λ_0 has multiplicity $m_0 \geq 2$, then the equation $(\mathbf{A} - \lambda_0 \mathbf{I})^2 \boldsymbol{\xi} = \mathbf{0}$ has at least two linearly independent solutions (notice that any eigenvector corresponding to λ_0 is a solution to this equation). Therefore, if $m_0 = 2$, both solutions generated by λ_0 can be obtained from solving $(\mathbf{A} - \lambda_0 \mathbf{I})^2 \boldsymbol{\xi} = \mathbf{0}$ [or perhaps only the system $(\mathbf{A} - \lambda_0 \mathbf{I})\boldsymbol{\xi} = \mathbf{0}$]. However, if $m_0 > 2$ then the solutions to $(\mathbf{A} - \lambda_0 \mathbf{I})^2 \boldsymbol{\xi} = \mathbf{0}$ may or may not give all m_0 solutions generated by λ_0. In particular, there may be a solution of the form

$$\mathbf{y}(t) = e^{\lambda_0 t}(\boldsymbol{\xi} + t\boldsymbol{\eta} + t^2\boldsymbol{\xi}) \tag{30}$$

It is routine to check that if $m_0 > 2$ and if $\boldsymbol{\xi}$ is a vector such that

$$(\mathbf{A} - \lambda_0 \mathbf{I})^3 \boldsymbol{\xi} = \mathbf{0} \quad \text{and} \quad (\mathbf{A} - \lambda_0 \mathbf{I})^2 \boldsymbol{\xi} \neq \mathbf{0} \tag{31}$$

then there is a solution to (1) of the form

$$\mathbf{y}(t) = e^{\lambda_0 t}\left[\boldsymbol{\xi} + t(\mathbf{A} - \lambda_0 \mathbf{I})\boldsymbol{\xi} + \frac{t^2}{2}(\mathbf{A} - \lambda_0 \mathbf{I})^2 \boldsymbol{\xi} \right] \tag{32}$$

If the solution vectors to (21), (26), and (31) do not generate m_0 linearly independent solutions to (1) [note that it must be the case that $m_0 > 3$ if the solutions of the form (23), (27), and (32) do not exhaust all solutions generated by λ_0], then the equation

$$(\mathbf{A} - \lambda_0 \mathbf{I})^4 \boldsymbol{\xi} = \mathbf{0} \qquad (\mathbf{A} - \lambda_0 \mathbf{I})^3 \boldsymbol{\xi} \neq \mathbf{0}$$

must be considered in order to determine a solution having the form

$$\mathbf{y}(t) = e^{\lambda_0 t}\left[\boldsymbol{\xi} + t(\mathbf{A} - \lambda_0 \mathbf{I})\boldsymbol{\xi} + \frac{t^2}{2!}(\mathbf{A} - \lambda_0 \mathbf{I})^2\boldsymbol{\xi} + \frac{t^3}{3!}(\mathbf{A} - \lambda_0 \mathbf{I})^3\boldsymbol{\xi}\right]$$

This process can be continued until all m_0 solutions generated by λ_0 are found. Also, this process is guaranteed to terminate in at most m_0 steps. In particular, the highest power of t that can possibly occur in a solution generated by λ_0 is $t^{m_0 - 1}$.

PROBLEMS

8.4-1. Compute the characteristic polynomial and the eigenvalues for the matrix in each of the following homogeneous systems. Moreover, for each real eigenvalue λ_0, determine all solutions of the form $\mathbf{y}(t) = e^{\lambda_0 t}\boldsymbol{\xi}$, where $\boldsymbol{\xi}$ is an appropriate vector.

(a) $\mathbf{y}' = \begin{pmatrix} -4 & 2 \\ -3 & 1 \end{pmatrix}\mathbf{y}$

(b) $\mathbf{y}' = \begin{pmatrix} -1 & 4 \\ 4 & -1 \end{pmatrix}\mathbf{y}$

(c) $\mathbf{y}' = \begin{pmatrix} -3 & -7 & -5 \\ 2 & 4 & 3 \\ 1 & 2 & 2 \end{pmatrix}\mathbf{y}$

(d) $\mathbf{y}' = \begin{pmatrix} 0 & 1 & 0 \\ 0 & 0 & 1 \\ 4 & -4 & 1 \end{pmatrix}\mathbf{y}$

(e) $\mathbf{y}' = \begin{pmatrix} -1 & 1 & 0 \\ 1 & -1 & 1 \\ 0 & 1 & -1 \end{pmatrix}\mathbf{y}$

(f) $\mathbf{y}' = \begin{pmatrix} 2 & 3 \\ -3 & 2 \end{pmatrix}\mathbf{y}$

8.4-2. Compute the characteristic polynomial and the eigenvalues for the matrix in each of the following homogeneous systems. Moreover, for each pair of complex conjugate eigenvalues $\lambda_0 = \alpha \pm i\beta$, determine two linearly independent real-valued solutions.

(a) $\mathbf{y}' = \begin{pmatrix} -1 & -4 \\ 4 & -1 \end{pmatrix}\mathbf{y}$

(b) $\mathbf{y}' = \begin{pmatrix} 0 & 1 & 0 \\ 0 & 0 & 1 \\ 4 & -4 & 1 \end{pmatrix}\mathbf{y}$ (See Problem 8.4-1*d*.)

(c) $\mathbf{y}' = \begin{pmatrix} 10 & 10 & 12 \\ -6 & -6 & -7 \\ -2 & -2 & -2 \end{pmatrix}\mathbf{y}$

8.4-3. Determine the general solution to each of the following homogeneous differential equations.

(a) $\mathbf{y}' = \begin{pmatrix} 1 & 1 & 1 \\ 0 & 2 & 2 \\ 0 & 0 & 3 \end{pmatrix} \mathbf{y}$

(b) $\mathbf{y}' = \begin{pmatrix} 0 & 1 & 0 \\ 0 & 0 & 1 \\ 4 & -4 & 1 \end{pmatrix} \mathbf{y}$ (See Problems 8.4-1*d* and 8.4-2*b*.)

(c) $\mathbf{y}' = \begin{pmatrix} 0 & 1 & 0 \\ 0 & 0 & 1 \\ 1 & 1 & -1 \end{pmatrix} \mathbf{y}$

(d) $\mathbf{y}' = \begin{pmatrix} 10 & 10 & 12 \\ -6 & -6 & -7 \\ -2 & -2 & -2 \end{pmatrix} \mathbf{y}$ (See Problem 8.4-2*c*.)

(e) $\mathbf{y}' = \begin{pmatrix} -3 & -7 & -5 \\ 2 & 4 & 3 \\ 1 & 2 & 2 \end{pmatrix} \mathbf{y}$ (See Problem 8.4-1*c*.)

(f) $\mathbf{y}' = \begin{pmatrix} 1 & 1 & 1 & 1 \\ 1 & 1 & 1 & 1 \\ 1 & 1 & 1 & 1 \\ 1 & 1 & 1 & 1 \end{pmatrix} \mathbf{y}$

(g) $\mathbf{y}' = \begin{pmatrix} -2 & 1 & 0 & 0 \\ -1 & -2 & 0 & 0 \\ 0 & 0 & 0 & 1 \\ 0 & 0 & -1 & 2 \end{pmatrix} \mathbf{y}$

(h) $\mathbf{y}' = \begin{pmatrix} 3 & 2 & 1 \\ -3 & -1 & -1 \\ 1 & 0 & 1 \end{pmatrix} \mathbf{y}$

8.4.4. (a) Determine the general solution to equation (24) in Example 8.4-8.

(b) Determine the general solution to equation (28) in Example 8.4-9.

*8.5 THE MATRIX EXPONENTIAL

In previous sections solutions to linear differential systems are described in terms of column vector–valued functions. However, in this section solutions are described in terms of matrix-valued functions, and these ideas are related to the notion of an exponential function for matrices. Suppose that for each pair i, j of integers in $\{1, \ldots, n\}$, y_{ij} is a function on $(-\infty, \infty)$. Now for each t

let $\mathbf{Y}(t)$ denote the $n \times n$ matrix whose ijth component is $y_{ij}(t)$: that is

$$\text{(1)} \qquad \mathbf{Y}(t) = \begin{pmatrix} y_{11}(t) & y_{12}(t) & \cdots & y_{1n}(t) \\ y_{21}(t) & y_{22}(t) & \cdots & y_{2n}(t) \\ \cdots & \cdots & \cdots & \cdots \\ y_{n1}(t) & y_{n2}(t) & \cdots & y_{nn}(t) \end{pmatrix}$$

or, more briefly, $\mathbf{Y}(t) = (y_{ij}(t))$. Then $\mathbf{Y}$ is called an $n \times n$ *matrix–valued function* on $(-\infty, \infty)$. The matrix-valued function $\mathbf{Y}$ is said to be *continuous* if every component y_{ij} is continuous and is said to be differentiable if every component y_{ij} is differentiable. Moreover, if $\mathbf{Y}$ is differentiable, then

$$\text{(2)} \qquad \mathbf{Y}'(t) \equiv \begin{pmatrix} y'_{11}(t) & y'_{12}(t) & \cdots & y'_{1n}(t) \\ y'_{21}(t) & y'_{22}(t) & \cdots & y'_{2n}(t) \\ \cdots & \cdots & \cdots & \cdots \\ y'_{n1}(t) & y'_{n2}(t) & \cdots & y'_{nn}(t) \end{pmatrix}$$

If both $\mathbf{X} = (x_{ij})$ and $\mathbf{Y} = (y_{ij})$ are differentiable functions on $(-\infty, \infty)$ and $\mathbf{Z}(t) \equiv \mathbf{X}(t) \cdot \mathbf{Y}(t)$, then $\mathbf{Z}(t)$ is also a differentiable function on $(-\infty, \infty)$ and the product rule for differentiation holds:

$$\text{(3)} \qquad \frac{d}{dt}[\mathbf{X}(t)\mathbf{Y}(t)] = \mathbf{X}'(t)\mathbf{Y}(t) + \mathbf{X}(t)\mathbf{Y}'(t)$$

The formula (3) can be established using the corresponding property for real-valued functions and the formula for matrix multiplication [see (7) in Section 8.1].

The type of equation under consideration in this section has the form

$$\text{(4)} \qquad \mathbf{Y}' = \mathbf{A}\mathbf{Y}$$

and is called a *first order matrix differential system.* A differentiable matrix-valued function $\mathbf{Y}$ on $(-\infty, \infty)$ is said to be a solution to (4) if $\mathbf{Y}'(t) = \mathbf{A} \cdot \mathbf{Y}(t)$ for all t. Note that if $\mathbf{Y}(t) \equiv 0$ on $(-\infty, \infty)$, then $\mathbf{Y}$ is a solution to (4) and is called the *trivial solution.*

EXAMPLE 8.5-1

Consider the matrix system

$$\text{(5)} \qquad \mathbf{Y}' = \begin{pmatrix} 1 & -3 \\ 2 & -4 \end{pmatrix}\mathbf{Y}$$

Define

$$\mathbf{Y}(t) = \begin{pmatrix} e^{-2t} & 3e^{-t} \\ e^{-2t} & 2e^{-t} \end{pmatrix}$$

for all t and note that if $\mathbf{A}$ is the 2×2 matrix in (5) then

$$\mathbf{AY}(t) = \begin{pmatrix} 1 & -3 \\ 2 & -4 \end{pmatrix} \begin{pmatrix} e^{-2t} & 3e^{-t} \\ e^{-2t} & 2e^{-t} \end{pmatrix} = \begin{pmatrix} -2e^{-2t} & -3e^{-t} \\ -2e^{-2t} & -2e^{-t} \end{pmatrix}$$

and

$$\mathbf{Y}'(t) = \begin{pmatrix} -2e^{-2t} & -3e^{-t} \\ -2e^{-2t} & 2e^{-t} \end{pmatrix}$$

Therefore, $\mathbf{Y}'(t) = \mathbf{AY}(t)$ and so $\mathbf{Y}$ is a solution to (5).

It is extremely helpful to consider the n vector-valued functions determined by the columns of the matrix-valued function $\mathbf{Y}$. So set

$$\mathbf{y}_1(t) \equiv \begin{pmatrix} y_{11}(t) \\ y_{21}(t) \\ \vdots \\ y_{n1}(t) \end{pmatrix}, \mathbf{y}_1(t) \equiv \begin{pmatrix} y_{12}(t) \\ y_{22}(t) \\ \vdots \\ y_{n2}(t) \end{pmatrix}, \ldots, \mathbf{y}_n(t) = \begin{pmatrix} y_{1n}(t) \\ y_{2n}(t) \\ \vdots \\ y_{nn}(t) \end{pmatrix}$$

Then

$$\mathbf{Y}(t) = (\mathbf{y}_1(t), \mathbf{y}_2(t), \ldots, \mathbf{y}_n(t)) \tag{6}$$

and

$$\mathbf{Y}'(t) = (\mathbf{y}'_1(t), \mathbf{y}'_2(t), \ldots, \mathbf{y}'_n(t)) \tag{7}$$

[see the notations (9) and (10) in Section 8.1]. Moreover, by formula (11) in Section 8.1,

$$\mathbf{AY}(t) = (\mathbf{Ay}_1(t), \mathbf{Ay}_2(t), \ldots, \mathbf{Ay}_n(t)) \tag{8}$$

Comparing formulas (7) and (8) gives immediately the following lemma:

□ **Lemma 1**

The matrix-valued function $\mathbf{Y} = (\mathbf{y}_1, \mathbf{y}_2, \ldots, \mathbf{y}_n)$ is a solution to (4) if and only if each one of the vector-valued functions $\mathbf{y}_1, \mathbf{y}_2, \ldots, \mathbf{y}_n$ is a solution to the vector system

$$\mathbf{y}' = \mathbf{Ay} \tag{9}$$

The problem now is to determine how to generate all the solutions to the matrix differential equation (4). It follows from Theorem 8.3-1 that the n vector-valued functions determined by the columns of $\mathbf{Y}$ generate the general solution to (9) if and only if they are linearly independent. Therefore, we make the following definition: A matrix solution $\mathbf{Y}$ to (4) is said to be a *fundamental matrix solution* if det $\mathbf{Y}(0) \neq 0$ [where det $\mathbf{Y}(0)$ is the determinant of the matrix $\mathbf{Y}(0)$]. According to Lemma 2 in Section 8.3, $\mathbf{Y} = (\mathbf{y}_1, \mathbf{y}_2, \ldots, \mathbf{y}_n)$ is a fundamental matrix solution to (4) if and only if the set of linear combinations

$$\mathbf{y} = c_1 \mathbf{y}_1 + \cdots + c_n \mathbf{y}_n \qquad c_1, \ldots, c_n \text{ are constants} \tag{10}$$

is the general solution to the vector system (9). If

$$\mathbf{c} \equiv \begin{pmatrix} c_1 \\ c_2 \\ \vdots \\ c_n \end{pmatrix}$$

it follows from the definition of multiplication of a matrix with a vector that

$$\begin{aligned} \mathbf{Y}(t)\mathbf{c} &= (\mathbf{y}_1(t), \mathbf{y}_2(t), \ldots, \mathbf{y}_n(t))\mathbf{c} \\ &= \mathbf{y}_1(t)c_1 + \mathbf{y}_2(t)c_2 + \cdots + \mathbf{y}_n(t)c_n \end{aligned}$$

Therefore, the following is valid:

(10′) The family of all $\mathbf{y}(t) \equiv \mathbf{Y}(t)\mathbf{c}$, where $\mathbf{c}$ is a constant vector, is precisely the same as the family (10).

Note in particular that if $\mathbf{Y}$ is a solution to (4) and $\mathbf{c}$ is a vector, then $\mathbf{y}(t) \equiv \mathbf{Y}(t)\mathbf{c}$ is a solution to the vector system (9). Combining the preceding comments gives the following important result:

■ **Theorem 8.5-1**

Suppose that $\mathbf{Y} = (\mathbf{y}_1, \mathbf{y}_2, \ldots, \mathbf{y}_n)$ is a matrix solution to (4). Then any one of the following statements implies the other three.

(i) $\mathbf{Y}$ is a fundamental matrix solution [i.e., det $\mathbf{Y}(0) \neq 0$].

(ii) The column functions $\mathbf{y}_1, \mathbf{y}_2, \ldots, \mathbf{y}_n$ are linearly independent.

(iii) The family of vector functions $\mathbf{y}(t) \equiv \mathbf{Y}(t)\mathbf{c}$, where $\mathbf{c}$ is a constant vector, is the general solution to (9).

(iv) det $\mathbf{Y}(t) \neq 0$ for all t.

EXAMPLE 8.5-2

Consider the 3×3 matrix system

$$\mathbf{Y}' = \begin{pmatrix} 4 & 6 & 6 \\ 1 & 3 & 2 \\ -1 & -4 & -3 \end{pmatrix} \mathbf{Y} \tag{11}$$

(see Example 8.4-3). A matrix solution to (11) is

$$\mathbf{Y}(t) = \begin{pmatrix} 0 & -6e^{-t} & 3e^{4t} \\ -e^t & -2e^{-t} & e^{4t} \\ e^t & 7e^{-t} & -e^{4t} \end{pmatrix}$$

(again see Example 8.4-3). Since

$$\det \mathbf{Y}(0) = \det \begin{pmatrix} 0 & -6 & 3 \\ -1 & -2 & 1 \\ 1 & 7 & -1 \end{pmatrix} = -21 - 6 + 6 + 6 = -15 \neq 0$$

it follows from Theorem 8.5-1 that $\mathbf{Y}$ is a fundamental matrix solution and that every solution $\mathbf{y}$ to the system

$$\mathbf{y}' = \begin{pmatrix} 4 & 6 & 6 \\ 1 & 3 & 2 \\ -1 & -4 & -3 \end{pmatrix} \mathbf{y} \tag{12}$$

is of the form

$$\mathbf{y}(t) = \begin{pmatrix} 0 & -6e^{-t} & 3e^{4t} \\ -e^t & -2e^{-t} & e^{4t} \\ e^t & 7e^{-t} & -e^{4t} \end{pmatrix} \begin{pmatrix} c_1 \\ c_2 \\ c_3 \end{pmatrix} \tag{13}$$

where c_1, c_2, and c_3 are constants.

Consider now the initial value problem associated with the vector system (9):

$$\mathbf{y}' = \mathbf{A}\mathbf{y} \qquad \mathbf{y}(0) = \boldsymbol{\eta} \tag{14}$$

If a fundamental matrix solution $\mathbf{Y}$ to (4) is known, then the solution to (14) must be of the form $\mathbf{y}(t) = \mathbf{Y}(t)\mathbf{c}$ for some constant vector $\mathbf{c}$. Since $\mathbf{y}(0) = \mathbf{Y}(0)\mathbf{c}$ and we want $\mathbf{y}(0) = \boldsymbol{\eta}$, the vector $\mathbf{c}$ has to be chosen so that $\mathbf{Y}(0)\mathbf{c} = \boldsymbol{\eta}$. Therefore, it is convenient to determine a fundamental matrix $\mathbf{Y}$ so that the algebraic system $\mathbf{Y}(0)\mathbf{c} = \boldsymbol{\eta}$ is simple to solve for the constant vector $\mathbf{c}$. The simplest such case is when $\mathbf{Y}(0) = \mathbf{I}$, where $\mathbf{I}$ is the identity matrix: $\mathbf{I} = (\delta_{ij})$ where $\delta_{ii} = 1$ and $\delta_{ij} = 0$ if $i \neq j$. For if $\mathbf{Y}(0) = \mathbf{I}$ we can simply take $\mathbf{c} = \boldsymbol{\eta}$ since $\mathbf{I}\mathbf{c} = \mathbf{c}$ for every constant matrix $\mathbf{c}$. The matrix solution $\mathbf{Y}$ to (4) such that $\mathbf{Y}(0) = \mathbf{I}$ is called the *matrix exponential solution to (4)* and the notation $\mathbf{Y}(t) = \mathbf{e}^{t\mathbf{A}}$ is used. Therefore,

$$\frac{d}{dt}\mathbf{e}^{t\mathbf{A}} = \mathbf{A}\mathbf{e}^{tA} \quad \text{and} \quad \mathbf{e}^0 = \mathbf{I} \tag{15}$$

Since $\det \mathbf{I} = 1 \neq 0$, the matrix exponential is a fundamental matrix solution to (4) and, moreover,

(16) $y(t) \equiv \mathbf{e}^{t\mathbf{A}}\boldsymbol{\eta}$ is the solution to the initial value problem (14) for every initial value $\boldsymbol{\eta}$.

EXAMPLE 8.5-3

Consider the 2 × 2 system

$$\mathbf{Y}' = \begin{pmatrix} 0 & 1 \\ 1 & 0 \end{pmatrix} \mathbf{Y} \tag{17}$$

It is easy to check that

$$\mathbf{Y}(t) \equiv \begin{pmatrix} e^t & -e^{-t} \\ e^t & e^{-t} \end{pmatrix}$$

is a fundamental matrix for (17). Since

$$\mathbf{Y}(0) = \begin{pmatrix} 1 & -1 \\ 1 & 1 \end{pmatrix} \neq \begin{pmatrix} 1 & 0 \\ 0 & 1 \end{pmatrix}$$

this fundamental matrix is not the exponential solution to (17). If

$$\mathbf{X}(t) = \begin{pmatrix} \dfrac{e^t + e^{-t}}{2} & \dfrac{e^t - e^{-t}}{2} \\ \dfrac{e^t - e^{-t}}{2} & \dfrac{e^t - e^{-t}}{2} \end{pmatrix} = \begin{pmatrix} \cosh t & \sinh t \\ \sinh t & \cosh t \end{pmatrix}$$

then $\mathbf{X}$ is also a fundamental matrix solution to (17). Since $\mathbf{X}(0) = \mathbf{I}$ it follows that $\mathbf{X}$ is the fundamental matrix solution to (17). Therefore,

$$\mathbf{exp}\left[t \begin{pmatrix} 0 & 1 \\ 1 & 0 \end{pmatrix}\right] = \begin{pmatrix} \cosh t & \sinh t \\ \sinh t & \cosh t \end{pmatrix}$$

for all numbers t. It is of interest to note that there is the following connection between the fundamental solution $\mathbf{Y}$ and the matrix exponential:

$$\mathbf{exp}\left[t \begin{pmatrix} 0 & 1 \\ 1 & 0 \end{pmatrix}\right] = \begin{pmatrix} e^t & -e^{-t} \\ e^t & e^{-t} \end{pmatrix} \begin{pmatrix} \frac{1}{2} & \frac{1}{2} \\ -\frac{1}{2} & \frac{1}{2} \end{pmatrix}$$

The principal concern now is constructing the matrix exponential solution from a known fundamental matrix solution $\mathbf{Y}$. A crucial point is to realize that the following is valid:

(18) If $\mathbf{Y}$ is a matrix solution to (4) and $\mathbf{C}$ is a constant matrix, then $\mathbf{X}(t) \equiv \mathbf{Y}(t)\mathbf{C}$ is also a matrix solution to (4).

The assertion (18) is not difficult to establish. Suppose that $\mathbf{Y}$ is a matrix solution to (4) and $\mathbf{X}(t) = \mathbf{Y}(t)\mathbf{C}$. Since $d/dt\ \mathbf{C} = \mathbf{0}$ it follows from the product rule [see (3)] that

$$\mathbf{X}'(t) = \mathbf{Y}'(t)\mathbf{C} + \mathbf{Y}(t) \cdot \mathbf{0} = \mathbf{Y}'(t)\mathbf{C}$$

and since $\mathbf{Y}'(t) = \mathbf{AY}(t)$ it follows that

$$\mathbf{X}'(t) = [\mathbf{AY}(t)]\mathbf{C} = \mathbf{A}[\mathbf{Y}(t)\mathbf{C}] = \mathbf{AX}(t)$$

so $\mathbf{X}$ is a matrix solution to (4) by definition (see Problem 8.1-3). Therefore, if $\mathbf{Y}$ is a fundamental matrix solution to (4) and $\mathbf{C}$ is a matrix such that $\mathbf{Y}(0)\mathbf{C} = \mathbf{I}$, then $\mathbf{e}^{t\mathbf{A}} = \mathbf{Y}(t)\mathbf{C}$ for all t. Therefore, $\mathbf{C} = \mathbf{Y}(0)^{-1}$ (see Section 8.2*c*), and we have the following result.

Theorem 8.5-3

Suppose that Y is *any* fundamental matrix solution to (4). Then

$$\mathbf{e}^{t\mathbf{A}} = \mathbf{Y}(t)\mathbf{Y}(0)^{-1} \tag{19}$$

for all t.

Since $\det \mathbf{Y}(0) \neq 0$ we know that $\mathbf{Y}(0)^{-1}$ exists, and we have by (18) that $\mathbf{X}(t) \equiv \mathbf{Y}(t)\mathbf{Y}(0)^{-1}$ is a matrix solution to (4). Since $\mathbf{X}(0) = \mathbf{Y}(0)\mathbf{Y}(0)^{-1} = \mathbf{I}$ by definition of inverse, it follows that (19) is correct.

EXAMPLE 8.5-4

Suppose that

$$\mathbf{A} = \begin{pmatrix} 4 & 6 & 6 \\ 1 & 3 & 2 \\ -1 & -4 & -3 \end{pmatrix}$$

Since

$$\mathbf{Y}(t) = \begin{pmatrix} 0 & -6e^{-t} & 3e^{4t} \\ -e^{t} & -2e^{-t} & e^{4t} \\ e^{t} & 7e^{t} & -e^{4t} \end{pmatrix}$$

is a fundamental matrix solution to $\mathbf{Y}' = \mathbf{AY}$ (see Example 8.5-2), it follows that

$$\mathbf{e}^{t\mathbf{A}} = \begin{pmatrix} 0 & -6\epsilon^{-t} & 3e^{4t} \\ -e^{t} & -2e^{-t} & e^{4t} \\ e^{t} & 7e^{-t} & -e^{4t} \end{pmatrix}\begin{pmatrix} 0 & -6 & 3 \\ -1 & -2 & 1 \\ 1 & 7 & -1 \end{pmatrix}^{-1}$$

From Example 8.5-2, $\det \mathbf{Y}(0) = -15$, and using (21) in Section 8.5-2 we have

$$\mathbf{Y}(0)^{-1} = -\tfrac{1}{15}\begin{pmatrix} -5 & 15 & 0 \\ 0 & -3 & -3 \\ -5 & -6 & -6 \end{pmatrix} = \begin{pmatrix} \frac{1}{3} & -1 & 0 \\ 0 & \frac{1}{5} & \frac{1}{5} \\ \frac{1}{3} & \frac{2}{5} & \frac{2}{5} \end{pmatrix}$$

and hence

$$\mathbf{e}^{t\mathbf{A}} = \begin{pmatrix} e^{4t} & \dfrac{6e^{4t} - 6e^{-t}}{5} & \dfrac{6e^{4t} - 6e^{-t}}{5} \\ \dfrac{e^{4t} - e^{t}}{3} & \dfrac{5e^{t} - 2e^{-t} + 2e^{4t}}{5} & \dfrac{2e^{4t} - 2e^{-t}}{5} \\ \dfrac{e^{t} - e^{4t}}{3} & \dfrac{7e^{-t} - 5e^{t} - 2e^{4t}}{5} & \dfrac{7e^{-t} - 2e^{4t}}{5} \end{pmatrix}$$

One of the advantages of the matrix exponential solution to (4) over other fundamental matrix solutions to (4) is that solutions to the initial value problem (14) can be trivially obtained from the matrix exponential [see (16)].

PROBLEMS

8.5-1. Determine the matrix exponential $\mathbf{e}^{t\mathbf{A}}$ for each of the following matrices $\mathbf{A}$:

(a) $\mathbf{A} = \begin{pmatrix} 1 & 2 \\ 2 & 1 \end{pmatrix}$

(b) $\mathbf{A} = \begin{pmatrix} 0 & 2 \\ -2 & 0 \end{pmatrix}$

(c) $\mathbf{A} = \begin{pmatrix} 1 & 1 \\ 1 & 1 \end{pmatrix}$

(d) $\mathbf{A} = \begin{pmatrix} 0 & 1 & 0 \\ 0 & 0 & 1 \\ 1 & -1 & 1 \end{pmatrix}$ (See Example 8.4-7.)

(e) $\mathbf{A} = \begin{pmatrix} -6 & -7 & -13 \\ 5 & 6 & 9 \\ 2 & 2 & 5 \end{pmatrix}$ (See Example 8.4-9.)

(f) $\mathbf{A} = \begin{pmatrix} 3 & 0 & -2 \\ 1 & 2 & -2 \\ 1 & 3 & -3 \end{pmatrix}$ (See Example 8.3-1.)

* **8.5-2.** Prove the product rule formula (3) for the differentiation of the product of two matrix functions.

(a) Suppose that $\mathbf{Y}$ is a differentiable matrix-valued function and that $\mathbf{X}(t) \equiv \mathbf{Y}(t)^2$ $[= \mathbf{Y}(t) \cdot \mathbf{Y}(t)]$ for all t. Show that $\mathbf{X}'(t) = \mathbf{Y}'(t)\mathbf{Y}(t) + \mathbf{Y}(t)\mathbf{Y}'(t)$. Give an example to show that $\mathbf{X}'(t)$ is not necessarily $2\mathbf{Y}(t)\mathbf{Y}'(t)$. [*Hint:* Determine a $\mathbf{Y}$ such that $\mathbf{Y}(t)\mathbf{Y}'(t) \neq \mathbf{Y}'(t)\mathbf{Y}(t)$.]

(b) If $\mathbf{X}$, $\mathbf{Y}$, and $\mathbf{Z}$ are each differentiable matrix-valued functions, determine a formula for $d/dt\,[\mathbf{X}(t)\mathbf{Y}(t)\mathbf{Z}(t)]$. Determine also a formula for $d/dt\,[\mathbf{X}(t)^3]$.

8.5-3. Suppose that $\mathbf{y}_j$ is a solution to the initial value problem $\mathbf{y}' = \mathbf{A}\mathbf{y}$, $\mathbf{y}(0) = \mathbf{e}_j$ for each $\mathbf{j} \in \{1, \ldots, n\}$. Show that $\mathbf{e}^{t\mathbf{A}} = (\mathbf{y}_1, \mathbf{y}_2, \ldots, \mathbf{y}_n)$.

*8.6 SOLUTION COMPUTATION USING MATRIX METHODS

Section 8.4 discusses a method for computing solutions to homogeneous linear systems using eigenvalues and eigenvectors. In this section we use properties of the matrix exponential solution in order to develop techniques for the computation of solutions. Suppose that $\mathbf{A}$ is an $n \times n$ matrix and consider the initial value vector equation

$$\mathbf{y}' = \mathbf{A}\mathbf{y} \qquad \mathbf{y}(0) = \boldsymbol{\eta} \tag{1}$$

and the matrix equation

$$\mathbf{Y}' = \mathbf{A}\mathbf{Y} \tag{2}$$

The matrix exponential $\mathbf{e}^{t\mathbf{A}}$ is the solution to the matrix equation (2) that equals the identity matrix $\mathbf{I}$ when $t = 0$. Once $\mathbf{e}^{t\mathbf{A}}$ is computed, the solution to the initial value problem (1) is $\mathbf{y}(t) = \mathbf{e}^{t\mathbf{A}}\boldsymbol{\eta}$ for all t. Therefore, the concern in this section is to indicate special techniques for computing $\mathbf{e}^{t\mathbf{A}}$.

For each $n \times n$ matrix $\mathbf{A}$ we define $\mathbf{A}^0 = \mathbf{I}$, $\mathbf{A}^1 = \mathbf{A}$, $\mathbf{A}^2 = \mathbf{A} \cdot \mathbf{A}$, and by induction on m, $\mathbf{A}^{m+1} = \mathbf{A}^m \cdot \mathbf{A}$ for each positive integer m. Moreover, if p is any polynomial, say

$$p(\lambda) = a_0 + a_1\lambda + a_2\lambda^2 + \cdots + a_m\lambda^m$$

then the matrix $\mathbf{p}(\mathbf{A})$ is defined by

$$\mathbf{p}(\mathbf{A}) = a_0\mathbf{I} + a_1\mathbf{A} + a_2\mathbf{A}^2 + \cdots + a_m\mathbf{A}^m$$

We want to indicate now that the matrix exponential $\mathbf{e}^{t\mathbf{A}}$ can be represented as an infinite series resembling the series for the number-valued exponential. Recall that if a is a real number then

$$e^{ta} = 1 + ta + \frac{t^2}{2!}a^2 + \frac{t^3}{3!}a^3 + \cdots \equiv \sum_{k=0}^{\infty} \frac{t^k}{k!}a^k \tag{3}$$

for all t. It is the case that for *any* $n \times n$ matrix $\mathbf{A}$, the matrix exponential $\mathbf{e}^{t\mathbf{A}}$ has this same form:

$$\mathbf{e}^{t\mathbf{A}} = \mathbf{I} + t\mathbf{A} + \frac{t^2}{2!}\mathbf{A}^2 + \frac{t^3}{3!}\mathbf{A}^3 + \cdots + \frac{t^m}{m!}\mathbf{A}^m + \cdots \tag{4}$$

or more concisely

$$\mathbf{e}^{t\mathbf{A}} = \sum_{k=0}^{\infty} \frac{t^k}{k!}\mathbf{A}^k \tag{4'}$$

Formula (4) will be used, but the fact that this series converges to $\mathbf{e}^{t\mathbf{A}}$ will not be established. Taking the derivative with respect to t of the right-hand side of (4) term by term (which is in fact valid), we obtain

$$\begin{aligned}\frac{d}{dt}\,\mathbf{e}^{t\mathbf{A}} &= 0 + \mathbf{A} + \frac{2t}{2!}\mathbf{A}^2 + \frac{3t^2}{3!}\mathbf{A}^3 + \frac{4t^3}{4!}\mathbf{A}^4 + \cdots \\ &= \mathbf{A} + t\mathbf{A}^2 + \frac{t^2}{2!}\mathbf{A}^3 + \frac{t^3}{3!}\mathbf{A}^4 + \cdots \\ &= \mathbf{A}\left(\mathbf{I} + t\mathbf{A} + \frac{t^2}{2!}\mathbf{A}^2 + \frac{t^3}{3!}\mathbf{A}^3 + \cdots\right) \\ &= \mathbf{A}\mathbf{e}^{t\mathbf{A}}\end{aligned}$$

Therefore the right-hand side of (4) defines a matrix solution to (2), and since it is clearly equal to the identity $\mathbf{I}$ when $t = 0$, we have at least formally established that the right-hand side of (4) does indeed represent the matrix exponential. There are some special cases where the series representation (4) can be helpful in the direct computation of $\mathbf{e}^{t\mathbf{A}}$, and a few examples are given here in order to illustrate some of these situations.

EXAMPLE 8.6-1

Suppose that λ is a number and $\mathbf{A} = \lambda\mathbf{I}$. Then

$\mathbf{A}^2 = \lambda^2\mathbf{I},\ \mathbf{A}^3 = \lambda^3\mathbf{I},\ \ldots,\ \mathbf{A}^m = \lambda^m\mathbf{I},\ \ldots$

and it follows from (4) and (3) that

$$\begin{aligned}\mathbf{e}^{t\lambda\mathbf{I}} &= \mathbf{I} + t\lambda\mathbf{I} + \frac{t^2}{2!}\lambda^2\mathbf{I} + \frac{t^3}{3!}\lambda^3\mathbf{I} + \cdots \\ &= \left(1 + t\lambda + \frac{t^2\lambda^2}{2!} + \frac{t^3\lambda^3}{3!} + \cdots\right)\mathbf{I} \\ &= e^{t\lambda}\mathbf{I}\end{aligned}$$

for all t.

EXAMPLE 8.6-2

Suppose that

$$\mathbf{A} = \begin{pmatrix} 0 & 1 & 0 \\ 0 & 0 & 1 \\ 0 & 0 & 0 \end{pmatrix}$$

Then

$$\mathbf{A}^2 = \begin{pmatrix} 0 & 0 & 1 \\ 0 & 0 & 0 \\ 0 & 0 & 0 \end{pmatrix} \qquad \mathbf{A}^3 = \begin{pmatrix} 0 & 0 & 0 \\ 0 & 0 & 0 \\ 0 & 0 & 0 \end{pmatrix}$$

and so $\mathbf{A}^4 = \mathbf{A} \cdot \mathbf{A}^3 = \mathbf{A} \cdot \mathbf{0} = \mathbf{0}$, $\mathbf{A}^5 = \mathbf{A} \cdot \mathbf{A}^4 = \mathbf{0}$, etc. Therefore, $\mathbf{A}^m = \mathbf{0}$ for all $m \geq 3$, and it follows from (4) that

$$\mathbf{e}^{t\mathbf{A}} = \mathbf{I} + t\mathbf{A} + \frac{t^2}{2!}\mathbf{A}^2$$

$$= \begin{pmatrix} 1 & 0 & 0 \\ 0 & 1 & 0 \\ 0 & 0 & 1 \end{pmatrix} + \begin{pmatrix} 0 & t & 0 \\ 0 & 0 & t \\ 0 & 0 & 0 \end{pmatrix} + \begin{pmatrix} 0 & 0 & \frac{t^2}{2} \\ 0 & 0 & 0 \\ 0 & 0 & 0 \end{pmatrix}$$

$$= \begin{pmatrix} 1 & t & \frac{t^2}{2} \\ 0 & 1 & t \\ 0 & 0 & 1 \end{pmatrix}$$

for all t.

Now suppose that λ is a real number and define $\mathbf{X}(t) = e^{-\lambda t}\mathbf{e}^{t\mathbf{A}}$ for all t. Using the product rule for differentiation and the fact that $d/dt\, \mathbf{e}^{t\mathbf{A}} = \mathbf{A}\mathbf{e}^{t\mathbf{A}}$,

$$\begin{aligned} \mathbf{X}'(t) &= e^{-\lambda t}\frac{d}{dt}\mathbf{e}^{t\mathbf{A}} - \lambda e^{-\lambda t}\mathbf{e}^{t\mathbf{A}} \\ &= e^{-\lambda t}\mathbf{A}\mathbf{e}^{t\mathbf{A}} - \lambda e^{-\lambda t}\mathbf{e}^{t\mathbf{A}} \\ &= \mathbf{A}(e^{-\lambda t}\mathbf{e}^{t\mathbf{A}}) - (e^{-\lambda t}\mathbf{e}^{t\mathbf{A}}) \\ &= (\mathbf{A} - \lambda\mathbf{I})(e^{-\lambda t}\mathbf{e}^{t\mathbf{A}}) \\ &= (\mathbf{A} - \lambda\mathbf{I})\mathbf{X}(t) \end{aligned}$$

Therefore, $\mathbf{X}' = (\mathbf{A} - \lambda\mathbf{I})\mathbf{X}$, $\mathbf{X}(0) = \mathbf{I}$, and it follows by definition that $\mathbf{X}(t) = \mathbf{e}^{t(\mathbf{A}-\lambda\mathbf{I})}$. Since $\mathbf{X}(t) = e^{-\lambda t}\mathbf{e}^{t\mathbf{A}}$ as well, it follows that

$$\mathbf{e}^{t\mathbf{A}} = e^{\lambda t}\mathbf{e}^{t(\mathbf{A}-\lambda\mathbf{I})} \qquad \text{for all numbers } t \text{ and } \lambda \tag{5}$$

From (5) it follows that $\mathbf{e}^{t\mathbf{A}}$ can be computed whenever $\mathbf{e}^{t(\mathbf{A}-\lambda\mathbf{I})}$ can be computed for some number λ.

EXAMPLE 8.6-3

Suppose that

$$\mathbf{A} = \begin{pmatrix} -3 & 1 & 0 \\ 0 & -3 & 1 \\ 0 & 0 & -3 \end{pmatrix}$$

Then

$$\mathbf{A} + 3\mathbf{I} = \begin{pmatrix} 0 & 1 & 0 \\ 0 & 0 & 1 \\ 0 & 0 & 0 \end{pmatrix}$$

and it follows from Example 8.6-2 that

$$\mathbf{e}^{t(\mathbf{A}+3\mathbf{I})} = \begin{pmatrix} 1 & t & \dfrac{t^2}{2} \\ 0 & 1 & t \\ 0 & 0 & 1 \end{pmatrix}$$

and so

$$\mathbf{e}^{t\mathbf{A}} = \begin{pmatrix} e^{-3t} & te^{-3t} & \dfrac{t^2e^{-3t}}{2} \\ 0 & e^{-3t} & te^{-3t} \\ 0 & 0 & e^{-3t} \end{pmatrix}$$

by formula (5).

8.6a Computation Using Cayley-Hamilton Theorem

Recall that the *characteristic polynomial* of the $n \times n$ matrix $\mathbf{A}$ is the nth order polynomial p_A defined by the formula

(6) $$p_A(\lambda) \equiv \det(\mathbf{A} - \lambda\mathbf{I})$$

[see (4) in Section 8.4]. Since p_A is a polynomial, the matrix $\mathbf{p}_A(\mathbf{A})$ is well-defined, and the *Cayley-Hamilton theorem* asserts that $\mathbf{p}_A(\mathbf{A})$ is the zero matrix for every matrix $\mathbf{A}$:

Theorem 8.6-1

Suppose that $\mathbf{A}$ is an $n \times n$ matrix and p_A is the characteristic polynomial of $\mathbf{A}$. Then $\mathbf{p}_A(\mathbf{A}) = \mathbf{0}$.

The proof of this theorem is omitted and we indicate only some of the applications of it to the computation of $\mathbf{e}^{t\mathbf{A}}$. Write p_A in the form

$$p_A(\lambda) = a_0 + a_1\lambda + \cdots + a_{n-1}\lambda^{n-1} + (-1)^n\lambda^n$$

[it follows directly from the definition that the coefficient of λ^n is $(-1)^n$]. Since $\mathbf{p}_A(\mathbf{A}) = \mathbf{0}$ and the coefficient of $\mathbf{A}^n$ in $\mathbf{p}_A(\mathbf{A})$ is nonzero, it follows that there are constants $\beta_{0,n}, \beta_{1,n}, \ldots, \beta_{n-1,n}$ such that

(7) $$\mathbf{A}^n = \beta_{0,n}\mathbf{I} + \beta_{1,n}\mathbf{A} + \cdots + \beta_{n-1,n}\mathbf{A}^{n-1}$$

Multiplying each side of equation (7) by $\mathbf{A}$ and substituting (7) for $\mathbf{A}^n$ in the resulting equation shows that

$$\mathbf{A}^{n+1} = \beta_{0,n+1}\mathbf{I} + \beta_{1,n+1}\mathbf{A} + \cdots + \beta_{n-1,n+1}\mathbf{A}^{n-1}$$

Continuing in this manner we find that

(8) $$\mathbf{A}^m = \beta_{0,m}\mathbf{I} + \beta_{1,m}\mathbf{A} + \cdots + \beta_{n-1,m}\mathbf{A}^{n-1} \qquad \text{for } m = n, n+1, \ldots$$

Therefore, *every* power of $\mathbf{A}$ is a linear combination of $\mathbf{I}$, $\mathbf{A}^1$, $\mathbf{A}^2, \ldots, \mathbf{A}^{n-1}$. The fact that (8) is true can be used to simplify the computation of higher powers of $\mathbf{A}$. It is extremely important to analyze the eigenvalues and eigenvectors of powers of $\mathbf{A}$ relative to those of $\mathbf{A}$ itself. Suppose that λ_i is an eigenvalue of $\mathbf{A}$ and $\boldsymbol{\eta}$ is a corresponding eigenvector: $\mathbf{A}\boldsymbol{\eta} = \lambda_i\boldsymbol{\eta}$. Then

$$\mathbf{A}^2\boldsymbol{\eta} = \mathbf{A}(\mathbf{A}\boldsymbol{\eta}) = \mathbf{A}(\lambda_i\boldsymbol{\eta}) = \lambda_i\mathbf{A}\boldsymbol{\eta} = \lambda_i^2\boldsymbol{\eta}$$
$$\mathbf{A}^3\boldsymbol{\eta} = \mathbf{A}(\mathbf{A}^2\boldsymbol{\eta}) = \mathbf{A}(\lambda_i^2\boldsymbol{\eta}) = \lambda_i^2\mathbf{A}\boldsymbol{\eta} = \lambda_i^3\boldsymbol{\eta}$$

and continuing in this manner, it follows that

(9) $$\text{If } \mathbf{A}\boldsymbol{\eta} = \lambda_i\boldsymbol{\eta} \qquad \text{then} \qquad \mathbf{A}^m\boldsymbol{\eta} = \lambda_i^m\boldsymbol{\eta} \qquad \text{for all } m = 2, 3, \ldots$$

Evaluating each side of (8) at $\boldsymbol{\eta}$ and using (9), it follows that

$$\lambda_i^m\boldsymbol{\eta} = \beta_{0,m}\boldsymbol{\eta} + \beta_{1,m}\lambda_i\boldsymbol{\eta} + \cdots + \beta_{n-1,m}\lambda_i^{m-1}\boldsymbol{\eta}$$

Since $\boldsymbol{\eta} \neq 0$ it follows that

(10) $$\lambda_i^m = \beta_{0,m} + \beta_{1,m}\lambda_i + \cdots + \beta_{n-1,m}\lambda_i^{m-1}$$

for each eigenvalue λ_i of $\mathbf{A}$. Therefore, if $\mathbf{A}$ has n distinct eigenvalues $\lambda_1, \ldots, \lambda_n$, then letting i take on values in $\{1, \ldots, n\}$ in (10) gives n equations to solve for the n unknowns $\beta_{0,m}, \ldots, \beta_{n-1,m}$. Notice that *it is not necessary to compute the eigenvectors.*

Using (8) it follows that if q is *any* polynomial there is a polynomial $\mathbf{q}_A$ *of degree less than* n such that $\mathbf{q}(\mathbf{A}) = \mathbf{q}_A(\mathbf{A})$. Using a limiting process in the series representation (4) of $\mathbf{e}^{t\mathbf{A}}$ it can be shown that for each number t there are numbers $\alpha_0(t), \alpha_1(t), \ldots, \alpha_{n-1}(t)$ such that

(11) $$\mathbf{e}^{t\mathbf{A}} = \alpha_0(t)\mathbf{I} + \alpha_1(t)\mathbf{A} + \cdots + \alpha_{n-1}(t)\mathbf{A}^{n-1}$$

Again suppose that λ_i is an eigenvalue of $\mathbf{A}$ and that $\boldsymbol{\eta}$ is a corresponding eigenvector. Using (9) and the series representation (4) it follows that

$$\begin{aligned}\mathbf{e}^{t\mathbf{A}}\boldsymbol{\eta} &= \boldsymbol{\eta} + t\mathbf{A}\boldsymbol{\eta} + \frac{t^2}{2!}\mathbf{A}^2\boldsymbol{\eta} + \frac{t^3}{3!}\mathbf{A}^3\boldsymbol{\eta} + \cdots \\ &= \boldsymbol{\eta} + t\lambda_i\boldsymbol{\eta} + \frac{t^2}{2!}\lambda_i^2\boldsymbol{\eta} + \frac{t^3}{3!}\lambda_i^3\boldsymbol{\eta} + \cdots \\ &= \left[1 + t\lambda_i + \frac{(t\lambda_i)^2}{2!}\lambda_i^2 + \frac{(t\lambda_i)^3}{3!} + \cdots\right]\boldsymbol{\eta} \\ &= e^{t\lambda_i}\boldsymbol{\eta}\end{aligned}$$

Substituting into (11) we have that

$$e^{t\lambda_i}\boldsymbol{\eta} = \alpha_0(t)\boldsymbol{\eta} + \alpha_1(t)\lambda_i\boldsymbol{\eta} + \cdots + \alpha_{n-1}(t)\lambda_i^{n-1}\boldsymbol{\eta}$$

and since $\boldsymbol{\eta} \neq \mathbf{0}$ the formula

$$e^{t\lambda_i} = \alpha_0(t) + \alpha_1(t)\lambda_i + \cdots + \alpha_{n-1}(t)\lambda_i^{n-1} \tag{12}$$

is valid for all t and each eigenvalue λ_i of $\mathbf{A}$.

If each of the n eigenvalues $\lambda_1, \lambda_2, \ldots, \lambda_n$ are distinct, then the n equations indicated by (12) are

$$\begin{aligned} \alpha_0(t) + \alpha_1(t)\lambda_1 + \cdots + \alpha_{n-1}(t)\lambda_1^{n-1} &= e^{t\lambda_1} \\ \alpha_0(t) + \alpha_1(t)\lambda_2 + \cdots + \alpha_{n-1}(t)\lambda_2^{n-1} &= e^{t\lambda_2} \\ \cdots\cdots\cdots\cdots\cdots\cdots\cdots\cdots\cdots & \\ \alpha_0(t) + \alpha_1(t)\lambda_n + \cdots + \alpha_{n-1}(t)\lambda_n^{n-1} &= e^{t\lambda_n} \end{aligned} \tag{13}$$

If $\lambda_i \neq \lambda_j$ for $i \neq j$ then equation (13) has unique solutions $\alpha_0(t), \alpha_1(t), \ldots, \alpha_{n-1}(t)$, and substituting for these values into (11) gives $\mathbf{e}^{t\mathbf{A}}$. Notice again that the eigenvectors do not need to be computed.

EXAMPLE 8.6-4

Suppose that

$$\mathbf{A} = \begin{pmatrix} 4 & 6 & 6 \\ 1 & 3 & 2 \\ -1 & -4 & -3 \end{pmatrix}$$

Then $\det(\mathbf{A} - \lambda\mathbf{I}) = -(\lambda - 1)(\lambda + 1)(\lambda - 4)$, and so $1, -1, 4$ are the eigenvalues (see Example 8.4-3). Substituting $n = 3$, $\lambda_1 = 1$, $\lambda_2 = -1$, and $\lambda_3 = 4$ in (13), we obtain the system

$$\begin{aligned} \alpha_0(t) + \alpha_1(t) + \alpha_2(t) &= e^t \\ \alpha_0(t) - \alpha_1(t) + \alpha_2(t) &= e^{-t} \\ \alpha_0(t) + 4\alpha_1(t) + 16\alpha_2(t) &= e^{4t} \end{aligned}$$

Subtracting each side of the first equation from the second, we obtain $2\alpha_1(t) = e^t - e^{-t}$, and hence

$$\alpha_1(t) = \tfrac{1}{2}e^t - \tfrac{1}{2}e^{-t}$$

Subtracting each side of the second equation from the third, we have

$$5\alpha_1(t) + 15\alpha_2(t) = e^{4t} - e^{-t}$$

and hence

$$\alpha_2(t) = \frac{e^{4t} - e^{-t}}{15} - \frac{e^t - e^{-t}}{6} = \tfrac{1}{15}e^{4t} + \tfrac{1}{10}e^{-t} - \tfrac{1}{6}e^t$$

Substituting for α_1 and α_2 in the first equation,

$$\begin{aligned} \alpha_0(t) &= e^t - \tfrac{1}{15}e^{4t} - \tfrac{1}{10}e^{-t} + \tfrac{1}{6}e^t - \tfrac{1}{2}e^t + \tfrac{1}{2}e^{-t} \\ &= \tfrac{2}{3}e^t + \tfrac{2}{5}e^{-t} - \tfrac{1}{15}e^{4t} \end{aligned}$$

Therefore, using (11)

$$\mathbf{e}^{t\mathbf{A}} = \alpha_0(t)\begin{pmatrix} 1 & 0 & 0 \\ 0 & 1 & 0 \\ 0 & 0 & 1 \end{pmatrix} + \alpha_1(t)\begin{pmatrix} 4 & 6 & 6 \\ 1 & 3 & 2 \\ -1 & -4 & -3 \end{pmatrix} + \alpha_2(t)\begin{pmatrix} 16 & 18 & 18 \\ 5 & 7 & 6 \\ -5 & -6 & -5 \end{pmatrix}$$

and substituting for $\alpha_0(t)$, $\alpha_1(t)$, $\alpha_2(t)$ and collecting terms, it follows that

$$\mathbf{e}^{t\mathbf{A}} = \begin{pmatrix} e^{4t} & \frac{6}{5}e^{4t} - \frac{6}{5}e^{-t} & \frac{6}{5}e^{4t} - \frac{6}{5}e^{-t} \\ \frac{1}{3}e^{4t} - \frac{1}{3}e^{t} & \frac{2}{5}e^{4t} + e^{t} - \frac{2}{5}e^{-t} & \frac{2}{5}e^{4t} - \frac{2}{5}e^{-t} \\ -\frac{1}{3}e^{4t} + \frac{1}{3}e^{t} & -\frac{2}{5}e^{4t} - e^{t} + \frac{7}{5}e^{-t} & -\frac{2}{5}e^{4t} + \frac{7}{5}e^{-t} \end{pmatrix}$$

This is the same expression that was obtained in Example 8.5-4.

▶ **EXAMPLE 8.6-5**

Suppose that

$$\mathbf{A} = \begin{pmatrix} 0 & -3 \\ 3 & 0 \end{pmatrix}$$

Then

$$\det(\mathbf{A} - \lambda\mathbf{I}) = \det\begin{pmatrix} -\lambda & -3 \\ 3 & -\lambda \end{pmatrix} = \lambda^2 + 9$$

and it follows that $\lambda = \pm 3i$ are the eigenvalues of $\mathbf{A}$. Setting $n = 2$ and $\lambda_i = 3i$ in (12), the equation

$$\alpha_0(t) + \alpha_1(t)3i = e^{3it} = \cos 3t + i \sin 3t$$

Setting real and imaginary parts equal, we immediately obtain

$$\alpha_0(t) = \cos 3t \qquad \text{and} \qquad \alpha_1(t) = \tfrac{1}{3}\sin 3t$$

Therefore, by formula (11),

$$\mathbf{e}^{t\mathbf{A}} = (\cos 3t)\begin{pmatrix} 1 & 0 \\ 0 & 1 \end{pmatrix} + (\tfrac{1}{3}\sin 3t)\begin{pmatrix} 0 & -3 \\ 3 & 0 \end{pmatrix}$$

$$= \begin{pmatrix} \cos 3t & -\sin 3t \\ \sin 3t & \cos 3t \end{pmatrix}$$

for all t.

Notice that when a pair of complex conjugate eigenvalues occur, we use only one of the equations in (13) and equate the corresponding real parts and the corresponding imaginary parts. Also, if the eigenvalues are all simple (and hence distinct), then (12) provides a system of n different equations [see (13)] that always has a unique set of solutions $\alpha_0, \alpha_1, \ldots, \alpha_{n-1}$. However, if any of the eigenvalues are multiple, then (12) does not provide enough equations to determine $\alpha_0, \alpha_1, \ldots, \alpha_{n-1}$. This procedure is easily modified to handle this case, however. If λ_i is an eigenvalue with multiplicity 2, then in addition to (12) the equation

$$te^{\lambda_i t} = \alpha_1(t) + 2\alpha_2(t)\lambda_i + \cdots + (n-1)\alpha_{n-1}(t)\lambda_i^{n-2} \tag{14}$$

is considered. Note that the left-hand side of (14) is the derivative of $\lambda \to e^{\lambda t}$ evaluated at $\lambda = \lambda_i$, and the right-hand side is the derivative of

$$\lambda \to \alpha_0(t) + \alpha_1(t)\lambda + \alpha_2(t)\lambda^2 + \cdots + \alpha_{n-1}(t)\lambda^{n-1}$$

evaluated at $\lambda = \lambda_i$. Of course, if the multiplicity is higher, then this procedure is continued. For example, if λ_i has multiplicity 3, then in addition to (12) and (14) the equation

$$t^2e^{\lambda_i t} = 2\alpha_2(t) + 6\alpha_3(t)\lambda_i + \cdots + (n-1)(n-2)\alpha_{n-1}(t)\lambda_i^{n-3} \tag{15}$$

is considered. This equation is obtained from the second derivative of the functions $\lambda \to e^{\lambda t}$ and

$$\lambda \to \alpha_0(t) + \alpha_1(t)\lambda + \cdots + \alpha_{n-1}(t)\lambda^{n-1}$$

evaluated at $\lambda = \lambda_i$.

EXAMPLE 8.6-6

Suppose that

$$\mathbf{A} = \begin{pmatrix} 0 & 1 \\ -1 & -2 \end{pmatrix}$$

Then

$$\begin{aligned} \det(\mathbf{A} - \lambda\mathbf{I}) &= \det\begin{pmatrix} -\lambda & 1 \\ -1 & -2-\lambda \end{pmatrix} = (-\lambda)(-2-\lambda) + 1 \\ &= \lambda^2 + 2\lambda + 1 = (\lambda+1)^2 \end{aligned}$$

and it follows that $\lambda = -1$ is an eigenvalue of $\mathbf{A}$ with multiplicity 2. From (12) and (14) the system

$$\begin{aligned} \alpha_0(t) - \alpha_1(t) &= e^{-t} \\ \alpha_1(t) &= te^{-t} \end{aligned}$$

is obtained. Therefore $\alpha_1(t) = te^{-t}$, $\alpha_0(t) = (1+t)e^{-t}$, and

$$\mathbf{e}^{t\mathbf{A}} = (1+t)e^{-t}\begin{pmatrix}1 & 0\\ 0 & 1\end{pmatrix} + te^{-t}\begin{pmatrix}0 & 1\\ -1 & -2\end{pmatrix}$$

$$= \begin{pmatrix}(1+t)e^{-t} & te^{-t}\\ -te^{-t} & (1-t)e^{-t}\end{pmatrix}$$

for all t.

PROBLEMS

8.6-1. Show that for each of the following matrices $\mathbf{A}$ there is a positive integer m and a number λ such that $(\mathbf{A} - \lambda\mathbf{I})^m = 0$. Then use (5) and the series representation (4) to compute $\mathbf{e}^{t\mathbf{A}}$ for all t.

(a) $\mathbf{A} = \begin{pmatrix}0 & 2 & 1\\ 0 & 0 & 3\\ 0 & 0 & 0\end{pmatrix}$

(b) $\mathbf{A} = \begin{pmatrix}3 & 0 & 0\\ 2 & 3 & 0\\ 1 & 0 & 3\end{pmatrix}$

(c) $\mathbf{A} = \begin{pmatrix}1 & 2 & 0 & 0\\ 0 & 1 & 0 & 0\\ 0 & 0 & 1 & 0\\ 0 & 0 & 1 & 1\end{pmatrix}$

(d) $\mathbf{A} = \begin{pmatrix}0 & 1 & 0\\ 0 & 0 & 0\\ 1 & 1 & 0\end{pmatrix}$

8.6-2. Use formula (11) [along with (12) and (14)] to compute $\mathbf{e}^{t\mathbf{A}}$ for each of the following matrices $\mathbf{A}$.

(a) $\mathbf{A} = \begin{pmatrix}4 & -2\\ 3 & -1\end{pmatrix}$

(b) $\mathbf{A} = \begin{pmatrix}1 & 2\\ -2 & 1\end{pmatrix}$

(c) $\mathbf{A} = \begin{pmatrix}-1 & 1\\ -4 & 3\end{pmatrix}$

(d) $\mathbf{A} = \begin{pmatrix}1 & 1 & 1\\ 2 & 2 & 2\\ 1 & 1 & 1\end{pmatrix}$

(e) $\mathbf{A} = \begin{pmatrix}-1 & 1 & 0\\ 1 & -1 & 1\\ 0 & 1 & -1\end{pmatrix}$

(f) $\mathbf{A} = \begin{pmatrix}-6 & -7 & -13\\ 5 & 6 & 9\\ 2 & 2 & 5\end{pmatrix}$

***8.6-3.** A matrix $\mathbf{A} = (a_{ij})$ is said to be *diagonal* if $a_{ij} = 0$ for $i \neq j$. If $\mathbf{A}$ is diagonal we write $\mathbf{A} = \operatorname{diag}(a_{11}, a_{22}, \ldots, a_{nn})$. Show that

$$\mathbf{e}^{t\operatorname{diag}(a_{11}, a_{22}, \ldots, a_{nn})} = \operatorname{diag}(e^{ta_{11}}, e^{ta_{22}}, \ldots, e^{ta_{nn}})$$

for all t.

***8.6-4.** Suppose that $\mathbf{A}$ is an $n \times n$ matrix and there is a real number α such that $\mathbf{A}^2 = \alpha\mathbf{A}$. Compute $\mathbf{e}^{t\mathbf{A}}$.

*8.7 SOLUTION REPRESENTATION FOR NONHOMOGENEOUS SYSTEMS

In this section some basic results for nonhomogeneous systems are indicated. As before, $\mathbf{A}$ is an $n \times n$ matrix, and

$$\mathbf{f}(t) = \begin{pmatrix} f_1(t) \\ f_2(t) \\ \vdots \\ f_n(t) \end{pmatrix}$$

is a continuous vector-valued function. We consider the nonhomogeneous system

$$\mathbf{y}' = \mathbf{A}\mathbf{y} + \mathbf{f}(t) \tag{1}$$

along with the corresponding homogeneous system

$$\mathbf{y}' = \mathbf{A}\mathbf{y} \tag{2}$$

According to Theorem 8.3-2, if $\mathbf{y}_p$ is a "particular" solution to (1), then the general solution to (1) is the set of all functions of the form $\mathbf{y} = \mathbf{y}_H + \mathbf{y}_p$ where $\mathbf{y}_H$ is a solution to the homogeneous system (2). Therefore, the concern here is to determine a particular solution to the nonhomogeneous system (1). It is of interest to observe that there is an analogy between the procedure for the system (1) and the one-dimensional nonhomogeneous system studied in Section 1.2.

The exponential matrix solution $\mathbf{e}^{t\mathbf{A}}$ is as described in Sections 8.5 and 8.6, and the following two properties are needed for the developments in this section.

□ **Lemma 1**

Suppose that $\mathbf{A}$ is an $n \times n$ matrix. Then

(i) $\mathbf{e}^{(t+s)\mathbf{A}} = \mathbf{e}^{t\mathbf{A}}\mathbf{e}^{s\mathbf{A}}$ for all numbers t and s, and

(ii) $\mathbf{A}\mathbf{e}^{t\mathbf{A}} = \mathbf{e}^{t\mathbf{A}}\mathbf{A}$ for all numbers t.

Part (ii) follows easily from the series representation (4) in Section 8.7. For note that

$$\begin{aligned} \mathbf{A}\mathbf{e}^{t\mathbf{A}} &= \mathbf{A}\left(\mathbf{I} + t\mathbf{A} + \frac{t^2}{2!}\mathbf{A}^2 + \cdots\right) \\ &= \mathbf{A} + t\mathbf{A}^2 + \frac{t^2}{2!}\mathbf{A}^3 + \cdots \\ &= \left(\mathbf{I} + t\mathbf{A} + \frac{t^2}{2!}\mathbf{A}^2 + \cdots\right)\mathbf{A} \\ &= \mathbf{e}^{t\mathbf{A}}\mathbf{A} \end{aligned}$$

and (ii) is seen to be true. Part (i) can be established using the uniqueness of solutions to the initial value problem

$$\mathbf{y}' = \mathbf{A}\mathbf{y} \qquad \mathbf{y}(0) = \boldsymbol{\eta} \tag{3}$$

For consider s as being constant and let $\boldsymbol{\xi}$ be any vector. Define

$$\mathbf{v}(t) \equiv \mathbf{e}^{t\mathbf{A}}\mathbf{e}^{s\mathbf{A}}\boldsymbol{\xi} \qquad \text{and} \qquad \mathbf{w}(t) \equiv \mathbf{e}^{(t+s)\mathbf{A}}\boldsymbol{\xi}$$

for all t. Then

$$\mathbf{v}'(t) = \left(\frac{d}{dt}\,\mathbf{e}^{t\mathbf{A}}\right)\mathbf{e}^{s\mathbf{A}}\boldsymbol{\xi} = \mathbf{A}\mathbf{e}^{t\mathbf{A}}\mathbf{e}^{s\mathbf{A}}\boldsymbol{\xi} = \mathbf{A}\mathbf{v}(t)$$

and

$$\mathbf{w}'(t) = \frac{d}{dt}\,(\mathbf{e}^{(t+s)\mathbf{A}}\boldsymbol{\xi}) = \frac{d}{dr}\,(\mathbf{e}^{r\mathbf{A}}\boldsymbol{\xi})\bigg|_{r=t+s}$$

$$= \mathbf{A}\mathbf{e}^{r\mathbf{A}}\boldsymbol{\xi}\bigg|_{r=t+s} = \mathbf{A}\mathbf{e}^{(t+s)\mathbf{A}}\boldsymbol{\xi} = \mathbf{A}\mathbf{w}(t)$$

Therefore, both $\mathbf{v}$ and $\mathbf{w}$ are solutions to (2), and since

$$\mathbf{v}(0) = \mathbf{I}\mathbf{e}^{s\mathbf{A}}\boldsymbol{\xi} = \mathbf{e}^{s\mathbf{A}}\boldsymbol{\xi}$$

and

$$\mathbf{w}(0) = \mathbf{e}^{(0+s)\mathbf{A}}\boldsymbol{\xi} = \mathbf{e}^{s\mathbf{A}}\boldsymbol{\xi}$$

we have that both $\mathbf{v}$ and $\mathbf{w}$ are solutions to (3) with $\boldsymbol{\eta} = \mathbf{e}^{s\mathbf{A}}\boldsymbol{\xi}$. By uniqueness of solutions to (3), $\mathbf{v}(t) \equiv \mathbf{w}(t)$ and hence

$$\mathbf{e}^{(t+s)\mathbf{A}}\boldsymbol{\xi} \equiv \mathbf{e}^{t\mathbf{A}}\mathbf{e}^{s\mathbf{A}}\boldsymbol{\xi}$$

for all numbers t and s and all vectors $\boldsymbol{\xi}$. Assertion (i) follows from this.

The fundamental result of this section is the *variation of constants formula* for systems. In fact the formula for systems is essentially the same as the one-dimensional case derived in the first chapter (see Theorem 1.2-1).

■ **Theorem 8.7-1**

Suppose that $\mathbf{A}$ and $\mathbf{f}$ are as above. Then for each vector $\boldsymbol{\eta}$ the solution $\mathbf{y}$ to (1) such that $\mathbf{y}(0) = \boldsymbol{\eta}$ can be written

$$\mathbf{y}(t) = \mathbf{e}^{t\mathbf{A}}\boldsymbol{\eta} + \int_0^t \mathbf{e}^{(t-s)\mathbf{A}}\mathbf{f}(s)\,ds \tag{4}$$

for all numbers t.

It follows from the product rule for differentiation that if $\mathbf{y}$ is the solution to (1) then

$$\frac{d}{dt}[\mathbf{e}^{-t\mathbf{A}}\mathbf{y}(t)] = \mathbf{e}^{-t\mathbf{A}}\mathbf{y}'(t) + \frac{d}{dt}(\mathbf{e}^{-t\mathbf{A}})\mathbf{y}(t)$$

$$= \mathbf{e}^{-t\mathbf{A}}\mathbf{y}'(t) - \mathbf{A}\mathbf{e}^{-t\mathbf{A}}\mathbf{y}(t)$$

However, $\mathbf{A}\mathbf{e}^{-t\mathbf{A}} = \mathbf{e}^{-t\mathbf{A}}\mathbf{A}$ by (ii) of Lemma 1, and it follows that

$$\frac{d}{dt}[\mathbf{e}^{-t\mathbf{A}}\mathbf{y}(t)] = \mathbf{e}^{-t\mathbf{A}}\mathbf{y}'(t) - \mathbf{e}^{-t\mathbf{A}}\mathbf{A}\mathbf{y}(t) \tag{5}$$

Therefore, if $\mathbf{y}$ is a solution to (1), then

$$\mathbf{y}'(t) - \mathbf{A}\mathbf{y}(t) = \mathbf{f}(t)$$

and multiplying each side of this equation by $\mathbf{e}^{-t\mathbf{A}}$, we obtain

$$\mathbf{e}^{-t\mathbf{A}}\mathbf{y}'(t) - \mathbf{e}^{-t\mathbf{A}}\mathbf{A}\mathbf{y}(t) = \mathbf{e}^{-t\mathbf{A}}\mathbf{f}(t)$$

By formula (5) we see that

$$\frac{d}{dt}[\mathbf{e}^{-t\mathbf{A}}\mathbf{y}(t)] = \mathbf{e}^{-t\mathbf{A}}\mathbf{f}(t)$$

Integrating each side of this equation from $s = 0$ to $s = t$ we have

$$\int_0^t \frac{d}{ds}[\mathbf{e}^{-s\mathbf{A}}\mathbf{y}(s)]\,ds = \int_0^t \mathbf{e}^{-s\mathbf{A}}\mathbf{f}(s)\,ds$$

and so

$$\mathbf{e}^{-t\mathbf{A}}\mathbf{y}(t) - \mathbf{e}^{-0\mathbf{A}}\mathbf{y}(0) = \int_0^t \mathbf{e}^{-s\mathbf{A}}\mathbf{f}(s)\,ds$$

But $\mathbf{y}(0) = \boldsymbol{\eta}$ and $e^{-0\mathbf{A}} = \mathbf{I}$, and it follows that

$$\mathbf{e}^{-t\mathbf{A}}\mathbf{y}(t) = \boldsymbol{\eta} + \int_0^t \mathbf{e}^{-s\mathbf{A}}\mathbf{f}(s)\,ds$$

By (i) of Lemma 1, $\mathbf{e}^{t\mathbf{A}}\mathbf{e}^{-t\mathbf{A}} = \mathbf{e}^{0\mathbf{A}} = \mathbf{I}$, so multiplying each side of the preceding equation by $\mathbf{e}^{t\mathbf{A}}$ leads to the expression

$$\mathbf{y}(t) = \mathbf{e}^{t\mathbf{A}}\boldsymbol{\eta} + \mathbf{e}^{t\mathbf{A}}\int_0^t \mathbf{e}^{-s\mathbf{A}}\mathbf{f}(s)\,ds$$

The formula (4) now follows from the fact that

$$\mathbf{e}^{t\mathbf{A}}\int_0^t \mathbf{e}^{-s\mathbf{A}}\mathbf{f}(s)\,ds = \int_0^t \mathbf{e}^{t\mathbf{A}}\mathbf{e}^{-s\mathbf{A}}\mathbf{f}(s)\,ds$$

$$= \int_0^t \mathbf{e}^{(t-s)\mathbf{A}}\mathbf{f}(s)\,ds$$

EXAMPLE 8.7-1

Consider the equation

$$\mathbf{y}' = \begin{pmatrix} 0 & 1 \\ -1 & 0 \end{pmatrix}\mathbf{y} + \begin{pmatrix} \sec t \\ 0 \end{pmatrix} \qquad \mathbf{y}(0) = \begin{pmatrix} 0 \\ 0 \end{pmatrix} \qquad 0 \le t < \frac{\pi}{2} \tag{6}$$

It is easy to check that

$$\mathbf{exp}\left[t\begin{pmatrix} 0 & 1 \\ -1 & 0 \end{pmatrix}\right] = \begin{pmatrix} \cos t & \sin t \\ -\sin t & \cos t \end{pmatrix}$$

and hence by formula (4) the solution to (6) is

$$\begin{aligned}
\mathbf{y}(t) &= \int_0^t \begin{pmatrix} \cos (t-s) & \sin (t-s) \\ -\sin (t-s) & \cos (t-s) \end{pmatrix}\begin{pmatrix} \sec s \\ 0 \end{pmatrix} ds \\
&= \int_0^t \begin{pmatrix} \cos (t-s) \sec s \\ -\sin (t-s) \sec s \end{pmatrix} ds \\
&= \int_0^t \begin{pmatrix} (\cos t \cos s + \sin t \sin s) \sec s \\ (-\sin t \cos s + \cos t \sin s) \sec s \end{pmatrix} ds \\
&= \int_0^t \begin{pmatrix} \cos t + \dfrac{\sin t \sin s}{\cos s} \\ -\sin t + \dfrac{\cos t \sin s}{\cos s} \end{pmatrix} ds \\
&= \begin{pmatrix} \cos t \displaystyle\int_0^t ds + \sin t \int_0^t \frac{\sin s}{\cos s} ds \\ -\sin t \displaystyle\int_0^t ds + \cos t \int_0^t \frac{\sin s}{\cos s} ds \end{pmatrix} \\
&= \begin{pmatrix} t \cos t - \ln (\cos t) \sin t \\ -t \sin t - \ln (\cos t) \cos t \end{pmatrix}
\end{aligned}$$

for all t in $[0, \pi/2)$.

If the nonhomogeneous term $\mathbf{f}$ is of a special form, then a technique similar to the method of undetermined coefficients can be used to obtain a particular solution (see Sections 2.4 and 7.3). As an illustration, suppose that γ is a real or complex number, that $\boldsymbol{\zeta}$ is a vector with real or complex components, and consider the equation

$$\mathbf{y}' = \mathbf{A}\mathbf{y} + e^{\gamma t}\boldsymbol{\zeta} \tag{7}$$

The procedure is to *assume that (7) has a particular solution of the form* $\mathbf{y}_p(t) \equiv e^{\gamma t}\boldsymbol{\xi}$ for some vector $\boldsymbol{\xi}$, and then determine such a vector $\boldsymbol{\xi}$ whenever possible. Since $\mathbf{y}_p'(t) \equiv \gamma e^{\gamma t}\boldsymbol{\xi}$ in such a case, substituting $\mathbf{y}_p$ into (7) shows that the equation

$$\gamma e^{\gamma t}\boldsymbol{\xi} = \mathbf{A}(e^{\gamma t}\boldsymbol{\xi}) + e^{\gamma t}\boldsymbol{\zeta}$$

must be satisfied. Since $e^{\gamma t}$ is a factor of both sides of this equation, it follows that $\boldsymbol{\xi}$ should satisfy the equation $\gamma\boldsymbol{\xi} - \mathbf{A}\boldsymbol{\xi} = \boldsymbol{\zeta}$. Therefore, we have the following result:

● **Proposition 1**

Suppose that γ is a real or complex number and that $\boldsymbol{\zeta}$ is a vector with real or complex components. If there exists any vector $\boldsymbol{\xi}$ such that

$$\gamma\boldsymbol{\xi} - \mathbf{A}\boldsymbol{\xi} = \boldsymbol{\zeta} \tag{8}$$

then $\mathbf{y}_p(t) \equiv e^{\gamma t}\boldsymbol{\xi}$ is a particular solution to (7).

If γ is not an eigenvalue of $\mathbf{A}$, then $\det(\gamma\mathbf{I} - \mathbf{A}) \neq 0$ and (8) always has a unique solution for every $\boldsymbol{\zeta}$. Note, however, that (8) may or may not have a solution if γ is an eigenvalue of $\mathbf{A}$. If $e^{\gamma t}\boldsymbol{\zeta}$ is complex-valued and $\mathbf{y}_p(t) \equiv e^{\gamma t}\boldsymbol{\xi}$ is a solution to (7), then $\mathbf{x}(t) \equiv \text{Re}\,(e^{\gamma t}\boldsymbol{\xi})$ and $\mathbf{y}(t) \equiv \text{Im}\,(e^{\gamma t}\boldsymbol{\xi})$ (see Section 8.4*b* for a discussion of vectors with complex components) are vectors with only real components for each t. Moreover, since the components of $\mathbf{A}$ are real,

$$\begin{aligned}\mathbf{x}'(t) &= \text{Re}\,[\mathbf{A}\mathbf{y}_p(t) + e^{\gamma t}\boldsymbol{\zeta}] \\ &= \text{Re}\,[\mathbf{A}\mathbf{y}_p(t)] + \text{Re}\,(e^{\gamma t}\boldsymbol{\zeta}) \\ &= \mathbf{A}\,\text{Re}\,[\mathbf{y}_p(t)] + \text{Re}\,(e^{\gamma t}\boldsymbol{\zeta})\end{aligned}$$

and it follows that

$$\mathbf{x}'(t) = \mathbf{A}\mathbf{x}(t) + \text{Re}\,(e^{\gamma t}\boldsymbol{\zeta}) \qquad \text{if } x(t) \equiv \text{Re}\,[\mathbf{y}_p(t)] \tag{9}$$

Similarly,

$$\mathbf{y}'(t) = \mathbf{A}\mathbf{y}(t) + \text{Im}\,(e^{\gamma t}\boldsymbol{\zeta}) \qquad \text{if } y(t) \equiv \text{Im}\,[\mathbf{y}_p(t)] \tag{10}$$

The preceding technique for determining particular solutions to (1) is illustrated with two examples.

▶ **EXAMPLE 8.7-2**

Consider the system

$$\mathbf{y}' = \begin{pmatrix} 1 & -2 & 0 \\ -2 & -1 & -2 \\ 1 & 1 & 6 \end{pmatrix}\mathbf{y} + \begin{pmatrix} e^{2t} \\ 0 \\ -e^{2t} \end{pmatrix} \tag{11}$$

Assuming that

$$\mathbf{y}_p(t) = e^{2t}\begin{pmatrix}\xi_1\\ \xi_2\\ \xi_3\end{pmatrix}$$

leads to the system

$$2\begin{pmatrix}\xi_1\\ \xi_2\\ \xi_3\end{pmatrix} - \begin{pmatrix}1 & -2 & 0\\ -2 & -1 & -2\\ 1 & 1 & 6\end{pmatrix}\begin{pmatrix}\xi_1\\ \xi_2\\ \xi_3\end{pmatrix} = \begin{pmatrix}1\\ 0\\ -1\end{pmatrix}$$

[see equation (8) in Proposition 1]. Therefore, the algebraic system

$$\begin{aligned}\xi_1 + 2\xi_2 \qquad\quad &= 1\\ 2\xi_1 + 3\xi_2 + 2\xi_3 &= 0\\ -\xi_1 - \ \xi_2 - 4\xi_3 &= -1\end{aligned}$$

needs to be solved. Using the first equation to eliminate ξ_1 from the second two,

$$\begin{aligned}\xi_1 + 2\xi_2 \qquad\quad &= 1\\ -\ \xi_2 + 2\xi_3 &= -2\\ \xi_2 - 4\xi_3 &= 0\end{aligned}$$

and then using the second to eliminate ξ_2 from the third,

$$\begin{aligned}\xi_1 + 2\xi_2 \qquad\quad &= 1\\ -\ \xi_2 + 2\xi_3 &= -2\\ -\ 2\xi_3 &= -2\end{aligned}$$

Therefore, $\xi_3 = 1$, $\xi_2 = 4$, and $\xi_1 = -7$, and it follows that

$$\mathbf{y}_p(t) \equiv e^{2t}\begin{pmatrix}-7\\ 4\\ 1\end{pmatrix}$$

is a particular solution to (11).

► **EXAMPLE 8.7-3**

Consider the system

$$\mathbf{y}' = \begin{pmatrix}0 & 1\\ 1 & 0\end{pmatrix}\mathbf{y} + \begin{pmatrix}\cos t\\ \sin t\end{pmatrix} \tag{12}$$

Since

$$\begin{aligned}e^{it}\begin{pmatrix}1\\ -i\end{pmatrix} &= (\cos t + i\sin t)\begin{pmatrix}1\\ -i\end{pmatrix}\\ &= \begin{pmatrix}\cos t\\ \sin t\end{pmatrix} + i\begin{pmatrix}\sin t\\ -\cos t\end{pmatrix}\end{aligned}$$

it follows that if $\mathbf{y}_p$ is a solution to

$$\mathbf{y}_p' = \begin{pmatrix} 0 & 1 \\ 1 & 0 \end{pmatrix} \mathbf{y}_p + e^{it} \begin{pmatrix} 1 \\ -i \end{pmatrix}$$

then $x_p(t) \equiv \text{Re}\,[y_p(t)]$ is a solution to (12)—see the system (9). Assuming that

$$\mathbf{y}_p = e^{it} \begin{pmatrix} \xi_1 \\ \xi_2 \end{pmatrix}$$

leads to the system

$$i \begin{pmatrix} \xi_1 \\ \xi_2 \end{pmatrix} - \begin{pmatrix} 0 & 1 \\ 1 & 0 \end{pmatrix} \begin{pmatrix} \xi_1 \\ \xi_2 \end{pmatrix} = \begin{pmatrix} 1 \\ -i \end{pmatrix}$$

or

$$\begin{aligned} i\xi_1 - \xi_2 &= 1 \\ -\xi_1 + i\xi_2 &= -i \end{aligned}$$

Therefore, $\xi_1 = 0$, $\xi_2 = -1$, so

$$\begin{aligned} \mathbf{y}(t) &= \text{Re}\,(\cos t + i \sin t) \begin{pmatrix} 0 \\ -1 \end{pmatrix} \\ &= \text{Re} \begin{pmatrix} 0 \\ -\cos t - i \sin t \end{pmatrix} \\ &= \begin{pmatrix} 0 \\ -\cos t \end{pmatrix} \end{aligned}$$

is a particular solution to (12). Since

$$\mathbf{exp} \left[t \begin{pmatrix} 0 & 1 \\ 1 & 0 \end{pmatrix} \right] = \begin{pmatrix} \cosh t & \sinh t \\ \sinh t & \cosh t \end{pmatrix}$$

for all t, it follows that the general solution to (12) can be written in the form

$$\begin{aligned} \mathbf{y}(t) &= \begin{pmatrix} \cosh t & \sinh t \\ \sinh t & \cosh t \end{pmatrix} \begin{pmatrix} c_1 \\ c_2 \end{pmatrix} + \begin{pmatrix} 0 \\ -\cos t \end{pmatrix} \\ &= \begin{pmatrix} c_1 \cosh t + c_2 \sinh t \\ c_1 \sinh t + c_2 \cosh t - \cos t \end{pmatrix} \end{aligned}$$

where c_1 and c_2 are arbitrary constants. The constants c_1 and c_2 can be determined so that any given initial value for (12) is satisfied.

PROBLEMS

8.7-1. Determine a particular solution to each of the following equations:

(a) $\mathbf{y}' = \begin{pmatrix} 0 & 1 \\ -1 & 0 \end{pmatrix} \mathbf{y} + \begin{pmatrix} \tan t \\ 0 \end{pmatrix} \qquad 0 \le t < \dfrac{\pi}{2}$

(b) $\mathbf{y}' = \begin{pmatrix} 0 & 1 \\ -1 & 0 \end{pmatrix}\mathbf{y} + \begin{pmatrix} 0 \\ \csc t \end{pmatrix} \qquad \frac{\pi}{2} \le t < \pi$

(c) $\mathbf{y}' = \begin{pmatrix} -2 & 5 \\ -1 & 2 \end{pmatrix}\mathbf{y} + \begin{pmatrix} 0 \\ \sec t \end{pmatrix} \qquad 0 \le t < \frac{\pi}{2}$

(d) $\mathbf{y}' = \begin{pmatrix} -2 & 5 \\ -1 & 2 \end{pmatrix}\mathbf{y} + e^t\begin{pmatrix} 1 \\ 2 \end{pmatrix}$

(e) $\mathbf{y}' = \begin{pmatrix} 0 & 1 \\ 1 & 0 \end{pmatrix}\mathbf{y} + \begin{pmatrix} \cos 2t \\ -\sin 2t \end{pmatrix}$

8.7-2. Determine first the general solution for each of the following systems and then determine a solution that satisfies the given initial data.

(a) $\mathbf{y}' = \begin{pmatrix} 1 & 2 \\ -2 & 1 \end{pmatrix}\mathbf{y} + \begin{pmatrix} e^t \\ -3e^t \end{pmatrix} \qquad \mathbf{y}(0) = \begin{pmatrix} -2 \\ 2 \end{pmatrix}$

(b) $\mathbf{y}' = \begin{pmatrix} -2 & 5 \\ -1 & 2 \end{pmatrix}\mathbf{y} + \begin{pmatrix} 0 \\ \sec t \end{pmatrix} \qquad \mathbf{y}(0) = \begin{pmatrix} 2 \\ 3 \end{pmatrix}$ (See Problem 8.7-1*c*.)

(c) $\mathbf{y}' = \begin{pmatrix} 2 & 1 & -2 \\ 0 & 3 & -2 \\ 3 & 1 & -3 \end{pmatrix}\mathbf{y} + \begin{pmatrix} -5e^{3t} \\ -6e^{3t} \\ -9e^{3t} \end{pmatrix} \qquad \mathbf{y}(0) = \begin{pmatrix} 0 \\ 0 \\ 0 \end{pmatrix}$

8.7-3. Suppose that $\mathbf{y}_1$ and $\mathbf{y}_2$ satisfy

$$\mathbf{y}_1' = \mathbf{A}\mathbf{y}_1 + \mathbf{f}_1(t) \qquad \text{and} \qquad \mathbf{y}_2' = \mathbf{A}\mathbf{y}_2 + \mathbf{f}_2(t)$$

respectively. If $\mathbf{y}(t) \equiv \mathbf{y}_1(t) + \mathbf{y}_2(t)$, show that $\mathbf{y}$ satisfies

$$\mathbf{y}' = \mathbf{A}\mathbf{y} + \mathbf{f}_1(t) + \mathbf{f}_2(t)$$

8.7-4. Using Problem 8.7-3 and Proposition 1, determine particular solutions to the following systems:

(a) $\mathbf{y}' = \begin{pmatrix} 1 & 2 \\ 2 & 1 \end{pmatrix}\mathbf{y} + \begin{pmatrix} e^t \\ 38 \sin t \end{pmatrix}$

(b) $\mathbf{y}' = \begin{pmatrix} 0 & 1 & 0 \\ 0 & 0 & 1 \\ 1 & -1 & 1 \end{pmatrix}\mathbf{y} + \begin{pmatrix} 2e^{-t} \\ e^{2t} \\ e^{-t} \end{pmatrix}$

***8.7-5.** Suppose that p is a real polynomial of degree $m \ge 0$ and consider the nonhomogeneous equation

$$\mathbf{y}' = \mathbf{A}\mathbf{y} + p(t)e^{\gamma t}\boldsymbol{\zeta} \tag{13}$$

We assume that a particular solution $\mathbf{y}_p$ to (13) has the form

$$\mathbf{y}_p(t) = e^{\gamma t}[\boldsymbol{\xi}_0 + \boldsymbol{\xi}_1 t + \cdots + \boldsymbol{\xi}_k t^k] \tag{14}$$

for some integer k. If γ is not an eigenvalue of $\mathbf{A}$, show that the system (13) has a solution of the form (14) with $k = m$.

***8.7-6.** Using Problem 8.7-5, determine particular solutions to each of the following systems

(a) $\mathbf{y}' = \begin{pmatrix} 0 & 4 \\ 4 & 0 \end{pmatrix} \mathbf{y} + \begin{pmatrix} t \\ -3 \end{pmatrix}$

(b) $\mathbf{y}' = \begin{pmatrix} 2 & -5 \\ 1 & -2 \end{pmatrix} \mathbf{y} + \begin{pmatrix} e^{2t} \\ te^{2t} \end{pmatrix}$

(c) $\mathbf{y}' = \begin{pmatrix} 4 & 6 & 6 \\ 1 & 3 & 2 \\ -1 & -4 & 2 \end{pmatrix} \mathbf{y} + \begin{pmatrix} 3 \\ 2-t \\ t \end{pmatrix}$

(d) $\mathbf{y}' = \begin{pmatrix} 10 & 10 & 12 \\ -6 & -6 & -7 \\ -2 & -2 & -2 \end{pmatrix} \mathbf{y} + \begin{pmatrix} t \\ t \\ 2t \end{pmatrix}$

[Assume $\mathbf{y}$ has the form (14) with $k = 2$, $\gamma = 0$.]

Remarks The study of systems of linear differential equations using matrix methods and terminology can be found in Hirsch and Smale [7] and Strang [13]. See also Petrovski [12].

Bibliography

1. Birkhoff, G., and G. C. Rota: *Ordinary Differential Equations*, Ginn, Boston, 1962.
2. Churchill, R. V.: *Operational Mathematics*, McGraw-Hill, New York, 1958.
3. Coddington, E. A.: *An Introduction to Ordinary Differential Equations*, Prentice-Hall, Englewood Cliffs, N.J., 1961.
4. Coddington, E. A., and N. Levinson: *Theory of Ordinary Differential Equations*, McGraw-Hill, New York, 1955.
5. Conte, S. E., and C. de Boor: *Elementary Numerical Analysis*, McGraw-Hill, New York, 1980.
6. Hildebrand, F. B.: *Advanced Calculus for Applications*, Prentice-Hall, Englewood Cliffs, N.J., 1962.
7. Hirsch, M. W., and S. Smale: *Differential Equations, Dynamical Systems, and Linear Algebra*, Academic Press, New York, 1974.
8. Ince, E. L.: *Ordinary Differential Equations*, Longmans, Green, London, 1929.
9. Kamke, E.: *Differentialgleichungen Loesungsmethoden und Loesungen*, Chelsea, New York, 1971.
10. Kreyszig, E.: *Advanced Engineering Mathematics*, Wiley, New York, 1972.
11. McCormick, J. M., and M. G. Salvadori: *Numerical Methods in Fortran*, Prentice-Hall, Englewood Cliffs, N.J., 1964.
12. Petrovski, I. G.: *Ordinary Differential Equations*, Prentice-Hall, Englewood Cliffs, N.J., 1966.
13. Strang, G.: *Linear Algebra and its Applications*, Academic Press, New York, 1976.
14. Struble, R. A.: *Nonlinear Differential Equations*, McGraw-Hill, New York, 1962.
15. Wylie, C. R., and L. C. Barrett: *Advanced Engineering Mathematics*, McGraw-Hill, New York, 1982.

Answers to Selected Exercises

CHAPTER 1

Section 1.1

1.1-2. (a) $y(t) = t^3/3 + 3t + c$

(b) $y(t) = te^{2t}/2 - e^{2t}/4 + c$

(c) $y(t) = (-\cos 3t)/3 + c$

(d) $y(t) = t/2 - (\sin 2t)/4 + c$

(e) $y(t) = [\ln (t^2 + 4)]/2 + c$

(g) $y(t) = 2t^{3/2}/3 - t + 2t^{1/2} - 2 \ln (\sqrt{t} + 1) + c$

1.1-3. (a) $y(t) = 3e^{2t} + 2 \quad (c = 3)$

(b) $y(t) = (2t + 7)^{-1/2} \quad (c = 7)$

(c) $y(t) = -(60 + 2e^t)^{1/2} \quad (c = 60)$

(d) $y(t) = te^{2t}/2 - e^{2t}/4 + 5 - e^2/4 \quad (c = 5 - e^2/4)$

(e) $y(t) = t/2 - (\sin 2t)/4 + 3 - \pi/12 + \sqrt{3}/8 \quad (c = 3 - \pi/12 + \sqrt{3}/8)$

1.1-4.

(a)

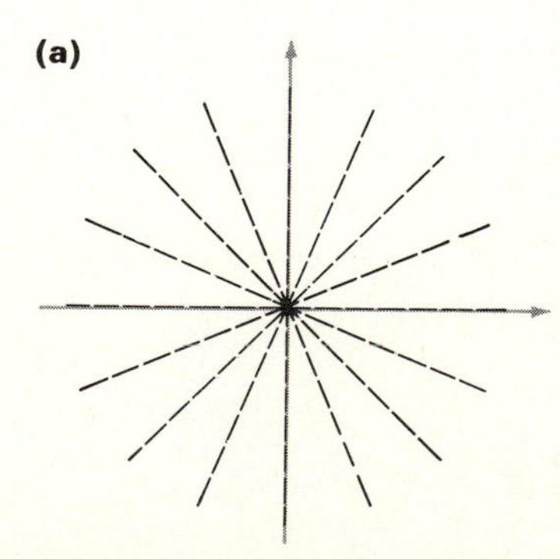

(b)

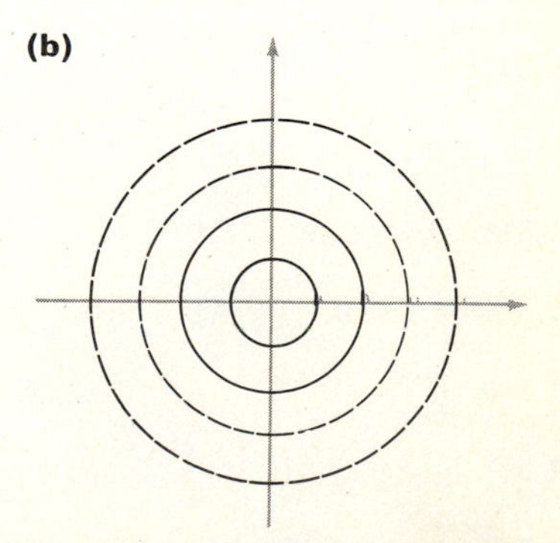

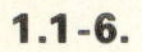

1.1-6.

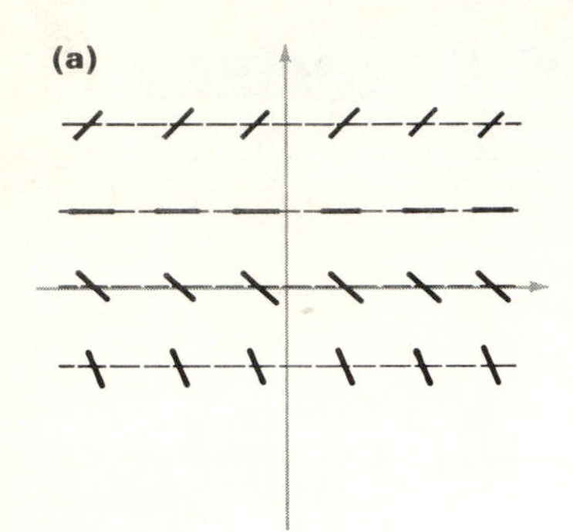

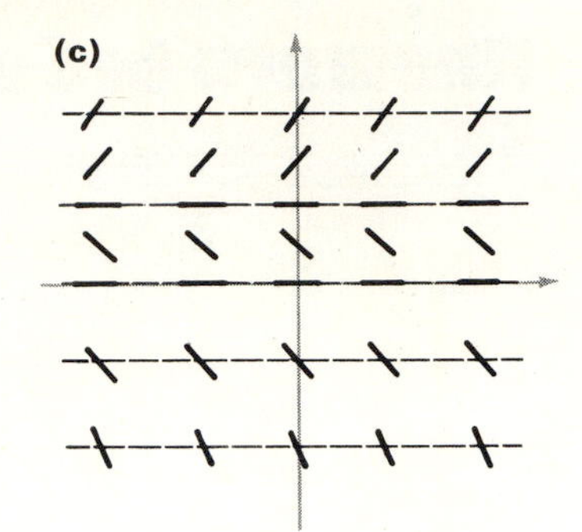

Section 1.2

1.2-1. **(a)** $y(t) = ce^{t + t^3/3}$
(b) $y(t) = ce^{-t}$
(c) $y(t) = ce^{2t} - e^{-3t}/5$
(d) $y(t) = ce^{2t} + te^{2t}$
(e) $y(t) = ce^{-t} + t - 1$
(f) $y(t) = ct^{-2} - t^{-1} \cos t + t^{-2} \sin t$
(g) $y(t) = t \sec t + c \sec t$
(h) $y(t) = (t^2 + 1)\{c + \arctan t + [\ln (t^2 + 1)]/2\}$

1.2-2. **(a)** $y(t) = 2e^t$
(b) $y(t) = e^{2t}/3$
(c) $y(t) = t^3/2 - 5t/2$
(d) $y(t) = \sin t$
(e) $y(t) = 6(t + 1)^2$

1.2-3. **(a)** $y(t) = (7 \cos^2 2t)/2 + \frac{1}{4}$
(b) $y(t) = (t \ln t + 1 - t)/(\ln t)$
(c) $y(t) = [-3t + 2 \ln (t + 1) + \frac{11}{6} - 2 \ln \frac{3}{2}][(1 + t)/(1 - t)]$

1.2-5. **(a)** $y = ce^{-\cos t}$ (c any constant)
(b) $y = 0$
(c) $y = (\cos \pi t + \pi \sin \pi t)/(1 + \pi^2)$
(d) No 2π-periodic solutions

Section 1.3

1.3-1. $T = (3 \ln 2)/(\ln \frac{5}{4}) \cong 9.32$. It takes $(3 \ln 4)/(\ln \frac{5}{4}) \cong 18.64$ years for 75 percent to disintegrate

1.3-2. $T = (6 \ln 2)/(\ln \frac{10}{9}) \cong 39.47$ months, $N'(18) = 109.35$ cts/min

1.3-3. $N(100)$ is $100 \times 2^{-1/5}\%$ ($\cong 87.05\%$) of N_0
$N(250)$ is $100 \times 2^{-1/2}\%$ ($\cong 70.71\%$) of N_0
$N(1000)$ is 25% of N_0.

1.3-4. $N(t) = N_0 e^{-(\ln N_0 - \ln N_1)(t - T_0)/(T_1 - T_0)}$

1.3-5. $Q(t) = 10 - 10e^{-t/5}$ and $t_1 = 5 \ln 2$

1.3-6. $Q(t) = 40e^{-t/20}$ and $t_1 = 20 \ln 10$

1.3-7. $b_5 = 0.1 \times (1 - e^{-6})$ and $b_{10} = 0.2 \times (1 - e^{-6})$

1.3-8. $Q(t) = 8000(20 + t)^{-2}$ $Q(40) = \frac{20}{9}$

1.3-9. $Q(t) = (64 - t)/2 - (64 - t)^2/128 \qquad Q_{max} = 8 = Q(32)$

1.3-10. $v(\infty) = -8$ and $v[(\ln 4)/4] = -6$

1.3-11. $v(t) = 32e^{-2t} - 16$, $y(t) = 16 - 16t - 16e^{-2t}$, and it rises to maximum height of $8(1 - \ln 2) \cong 2.455$ in $t = (\ln 2)/2 \cong 0.3466$

1.3-12. $t = 2v_0/g$

1.3-15. $\bar{v}(t) = v_0 - g(t - t_0)$, which is indeed the solution to (8) with $\rho = 0$. Also, $\bar{y}(t) = -g(t - t_0)^2/2 + v_0(t - t_0) + y_0$, and hence $\bar{y}' = \bar{v}$

1.3-16. $t_m \to v_0/g$ and $y_m \to v_0^2/2g$ as $\rho \to 0+$

1.3-17. **(a)** $T(t) = S_0 + (T_0 - S_0)e^{-\rho(t - t_0)}$
(b) $T(16) = 97.5°$ $\quad t_{60} = (8 \ln 4)/(\ln \frac{4}{3}) \cong 38.55$
(c) $S_0 = 83.333°$ and $T(3) = 84.4°$

Section 1.4

1.4-1. **(a)** $-y^{-1} = t^2 - \frac{1}{4}$
(b) $\ln |y^2 - 1| = 2 \ln |t| + \ln 3 - 1 + t^2$
(c) $\ln\left(\dfrac{|y|}{\sqrt{1 + y^2}}\right) = e^t - 1 - \ln \sqrt{2}$
(d) $\sin y = \sin t + \frac{1}{2}$

1.4-2. **(a)** $y(t) = \dfrac{4}{1 - 4t^2} \qquad -\frac{1}{2} < t < \frac{1}{2}$
(b) $y(t) = \dfrac{2}{\sqrt{8t^2 + 1}} \qquad -\infty < t < \infty$
(c) $y(t) = -\sqrt{2 \sin t + 16} \qquad -\infty < t < \infty$
(d) $y(t) = \sqrt{2 \sin t + 1} \qquad -\dfrac{\pi}{6} < t < \dfrac{7\pi}{6}$
(e) $y(t) = 2 \tan 2t \qquad -\dfrac{\pi}{4} < t < \dfrac{\pi}{4}$

1.4-3. **(a)** $y(t) = y_0 e^{3(t^{2/3} - 1)/2} \qquad y(t) \to y_0 e^{-3/2}$ as $t \to 0+$
(b) $y(t) = y_0 t \qquad y(t) \to 0$ as $t \to 0+$
(c) $y(t) = e^{(1 - t^{-2})/2} \qquad y(t) \to 0$ as $t \to 0+$

1.4-4. **(a)** $y(t) = \dfrac{-2}{\sqrt{8t + 1}} \qquad -\frac{1}{8} < t < \infty$
(b) $y(t) = \dfrac{1}{1 + e^t} \qquad -\infty < t < \infty$
(c) $y(t) = \arctan t \qquad -\infty < t < \infty$
(d) $y(t) = \arctan (t + 1) \qquad -\infty < t < \infty$
(e) $y(t) = \dfrac{3 - e^{4t}}{1 + e^{4t}} \qquad -\infty < t < \infty$
(f) $y(t) = \dfrac{5e^{4t} + 3}{1 - 5e^{4t}} \qquad -\dfrac{\ln 5}{4} < t < \infty$

(h) $y(t) = 2\tan\left(2t + \frac{\pi}{4}\right) - 1 \qquad \frac{-3\pi}{8} < t < \frac{\pi}{8}$

(i) $y(t) = \dfrac{-3 - \sqrt{1 + 8e^t}}{2} \qquad -\infty < t < \infty$

1.4-5. If $y' = -y^\alpha$, $y(0) = y_0 > 0$ where $0 < \alpha < 1$, then

$$y(t) = \begin{cases} [y_0^{1-\alpha} - (1-\alpha)t]^{(1-\alpha)^{-1}} & \text{if } t < \dfrac{y_0^{1-\alpha}}{1-\alpha} \\ 0 & \text{if } t > \dfrac{y_0^{1-\alpha}}{1-\alpha} \end{cases}$$

1.4-6. **(a)** $y(t) = \tan(t + \pi/4) - t - 1 \qquad -5\pi/4 < t < \pi/4$

(b) $y(t) = t + 1 \qquad -\infty < t < \infty$

1.4-7. **(a)** $x(t) = 3e^{7t} - 2 \qquad -\infty < t < \infty$

(b) $r(s) = -(4 - 2\sin s)^{-1/2} \qquad -\infty < s < \infty$

(c) $r(s) = (1 - 2\sin s)^{-1/2} \qquad -7\pi/6 < s < \pi/6$

(d) $P(u) = (u^2 \ln u)/2 - u^2/4 - e^2/4 \qquad 0 < u < \infty$

(e) $u(t) = t^{-1} \qquad 0 < t < \infty$

(f) $Q(r) = [-\ln(3 - 2e^r)]/2 \qquad -\infty < r < \ln\frac{3}{2}$

(g) $y(t) = t\sin t - (\pi \sin t)/2 \qquad 0 < t < \pi$

(h) $y(x) = -1 + \sqrt{x^2 + 2x + 16} \qquad -\infty < x < \infty$

Equations (a), (d), and (e) are both linear and separable.

Section 1.5

1.5-1. **(a)**

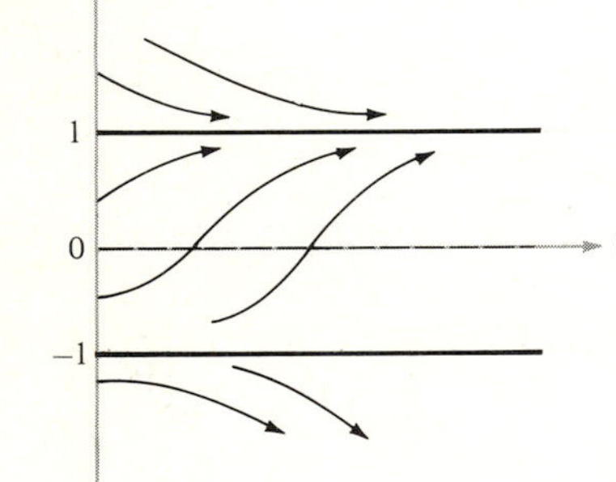

CP = $\{1, -1\}$
Convexity change: $y = 0$

(b)

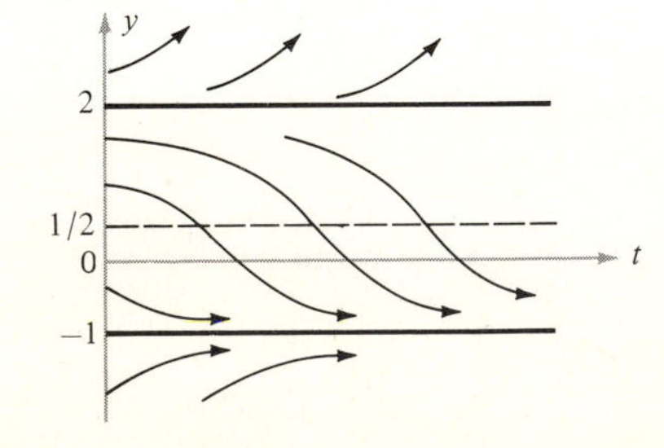

CP = $\{2, -1\}$
Convexity change: $y = \frac{1}{2}$

(d)

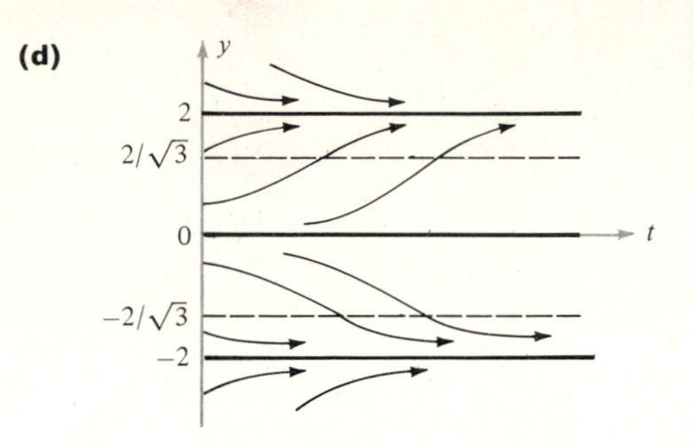

$\text{CP} = \{2, 0, -2\}$

Convexity change: $y = \pm \dfrac{2}{\sqrt{3}}$

(f)

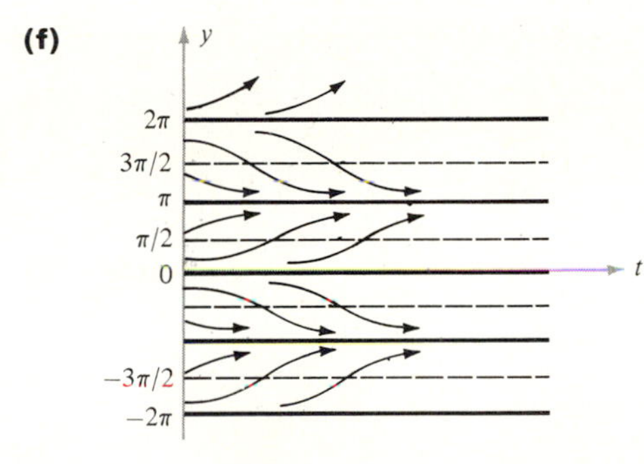

$\text{CP} = \{n\pi : n = 0, \pm 1, \pm 2, \ldots\}$

Convexity change: $y = \pm \dfrac{\pi}{2}, \pm \dfrac{3\pi}{2}, \pm \dfrac{5\pi}{2}, \ldots$

(h)

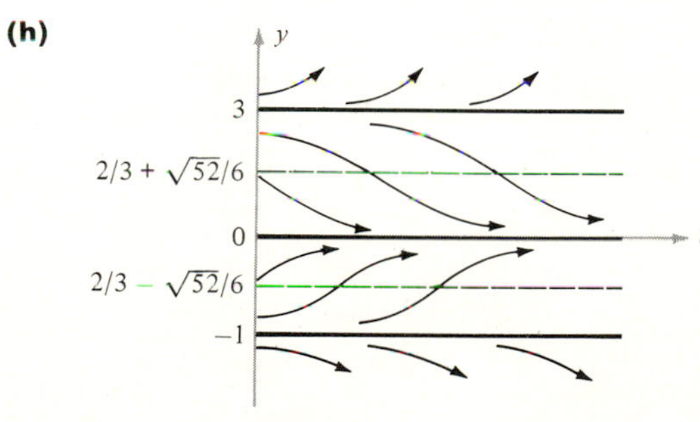

$\text{CP} = \{3, 0, -1\}$

Convexity change: $y = (4 \pm \sqrt{52})/6$

(k)

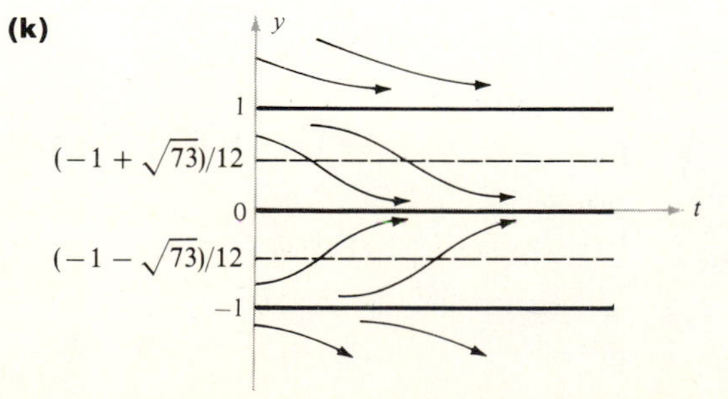

$\text{CP} = \{1, 0, -1\}$
Convexity change: $y = (-1 \pm \sqrt{73})/12$

1.5-2. (a)

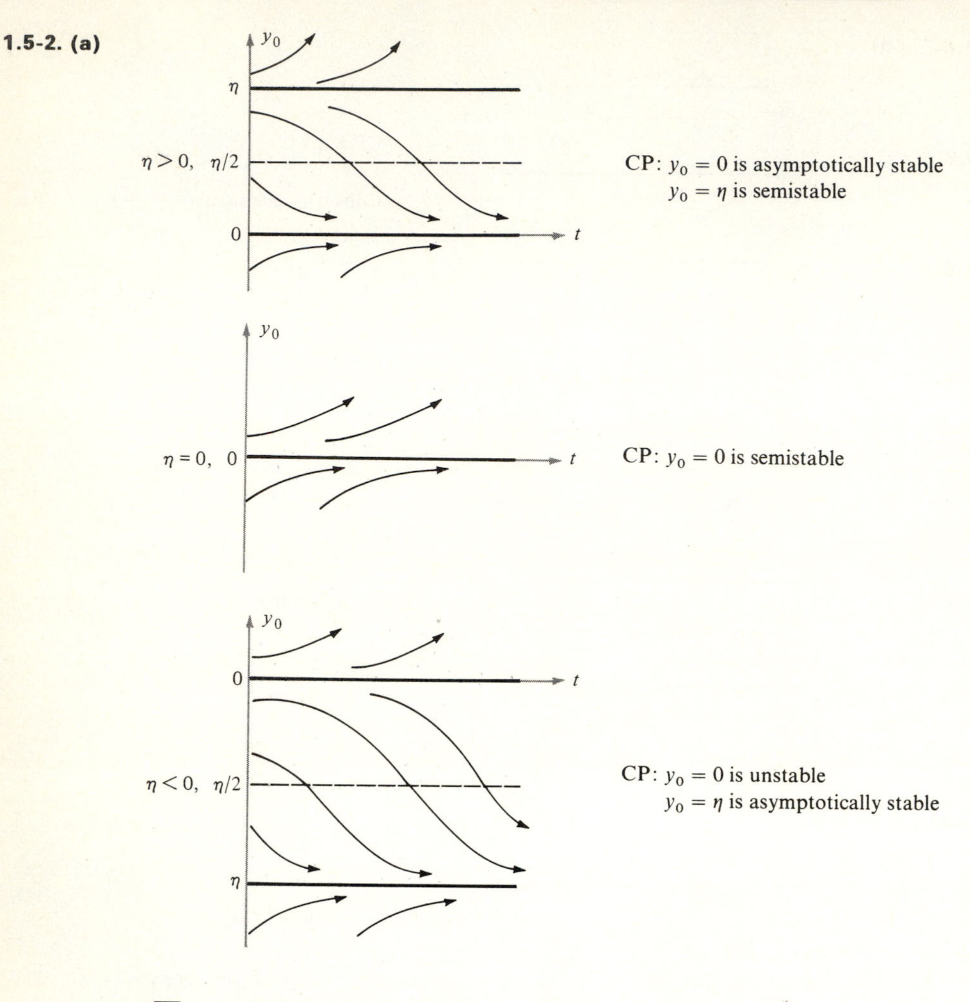

1.5-3. $v(t) = \sqrt{\dfrac{mg}{\rho}}\left(\dfrac{1 - e^{2t\sqrt{mg/\rho}}}{1 + e^{2t\sqrt{mg/\rho}}}\right)$ and $v(t) \to -\sqrt{\dfrac{mg}{\rho}}$ as $t \to +\infty$

1.5-5. (a) If n is odd, $z = 0$ is unstable, and if n is even, then $z = 0$ is semistable.

(b) If n is odd, then $z = 0$ is stable, and if n is even, then $z = 0$ is semistable.

Section 1.6

1.6-1. (a) $y(t) = \dfrac{2t^2}{3 - 2t} \qquad 0 < t < \frac{3}{2}$

(b) $y(t) = t(\frac{5}{4} - t)^{1/2} \qquad 0 < t < \frac{5}{4}$

(c) $y^2 + 2ty - 14 - t^2 = 0$ and explicitly
$y(t) = -t + (2t^2 + 14)^{1/2} \qquad t > 0$

(d) $y(t) = \dfrac{2t^3}{32 - t^2} \qquad 0 < t < 4\sqrt{2}$

(e) $y(t) = (6t + t^2)^{1/2} \qquad t > 0$

1.6-2. (a) $y = -\sqrt{e^{2t} - t^2}$ $\quad -\alpha < t < \infty$ $\quad$ where $\alpha > 0$ is such that $e^{-2\alpha} = \alpha^2$

(b) $y = \exp[(e^{2t} - 2t - 1)/4]$

(c) $y = \arcsin(1 - e^{t^2})$ $\quad -\sqrt{\ln 2} < t < \sqrt{\ln 2}$

1.6-3. (a) $y = (1 - 8e^{t^2}/9)^{-1/2}$ $\quad -\sqrt{\ln \frac{9}{8}} < t < \sqrt{\ln \frac{9}{8}}$

(b) $y = (1 - t)^{-1}$ $\quad -\infty < t < 1$

(c) $y = (2t^{-1}/3 - 5t^2/12)^{1/2}$ $\quad 0 < t < 2/\sqrt[3]{5}$

1.6-4. (a) $y^2 + 4ty - t^2 = c$ defines y implicitly

(b) $y = \frac{3}{2}t \tan\left(\frac{\sqrt{3}}{2} \ln t + c\right) - \frac{1}{2}t$ $\quad t > 0$

1.6-5. (b) $|(y + t)^3| = c(y + 2t - 1)^4$ $\quad$ defines y implicitly

1.6-7. (a) $u(s) = \dfrac{1}{1 - \ln s}$ $\quad 0 < s < e$

(b) $p(t) = \dfrac{-t + \sqrt{3t^2 + 4}}{2}$ $\quad t > 0$

(c) $x(r) = \dfrac{1}{1 - r + e^{-r}}$ $\quad r > 0$ $\quad$ and $\quad 1 + e^{-\gamma} > \gamma$

(d) $Q(s) = e^s - (e^s + e^2)/s$ $\quad s > 0$

(e) $u(t) = -(2e^{2t} - e^t)^{1/2}$ $\quad -\ln 2 < t < \infty$ $\quad$ (use the transformation $v = u^2$)

(f) $u(v) = v\sqrt{9v^2 - 4}$ $\quad \frac{2}{3} < v < \infty$

(g) $T(s) = 2s^2 - 153/s^2$ $\quad 0 < s < \infty$

(h) $y(x) = \sqrt{6x^2 + 3}$ $\quad -\infty < x < \infty$

Section 1.7

1.7-1. (a) Exact, $U(x, y) = x^3/3 + (\sin 2y)/2$

(b) Exact, $U(x, y) = x^2y^3 - y^2x$

(c) Not exact but variables separate and $U(x, y) = [\ln(1 + x^2)]/2 + [\ln(1 + y^3)]/3$

(d) Not exact but variables separate and $U(x, y) = \ln|y| + y + x$

(e) Exact, $U(x, y) = -(y^2 + ye^x + \cos x)$

(f) Not exact but variables separate and $U(x, y) = xe^x - e^x + \ln|y| - y$

1.7-2. (a) $U(x, y) = \frac{1}{2}x^2y^2 + \frac{1}{2}x^4$ $\quad$ (take $\mu = x$)

(b) $U(x, y) = \frac{1}{2}x^2 + \dfrac{x}{y} - \dfrac{1}{y}$ $\quad$ (take $\mu = y^{-3}$)

(c) $U(x, y) = 2\ln x - \dfrac{y^2}{2x^2}$ $\quad$ (take $\mu = x^{-3}$)

(d) $U(x, y) = -4x^{-3/4}y^{9/4}/3 + 4x^{5/4}y^{5/4}/5$ $\quad$ (take $\mu = x^{-7/4}y^{1/4}$)

1.7-4. (a) $\mu = e^x$ $\quad$ and $\quad U(x, y) = (y - x + 1)e^x$

(b) $\mu = \sec^3 x$ $\quad$ and $\quad U(x, y) = y\sec^2 x$

(c) $\mu = y^{-2}$ $\quad$ and $\quad U(x, y) = y + x^2/y$

1.7-5. (a) $y = cx$ $\qquad$ **(d)** $y^2 = -2x + c$

(b) $|y| = cx^4$ $\qquad$ **(e)** $x^2 + 3y^2 = c$

(c) $3x^2y + y^3 = c$ $\qquad$ **(f)** $xy = c$

1.7-6. **(a)** $r^2 - \theta + \ln|\theta + 1| = c \qquad r = (\theta - \ln|\theta + 1| + 2)^{1/2}$

(b) $uv^2 - v \tan u = c$

(c) $u = \ln 2 - \ln(c + e^{-t^2}) \qquad u = \ln 2 - \ln(1 + e^{-t^2})$

(d) $(r + 2\theta)\theta^2 = c(r + \theta) \qquad r = \dfrac{4\theta - 6\theta^3}{3\theta^2 - 4}$

(e) $y^2 + x^3y - 3x^2 = c \qquad y = -x^3/2 + \sqrt{x^6 + 12x^2 + 12}/2$

(f) $\tan^2 P + s^2 = c$

(g) $y = 2x^2 + cx^{-2}$

(h) $q = 4[1 + (c - 1)e^{-t/2}]^{-2} \qquad q(t) = 4(1 + e^{-t/2})^{-2}$

(i) $qe^{p^2} = c(q - 3)$

(j) $Q = c \sec^3 t + \tan t \sec^2 t - \frac{1}{3}\tan^3 t$

(k) $e^u = ce^t + e^{3t}/2 \qquad u(t) = \ln(e^t + e^{3t}) - \ln 2$

(l) $s^2 - x^2 = cs$

Section 1.8

1.8-1. Increases threefold in $\dfrac{34 \ln 3}{\ln 2} \cong 53.9$ years and fivefold in $\dfrac{34 \ln 5}{\ln 2} \cong 78.9$ years.

1.8-2. $p(1985) \cong 8 \times 10^6$, and if $T = 1988.9$ then $p(T) \cong 10^7$.

1.8-4. If $p_0 > h/a$, $p(t)$ increases without bound as $t \to +\infty$; if $p_0 = h/a$, then $p(t) \equiv h/a$; and if $p_0 < h/a$, then $p(t) = 0$ in a finite length of time.

1.8-6. **(a)**

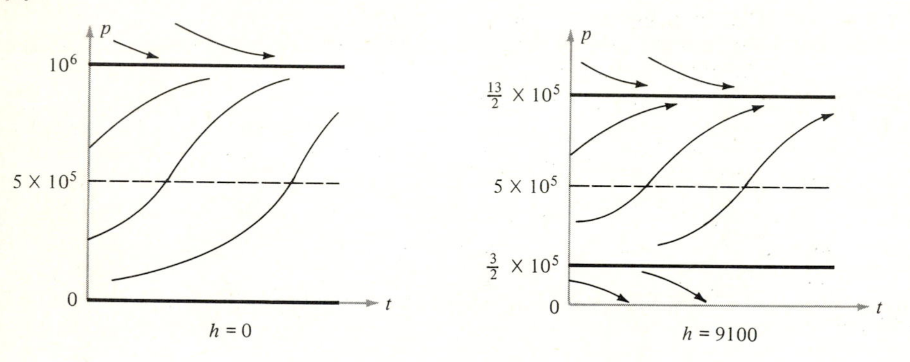

(b) $h^* = 10^4$, and if $h = 10^4$, $p' = -4 \times 10^{-8}(p - 5 \times 10^5)^2$, and $p(t) = 5 \times 10^5 + [(p_0 - 5 \times 10^5)^{-1} + 4 \times 10^{-8}t]^{-1}$ so long as $p(t)$ remains positive (and $p_0 \neq 5 \times 10^5$).

1.8-7. **(a)**

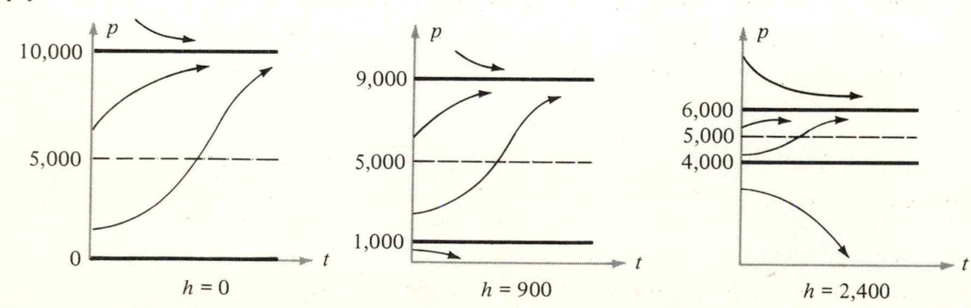

(b) $h_0 = 1600$ and $\lim_{t\to+\infty} p(t) = 8000$

(c) $h^* = 2500$ and $p(t) = \dfrac{10{,}000p_0 + 5000(p_0 - 5000)t}{10{,}000 + (p_0 - 5000)t}$

1.8-8. He should wait approximately 3.28 months.

1.8-9. $c = e^{-1}$

1.8-10. (c) $p(t) = [\frac{1}{4} + (p_0^{-2} - \frac{1}{4})e^{-2t}]^{-1/2}$
(d) $p(t) = [\frac{1}{2} + (p_0^{-1/2} - \frac{1}{2})e^{(-1/2)t}]^{-2}$

Section 1.9

1.9-1. (a) **(i)** $y = \dfrac{a_0 b_0 e^{k(b_0 - a_0)t} - a_0 b_0}{b_0 e^{k(b_0 - a_0)t} - a_0}$

(ii) $y = \dfrac{a_0^2 kt}{a_0 kt + 1}$

(b) $k = \dfrac{1}{(b_0 - a_0)t_1} \ln\left[\dfrac{a_0(b_0 - y_1)}{b_0(a_0 - y_1)}\right]$ if $a_0 \neq b_0$ and

$k = \dfrac{y_1}{a_0 t_1(a_0 - y_1)}$ if $a_0 = b_0$

1.9-2. (a) 75% **(b)** $33\frac{1}{3}\%$

Section 1.10

1.10-1. $ev(0) = \sqrt{2Rg}$ $ev(3R) = \frac{1}{2}\sqrt{2Rg}$ $ev(15R) = \frac{1}{4}\sqrt{2Rg}$ and $ev(99R) = \frac{1}{10}\sqrt{2Rg}$

1.10-2. (a) $v(y) = R\sqrt{\dfrac{2g}{R + y}} = ev(y)$
(b) $y(t) = (R^{3/2} + 3Rt\sqrt{2g}/2)^{2/3} - R$
(c) $t_R = \dfrac{2(2\sqrt{2} - 1)\sqrt{R}}{3\sqrt{2g}}$ $t_{2R} = \dfrac{2(3\sqrt{3} - 1)\sqrt{R}}{3\sqrt{2g}}$ $t_{3R} = \dfrac{14\sqrt{R}}{3\sqrt{2g}}$

1.10-3. (a) $H_{max}(\frac{1}{2}) = R/3$ $H_{max}(\frac{1}{4}) = R/15$ $H_{max}(\frac{1}{10}) = R/99$
(b) $v_0^2/2g$ (This is the same as in Problem 1.3-12.)

1.10-4. (a) $v(y_0) = -R\left[2g\left(\dfrac{1}{R} - \dfrac{1}{R + y_0}\right)\right]^{1/2}$ $v(y_0) \to -\sqrt{2Rg}$ as $y_0 \to \infty$
(b) $Rt\left(\dfrac{2g}{R + y_0}\right)^{1/2} = [(R + y)(y_0 - y)]^{1/2} + (R + y_0) \arcsin\left(\dfrac{y_0 - y}{R + y_0}\right)^{1/2}$
(c) $T = \sqrt{\dfrac{y_0(R + y_0)}{2gR}} + \sqrt{\dfrac{R + y_0}{R^2 2g}} \cdot \arcsin\sqrt{\dfrac{y_0}{R + y_0}}$

1.10-5. If $\alpha > \gamma$, then $P^* = [-(\beta + \delta) + \sqrt{(\beta + \delta)^2 + 4\varepsilon(\alpha - \gamma)}]/2\varepsilon$ is critical point.
If $\alpha = \gamma$, then $P^* = 0$ is critical point.
If $\alpha < \gamma$, then no nonnegative critical point.

CHAPTER 2

Section 2.1

2.1-1. (a) $y(t) = 3t^2 + c_1 t + c_2$

(b) $y(t) = te^{2t}/4 - e^{2t}/4 + c_1 t + c_2$

(c) $y_{\text{gen}}(t) = (-\cos 3t)/9 + c_1 t + c_2$
$y(t) = (-\cos 3t)/9 - 2t + 2\pi - \frac{26}{9}$

(d) $y(t) = [t \arctan (t/2)]/2 - [\ln (1 + t^2/4)]/2 + c_1 t + c_2$

(e) $y_{\text{gen}}(t) = e^{-t} + t^3/2 + c_1 t + c_2$
$y(t) = e^{-t} + t^3/2 + 3$

2.1-2. (a) $c_1 = 2, c_2 = 1$

(b) $c_1 = \frac{2}{3}, c_2 = \frac{32}{3}$

(c) $c_1 = -1, c_2 = -1$

(d) $c_1 = \frac{3}{4}, c_2 = \frac{5}{4}$

2.1-3. (a) $y = t^3/6 - t/6$

(b) $y = -\sin t - t$

(c) No solution

(d) $y = -\cos t + c$

2.1-4. (a) $y(t) = \ln (t + \frac{1}{2}) + 1 + \ln 2 \qquad t > -\frac{1}{2}$

(b) $y(t) = 2te^t - e^t$

(c) $y(t) = \dfrac{2}{1 - 2t} \qquad t < \frac{1}{2}$

(d) $y(t) = -1$

(e) $y(t) = \dfrac{6 \cdot 3^{1/3}}{(6^{1/2} - 3^{1/6}t)^2} \qquad t < 6^{1/2}/3^{1/6}$

2.1-5. (a) $|y| = c_2|\cos (-kt + c_1)|$ for $y'' = -k^2 y$ and
$|y| = c_2|\cosh (kt + c_1)|$ for $y'' = k^2 y$

(b) $|y| = c_2 e^{-2t}|(e^{-3t} - c_1)|$

Section 2.2

2.2-1. (a) $W = e^{3t}$, independent

(b) $W = 2$, independent

(c) $W = 0$, dependent

(d) $W = 0$, independent

(e) $W = e^{3t}$, independent

(f) $W = 3 \sin 2t \cos 3t - 2 \sin 3t \cos 2t$, independent

(g) $W = -4$, independent

(h) $W = (t + 1)^{-2}t^{-2}$, independent

(i) $W = 0$, dependent (recall that $\sin 2t = 2 \sin t \cos t$)

2.2-2. (a) $c_1 = \frac{1}{2}, c_2 = \frac{1}{2}$

(b) $c_1 = 1, c_2 = \frac{1}{3}$

(c) $c_1 = 4, c_2 = \frac{1}{4}$

(d) $c_1 = 1, c_2 = 2$

(e) $c_1 = \frac{1}{3}, c_2 = -\frac{1}{3}$

(f) $c_1 = 1, c_2 = 0$

(g) $c_1 = \sqrt{3} - 2, c_2 = 2\sqrt{3} - 1$

2.2-4. The coefficient t^2 of y'' is 0 at 0, and Problem 2.2-3 is not valid on any interval containing 0.

2.2-5. (a) Since $W[y_1, y_2; t_0] = 0$, they are linearly dependent.

(b) In either case they are linearly independent whenever y_1 is not the trivial solution (if $y_1 \equiv 0$, they are linearly dependent). The crucial issue is that if y_1 is nontrivial then both $y_1(t_0)$ and $y_1'(t_0)$ cannot equal zero, and hence $W[y_1, y_2; t_0] \neq 0$.

(c) Since $W[x_1, x_2; t] = -2W[y_1, y_2; t]$, x_1, x_2 are linearly independent if and only if y_1, y_2 are linearly independent.

2.2-6. Since $y_2(t) = \begin{cases} t^3 & \text{if } t \geq 0 \\ -t^3 & \text{if } t < 0 \end{cases}$ we have $y_2'(t) = \begin{cases} 3t^2 & \text{if } t \geq 0 \\ -3t^2 & \text{if } t < 0 \end{cases}$

Therefore, $y_2'(t) = 3t\,|t|$ and $W[t^3, t^2\,|t|; t] \equiv 0$.

Section 2.3

2.3-1. (a) $y = c_1 e^{2t} + c_2 e^{-4t}$

(b) $y = c_1 e^{3t} + c_2 e^{-2t}$

(c) $y = c_1 + c_2 e^{-3t}$

(d) $y = (c_1 + c_2 t)e^{2t}$

(e) $y = c_1 e^{\sqrt{3}t} + c_2 e^{-\sqrt{3}t}$

(f) $y = (c_1 \cos 2t + c_2 \sin 2t)e^t$

(g) $y = c_1 e^t + c_2 e^{-3t/2}$

(h) $y = \left(c_1 \cos \frac{\sqrt{7}}{2} t + c_2 \sin \frac{\sqrt{7}}{2} t\right) e^{3t/2}$

(i) $y = c_1 \cos 2\sqrt{2}t + c_2 \sin 2\sqrt{2}t$

(j) $y = (c_1 + c_2 t)e^{3t/2}$

(k) $y = c_1 e^{2t} + c_2 e^{-2t/3}$

(l) $y = (c_1 + c_2 t)e^{-3t}$

(m) $y = \left(c_1 \cos \frac{\sqrt{3}}{2} t + c_2 \sin \frac{\sqrt{3}}{2} t\right) e^{-t/2}$

2.3-2. (a) $y = \frac{1}{2} \cos 2t + \frac{\sqrt{3}}{2} \sin 2t$

(b) $y = 3e^t$

(c) $y = \left(\frac{5\sqrt{3}}{3} \sin \frac{\sqrt{3}}{2} t - \cos \frac{\sqrt{3}}{2} t\right) e^{t/2}$

(d) $y = 8e^{t-1} - 5te^{t-1}$

(e) $y = e^{2t}/4$

2.3-3. (a) $\eta < 1 : y = c_1 e^{(-1+\sqrt{1-\eta})t} + c_2 e^{(-1-\sqrt{1-\eta})t}$
$\eta = 1 : y = c_1 e^{-t} + c_2 te^{-t}$
$\eta > 1 : y = [c_1 \cos(\sqrt{\eta - 1}t) + c_2 \sin(\sqrt{\eta - 1}t)]e^{-t}$
All solutions converge to zero as $t \to \infty$ only in case $\eta > 0$.

(b) $|\eta| > 1 : y = c_1 e^{(-2\eta - 2\sqrt{\eta^2 - 1})t} + c_2 e^{(-2\eta + 2\sqrt{\eta^2 - 1})t}$
$\eta = \pm 1 : y = c_1 e^{-2t} + c_2 te^{-2t}$
$|\eta| < 1 : y = [c_1 \cos(2\sqrt{1 - \eta^2}t) + c_2 \sin(2\sqrt{1 - \eta^2}t)]e^{-2\eta t}$
All solutions converge to zero as $t \to \infty$ only in case $\eta > 0$.

(c) $\eta < 0$ or $\eta > 1 : y = c_1 e^{(-\eta + \sqrt{\eta^2 - \eta})t} + c_2 e^{(-\eta - \sqrt{\eta^2 - \eta})t}$
$0 < \eta < 1 : y = [c_1 \cos(\sqrt{\eta - \eta^2}t) + c_2 \sin(\sqrt{\eta - \eta^2}t)]e^{-\eta t}$
$\eta = 0 : y = c_1 + c_2 t \qquad \eta = 1 : y = c_1 e^{-t} + c_2 te^{-t}$
All solutions converge to zero as $t \to \infty$ only in case $\eta > 0$.

(d) $\eta > 0$ or $\eta < -4 : y = e^{\eta t/2}(c_1 e^{\sqrt{\eta^2 + 4\eta}t/2} + c_2 e^{-\sqrt{\eta^2 + 4\eta}t/2})$
$-4 < \eta < 0 : y = e^{\eta t/2}\left(c_1 \cos \frac{-\sqrt{\eta^2 - 4\eta}t}{2} + c_2 \sin \frac{-\sqrt{\eta^2 - 4\eta}t}{2}\right)$
$\eta = 0 : y = c_1 + c_2 t \qquad \eta = -4 : y = c_1 e^{-2t} + c_2 te^{-2t}$
All solutions converge to zero as $t \to \infty$ only in case $\eta < 0$.

2.3-5. The family $y = \bar{c}_1 e^{\alpha t} \cos(\beta t + \bar{c}_2)$ is also the general solution.

2.3-6. The families $y = k_1 e^{\alpha t} \sinh(\beta t + k_2)$ and $y = \bar{k} e^{\alpha t} \cosh(\beta t + \bar{k}_2)$ are both families of solutions, but neither family is the general solution.

2.3-7. If $b^2 - 4ac < 0$ then there is an infinite number of zeros, and if $b^2 - 4ac > 0$ and y is nontrivial, there is either 1 zero or no zeros for y.

Section 2.4

2.4-1.
(a) $y_p = t/2 - \frac{1}{3}$
(b) $y_p = -\sin 2t$
(c) $y_p = e^t/3$
(d) $y_p = t^3/3 + 8t/9 + \frac{26}{27}$
(e) $y_p = e^{-2t}/5 + 6t$
(f) $y_p = -e^t(\cos t + \sin t)\frac{1}{2}$

2.4-2.
(a) $y_p = \frac{1}{3} - te^{3t}/2$
(b) $y_p = t^3 e^{-t}/6$
(c) $y_p = e^t(2\cos 2t + \sin 2t)/20$
(d) $y_p = (t \cosh t)/2 \quad [= t(e^t + e^{-t})/4]$
(e) $y_p = te^{-t} + t^3/3 - t^2 + 3t$
(f) $y_p = -t(\sin 2t + \cos 2t)/4$

2.4-3.
(a) $y_p = (d_1 t^3 + d_2 t^2 + d_3 t)e^{2t} + e^{2t}(d_4 \cos 2t + d_5 \sin 2t) + d_6 \cos 2t + d_7 \sin 2t + d_8$

(b) $y_p = (d_1 t^2 + d_2 t)\cos 3t + (d_3 t^2 + d_4 t)\sin 3t + d_5 t \cos 3t + d_6 t \sin 3t + e^t(d_7 \cos 3t + d_8 \sin 3t)$

(c) $y_p = (d_1 t^4 + d_2 t^3 + d_3 t^2)e^{2t} + (d_4 t + d_5)e^t \cos 3t + (d_6 t + d_7)e^t \sin 3t$

(d) $y_p = (d_1 t^3 + d_2 t^2 + d_3 t)e^t \cos t + (d_4 t^3 + d_5 t^2 + d_6 t)e^t \sin t + (d_7 t^2 + d_8 t + d_9)e^t + d_{10} t^2 + d_{11} t + d_{12}$

2.4-4.
(a) $y = 9e^{3t}/16 + 23e^{-t}/16 - 2 - te^{-t}/4$
(b) $y = 5e^{2t}/4 - e^{-2t}/36 - (3t + 2)(e^t/9)$
(c) $y = 2e^{-2t}/3 + 4e^t/3 - 2\cos t + \sin t$

2.4-5. $\{1, \cos 2\beta t, \sin 2\beta t)\}$ is a differential family for $\sin^2 \beta t$
(a) $\{\cos 4t, \sin 4t\}$
(b) $\{\sin t, \cos t, \sin 3t, \cos 3t\}$

2.4-6.
(a) $y = c_1 e^{2t} + c_2 e^{-2t} + (\cos 2t)/16 - \frac{1}{8}$
(b) $y = c_1 \cos 2t + c_2 \sin 2t - (t \sin 2t)/8 + \frac{1}{8}$
(c) $y = c_1 e^{2t} + c_2 e^{-2t} - (11 \sin^3 t)/65 - (6 \sin t \cos^2 t)/65$

2.4-10. $y_p = (At \sin \omega t)/2\omega \quad$ if $\omega = \beta \quad$ and
$y_p = (A \cos \beta t)/(\omega^2 - \beta^2) \quad$ if $\omega \neq \beta$

Section 2.5

2.5-1.
(a) $y_p = (-\cos t) \ln |\sec t + \tan t|$
(b) $y_p = -3t^2 e^{-t}/4 + t^2 e^{-t} \ln t$
(c) $y_p = (\cos 2t)(\ln |\cos 2t|)/4 + (t \sin 2t)/2$
(d) $y_p = e^{-t} \arctan e^t - [e^{-2t} \ln(1 + e^{2t})]/2$
(e) $y_p = -e^t \cos t \ln |\sec t + \tan t|$
(f) $y_p = [e^t \ln(1 + e^{-t}) - e^{-t} \ln(1 + e^t) - 1)]/2$
(g) $y_p = (\sin t) \ln(1 + \sin t) + 1 - t \cos t + \cos t \tan t$

2.5-2. **(a)** $y = c_1 t + c_2 t \ln t + t(\ln t)^2$

(b) $y = c_1 e^t + c_2 e^t \ln t + te^t$

(c) $y = c_1 t + c_2 t^2 - t \ln t - t(\ln t)^2/2$

2.5-9. **(a)** $y(t) \to 0$ as $t \to +\infty$

(b) $y(t) \to \frac{1}{c}$ as $t \to +\infty$

(c) $y(t) \to \frac{2}{c}$ as $t \to +\infty$

(d) $y(t)$ bounded on $[0, \infty)$

(e) $y(t) \to \frac{2}{c}$ as $t \to \infty$

2.5-10. **(a)** $s = c_1 e^{2t} + c_2 te^{2t} - 2t^2 e^{2t} + \frac{1}{4}$

(b) $r = c_1 e^{\theta} + c_2 \theta e^{\theta} - e^{\theta} \ln |1 - \theta|$

(c) $z = c_1 \cos x + c_2 \sin x + (\sec x \tan x)/3$

(d) $u = c_1 \cos s + c_2 \sin s + e^s/2 - (s \cos s)/2$

(e) $\phi = c_1 e^s + c_2 e^{2s} + (e^s + e^{2s}) \ln (1 + e^{-s})$

Section 2.6

2.6-1. Equation is $y'' + 16y = 0$, $y(0) = y_0$, $y'(0) = v_0$

(a) $y(t) = 3 \cos 4t$

(b) $y(t) = -\cos 4t + \sqrt{3} \sin 4t$, $A = 2$, $f = 2/\pi$, and $y(T) = 0$ when $T = \pi/24$

(c) $v_0 = \pm 4\sqrt{5}$

2.6-2. **(a)** $y(t) = \cos (8\sqrt{2}t) + \frac{\sqrt{2}}{8} \sin (8\sqrt{2}t)$

(b) $y(t) = \cos (8\sqrt{2}t) - \frac{\sqrt{2}}{8} \sin (8\sqrt{2}t)$

(c) $y(t) = -\frac{\sqrt{2}}{4} \sin (8\sqrt{2}t)$

2.6-3. Equation is $\frac{1}{2}y'' + \mu y' + 8y = 0$, $y(0) = y_0$, $y'(0) = v_0$

(a) $y = (e^{-2t} \sin 2\sqrt{3}t)/\sqrt{3}$ if $\mu = 2$

$y = 2te^{-4t}$ if $\mu = 4$

$y = e^{-2t}/3 - e^{-8t}/3$ if $\mu = 5$

(b) $y = e^{-2t}[-6 \cos (2\sqrt{3}t) - 2\sqrt{3} \sin (2\sqrt{3}t)]/3$ if $\mu = 2$

$y = -2e^{-4t} - 8te^{-4t}$ if $\mu = 4$

$y = -8e^{-2t}/3 + 2e^{-8t}/3$ if $\mu = 5$

2.6-4. Equation is $\frac{1}{2}y'' + \mu y' + 2y = 0$, $y(0) = y_0$, $y'(0) = v_0$

(a) $y = -e^{-t}[\cos \sqrt{3}t) + (\sin \sqrt{3}t)/\sqrt{3}]$

$$= -\frac{2}{\sqrt{3}} e^{-t} \sin \left(\sqrt{3}t + \frac{\pi}{3}\right), \ t_0 = \frac{2\pi\sqrt{3}}{9}, \ t_1 = \frac{\pi\sqrt{3}}{3}$$

(b) $y = [(v_0 - 2)t - 1]e^{-2t}$, $v^* = 2$, $T_{v_0} = \frac{1}{v_0 - 2}$, $S_{v_0} = \frac{v_0}{2(v_0 - 2)}$

(c) $y = (v_0 - 4)(e^{-t}/3) + (1 - v_0)(e^{-4t}/3)$
If $0 \le v_0 \le 4$ then $y(t) < 0$ for all $t \ge 0$;
if $v_0 > 4$ then $y(t) < 0$ for $0 \le t < T_{v_0}$;
and $y(t) > 0$ for $t > T_{v_0}$ when $T_{v_0} = \frac{1}{3} \ln \frac{v_0 - 1}{v_0 - 4}$;
unique maximum occurs at $S_{v_0} = \frac{1}{3} \ln \left(4 \frac{v_0 - 1}{v_0 - 4} \right)$

2.6-6. (a) $y(t) = \frac{A}{4 - \omega^2} [\sin \omega t - (\omega \sin 2t)/2]$

(b) $y = (\sin 2t)/2 - (t \cos 2t)/4 \quad \text{if } \mu = 0$

$$y = e^{-\mu t/2}\left[\frac{1}{2\mu} \cos\left(\frac{\sqrt{16 - \mu^2}}{2} t\right) + \frac{1}{2\sqrt{16 - \mu^2}} \sin\left(\frac{\sqrt{16 - \mu^2}}{2} t\right)\right] - \frac{1}{2\mu} \cos 2t \quad \text{if } 0 < \mu < 4$$

(c) $y = 32 \sin \frac{t}{8} \sin \frac{17t}{8}$

Section 2.7

2.7-1. (a) $y = c_1 t^2 + c_2 t^{-2}$
(b) $y = c_1 t \cos (\ln t) + c_2 t \sin (\ln t)$
(c) $y = c_1 t + c_2 t^2 + t \ln t + t^{-1}/3$
(d) $y = c_1 t^{-1} + c_2 t^{-1} \ln t + (\ln t)^2 - 4 \ln t + 6$

2.7-2. (a) $y = 2\sqrt{t} \cos [(\sqrt{3} \ln t)/2] - 4\sqrt{t} \sin [(\sqrt{3} \ln t)/2]/\sqrt{3}$
(b) $y = t - 1 + 2t^2 \ln t$

2.7-4. (a) $y = c_1 \cos [\ln (t + 1)] + c_2 \sin [\ln (t + 1)] + (t + 1)^2/5 - t$
(b) $y = c_1(3t - 1)^{5/3} + c_2(3t - 1)^{-1/3} - 3t/4 - \frac{23}{20}$

2.7-5. $y = c_1 \cos (t^2/2) + c_2 \sin (t^2/2)$

2.7-6. $y = e^{-e^t/2}[c_1 \cos (\sqrt{3}e^t/2) + c_2 \sin (\sqrt{3}e^t/2)] + e^t - 1$

2.7-7. $y = c_1 e^{2/t} + c_2 e^{-2/t}$

2.7-9. (a) $x = c_1 \cos t + c_2 \sin t - \cos t \ln |\sec t + \tan t|$
(b) $r = c_1 \cos (\theta/2) + c_2 \sin (\theta/2) - 2 \cos \theta$
(c) $y = c_1 x + \frac{c_2}{x} - \frac{1}{2} - \frac{x}{2} \ln |x| + \left(\frac{x}{2} - \frac{1}{2x}\right) \ln |1 + x|$
(d) $\phi = c_1 e^{2z} + c_2 ze^{2z} - z^2e^{2z}/2 - 2$
(e) $r = \frac{c_1 \cos (\ln \theta) + c_2 \sin (\ln \theta)}{\theta^2} + \frac{1}{5}$
(f) $\phi = c_1 e^{-t} + c_2 te^{-t} - e^{-t} \ln |1 - e^{-t}|$
(g) $\psi = c_1 t^2 + \frac{c_2}{t} - 2 \ln t + 1$
(h) $\phi = c_1 x^2 + c_2 x^{-4} + x^2$

Section 2.8

2.8-1. (a) $\frac{1}{10}Q'' + 4Q' + 200Q = E(t), Q(0) = 2, Q'(0) = 0$

(b) $Q(t) = e^{-20t}(2\cos 40t + \sin 40t)$
$I(t) = -100e^{-20t}\sin 40t$

(c) $Q(t) = e^{-20t}(38\cos 40t + 19\sin 40t)/20 + \frac{1}{10}$
(steady state is $\frac{1}{10}$)

(d) $Q(t) = e^{-20t}(156\cos 40t + 77\sin 40t)/80 + (2\cos 20t + \sin 20t)/40$
[steady state is $(2\cos 20t + \sin 20t)/40$]

2.8-2. (a) $Q(t) = e^{-5t/2}(-240\sin 5t - 800\cos 5t)/93 + 800e^{-t}/93$
$I(t) = e^{-5t/2}(800\cos 5t + 4600\sin 5t)/93 - 800e^{-t}/93$

(b) $Q(t) = e^{-5t/2}(-16\cos 5t - 72\sin 5t)/17 + (16\cos 5t + 64\sin 5t)/17$
$I(t) = 5e^{-5t/2}(-64\cos 5t + 52\sin 5t)/17 + 5(64\cos 5t - 16\sin 5t)/17$

2.8-3. Underdamped if $R^2 < 4L/C$, overdamped if $R^2 > 4L/C$, and critically damped if $R^2 = 4L/C$.

2.8-4. (a) $Q(t) = e^{-14t}(c_1\cos 7t + c_2\sin 7t) + 1/5$

(b) $Q(t) = e^{-14t}(c_1\cos 7t + c_2\sin 7t) + (\cos 7t + \sin 7t)/8$

(c) $Q(t) = e^{-14t}(c_1\cos 7t + c_2\sin 7t) + (\sin 7t - \cos 7t)/8$

2.8.5. $T = 2\pi/5, y(t) = (5\cos 5t + 4\sin 5t)/60$

2.8-6. $T = \sqrt{2}\pi/5, y(t) = [2\cos(5\sqrt{2}t) + \sqrt{2}\sin(5\sqrt{2}t)]/30$

2.8-7. $\rho_1 = 15.625$ for 2 times and $\rho_1 = 6.9444\cdots$ for 3 times

2.8-8. (a) $H > W/S_1S_2\rho$

(c) $T = \dfrac{2\pi\sqrt{W}}{\sqrt{S_1S_2\rho g}}$ $\quad A = \sqrt{y_0^2 + \dfrac{v_0^2W}{S_1S_2\rho g}}$

(d) $T_2 = T/\sqrt{2}, T_{2,2} = T/2$, and $T_{3,3} = T/3$

2.8-9. $y'' + \dfrac{\sqrt{3}L^2\rho g}{4W}y = 0, y(0) = y_0, y'(0) = v_0$

$$T = \frac{4\pi\sqrt{W}}{\sqrt[4]{3}\sqrt{\rho g L}}$$

CHAPTER 3

Section 3.1

3.1-1. (a) $\dfrac{2}{s^2+4}$ $\quad s > 0$

(b) $\dfrac{4}{(2s-1)^2}$ $\quad s > \frac{1}{2}$

(c) $\dfrac{2}{s^3}$ $\quad s > 0$

(d) $\dfrac{1-e^{-s}}{s}$ $\quad s > 0$

(e) $\dfrac{1+e^{-\pi s}}{s^2+1}$ $\quad s > 0$

3.1-2. **(a)** $\dfrac{2}{(s+4)^3}$

(b) $\dfrac{4s}{(s^2+4)^2}$

(c) $\dfrac{s^2-1}{(s^2+1)^2}$

(d) $\dfrac{2s+2}{(s^2+2s+2)^2}$

(e) $\dfrac{1}{2s}+\dfrac{s/2}{s^2+4}$

(f) $\dfrac{1}{2}\left(\dfrac{s-2}{(s-2)^2+1}\right)-\dfrac{1}{2}\left(\dfrac{s+2}{(s+2)^2+1}\right)$

(g) $\dfrac{1}{(s+1)^2}-\dfrac{3(s-1)}{(s-1)^2-4}$

(h) $\dfrac{2s(s^2-3)}{(s^2+1)^3}$

(i) $\dfrac{16(3s^2-16)}{(s^2+16)^3}$

(j) $\dfrac{1}{2s}-\dfrac{1}{2}\dfrac{s}{s^2+9}$

Section 3.2

3.2-1. **(a)** $2\cos 3t-(\sin 3t)/3$

(b) $5t^2e^t/2$

(c) $e^{-t}\cos(\sqrt{5}t)-\dfrac{1}{\sqrt{5}}e^{-t}\sin(\sqrt{5}t)$

(d) $e^{-2t}(\sinh 3t)/3$

(e) $t-\sin t$

(f) $(e^t-\cos t-\sin t)/2$

(g) $(e^t+e^{-t}-2)/2$

(h) $1-e^t+3te^t$

(i) $[1-e^{-t}\cos(\sqrt{3}t)-e^{-t}\sin(\sqrt{3}t)]/12$

3.2-2. **(a)** $e^{-2t}-2te^{-2t}+t^2e^{-2t}/2$

(b) $(\sin 2t-2t\cos 2t)/16$

(c) $(\sinh t-\sin t)/2$

(d) $t\sin t$

(e) $(t\cos t-\sin t)/2$

3.2-6. **(a)** $\dfrac{\pi}{2}-\arctan s$ **(b)** $\ln\dfrac{s+1}{s}$ **(c)** $\ln\dfrac{s}{\sqrt{s^2+1}}$

3.2-7. **(a)** $t^{-1}\sin t$ **(b)** $t^{-1}(e^t-e^{-t})$

3.2-8. **(a)** $\dfrac{1}{s^2+1}\dfrac{1+e^{-\pi s}}{1-e^{-\pi s}}$

(b) $\dfrac{1}{s}\dfrac{1-e^{-s}}{1+e^{-s}}$

(c) $\dfrac{1-e^{-s}-se^{-s}}{s^2-s^2e^{-s}}$

Section 3.3

3.3-1. **(a)** $y=\sinh t-1=e^t/2-e^{-t}/2-1$

(b) $y=(\cos t)/3-(\cos 2t)/3$

(c) $y=e^t/2+e^{3t}/10-3e^{-2t}/5$

(d) $y=(1-e^{3t}\cos t+3e^{3t}\sin t)/10$

(e) $y=(e^{-3t}-e^{2t}+5te^{2t})/25$

3.3-2. **(a)** $(1-9e^{-2s}+9e^{-\pi s})/s$

(b) $(-2+4e^{-s}-4e^{-2s})/s$

(c) $(1-2e^{-s}+4e^{-3s})/s$

(d) $(6-2e^{-\pi s}-4e^{-2\pi s})/s$

3.3-3. **(a)** $1/s + (1 + 2s)e^{-3s}/s^2$
(b) $(s/2 - \sqrt{3}/2)/(s^2 + 4)$
(c) $5^{2-s}/(s - 2)$
(d) $e^{-2s}(2 - 4s + 4s^2)/s^3$
(e) $1/s^2 + (\frac{1}{2} + \sqrt{3}s/2)/(s^2 + 1)$

3.3-4. **(a)** $y(t) = u_1(t)[\sin(2t - 2)]/2$
(b) $y(t) = u_4(t)(e^{2t-8} - e^{-t+4})/3$
(c) $y(t) = u_{\ln 2}(t)(e^{-t} + e^t/4 - 1)$
(d) $y(t) = u_3(t)(t - 3)^2 e^{(t-3)/2}/16$

3.3-6. **(a)** $y(t) = u_2(t)(1 - e^{t-2}/2 - e^{-t+2}/2)$
(b) $y(t) = 1 - \cos t + u_3(t)[\cos(t - 3) - 1]$
(c) $y(t) = e^{3t}/4 - e^{2t}/3 + e^{-t}/12 + u_1(t)(e^{2t-2} + 2e^{-t+2} - 3)/6$
(d) $y(t) = e^{-t}/4 - e^t/4 + te^t/2 - u_2(t)(e^{-t+4}/4 - 5e^t/4 + te^t/2)$

5.3-7. **(b)** $\mathscr{L}\{f(t)\}(s) = \dfrac{1 - e^{-s}}{s(1 + e^{-s})}$ same as Problem 3.2-8*b*

Section 3.4

3.4-1. **(a)** $y(t) = (9 + 4\cos t - 13\cos 2t)/12$
(b) $y(t) = (9e^{-3t} - 25e^{-t} + 20 - 4\cos t + 2\sin t)/20$
(c) $y(t) = 2 - 2\cos t + u_1(t)\sin(t - 1)$
(d) $y(t) = 2 - 2e^{-t} - 2te^{-t}$
$-u_5(t)[2 - 2e^{-(t-5)} - 2(t - 5)e^{-(t-5)}] + u_6(t)(t - 6)e^{-(t-6)}$

3.4-2. **(a)** $y(t) = e^{-t} - e^{-2t} + 3u_2(t)(e^{-t+2} - e^{-2t+4})$
Solution on $[0, 2)$ to hom. eq. with IC: $y(t) = 0$, $y'(0) = 1$
Solution on $[2, \infty)$ to hom. eq. with IC: $y(2) = -e^{-4} + e^{-1}$, $y'(2) = 2e^{-4} - e^{-2} + 5$
$y'(t)$ has jump discontinuity at $t = 2$

(b) $y(t) = 2e^{-t}\sin t - u_1(t)[e^{-t+1}\sin(t - 1)]$
Solutions on $[0, 1)$ to hom. eq. with IC: $y(0) = 0$, $y'(0) = 2$
Solution on $[1, \infty)$ to hom. eq. with IC: $y(1) = 2e^{-1}\sin 1$, $y'(1) = 2e^{-1}(\cos 1 - \sin 1) - 1$
$y'(t)$ has jump discontinuity at $t = 1$

(c) $y(t) = \sin t + u_2(t)\sin(t - 2)$
Solution on $[0, 2)$ to hom. eq. with IC: $y(0) = 0$, $y'(0) = 1$
Solution on $[2, \infty)$ to hom. eq. with IC: $y(2) = \sin 2$, $y'(2) = \cos 2 + 1$
$y'(t)$ has jump discontinuity at $t = 2$

(d) $y(t) = te^{-2t} - u_1(t)(t - 1)e^{-2t+2}$
Solution on $[0, 1)$ to hom. eq. with IC: $y(0) = 0$, $y'(0) = 1$
Solution on $[1, \infty)$ to hom. eq. with IC: $y(1) = e^{-2}$, $y'(1) = -e^{-2} - 1$
$y'(t)$ has jump discontinuity at $t = 1$

(e) $y(t) = 4e^{-2t} - 3e^{-t}$
Solution on $[0, \infty)$ to hom. eq. with IC: $y(0) = 1$, $y'(0) = -5$

CHAPTER 4

Section 4.1

4.1-1. **(a)** $\rho = \frac{1}{2}$
(b) $\rho = 1$
(c) $\rho = \infty$
(d) $\rho = 2$
(e) $\rho = \infty$
(f) $\rho = 1$

4.1-2. (a) $\dfrac{t}{1-t^2}$ $|t|<1$

(b) $\cos\sqrt{t}-1$ $t>0$

(c) $\dfrac{e^{-t}-1+t}{t}$ $t\neq 0$

(d) $\dfrac{4}{1-2t}$ $|t|<\frac{1}{2}$

(e) $\cos t^2$

(f) $t(e^{t^2}-1)$

4.1-3. (a) $t=\pm 1$ are regular singular points

(b) $t=0, \pm 1$ are regular singular points

(c) $t=\pm 1$ are regular singular points
$t=0$ is an irregular singular point

(d) $t=0$ is an irregular singular point

(e) $t=2, -1$ are regular singular points

4.1-5. (b) **(i)** $t=n\pi, n=0, \pm 1, \pm 2, \ldots$ are regular singular points

(ii) $t=2n\pi, n=0, \pm 1, \pm 2, \ldots$ are regular singular points

(iii) $t=0$ is a regular singular point

(iv) $t=0, -1, \pm 2, \pm 3, \ldots$ are regular singular points
$t=1$ is an irregular singular point

4.1-6. (a) $2+t+t^2=4+3(t-1)+(t-1)^2=2-(t+1)+(t+1)^2$

(b) $t^5=1+5(t-1)+10(t-1)^2+10(t-1)^3+5(t-1)^4+(t-1)^5$
$=32+80(t-2)+80(t-2)^2+40(t-2)^3+10(t-2)^4+(t-2)^5$

(c) $4-t^3=3-3(t-1)-3(t-1)^2-(t-1)^3$
$=12-12(t+2)+6(t+2)^2-(t+2)^3$

Section 4.2

4.2-1. (a) $y(t)=e^{2t}=\displaystyle\sum_{n=0}^{\infty}\frac{2^n t^n}{n!}$

(b) $y(t)=\sin 2t=\displaystyle\sum_{n=0}^{\infty}\frac{(-1)^n 2^{2n+1}}{(2n+1)!}t^{2n+1}$

(c) $y(t)=-2e^{3t}=\displaystyle\sum_{n=0}^{\infty}\frac{(-2)3^n t^n}{n!}$

(d) $y(t)=\cosh 2t=\displaystyle\sum_{n=0}^{\infty}\frac{4^n}{(2n)!}t^{2n}$ $\left(=\dfrac{e^{2t}+e^{-2t}}{2}\right)$

4.2-2. (a) $a_n=-\dfrac{a_{n-2}}{n-1}$ $n=2, 3, \ldots$

$$a_{2k}=\frac{(-1)^k a_0}{1\cdot 3\cdot 5\cdot\cdots\cdot(2k-1)} \qquad a_{2k+1}=\frac{(-1)^k a_1}{2\cdot 4\cdot\cdots\cdot(2k)} \qquad k=1, 2, \ldots$$

(b) $a_n=\dfrac{2(n-3)a_{n-2}}{n(n-1)}$ $n=2, 3, \ldots$

$$a_{2k}=\frac{2^k(2k-3)(2k-5)\cdots(-1)}{(2k)!}a_0 \qquad a_{2k+1}=0 \qquad k=1, 2, \ldots$$

(c) $a_n=\dfrac{(n-3)a_{n-2}}{n-1}$ $n=2, 3, \ldots$

$$a_{2k}=\frac{-a_0}{2k-1} \qquad a_{2k+1}=0 \qquad k=1, 2, \ldots$$

(d) $a_n = \dfrac{(n-1)(n-3)}{2n(n-1)} a_{n-2} \qquad n = 2, 3, \ldots$

$a_2 = \dfrac{a_0}{2} \qquad a_n = 0 \qquad n = 3, 4, \ldots$

(e) $a_n = -\dfrac{(n-3)}{n(n-1)} [(n-4)a_{n-2} + 2(n-1)a_{n-1}] \qquad n = 2, 3, \ldots$

$a_2 = a_1 - a_0 \qquad a_n = 0 \qquad n = 3, 4, \ldots$

4.2-3. (a) $a_0 = 1,\ a_1 = 0,\ a_2 = -\frac{3}{2},\ a_n = \dfrac{a_{n-3} - 3a_{n-2}}{n(n-1)} \qquad n = 3, 4, \ldots$

$a_3 = \frac{1}{6},\ a_4 = \frac{3}{8}$
$y(t) \cong 1 - 3t^2/2 + t^3/6 + 3t^4/8$

(b) $a_0 = 0,\ a_1 = 1,\ a_2 = 0,\ a_n = \dfrac{-3a_{n-2} + a_{n-3}}{n(n-1)} \qquad n = 3, 4, \ldots$

$a_3 = \frac{1}{6},\ a_4 = \frac{1}{12},\ a_5 = \frac{3}{40}$
$y(t) \cong t - t^3/6 + t^4/12 + 3t^5/40$

(c) $a_n = -\dfrac{a_{n-4}}{n(n-1)} \qquad n = 4, 5, \ldots$

$a_0 = 1,\ a_1 = 2,\ a_2 = 0,\ a_3 = 0,\ a_4 = -\frac{1}{12},\ a_5 = -\frac{1}{10}$
$y(t) \cong 1 + 2t - t^4/12 - t^5/10$

(d) $a_n = \dfrac{(n-1)(n-2)a_{n-1} + a_{n-2}}{2n(n-1)} \qquad n = 2, 3, \ldots$

$a_0 = 1,\ a_1 = -3,\ a_2 = \frac{1}{4},\ a_3 = -\frac{5}{24}$
$y(t) \cong 1 - 3(t-1) + (t-1)^2/4 - 5(t-1)^3/24$

(e) $a_n = \dfrac{a_{n-3} - 2a_{n-2}}{n(n-1)} \qquad n = 3, 4, \ldots$

$a_0 = 3,\ a_1 = 4,\ a_2 = -3,\ a_3 = -\frac{5}{6}$
$y(t) \cong 3 + 4(t+2) - 3(t+2)^2 - 5(t+2)^3/6$

(f) $a_n = -\dfrac{(n-1)(2n-3)a_{n-1} + [(n-2)(n-3) + 1]a_{n-2}}{n(n-1)} \qquad n = 2, 3, \ldots$

$a_0 = 2,\ a_1 = -1,\ a_2 = -\frac{1}{2},\ a_3 = \frac{2}{3}$
$y(t) \cong 2 - (t-1) - (t-1)^2/2 + 2(t-1)^3/3$

4.2-4. (a) $c_n = \dfrac{c_{n-3} + 1}{n(n-1)} \qquad n = 3, 4, \ldots$

$c_0 = 0,\ c_1 = 0,\ c_2 = \frac{1}{2},\ c_3 = \frac{1}{6},\ c_4 = \frac{1}{12},\ c_5 = \frac{3}{40}$
$y(t) \cong t^2/2 + t^3/6 + t^4/12 + 3t^5/40$

(b) $c_n = \dfrac{(n-3)c_{n-3} - c_{n-2}}{n(n-1)} \qquad n = 3, 4, \ldots$

$c_0 = 0,\ c_1 = 0,\ c_2 = \frac{1}{2},\ c_3 = \frac{1}{6},\ c_4 = \frac{7}{24},\ c_5 = \frac{3}{40}$
$y(t) \cong t^2/2 + t^3/6 + 7t^4/24 + 3t^5/40$

(c) $c_n = \dfrac{(n-3)c_{n-3} - c_{n-2}}{n(n-1)} \qquad n = 3, 4, \ldots$

$c_0 = 0,\ c_1 = 0,\ c_2 = \frac{1}{2},\ c_3 = 0,\ c_4 = -\frac{7}{24},\ c_5 = \frac{1}{20},\ c_6 = \frac{1}{720}$
$y(t) \cong t^2/2 - 7t^4/24 + t^5/20 + 7t^6/720$

4.2-5. The recursion formula is

$$a_n = -2\frac{n-2-p}{n(n-1)}a_{n-2} \qquad n = 2, 3, \ldots$$

Therefore a_{2k} is proportional to a_0 and a_{2k+1} to a_1 for $k = 1, 2, \ldots$. Moreover, if $n = p + 2$ then $a_n = 0$, and hence $a_{n+2k} = 0$ for $k = 0, 1, 2, \ldots$.
If $p = 0$, $y(t) \equiv 1$ is a polynomial solution.
If $p = 1$, $y(t) \equiv t$ is a polynomial solution.
If $p = 2$, $y(t) = 2t^2 + 1$ is a polynomial solution.
If $p = 3$, $y(t) = \frac{2}{3}t^3 + t$ is a polynomial solution.

Section 4.3

4.3-1. (a) $3\sigma^2 + 4\sigma + 1 = 0 \qquad \sigma = 1, -\frac{1}{3}$ [about $t_0 = 0$]

$$y_1(t) = \sum_{n=0}^{\infty} a_n t^{n-1/3} \qquad y_2(t) = \sum_{n=0}^{\infty} b_n t^{n-1}$$

(b) $\sigma^2 - \frac{1}{4} = 0 \qquad \sigma = \pm\frac{1}{2}$ (about $t_0 = 0$)

$$y_1(t) = \sum_{n=0}^{\infty} a_n t^{n+1/2} \qquad y_2(t) = \sum_{n=0}^{\infty} b_n t^{n-1/2} + Ky_1 \ln t$$

(c) $\sigma^2 - \sigma/2 = 0 \qquad \sigma = 0, \frac{1}{2}$ (about $t_0 = -2$)

$$y_1(t) = \sum_{n=0}^{\infty} a_n(t+2)^{n+1/2} \qquad y_2(t) = \sum_{n=0}^{\infty} b_n(t+2)^n$$

(d) $\sigma^2 = 0 \qquad \sigma = 0$ (about $t_0 = 0$)

$$y_1(t) = \sum_{n=0}^{\infty} a_n t^n \qquad y_2(t) = \sum_{n=0}^{\infty} b_n t^n + Ky_1(t) \ln t$$

and

$\sigma^2 + 5\sigma = 0 \qquad \sigma = 0, -5$ (about $t_0 = 1$)

$$y_1(t) = \sum_{n=0}^{\infty} a_n(t-1)^n \qquad y_2(t) = \sum_{n=0}^{\infty} b_n(t-1)^{n-5} + Ky_1(t) \ln (t-1)$$

(e) $\sigma^2 = 0 \qquad \sigma = 0$ (about $t_0 = -0$)

$$y_1(t) = \sum_{n=0}^{\infty} a_n(t+1)^n \qquad y_2(t) = \sum_{n=0}^{\infty} b_n(t+1)^n + Ky_1(t) \ln (t+1)$$

4.3-2. (a) Indicial equation is $2\sigma(2\sigma - 1) = 0$, so $\sigma = 0, \frac{1}{2}$. If $\sigma = 0$, recursion formula is

$$a_n = \frac{a_{n-1}}{2n(2n-1)} \quad \text{and} \quad a_n = \frac{a_0}{(2n)!} \quad \text{for } n = 1, 2, \ldots$$

If $\sigma = \frac{1}{2}$, recursion formula is

$$a_n = \frac{a_{n-1}}{(2n+1)(2n)} \quad \text{and} \quad a_n = \frac{a_0}{(2n+1)!} \quad n = 1, 2, \ldots$$

(b) Indicial equation is $\sigma^2 = 0$, so $\sigma = 0$. Recursion formula is

$$a_n = \frac{a_{n-1}}{n^2} \quad \text{and} \quad a_n = \frac{a_0}{(n!)^2} \quad \text{for } n = 1, 2, \ldots$$

(c) Indicial equation is $\sigma^2 - \sigma = 0$, so $\sigma = 0, 1$. If $\sigma = 1$, recursion formula is

$$a_n = \frac{-a_{n-1}}{(n+1)n} \quad \text{and} \quad a_n = \frac{(-1)^n a_0}{(n+1)!n!} \quad \text{for } n = 1, 2, \ldots$$

No other solutions of indicated form exists.

(d) Indicial equation is $\sigma^2 - 1 = 0$, so $\sigma = \pm 1$.

$$\text{If } \sigma = 1,\ a_n = \frac{-a_{n-1}}{n+2} \quad \text{and} \quad a_n = \frac{(-1)^n a_0}{(n+2)!}$$

$$\text{If } \sigma = -1,\ a_n = -\frac{a_{n-1}}{n} \quad \text{and} \quad a_n = \frac{(-1)^n a_0}{n!}$$

(e) Indicial equation is $\sigma^2 = 0$, so $\sigma = 0$. Recursion formula is

$$a_1 = 0 \qquad a_n = \frac{-a_{n-2}}{n^2} \qquad \text{for } n = 2, 3, \ldots$$

$$\text{Therefore, } a_{2k+1} = 0, \qquad a_{2k} = \frac{(-1)^k a_0}{(2^k k!)^2} \qquad k = 1, 2, \ldots$$

No other solutions of this form exist.

(f) Indicial equation is $\sigma^2 = 0$, so $\sigma = 0$. Recursion formula is $a_n = a_{n-1}$, so $a_n = a_0$ for $n = 1, 2, 3, \ldots$. No other solutions of this form exist.

(g) Indicial equation is $\sigma^2 = 0$, so $\sigma = 0$. Recursion formula is

$$a_n = \frac{n+4}{n} a_{n-1} \qquad \text{for } n = 1, 2, \ldots$$

Therefore,

$$a_n = \frac{(n+4)!}{n!4!} a_0 \qquad \text{for } n = 1, 2, \ldots$$

No other solutions of this form exist.

4.3-3. (c) $L_1(t) = 1 - t$, $L_2(t) = 1 - 2t + t^2$, and $L_3(t) = 1 - 3t + 3t^2/2 - t^3/6$

Section 4.4

4.4-1. (a) ∞ is a regular singular point and $\sigma = 0, 1$ are indicial roots at ∞. If $\sigma = 1$,

$$a_n = \frac{a_{n-1}}{(n+1)n} = \frac{a_0}{(n+1)!n!} \qquad \text{for } n = 1, 2, \ldots \qquad \text{so}$$

$$y(t) = t^{-1} \sum_{n=0}^{\infty} \frac{t^{-n}}{(n+1)!n!} \qquad \text{for } t > 0$$

is a solution (no further solutions for $\sigma = 0$).

(b) ∞ is an ordinary point, and

$$y(t) = a_0(1 - t^{-2}) + a_1 \left[\sum_{k=1}^{\infty} \frac{(2k-3)(2k-7)\cdots(-1)}{(2k+1)!} t^{-k} + t^{-1} \right]$$

is a general solution.

(c) ∞ is a regular singular point, and

$$y(t) = a_0 \sum_{n=0}^{\infty} t^{-n}/n! \qquad (= a_0 e^{1/t})$$

are all solutions of the indicated form.

4.4-2. (a) Indicial equation is $2\sigma^2 + \sigma = 0$, so $\sigma = 0, -\frac{1}{2}$ are indicial roots at ∞. If $\sigma = 0$,

$$a_n = \frac{-a_{n-1}}{n(2n+1)} \quad \text{for } n = 1, 2, \ldots \quad \text{and}$$

$a_1 = -a_0/3,\ a_2 = a_0/30,\ a_3 = -a_0/630.$
If $\sigma = -\frac{1}{2}$,

$$a_n = \frac{-a_{n-1}}{n(2n-1)} \quad \text{for } n = 1, 2, \ldots \quad \text{and}$$

$a_1 = -a_0,\ a_2 = a_0/6,\ a_3 = -a_0/90.$

(b) Indicial equation is $\sigma^2 - \sigma - 6 = 0$, so $\sigma = 3, -2$ are indicial roots at ∞. If $\sigma = 3$, $a_1 = 0$, and

$$a_n = \frac{(n+1)(n+2)}{n(n+5)} a_{n-2} \quad \text{for } n = 2, 3, \ldots$$

and $a_1 = 0$, $a_2 = -a_0/7$, $a_3 = 0$.
If $\sigma = -2$, $a_1 = 0$ and $n(n-5)a_n = (n-4)(n-3)a_{n-2}$ for $n = 2, 3, \ldots$, and $a_1 = 0$, $a_2 = -a_0/3$, $a_3 = 0$.

Section 4.5

4.5-5. (a) $y(t) = c_1 J_{1/2}(3t^2) + c_2 J_{-1/2}(3t^2)$
(b) $y(t) = c_1 J_{2/3}(t^2) + c_2 J_{-2/3}(t^2)$
(c) $y(t) = c_1 t^2 J_2(t)$
(d) $y(t) = c_1 t^{1/2} J_{1/3}(2t^{3/2}/3) + c_2 t^{1/2} J_{-1/3}(2t^{3/2}/3)$
(e) $y(t) = c_1 t J_{1/2}(t^2) + c_2 t J_{-1/2}(t^2)$

4.5-6. $P_5(t) = (63t^5 - 70t^3 + 15t)/8$ and $P_6(t) = (231t^6 - 315t^4 + 105t^2 - 5)/16$

4.5-11. (a) $y = c_1 F\left(\frac{\sqrt{33}-1}{16}, \frac{3-\sqrt{33}}{16}, \frac{1}{2}; t\right) + c_2 t^{1/2} F\left(\frac{\sqrt{33}+7}{16}, \frac{11-\sqrt{33}}{16}, \frac{3}{2}; t\right)$

(b) $y = c_1 F(1 + \sqrt{3}/2, 1 - \sqrt{3}/2, 3/2; t) + c_2 t^{-1/2} F\left(\frac{1+\sqrt{3}}{2}, \frac{1-\sqrt{3}}{2}, \frac{1}{2}; t\right)$

(c) $y = c_1 F(1, 2, 2; t)[= c_1(1-t)^{-1}]$

CHAPTER 5

Section 5.1

5.1-1. (a) $x = 3e^{2t}$
$y = e^{2t} - e^{-t}$

(b) $x = (t+1)^{-1}$
$y = 2t + 3 + \ln(t+1)$

(c) $x = t^2 + t$
$y = 4e^t - t^2 - 3t - 4$

(d) $x = (2t+1)^{-1/2}$
$y = 6[6t + 2 + 3\ln(2t+1)]^{-1}$

5.1-2. (a) $c_1 = 2, c_2 = 1$
(b) $c_1 = -3, c_2 = 4$
(c) $c_1 = \frac{1}{2} - \sqrt{3}, c_2 = 1 + \sqrt{3}/2$
(d) $c_1 = 1, c_2 = -1$

5.1-3. (a) $x = \cosh 2t$, $y = 2 \sinh 2t$
(b) $x = -3\cos 3t + (2\sin 3t)/3$
$y = 9\sin 3t + 2\cos 3t$
(c) $x = (e^{3t} - e^{-t})/2$
$y = (3e^{3t} + e^{-t})/2$

Section 5.2

5.2-1. (a) Independent **(c)** Dependent
(b) Independent **(d)** Independent

5.2-2. (a) $x' = y$, $y' = -4x + 3$ $\quad$ $x = c_1 \cos 2t + c_2 \sin 2t + \frac{3}{4}$, $y = -2c_1 \sin 2t + 2c_2 \cos 2t$

(b) $x' = y$, $y' = -2x - 2y + e^{2t}$ $\quad$ $x = e^{-t}(c_1 \cos t + c_2 \sin t) + e^{2t}/10$, $y = e^{-t}[(c_2 - c_1)\cos t - (c_2 + c_1)\sin t] + e^{2t}/5$

(c) $x' = y$, $y' = -4x + 4y$ $\quad$ $x = c_1 e^{2t} + c_2 te^{2t}$, $y = (2c_1 + c_2)e^{2t} + 2c_2 te^{2t}$

(d) $y' = y$, $y' = 2x + t + 3$ $\quad$ $x = c_1 e^{\sqrt{2}t} + c_2 e^{-\sqrt{2}t} - t/2 - \frac{3}{2}$, $y = \sqrt{2}c_1 e^{\sqrt{2}t} - \sqrt{2}c_2 e^{-\sqrt{2}t} - \frac{1}{2}$

Section 5.3

5.3-1. (a) $x = c_1 \cos \sqrt{2}t + c_2 \sin \sqrt{2}t$
$y = (-c_1 \sin \sqrt{2}t)/\sqrt{2} + (c_2 \cos \sqrt{2}t)/\sqrt{2}$

(b) $x = c_1 e^{\sqrt{2}t} + c_2 e^{-\sqrt{2}t}$
$y = c_1 e^{\sqrt{2}t}/\sqrt{2} - c_2 e^{-\sqrt{2}t}/\sqrt{2}$

(c) $x = c_1 e^{6t} + c_2 te^{6t}$
$y = (c_1 + c_2)e^{6t} + c_2 te^{6t}$

(d) $x = c_1 e^t \cos 4t + c_2 e^t \sin 4t$
$y = -c_2 e^t \cos 4t + c_1 e^t \sin 4t$

(e) $x = c_1 e^{3t} + c_2 e^{-t}$
$y = c_1 e^{3t} - c_2 e^{-t}$
$x = e^{-t}$, $y = -e^{-t}$

(f) $x = c_1 + c_2 e^t$
$y = c_2 e^t$
$x = -5 + 2e^t$, $y = 2e^t$

(g) $x = c_1 e^t \cos 2t + c_2 e^t \sin 2t$
$y = (c_1 + c_2)e^t(\cos 2t)/2 + (-c_1 + c_2)e^t(\sin 2t)/2$
$x = -2e^t \cos 2t + 4e^t \sin 2t$
$y = e^t \cos 2t + 3e^t \sin 2t$

5.3-2. (a) $x = c_1 \cos \sqrt{2}t + c_2 \sin \sqrt{2}t + 3$
$y = (-c_1 \sin \sqrt{2}t)/\sqrt{2} + (c_2 \cos \sqrt{2}t)/2 - \frac{1}{2}$

(b) $x = c_1 e^{\sqrt{2}t} + c_2 e^{-\sqrt{2}t} - \frac{1}{2}$
$y = c_1 e^{\sqrt{2}t}/\sqrt{2} - c_2 e^{-\sqrt{2}t}/\sqrt{2} - t/2$

(c) $x = c_1 + c_2 e^{2t} + t/2$
$y = -c_1 + c_2 e^{2t} - t/2 - \frac{1}{2}$

(d) $x = c_1 e^{3t} + c_2 e^{-t} - \frac{2}{3}$
$y = c_1 e^{3t} - c_2 e^{-t} + \frac{1}{3} - e^t/2$
$x = 5e^{3t}/12 + e^{-t}/4 - \frac{2}{3}$
$y = 5e^{3t}/12 - e^{-t}/4 + \frac{1}{3} - e^t/2$

5.3-3. (x_0, y_0) must satisfy $3y_0 + 2x_0 = 0$

5.3-4. (x_0, y_0) must satisfy $y_0 + 2x_0 = 1$. Moreover,

$$\lim_{t\to\infty} x(t) = 4 \qquad \text{and} \qquad \lim_{t\to\infty} y(t) = -9$$

5.3-6. (a) $Q_1(t) = 25 + 25e^{-2t/15} \to 25$ as $t \to \infty$
$Q_2(t) = 75 - 25e^{-2t/15} \to 75$ as $t \to \infty$

(b) $Q_1(t) = (25 + 75e^{-8t/35})/2 \to \frac{25}{2}$ as $t \to \infty$
$Q_2(t) = (175 - 75e^{-8t/35})/2 \to \frac{175}{2}$ as $t \to \infty$

5.3-7. $Q_1(\infty) = \dfrac{V_1(A_1 + A_2)}{V_1 + V_2}$ and $Q_2(\infty) = \dfrac{V_2(A_1 + A_2)}{V_1 + V_2}$

5.3-8. $Q_1(t) = (A_1 - \beta V_1)e^{-rt/V_1} + \beta V_1$ and if $V_1 \neq V_2$, then

$$Q_2(t) = \frac{(A_1 - \beta V_1)V_2}{V_1 - V_2} e^{-rt/V_1} + \left[A_2 - \beta V_2 - \frac{(A_1 - \beta V_2)V_2}{V_2 - V_2} \right] e^{-rt/V_1} + \beta V_2$$

and if $V_1 = V_2$, then

$$Q_2(t) = \left(A_2 - \beta V_2 + \frac{A_1 - \beta V_1}{V_1} t \right) e^{-rt/V_2} + \beta V_2$$

Moreover, $Q_1(\infty)$, $Q_2(\infty)$ both exist, with $Q_1(\infty) = \beta V_1$ and $Q_2(\infty) = \beta V_2$.

5.3-9. (a) $Q_1 = 10e^{-11t/60}(e^{\sqrt{101}t/60} + e^{-\sqrt{101}t/60}) + 30$
$Q_2 = e^{-11t/60}[(-1 + \sqrt{101})e^{\sqrt{101}t/60} + (-1 - \sqrt{101})e^{-\sqrt{101}t/60}] + 30$
$Q_1(t) \to 30$ and $Q_2(t) \to 30$ as $t \to \infty$

(b) $Q_1 = 4e^{-t/10}(e^{\sqrt{2}t/20} + e^{-\sqrt{2}t/120}] + 2$
$Q_2 = 4\sqrt{2}e^{-t/10}(e^{\sqrt{2}t/20} - e^{-\sqrt{2}t/20}) + 4$
$Q_1(t) \to 2$ and $Q_2(t) \to 4$ as $t \to \infty$

Section 5.4

5.4-1. (a) CP $= \{(2, -1)\}$

IC: $(x - 2)^2 + \dfrac{(y + 1)^2}{4} = c$

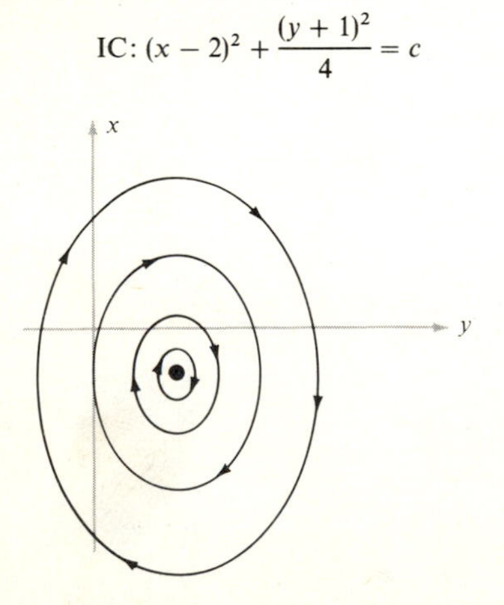

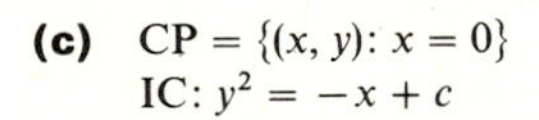

(c) CP $= \{(x, y) \colon x = 0\}$
IC: $y^2 = -x + c$

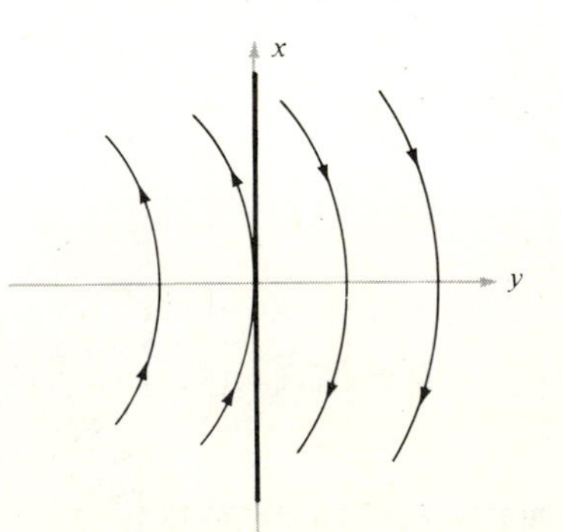

(f) $\mathrm{CP} = \{(x, y): x^2 + y^2 = 1\}$
$\mathrm{IC}: y^2 = -2x + c$

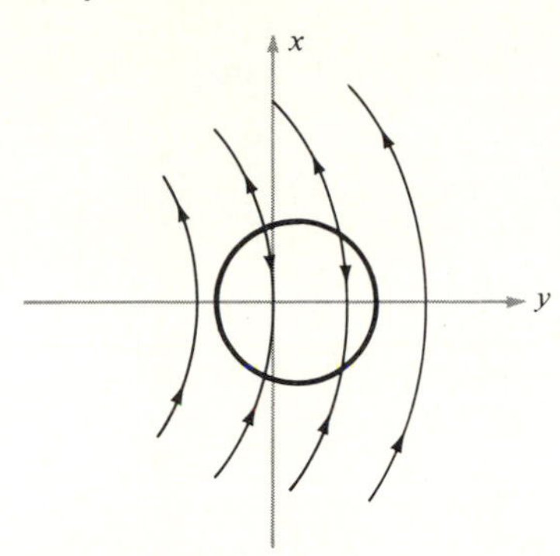

(h) $\mathrm{CP} = \{(x, y): x = 0\}$
$\mathrm{IC}: y^2 = c|x| - 1 \qquad c > 0$

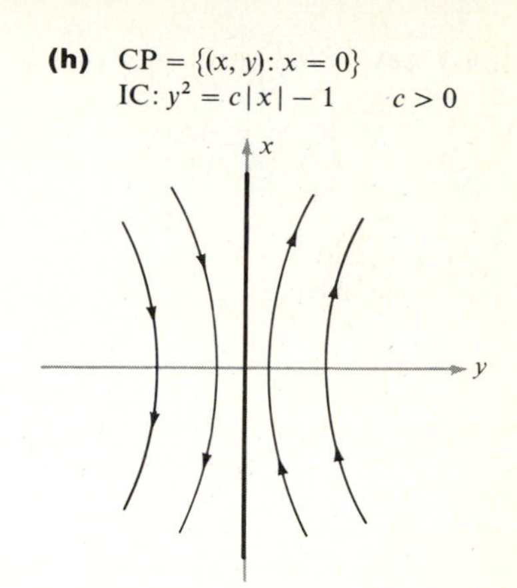

5.4-2. (a) $\mathrm{CP} = \{(1, 0)\} \cup \{(x, y): x = y\}$
$\mathrm{IC}: (x - 1)^2 + y^2 = k \qquad \text{where } k = (x_0 - 1)^2 + y_0^2$

(b) $\{(x_0, y_0): (x_0 - 1)^2 + y_0^2 = 5 \text{ and}, (x_0, y_0) \neq (-1, 1)\}$

(c) $x_\infty = y_\infty = (1 + \sqrt{15})/2 \qquad \text{if } x_0 = 3, y_0 = 2$
$x_\infty = y_\infty = (1 + \sqrt{7})/2 \qquad \text{if } x_0 = -1, y_0 = 0$

5.4-3. (a) $\mathrm{CP} = \{(x, y): y = 0\}$
$\mathrm{IC}: y = x^2 + c \qquad \text{where } c = y_0 - x_0^2$

(b) $x_\infty = -1, y_\infty = 0 \qquad \text{if } x_0 = 0, y_0 = -1$
$x_\infty = -1, y_\infty = 0 \qquad \text{if } x_0 = -2, y_0 = 3$
$x_\infty = -\sqrt{5}, y_\infty = 0 \qquad \text{if } x_0 = 2, y_0 = -1$

(c) $\{(x_0, y_0): y_0 = x_0^2 \text{ and } x_0 \leq 0\}$

(d) Trajectory is $\{(x, y): y = x^2, -1 \leq x < 0\}$ if $x_0 = -1, y_0 = 1$
Trajectory is $\{(x, y): y = x^2 + 1, x \geq 0\}$ if $x_0 = 0, y_0 = 1$
Trajectory is $\{(x, y): y = x^2 - 1, -1 < x < 0\}$ if $x_0 = 0, y_0 = 1$

5.4-5. Both plots are the same.

5.4-6. $\mathrm{CP} = \{(x, y): y = 0, x \geq 0\}$
$\mathrm{IC}: y + x + \beta/\alpha x = c \qquad \text{where } c = y_0 + x_0 + \beta/\alpha x_0$
$x_\infty = (c - \sqrt{c^2 - 4\beta/\alpha})/2, y_\infty = 0$

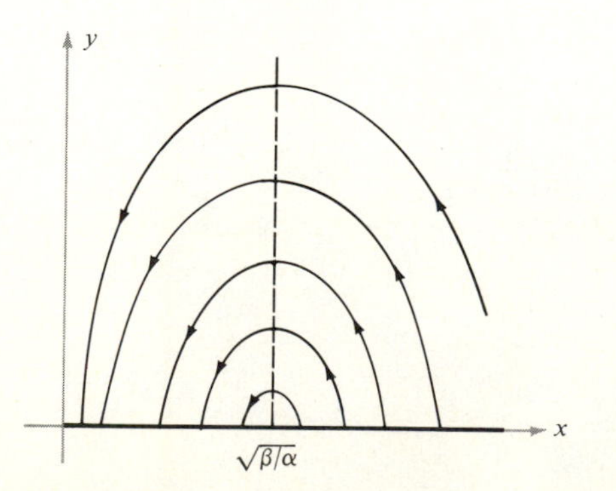

5.4-7. (a) $x' = y$
$y' = -g$
IC: $y^2 = -2g(x - c)$

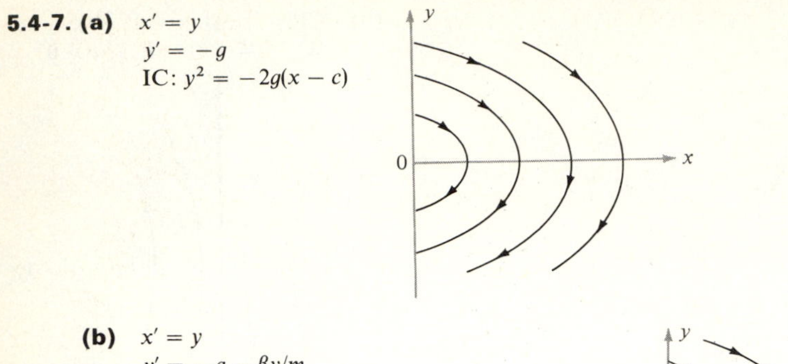

(b) $x' = y$
$y' = -g - \beta y/m$
IC: $x = (m^2g/\beta^2)\ln|\beta y/m + g| - my/\beta + c$

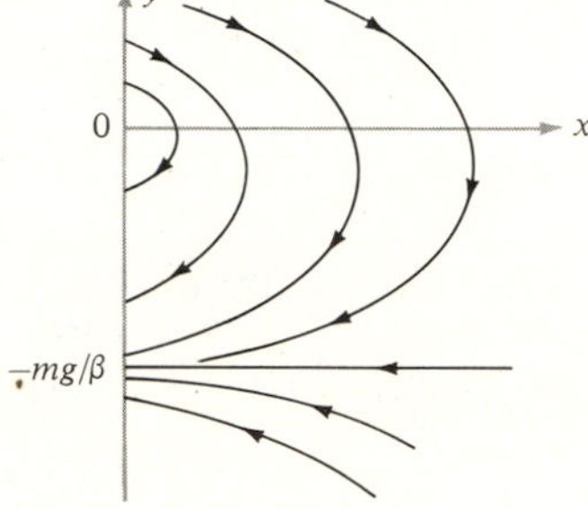

Section 5.5

5.5-1. (a) $\text{CP} = \{(x, y): x = 1\} \cup \{(0, 0)\}$
IC: $x^2 + y^2 = c$
If $0 < x(0)^2 + y(0)^2 < 1$, then the solution x, y is nonconstant and periodic. Otherwise, $x(t)$, $y(t)$ converges to a critical point as $t \to \infty$.

(b) $\text{CP} = \{(x, y): y = x^2\} \cup \{(0, -1)\}$
IC: $x^2/6 + (y + 1)^2 = c$
If $0 < x(0)^2/6 + [y(0) + 1]^2 < 1$, then the solution x, y is nonconstant and periodic. Otherwise, $x(t)$, $y(t)$ converges to a critical point as $t \to \infty$.

(c) $\text{CP} = \{(x, y): x = 0 \text{ or } y = -1\}$
IC: $y = -x^3 + c$
There are no nonconstant periodic solutions and $x(t)$, $y(t)$ converges to a critical point only in case $x(0) \leq 0$ or $y(0) \geq -1$.

5.5-2. (a) $x' = y$
$y' = -x^5$
$\text{CP} = \{(0, 0)\}$
IC: $x^6/x + y^2 = c$
$y_{\max} = A^3/\sqrt{3}$

(b) $x' = y$
$y' = -x|x|$
$\text{CP} = \{(0, 0)\}$
IC: $\dfrac{|x|x^2}{3} + \dfrac{y^2}{2} = c$
$y_{\max} = A^{3/2}\sqrt{\tfrac{3}{2}}$

(c) $x' = y$
$y' = -x(1 + x^2)^{-1}$
$\text{CP} = \{(0, 0)\}$
$y^2 + \ln(1 + x^2) = c$
$y_{\max} = \sqrt{\ln(1 + A^2)}$

5.5-4. (a) $\bar{U}(x, y) = gx^2/2l + y^2/2$ is a first integral, and the curves $\bar{U}(x, y) = k$ are ellipses centered at the origin.

(b) Same as the level curves $\bar{U}(x, y) = k$.

(c) $x = c_1 \cos(t\sqrt{g/l}) + c_2 \sin(t\sqrt{g/l})$
$y = c_2\sqrt{g/l} \cos(t\sqrt{g/l}) - c_1\sqrt{g/l} \sin(t\sqrt{g/l})$
$T = 2\pi\sqrt{l/g}$ so T increases as l increases, which is the same as in Problem 5.5-2.

Section 5.6

5.6-1. (a) Center point with counterclockwise motion around origin. Not asymptotically stable.

(b) Asymptotically stable node.

(c) Center point with counterclockwise motion around origin. Not asymptotically stable.

(d) Unstable node.

(e) Asymptotically stable focal point with clockwise motion about origin.

(f) Saddle point.

5.6-2. If $\mu = 0$, the origin is a center point with clockwise motion; if $0 < \mu < 2$, the origin is an asymptotically stable focal point with clockwise motion; and if $\mu \geq 2$, the origin is an asymptotically stable node (note that system is critically damped when $\mu = 2$).

5.6-3. (a) If $\eta < 0$, origin is asymptotically stable focal point; if $0 < \eta < 1$, origin is asymptotically stable node; if $\eta = 1$, there is a line of critical points through origin; and if $\eta > 1$, the origin is a saddle point.

5.6-4. (a) Center point

(b) Asymptotically stable focal point

(c) Asymptotically stable nodal point

(d) Unstable nodal point

(e) Saddle point

Section 5.7

5.7-1. (a) CP = $\{(-3, 2)\}$
Linearization about $(-3, 2)$ is $x' = 2y$, $y' = x$
$(-3, 2)$ is a saddle point

(b) CP = $\{(0, 0), (2, 4)\}$
Linearization about $(0, 0)$ is $x' = -2x$, $y' = -y$
$(0, 0)$ is asymptotically stable double tangent node
Linearization about $(2, 4)$ is $x' = 2x$, $y' = 4x - y$
$(2, 4)$ is a saddle point

(c) CP = $\{(0, 1), (1, 3)\}$
Linearization about $(0, 1)$ is $x' = -2x + y$, $y' = -2x$
$(0, 1)$ is asymptotically stable spiral point (with clockwise spiral)
Linearization about $(1, 3)$ is $x' = -2x + y$, $y' = y$
$(1, 3)$ is a saddle point

5.7-2. (a) If $\alpha > 3$, origin is asymptotically stable double tangent node; if $-1 < \alpha < 3$, origin is asymptotically stable spiral point; and if $\alpha < -1$, origin is saddle point.

(b) If $\alpha > 1$, origin is saddle point; if $0 < \alpha < 1$, origin is asymptotically stable double-tangent node; and if $\alpha < 0$, origin is asymptotically stable spiral point.

Section 5.8

5.8-1. (a) $\bar{p}(t) = A \sin(\sqrt{\alpha\gamma}\, t + \phi) \qquad \bar{q}(t) = -A \dfrac{\delta}{\beta}\sqrt{\dfrac{\alpha}{\gamma}} \cos(\sqrt{\alpha\gamma}\, t + \phi)$

where A and ϕ are arbitrary constants.

(b) $\beta^2\gamma\bar{q}^2 + \delta^2\alpha\bar{p}^2 = c$

(c) Same family as in part *b*.

5.8-2. Same as predator-prey model, with prey replaced by susceptibles and predators replaced by infectives. $\text{CP} = \{(0, 0), (\beta/\alpha, \gamma/\alpha)\}$, and all other solutions with positive initial data are periodic.

Section 5.9

5.9-1. (a) $x^4/4 + y^2/2 = E$
$A^4 = 4E \qquad$ for all E

(b) $(\arctan x^2)/2 + y^2/2 = E$
$\arctan 2E = A^2 \qquad 0 < E < \pi/4$

(c) $x^2|x|/3 + y^2/2 = E$
$A^3 = 3E \qquad$ for all E

(d) $x^6/6 + y^2/2 = E$
$A^6 = 6E \qquad$ for all E

(e) $(1 - e^{-x^2})/2 + y^2/2 = E$
$A = [-\ln(1 - 2E)]^{1/2} \qquad 0 < E < 1/2$

5.9-2. In Example 5.9-2, $A^2 = (-k + \sqrt{k^2 + 4lE})/l$, and in Example 5.9-3, $A^2 = (k + \sqrt{k^2 - 4lE})/l$.

CHAPTER 6

Section 6.1

6.1-1. (a) $\phi_4(t) = \frac{1}{2} - t/8 + 35t^2/64 - 111t^3/768$

(b) $\phi_4(t) = t^3/3$

(c) $\phi_4(t) = t + t^3/6$

(d) $\phi_4(t) = (t - 1) + (t - 1)^2$

6.1-2. (a) $\psi_3(t) = \frac{1}{4} + t/8 - t^2/64 \qquad \phi_3(t) = \frac{1}{2} + 3t/8 + 7t^2/64$

(b) $\psi_3(t) = 1 + t/2 + t^2/8 \qquad \phi_3(t) = \frac{1}{2} + t^2/8$

6.1-3. (a) $\phi_5(t) = \frac{1}{2} - 3t^2/16 + 3t^4/768$

(b) $\phi_5(t) = 1 - 3t - t^3/6 + t^4/4$

(c) $\phi_5(t) = t/4 - t^2/8 + t^4/96$

6.1-4. (a) $\phi_2(t) = \frac{1}{2} + 7t/8 \qquad -1.155t^2 \le y(t) - \phi_2(t) \le -0.066t^2 \qquad$ for $0 \le t \le \frac{1}{2}$

(b) $\phi_2(t) \equiv 0 \qquad 0.5t^2 \le y(t) \le 1.455t^2$

(c) $\phi_2(t) = 1 + e^{-1}t \cong 1 + 0.3679t$
$-3.01t^2 \le y(t) - \phi_2(t) \le 4t^2$

Section 6.2

6.2-1. (a)

t_i	$y_i\ \{=0.6y_{i-1}\}$	$y(t_i)\ \{=e^{-4t_i}\}$
0.1	0.6000	0.6703
0.2	0.3600	0.4493
0.3	0.2160	0.3012
0.4	0.1296	0.2019
0.5	0.0777	0.1353

(b)

t_i	$y_i\ \{=0.8y_{i-1}\}$	$y(t_i)\ \{=e^{-2t_i}\}$
0.1	0.8000	0.8187
0.2	0.6400	0.6703
0.3	0.5120	0.5488
0.4	0.4096	0.4493
0.5	0.3277	0.3679

(c)

t_i	$y_i\ \{=1.2y_{i-1} - 0.1\}$	$y(t_i)\ \{=\frac{1}{2} + \frac{1}{2}e^{2t_i}\}$
0.1	1.100	1.111
0.2	1.220	1.246
0.3	1.364	1.411
0.4	1.537	1.613
0.5	1.744	1.859

(d)

t_i	$y_i\ \{=y_{i-1} + 0.02i\ (y_{i-1})^2\}$	$y(t_i)\left\{=\frac{1}{1-t_i^2}\right\}$
0.1	1.020	1.010
0.2	1.062	1.042
0.3	1.130	1.099
0.4	1.232	1.190
0.5	1.384	1.333

(e)

t_i	$y_i\ \{=y_{i-1} + 0.1 + 0.1(y_{i-1})^2\}$	$y(t_i)\ \{=\tan t_i\}$
0.1	0.1000	0.1003
0.2	0.2010	0.2027
0.3	0.3050	0.3093
0.4	0.4143	0.4228
0.5	0.5315	0.5463

(f)

t_i	$y_i\ \{=1.1y_{i-1} + 0.1(y_{i-1})^2\}$	$y(t_i)\left\{=\frac{1}{1+e^{-t_i}}\right\}$
0.1	0.5250	0.5250
0.2	0.5499	0.5498
0.3	0.5747	0.5744
0.4	0.5991	0.5987
0.5	0.6231	0.6225

6.2-2. (a) $y(\frac{1}{2}) = 2$

Euler's method: $h = \frac{1}{8}$, $y_4 = 1.766$; $h = \frac{1}{12}$, $y_6 = 1.825$

Taylor's order 3: $h = \frac{1}{4}$, $y_2 = 1.885$; $h = \frac{1}{2}$, $y_3 = 1.933$

(b) $y(\frac{1}{2}) = \frac{2}{3} \cong 0.6667$
Euler's method: $h = \frac{1}{8}$, $y_4 = 0.6416$; $h = \frac{1}{12}$, $y_6 = 0.6506$
Taylor's order 3: $h = \frac{1}{8}$, $y_2 = 0.6139$; $h = \frac{1}{12}$, $y_3 = 0.6231$

6.2-3. $\lim_{n\to\infty} (1 + a/n)^n = \lim_{x\to\infty} e^{x \ln (1 + a/x)} = e^a$
(Use L'Hospital's rule)

6.2-4. (a) $y_1 = 1.01$, $y_2 = 1.0407$
(b) $y_1 = 0.47652$, $y_2 = 0.45710$

6.2-5. (a)

	Euler		Taylor (order 3)	
t_i	x_i	y_i	x_i	y_i
0.1	0.9000	−1.000	0.9250	−0.7300
0.2	0.8500	−0.4060	0.8943	−0.4911

Exact solutions to five digits: $x(0.1) = 0.92405$, $y(0.1) = -0.72515$
$x(0.2) = 0.89417$, $y(0.2) = -0.48196$

(b)

	Euler		Taylor (order 3)	
t_i	x_i	y_i	x_i	y_i
0.1	0.9500	3.225	0.9456	3.2353
0.2	0.8918	3.471	0.8832	3.4942

Section 6.3

6.3-1. (a) $y_1 = 0.6800$, $y_2 = 0.4624$, $y_3 = 0.3144$, $y_4 = 0.2138$
(b) $y_1 = 0.8200$, $y_2 = 0.6724$, $y_3 = 0.5512$, $y_4 = 0.4521$
(c) $y_1 = 0.5250$, $y_2 = 0.5498$, $y_3 = 0.5744$, $y_4 = 0.5986$
(d) $y_1 = 0.0005$, $y_2 = 0.0030$, $y_3 = 0.0095$, $y_4 = 0.0220$

6.3-2. (a) $h = \frac{1}{4}$: $y_1 = 1.3332$, $y_2 = 1.9988$
$h = \frac{1}{6}$: $y_1 = 1.2000$, $y_2 = 1.5000$, $y_3 = 1.9997$

(b) $h = \frac{1}{4}$: $y_1 = 0.8000$, $y_2 = 0.6667$
$h = \frac{1}{6}$: $y_1 = 0.8571$, $y_2 = 0.7500$, $y_3 = 0.6667$

6.3-3. (a) $y_1 = 0.6800$, $y_2 = 0.4624$, $y_3 = 0.3144$, $y_4 = 0.2138$
(b) $y_1 = 0.8200$, $y_2 = 0.6724$, $y_3 = 0.5512$, $y_4 = 0.4521$
(c) $y_1 = 0.5250$, $y_2 = 0.5498$, $y_3 = 0.5745$, $y_4 = 0.5987$
(d) $y_1 = 0.0003$, $y_2 = 0.0025$, $y_3 = 0.0088$, $y_4 = 0.0210$

6.3-7. (a) $x_1 = 0.9241$, $y_1 = -0.7252$
$x_2 = 0.8912$, $y_2 = -0.4820$

(b) $x_1 = 0.9457$, $y_1 = 3.2358$
$x_2 = 0.8834$, $y_2 = 3.4952$

6.3-8. (a)

Euler ($h = 0.1$)	Runge-Kutta ($h = 0.2$)
$x_1 = 0.1000,\ y_1 = 0.9800$	
$x_2 = 0.1960,\ y_2 = 0.9204$	$x_1 = 0.1947,\ y_1 = 0.9211$
$x_3 = 0.2814,\ y_3 = 0.8236$	
$x_4 = 0.3608,\ y_4 = 0.6935$	$x_2 = 0.3586,\ y_2 = 0.6968$

(b)

Euler ($h = 0.1$)	Runge-Kutta ($h = 0.2$)
$x_1 = 0.0995,\ y_1 = 0.0010$	
$x_2 = 0.0980,\ y_2 = -0.0199$	$x_1 = 0.0980,\ y_1 = -0.0918$
$x_3 = 0.0955,\ y_3 = -0.0296$	
$x_4 = 0.0921,\ y_4 = -0.0389$	$x_2 = 0.0921,\ y_2 = -0.0389$

(c)

Euler ($h = 0.1$)	Runge-Kutta ($h = 0.2$)
$x_1 = 0.0995,\ y_1 = -0.0094$	
$x_2 = 0.0981,\ y_2 = -0.0178$	$x_1 = 0.0982,\ y_1 = -0.0178$
$x_3 = 0.0959,\ y_3 = -0.0253$	
$x_4 = 0.0931,\ y_4 = -0.0318$	$x_2 = 0.0931,\ y_2 = -0.0318$

Section 6.4

6.4-1. (a) Improved Euler: $y_1 = 1.1025$, $y_2 = 1.2155$, $y_3 = 1.3401$
Nystrum-Trap.: $y_2 = 1.2186$, $y_3 = 1.3469$

(b) Improved Euler: $y_1 = 1.0100$, $y_2 = 1.0414$, $y_3 = 1.0984$
Nystrum-Trap.: $y_2 = 1.0419$, $y_3 = 1.0999$

(c) Improved Euler: $y_1 = 0.9050$, $y_2 = 0.8190$, $y_3 = 0.7412$
Nystrum-Trap.: $y_2 = 0.8188$, $y_3 = 0.7408$

(d) Improved Euler: $y_1 = 0.1005$, $y_2 = 0.2030$, $y_3 = 0.3098$
Nystrum-Trap.: $y_2 = 0.2030$, $y_3 = 0.3099$

6.4-2. (a) $y_2 = 1.2186$, $y_3 = 1.3468$

(b) $y_2 = 1.0419$, $y_3 = 1.0999$

(c) $y_2 = 0.8186$, $y_3 = 0.7407$

(d) $y_2 = 0.2031$, $y_3 = 0.3099$

6.4-5. (a) Runge-Kutta: $y_1 = 0.670400$, $y_2 = 0.449436$, $y_3 = 0.301302$
Adams-Moulton: $y_4 = 0.201921$, $y_5 = 0.135332$
Milne: $y_4 = 0.201921$, $y_5 = 0.135402$
Hamming: $y_4 = 0.201934$, $y_5 = 0.135336$

(b) Runge-Kutta: $y_1 = 0.913794$, $y_2 = 0.851191$, $y_3 = 0.807622$
Adams-Moulton: $y_4 = 0.779798$, $y_5 = 0.765267$
Milne: $y_4 = 0.779804$, $y_5 = 0.765280$
Hamming: $y_4 = 0.779799$, $y_5 = 0.765267$

(c) Runge-Kutta: $y_1 = 0.524979$, $y_2 = 0.549834$, $y_3 = 0.574443$
Adams-Moulton: $y_4 = 0.598688$, $y_5 = 0.622460$
Milne: $y_4 = 0.598688$, $y_5 = 0.622460$
Hamming: $y_4 = 0.598688$, $y_5 = 0.622460$

CHAPTER 7

Section 7.1

7.1-1. (a) $y(t) = 4 - e^{-t}/3 - 5e^{2t}/3$

(b) $y(t) = e^{-t}/2 + e^{-2t}(3 \cos t + 3 \sin t)/2$

(c) $y(t) = 2 \cos 2t + t \sin 2t$

7.1-2. (a) $y(t) = (\cos 2t)/8 + 7/8 + 7t^2/4$

(b) $y(t) = -e^t/2 + e^{-t}/2 + t + 1$

(c) $y(t) = -4 + 5e^t - 4te^t + t^2/2 - 2t$

Section 7.2

7.2-1. (a) $\lambda = (-1 \pm \sqrt{3}\,i)/2$ have multiplicity 3 and
$y = e^{-t/2}[(c_1 + c_2 t + c_3 t^2) \cos (\sqrt{3}\,t/2) + (c_4 + c_5 t + c_6 t^2) \sin (\sqrt{3}\,t/2)]$

(b) $\lambda = \pm 3i$ have multiplicity 2, $\lambda = 2$ has multiplicity 3, and
$y = (c_1 + c_2 t) \cos 3t + (c_3 + c_4 t) \sin 3t + (c_5 + c_6 t + c_7 t^2)e^{2t}$

(c) $\lambda = -1$ is simple, $\lambda = 1$ has multiplicity 3, $\lambda = (-1 \pm \sqrt{3}\,i)/2$ have multiplicity 2, and
$y = c_1 e^{-t} + (c_2 + c_3 t + c_4 t^2)e^t$
$\quad + e^{-t/2}[(c_5 + c_6 t) \cos (\sqrt{3}\,t/2) + (c_7 + c_8 t) \sin (\sqrt{3}\,t/2)]$

(d) $\lambda = 1$ is simple, $\lambda = 2$ has multiplicity 2, $\lambda = 3$ has multiplicity 3, and
$y = c_1 e^t + (c_2 + c_3 t)e^{2t} + (c_4 + c_5 t + c_6 t^2)e^{3t}$

7.2-2. (a) $y = c_1 e^{2t} + c_2 e^{-2t} + c_3 e^{\sqrt{2}t} + c_4 e^{-\sqrt{2}t}$

(b) $y = (c_1 + c_2 t)e^t + e^{-t/2}[(c_3 + c_4 t) \cos (\sqrt{3}\,t/2) + (c_5 + c_6 t) \sin (\sqrt{3}\,t/2)]$

(c) $y = c_1 e^{\sqrt{2}t} + c_2 e^{-\sqrt{2}t} + c_3 \cos \sqrt{2}\,t + c_4 \sin \sqrt{2}\,t$

(d) $y = c_1 e^t + e^{-t/2}[c_2 \cos (\sqrt{3}\,t/2) + c_3 \sin (\sqrt{3}\,t/2)] + c_4 e^{-2t}$
$\quad + e^t(c_5 \cos \sqrt{3}\,t + c_6 \sin \sqrt{3}\,t)$

7.2-3. (a) $y = c_1 e^{-t} + e^{-t/2}[c_2 \cos (\sqrt{7}\,t/2) + c_3 \sin (\sqrt{7}\,t/2)]$

(b) $y = c_1 + (c_2 + c_3 t + c_4 t^2)e^t$

(c) $y = c_1 e^{t/2} + e^{-t/2}[c_2 \cos (\sqrt{3}\,t/2) + c_3 \sin (\sqrt{3}\,t/2)]$

(d) $y = c_1 e^{-t/2} + c_2 e^{-3t} + c_3 e^{\sqrt{2}t} + c_4 e^{-\sqrt{2}t}$

7.2-4. (a) $y = c_1 e^t + c_2 e^{-t} + c_3 e^{\sqrt{2}t} + c_4 e^{-\sqrt{2}t}$

(b) $y = c_1 e^{-2t} + e^t(c_2 \cos \sqrt{3}\,t + c_3 \sin \sqrt{3}\,t)$

(c) $y = c_1 e^{2t} + e^{-t}(c_2 \cos t + c_3 \sin t)$

(d) $y = c_1 e^t + e^{-t/3}[c_2 \cos (\sqrt{2}\,t/3) + c_3 \sin (\sqrt{2}\,t/3)]$

(e) $y = e^{-t}[(c_1 + c_2 t) \cos \sqrt{3}\,t + (c_3 + c_4 t) \sin \sqrt{3}\,t] + (c_5 + c_6 t)e^{2t}$

(f) $y = c_1 e^{\sqrt{3}t} + c_2 e^{-\sqrt{3}t} + c_3 \cos \sqrt{2}\,t + c_4 \sin \sqrt{2}\,t$

(g) $y = c_1 e^t + c_2 e^{2t} + c_3 e^{3t}$

7.2-5. (a) $y = 1 - 2t + t^2$

(b) $y = 3e^t + (2 \cos 2t)/5 - (3 \sin 2t)/10$

(c) $y = e^{-t}/4 + (-1 + 2t)(e^t/4)$

(d) $y = -e^t/4 - 3e^{-t}/4 + \cos t + (\sin t)/2$

Section 7.3

7.3-1. (a) $y = c_1e^t + c_2e^{-t} + c_3 \cos t + c_4 \sin t + e^{2t}/7$

(b) $y = c_1e^{-t} + e^t(c_2 \cos t + c_3 \sin t) + 3t^2/2 + \frac{3}{2}$

(c) $y = c_1e^{-t} + c_2e^{-2t} + c_3e^{-3t} + e^t/24 + t/6 - \frac{11}{36}$

(d) $y = c_1e^t + c_2e^{-t} + c_3 \cos t + c_4 \sin t - t + te^{-t}/4$

7.3-2. (a) $y_p = (e^t + 1)^2/4 - (e^t + 1) + [\ln (e^t + 1)]/2$
$+ [(e^t + 1) - \ln (e^t + 1)]e^t + \ln (e^t + 1)\,(e^{2t}/2)$

(b) $y_p = (-t \cos t)/2 + [(\sin t)\ln |\cos t|]/2 + (\ln|\sec t + \tan t|)/2$

7.3-3. (a) $y_p = d_1 + d_2t + d_3t^2 + (d_4t^3 + d_5t^2) \cos t + (d_6t^3 + d_7t^2) \sin t$

(b) $y_p = d_1t^3 + d_2t^2 + d_3te^t$

(c) $y_p = d_1 \sin t + d_2 \cos t + te^{-t}(d_3 \sin t + d_4 \cos t)$

(d) $y_p = (d_1t^2 + d_2t)e^t + d_3 \cos 3t + d_4 \sin 3t + d_5t \sin t + d_6t \cos t$

7.3-4. If $r(\alpha) = r'(\alpha) = 0$ and $r''(\alpha) \neq 0$, then $y_p = t^2e^{\alpha t}/r''(\alpha)$; and if $r(\alpha) = r'(\alpha) = r''(\alpha) = 0$, then $r'''(\alpha) = 6a_3$ and $y_p = t^3e^{\alpha t}/6a_3$.

Section 7.4

7.4-1. (a) $x = c_1e^{-t} + e^{t/2}[c_2 \cos (\sqrt{3}t/2) + c_3 \sin (\sqrt{3}t/2)]$
$y = c_1e^{-t} + \frac{1}{2}e^{t/2}[(-c_2 + \sqrt{3}c_3) \cos (\sqrt{3}t/2) - (\sqrt{3}c_2 + c_3) \sin (\sqrt{3}t/2)]$

(b) $x = c_1e^{-2t} + c_2e^{3t} + c_3e^{-3t}$
$y = -c_1e^{-2t} + 4c_2e^{3t} - 2c_3e^{-3t}$

(c) $x_p = 2t + 2$
$y_p = -t^2$ (Homogeneous solution is in part *a*)

7.4-2. (a) $x = -12e^t + 14 \sin t + 13 \cos t$
$y = 12e^t - 3 \sin t - 8 \cos t$

(b)
$$x = \frac{(4\sqrt{2} + 3)e^{\sqrt{2}t} + (4\sqrt{2} - 3)e^{-\sqrt{2}t}}{8\sqrt{2}} + \frac{1}{3} \cos \sqrt{2}t - \frac{7}{12\sqrt{2}} \sin \sqrt{2}t - \frac{4}{3}e^t$$
$$y = \frac{(4\sqrt{2} + 3)e^{\sqrt{2}t} + (4\sqrt{2} - 3)e^{-\sqrt{2}t}}{16\sqrt{2}} - \frac{1}{6} \cos \sqrt{2}t - \frac{7}{24\sqrt{2}} \sin \sqrt{2}t - \frac{1}{3}e^t + \frac{1}{4}t$$

7.4-3. (a) $x = c_1 \cos 2t + c_2 \sin 2t + c_3 \cos t + c_4 \sin t$
$y = -c_1 \cos 2t - c_2 \sin 2t - 2c_3 \cos t - 2c_4 \sin t$
and $1/\pi$ and $1/2\pi$ are the natural frequencies.

(b) When frequency is $1/\pi$, then $x = -y$ (that is, $c_3 = c_4 = 0$ in part *a*), so m_1 is always the same distance from equilibrium as m_2 and in the opposite direction. When frequency is $1/2\pi$, $y = -2x$ (that is, $c_1 = c_2 = 0$ in part *a*), so m_2 is twice the distance from equilibrium as m_1 and in the opposite direction.

(c) $x = \cos 2t - \sin 2t - \cos t$
$t = -\cos 2t + \sin 2t + 2 \cos t$

7.4-4. (a) The natural frequencies are $\omega_1/2\pi$ and $\omega_2/2\pi$, where
$$\omega_1^2 = \frac{k_1 + 2k_2 + \sqrt{k_1^2 + 4k_2^2}}{8} \qquad \omega_2^2 = \frac{k_1 + 2k_2 - \sqrt{k_1^2 + 4k_2^2}}{8}$$

(b) The natural frequencies are $\omega_1/2\pi$ and $\omega_2/2\pi$, where

$$\omega_1^2 = \frac{2m_2 + m_1 + \sqrt{4m_2^2 + m_1^2}}{m_1 m_2} \qquad \omega_2^2 = \frac{2m_2 + m_1 - \sqrt{4m_2^2 + m_1^2}}{m_1 m_2}$$

7.4-5. (a) The natural frequencies are $\sqrt{k/m}/2\pi$ and $\sqrt{3k/m}/2\pi$.

(b) $x = c_1 \cos \sqrt{2}t + c_2 \sin \sqrt{2}t + c_3 \cos \sqrt{6}t + c_4 \sin \sqrt{6}t$
$y = c_1 \cos \sqrt{2}t + c_2 \sin \sqrt{2}t - c_3 \cos \sqrt{6}t - c_4 \sin \sqrt{6}t$ and
$\omega_1 = \sqrt{2}/2\pi$ and $\omega_2 = \sqrt{6}/2\pi$ are the natural frequencies.

7.4-6. (b) $y = x \tan \alpha - (x^2 g \sec^2 \alpha)/2v_0^2$, which is a parabola

(c) Time of flight is $(2v_0 \sin \alpha)/g$

(d) $R = (v_0^2 \sin 2\alpha)/g$ and R is maximum when $\alpha = \pi/4$

(e) $R = R_{max}/2$ when $\alpha = \pi/12$ and $\alpha = 5\pi/12$

(f) Maximum height is $(v_0^2 \sin^2 \alpha)/2g$

CHAPTER 8

Section 8.1

8.1-1. (a) $\begin{pmatrix} -6 \\ -4 \end{pmatrix}$ **(b)** $\begin{pmatrix} 3 \\ -8 \\ 2 \end{pmatrix}$ **(c)** $\begin{pmatrix} 2 \\ 16 \\ 4 \end{pmatrix}$ **(d)** $\begin{pmatrix} 0 \\ 0 \\ 0 \end{pmatrix}$

(e) $\begin{pmatrix} 0 \\ 1 \\ -2 \\ 6 \end{pmatrix}$ **(f)** $\begin{pmatrix} 6 & 8 & 5 \\ 6 & 6 & -3 \\ 6 & 10 & 4 \end{pmatrix}$ **(g)** $\begin{pmatrix} 6 & 0 & 6 & 4 \\ 5 & 2 & 1 & -1 \\ 9 & 5 & 5 & 3 \\ -8 & -4 & -4 & 2 \end{pmatrix}$

8.1-2. (a) $\begin{pmatrix} 5 \\ 4 \\ -5 \end{pmatrix}$ **(b)** $\begin{pmatrix} -7 & -8 & -3 \\ 1 & -9 & -7 \\ -8 & -1 & -7 \end{pmatrix}$ **(c)** $\begin{pmatrix} -1 & -1 & 1 \\ 3 & -3 & 1 \\ 0 & 2 & 0 \end{pmatrix}$ **(d)** $\begin{pmatrix} 4 \\ -4 \\ 18 \end{pmatrix}$

8.1-4. (a) $\begin{pmatrix} y_1 \\ y_2 \end{pmatrix}' = \begin{pmatrix} 6 & -2 \\ 0 & 1 \end{pmatrix}\begin{pmatrix} y_1 \\ y_2 \end{pmatrix} + \begin{pmatrix} te^t \\ -7e^{2t} \end{pmatrix}$

(b) $\begin{pmatrix} y_1 \\ y_2 \\ y_3 \end{pmatrix}' = \begin{pmatrix} 0 & 1 & -1 \\ 0 & 2 & 1 \\ -1 & 1 & -1 \end{pmatrix}\begin{pmatrix} y_1 \\ y_2 \\ y_3 \end{pmatrix} + \begin{pmatrix} \sin 2t \\ 2e^{-t} \\ -\cos 2t \end{pmatrix}$

(c) $\begin{pmatrix} y_1 \\ y_2 \\ y_3 \end{pmatrix}' = \begin{pmatrix} 0 & 0 & 3 \\ -4 & 1 & 0 \\ 6 & 0 & 0 \end{pmatrix}\begin{pmatrix} y_1 \\ y_2 \\ y_3 \end{pmatrix}$

(d) $\begin{pmatrix} y_1 \\ y_2 \\ y_3 \\ y_4 \end{pmatrix}' = \begin{pmatrix} 1 & 1 & 1 & 1 \\ 1 & 1 & 1 & 1 \\ 0 & 0 & 1 & -1 \\ 1 & 0 & -1 & 0 \end{pmatrix}\begin{pmatrix} y_1 \\ y_2 \\ y_3 \\ y_4 \end{pmatrix} + \begin{pmatrix} e^t \\ -e^{-t} \\ e^t \\ -e^{-t} \end{pmatrix}$

8.1-5. (b) $d_1 = -3, d_2 = 4$ **(c)** $c = -\frac{1}{3}$

8.1-6. (a) $\begin{pmatrix} y_1 \\ y_2 \\ y_3 \end{pmatrix}' = \begin{pmatrix} 0 & 1 & 0 \\ 0 & 0 & 1 \\ 1 & -1 & 1 \end{pmatrix}\begin{pmatrix} y_1 \\ y_2 \\ y_3 \end{pmatrix} + \begin{pmatrix} 0 \\ 0 \\ e^{2t} \end{pmatrix}$

(b) $\begin{pmatrix} y_1 \\ y_2 \\ y_3 \\ y_4 \end{pmatrix}' = \begin{pmatrix} 0 & 1 & 0 & 0 \\ 0 & 0 & 1 & 0 \\ 0 & 0 & 0 & 1 \\ -1 & 0 & 2 & 0 \end{pmatrix} \begin{pmatrix} y_1 \\ y_2 \\ y_3 \\ y_4 \end{pmatrix} + \begin{pmatrix} 0 \\ 0 \\ 0 \\ e^{-t}+1 \end{pmatrix}$

(c) $\begin{pmatrix} y_1 \\ y_2 \\ y_3 \end{pmatrix}' = \begin{pmatrix} 0 & 1 & 0 \\ 0 & 0 & 1 \\ 8 & 0 & 0 \end{pmatrix} \begin{pmatrix} y_1 \\ y_2 \\ y_3 \end{pmatrix}$

Section 8.2

8.2-1. (a) Linearly independent

(b) Linearly dependent: $2\begin{pmatrix} 1 \\ -1 \end{pmatrix} + 7\begin{pmatrix} 1 \\ 2 \end{pmatrix} - 3\begin{pmatrix} 3 \\ 4 \end{pmatrix} = \begin{pmatrix} 0 \\ 0 \end{pmatrix}$

(c) Linearly independent

(d) Linearly dependent: $\begin{pmatrix} 2 \\ 4 \\ 0 \end{pmatrix} + 4\begin{pmatrix} 3 \\ -1 \\ 7 \end{pmatrix} - 14\begin{pmatrix} 1 \\ 0 \\ 2 \end{pmatrix} = \begin{pmatrix} 0 \\ 0 \\ 0 \end{pmatrix}$

(e) Linearly dependent: $\begin{pmatrix} 1 \\ 2 \\ 3 \end{pmatrix} + 0\begin{pmatrix} 1 \\ 1 \\ 2 \end{pmatrix} - 2\begin{pmatrix} 2 \\ 2 \\ 2 \end{pmatrix} + \begin{pmatrix} 3 \\ 2 \\ 1 \end{pmatrix} = \begin{pmatrix} 0 \\ 0 \\ 0 \end{pmatrix}$

(f) Linearly dependent: $\begin{pmatrix} 1 \\ 1 \\ 0 \\ 0 \end{pmatrix} + 0\begin{pmatrix} 0 \\ 1 \\ 1 \\ 0 \end{pmatrix} + \begin{pmatrix} 0 \\ 0 \\ 1 \\ 1 \end{pmatrix} - \begin{pmatrix} 1 \\ 1 \\ 1 \\ 1 \end{pmatrix} = \begin{pmatrix} 0 \\ 0 \\ 0 \\ 0 \end{pmatrix}$

8.2-2. (a) $x_1 = -\frac{5}{4}, x_2 = -\frac{9}{8}, x_3 = \frac{1}{4}$

(b) $x_1 = -k, x_2 = k, x_3 = k$, k any constant

(c) $x_1 = -k, x_2 = 4k, x_3 = k$, k any constant

(d) $x_1 = -\frac{17}{5}, x_2 = \frac{3}{5}, x_3 = 2, x_4 = 1$

8.2-3. (a) $\det \mathbf{A} = 18$ **(b)** $\det \mathbf{A} = -18$

(c) $\det \mathbf{A} = 0$ and $\mathbf{Ax} = \boldsymbol{\theta}$ if $\mathbf{x} = k\begin{pmatrix} 1 \\ 1 \\ 1 \end{pmatrix}$, k a constant

(d) $\det \mathbf{A} = 0$ and $\mathbf{Ax} = \mathbf{0}$ if $\mathbf{x} = k\begin{pmatrix} 1 \\ 0 \\ -1 \end{pmatrix} + l\begin{pmatrix} 0 \\ 1 \\ -1 \end{pmatrix}$, k, l constants

Section 8.3

8.3-1. (a) $\mathbf{y}_1$ and $\mathbf{y}_2$ generate the general solution.

(b) $\mathbf{y}_1$ and $\mathbf{y}_2$ generate the general solution.

(c) $\mathbf{y}_1, \mathbf{y}_2, \mathbf{y}_3$ generate the solutions $\mathbf{y}$ to the equation that satisfy $\mathbf{y}(0) = \boldsymbol{\eta}$ where $\eta_1 = \eta_3$.

(d) $\mathbf{y}_1, \mathbf{y}_2, \mathbf{y}_3$ generate the general solution.

(e) $\mathbf{y}_1, \mathbf{y}_2, \mathbf{y}_3$ generate the solutions $\mathbf{y}$ to the equation that satisfy $\mathbf{y}(0) = \boldsymbol{\eta}$ where $\eta_2 = -\eta_3$.

8.3-2. (a) $\mathbf{y}(t) = \begin{pmatrix} 2e^{2t} - 3e^{-2t} + e^t \\ 4e^{2t} + 6e^{-2t} + e^{2t} \end{pmatrix}$ (that is, $c_1 = 2$, $c_2 = -3$)

(b) $\mathbf{y}(t) = \begin{pmatrix} 2e^{2t} + e^t + 1 \\ e^{2t} + e^t + 1 \\ e^t + te^t + 1 \end{pmatrix}$ (that is, $c_1 = c_2 = c_3 = 1$)

8.3-3. (a) $\mathbf{y} = \begin{pmatrix} c_1 e^t \\ c_2 e^{-t} \\ c_3 e^{-2t} - c_1 e^{-t} \end{pmatrix}$

(b) $\mathbf{y} = \begin{pmatrix} c_1 e^{2t} \\ c_1 te^{2t} + c_2 e^{2t} \\ c_1 te^{2t} + c_3 e^{2t} - (2t+1)/4 \end{pmatrix}$

(c) $\mathbf{y} = \begin{pmatrix} c_1 e^t \\ (c_1 t + c_2)e^t \\ (c_1 t^2/2 + c_2 t + c_1 t + c_3)e^t \end{pmatrix}$

(d) $\mathbf{y} = \begin{pmatrix} c_1 \\ c_2 e^{2t} \\ c_3 e^{-2t} + c_1/2 \\ c_4 e^{2t} + c_2 te^{2t} - \frac{1}{2} \end{pmatrix}$

Section 8.4

8.4-1. (a) $\mathbf{y} = c_1 e^{-t}\begin{pmatrix} 2 \\ 3 \end{pmatrix} + c_2 e^{-2t}\begin{pmatrix} 1 \\ 1 \end{pmatrix}$

$p(\lambda) = \lambda^2 + 3\lambda + 2$; $\lambda = -1, -2$ are eigenvalues

(b) $\mathbf{y} = c_1 e^{3t}\begin{pmatrix} 1 \\ 1 \end{pmatrix} + c_2 e^{-5t}\begin{pmatrix} 1 \\ -1 \end{pmatrix}$

$p(\lambda) = \lambda^2 + 2\lambda - 15$; $\lambda = 3, -5$ are eigenvalues

(c) $\mathbf{y} = c_1 e^t \begin{pmatrix} -3 \\ 1 \\ 1 \end{pmatrix}$

$p(\lambda) = -\lambda^3 + 3\lambda^2 - 3\lambda + 1 = -(\lambda - 1)^3$; $\lambda = 1, 1, 1$ are eigenvalues

(d) $\mathbf{y} = c_1 e^t \begin{pmatrix} 1 \\ 1 \\ 1 \end{pmatrix}$

$p(\lambda) = -\lambda^3 + \lambda^2 - 4\lambda + 4$; $\lambda = 1, \pm 2i$ are eigenvalues

(e) $\mathbf{y} = c_1 e^{-t}\begin{pmatrix} 1 \\ 0 \\ -1 \end{pmatrix} + c_2 e^{-t(1+\sqrt{2})}\begin{pmatrix} 1 \\ -\sqrt{2} \\ 1 \end{pmatrix} + c_3 e^{-t(1-\sqrt{2})}\begin{pmatrix} 1 \\ \sqrt{2} \\ 1 \end{pmatrix}$

$p(\lambda) = -\lambda^3 - 3\lambda^2 - \lambda + 1$; $\lambda = -1, -1 \pm \sqrt{2}$ are eigenvalues

(f) $p(\lambda) = \lambda^2 - 4\lambda + 13$; $\lambda = 2 \pm 3i$ are eigenvalues (no real eigenvalues)

8.4-2. (a) $p(\lambda) = (\lambda + 1)^2 + 4$; $\lambda = -1 \pm 2i$ are eigenvalues

$\mathbf{y} = c_1 e^{-t}\begin{pmatrix} \cos 2t \\ \sin 2t \end{pmatrix} + c_2 e^{-t}\begin{pmatrix} \sin 2t \\ -\cos 2t \end{pmatrix}$

(b) $p(\lambda) = -\lambda^3 + \lambda^2 - 4\lambda + 4 = -(\lambda - 1)(\lambda^2 + 4)$, $\lambda = 1$, $\pm 2i$ are eigenvalues

$$\mathbf{y} = c_1 \begin{pmatrix} \sin 2t \\ 2\cos 2t \\ -4\sin 2t \end{pmatrix} + c_2 \begin{pmatrix} \cos 2t \\ -2\sin 2t \\ -4\cos 2t \end{pmatrix}$$

(c) $p(\lambda) = -\lambda^3 + 2\lambda^2 - 2\lambda$; $\lambda = 0, 1 \pm i$ are eigenvalues

$$\mathbf{y} = c_1 \begin{pmatrix} 8e^t \cos t - 3e^t \sin t \\ -5e^t \cos t - 2e^t \sin t \\ -2e^t \cos t \end{pmatrix} + c_2 \begin{pmatrix} 3e^t \cos t + 8e^t \sin t \\ 2e^t \cos t - 5e^t \sin t \\ -2e^t \sin t \end{pmatrix}$$

8.4-3. (a) $$\mathbf{y}(t) = \begin{pmatrix} c_1 e^t - c_2 e^{2t} - 3c_3 e^{3t}/2 \\ c_2 e^{2t} + 2c_3 e^{3t} \\ c_3 e^{3t} \end{pmatrix}$$

(b) $$\mathbf{y}(t) = \begin{pmatrix} c_1 e^t + c_2 \sin 2t + c_3 \cos 2t \\ c_1 e^t + 2c_2 \cos 2t - 2c_3 \sin 2t \\ c_1 e^t - 4c_2 \sin 2t - 2c_3 \cos 2t \end{pmatrix}$$

(c) $$\mathbf{y}(t) = \begin{pmatrix} c_1 e^t + c_2 e^{-t} + c_3 te^{-t} \\ c_1 e^t - c_2 e^{-t} + c_3 e^{-t} - c_3 te^{-t} \\ c_1 e^t + c_2 e^{-t} - 2c_3 e^{-t} + c_3 te^{-t} \end{pmatrix}$$

(d) $$\mathbf{y}(t) = \begin{pmatrix} c_1 + (8c_2 + 3c_3)e^t \cos t + (8c_3 - 3c_2)e^t \sin t \\ -c_1 - (5c_2 - 2c_3)e^t \cos t - (2c_2 + 5c_3)e^t \sin t \\ -2c_2 e^t \cos t - 2c_2 e^t \sin t \end{pmatrix}$$

(e) $$\mathbf{y} = c_1 e^t \begin{pmatrix} -3 \\ 1 \\ 1 \end{pmatrix} + c_2 e^t \begin{pmatrix} 1 + t \\ -t \\ -1 \end{pmatrix} + c_3 e^t \begin{pmatrix} -2t \\ 1 \\ -1 + t \end{pmatrix}$$

(f) $$\mathbf{y} = c_1 \begin{pmatrix} 1 \\ -1 \\ 0 \\ 0 \end{pmatrix} + c_2 \begin{pmatrix} 0 \\ 1 \\ -1 \\ 0 \end{pmatrix} + c_2 \begin{pmatrix} 0 \\ 0 \\ 1 \\ -1 \end{pmatrix} + c_4 e^{4t} \begin{pmatrix} 1 \\ 1 \\ 1 \\ 1 \end{pmatrix}$$

(g) $$\mathbf{y} = c_1 \begin{pmatrix} e^{-2t} \cos t \\ -e^{-2t} \sin t \\ 0 \\ 0 \end{pmatrix} + c_2 \begin{pmatrix} e^{-2t} \sin t \\ e^{-2t} \cos t \\ 0 \\ 0 \end{pmatrix} + c_3 \begin{pmatrix} 0 \\ 0 \\ e^t \\ e^t \end{pmatrix} + c_4 \begin{pmatrix} 0 \\ 0 \\ te^t \\ (1 + t)e^t \end{pmatrix}$$

8.4-4. (a) $$\mathbf{y} = c_1 e^{-t} \begin{pmatrix} 3 \\ -1 \\ -2 \end{pmatrix} + c_2 e^{2t} \begin{pmatrix} 1 \\ 0 \\ 0 \end{pmatrix} + c_3 e^{2t} \begin{pmatrix} 0 \\ 1 \\ 1 \end{pmatrix}$$

Section 8.5

8.5-1. (a) $$\mathbf{e}^{t\mathbf{A}} = \begin{pmatrix} e^t \cosh 2t & e^t \sinh 2t \\ e^t \sinh 2t & e^t \cosh 2t \end{pmatrix}$$

(b) $$\mathbf{e}^{t\mathbf{A}} = \begin{pmatrix} \cos 2t & \sin 2t \\ -\sin 2t & \cos 2t \end{pmatrix}$$

(c) $$\mathbf{e}^{t\mathbf{A}} = \begin{pmatrix} (1 + e^{2t})/2 & (e^{2t} - 1)/2 \\ (e^{2t} - 1)/2 & (1 + e^{2t})/2 \end{pmatrix}$$

(d) $$e^{tA} = \begin{pmatrix} \dfrac{e^t - \sin t + \cos t}{2} & \sin t & \dfrac{e^t - \sin t - \cos t}{2} \\ \dfrac{e^t - \cos t - \sin t}{2} & \cos t & \dfrac{e^t - \cos t + \sin t}{2} \\ \dfrac{e^t + \sin t - \cos t}{2} & -\sin t & \dfrac{e^t + \sin t + \cos t}{2} \end{pmatrix}$$

(e) $$e^{tA} = \begin{pmatrix} 3e^t + (1-t)e^t - 3e^{3t} & 2e^t + (1-t)e^t - 3e^{3t} & 5e^t + (1-t)e^t - 6e^{3t} \\ -3e^t + (1+t)e^t + 2e^{3t} & -2e^t + (1+t)e^t + 2e^{3t} & -5e^t + (1+t)e^t + 4e^{3t} \\ e^{3t} - 1 & e^{3t} - 1 & 2e^{3t} - 1 \end{pmatrix}$$

Section 8.6

8.6-1. (a) $$e^{tA} = \begin{pmatrix} 1 & 2t & t + 3t^2 \\ 0 & 1 & 3t \\ 0 & 0 & 1 \end{pmatrix}$$

(b) $$e^{tA} = \begin{pmatrix} e^{3t} & 0 & 0 \\ 2te^{3t} & e^{3t} & 0 \\ te^{3t} & 0 & e^{3t} \end{pmatrix}$$

(c) $$e^{tA} = \begin{pmatrix} e^t & 2te^t & 0 & 0 \\ 0 & e^t & 0 & 0 \\ 0 & 0 & e^t & 0 \\ 0 & 0 & te^t & e^t \end{pmatrix}$$

(d) $$e^{tA} = \begin{pmatrix} 1 & t & 0 \\ 0 & 1 & 0 \\ t & t + t^2 & 1 \end{pmatrix}$$

8.6-2. (a) $$e^{tA} = \begin{pmatrix} \dfrac{3e^{2t} - e^t}{2} & \dfrac{e^t - e^{2t}}{2} \\ \dfrac{3e^{2t} - 3e^t}{2} & \dfrac{3e^t - e^{2t}}{2} \end{pmatrix}$$

(b) $$e^{tA} = \begin{pmatrix} e^t \cos 2t & -e^t \sin 2t \\ e^t \sin 2t & e^t \cos 2t \end{pmatrix}$$

(c) $$e^{tA} = \begin{pmatrix} e^t - 2te^t & te^t \\ -4te^t & e^t + 2te^t \end{pmatrix}$$

(d) $$e^{tA} = \begin{pmatrix} \dfrac{3 + e^{4t}}{4} & \dfrac{-1 + e^{4t}}{4} & \dfrac{-1 + e^{4t}}{4} \\ \dfrac{-2 + e^{4t}}{4} & \dfrac{2 + 2e^{4t}}{4} & \dfrac{-2 + 2e^{4t}}{4} \\ \dfrac{-1 + e^{4t}}{4} & \dfrac{-1 + e^{4t}}{4} & \dfrac{3 + e^{4t}}{4} \end{pmatrix}$$

(e) $$\mathbf{e}^{t\mathbf{A}} = \begin{pmatrix} \dfrac{2 + e^{\sqrt{2}t} + e^{-\sqrt{2}t}}{4} & \dfrac{\sqrt{2}e^{\sqrt{2}t} - \sqrt{2}e^{-\sqrt{2}t}}{4} & \dfrac{-2 + e^{\sqrt{2}t} + e^{-\sqrt{2}t}}{4} \\ \dfrac{\sqrt{2}e^{\sqrt{2}t} - \sqrt{2}e^{-\sqrt{2}t}}{4} & \dfrac{2e^{\sqrt{2}t} + 2e^{\sqrt{2}t}}{4} & \dfrac{\sqrt{2}e^{\sqrt{2}t} - \sqrt{2}e^{\sqrt{2}t}}{4} \\ \dfrac{-2 + e^{\sqrt{2}t} + e^{-\sqrt{2}t}}{4} & \dfrac{\sqrt{2}e^{\sqrt{2}t} - \sqrt{2}e^{-\sqrt{2}t}}{4} & \dfrac{2 + e^{\sqrt{2}t} + e^{-\sqrt{2}t}}{4} \end{pmatrix}$$

8.6-4. If $\mathbf{A}^2 = \alpha\mathbf{A}$ then $\mathbf{e}^{t\mathbf{A}} = \mathbf{I} + t\mathbf{A} + \left(\dfrac{e^{\alpha t} - \alpha t - 1}{\alpha}\right)\mathbf{A}$ where $\left(\dfrac{e^{\alpha t} - t - 1}{\alpha}\right) \equiv 0$ if $\alpha = 0$.

Section 8.7

8.7-1. (a) $$\mathbf{y}_p(t) = \begin{pmatrix} 1 - (\sin t)(\ln|\sec t + \tan t|) \\ (\cos t)(\ln|\sec t + \tan t|) \end{pmatrix}$$

(b) $$\mathbf{y}_p(t) = \begin{pmatrix} (\sin t)[\ln(\sin t)] - t\cos t \\ (\cos t)[\ln(\sin t)] + t\sin t \end{pmatrix}$$

(c) $$\mathbf{y}_p(t) = \begin{pmatrix} t\sin t + (\cos t)[\ln(\cos t) \\ t\cos t - (\sin t)[\ln(\cos t) \end{pmatrix}$$

(d) $$\mathbf{y}_p(t) = e^t\begin{pmatrix} \frac{9}{2} \\ \frac{5}{2} \end{pmatrix}$$

(e) $$\mathbf{y}_p(t) = \operatorname{Re}\left[e^{2it}\begin{pmatrix} -3i/5 \\ \frac{1}{5} \end{pmatrix}\right] = \begin{pmatrix} (3\sin 2t)/5 \\ (\cos 2t)/5 \end{pmatrix}$$

8.7-2. (a) $$\mathbf{y}_g = c_1\begin{pmatrix} e^t\cos 2t \\ -e^t\sin 2t \end{pmatrix} + c_2\begin{pmatrix} e^t\sin 2t \\ e^t\cos 2t \end{pmatrix} - e^t\begin{pmatrix} \frac{3}{2} \\ \frac{1}{2} \end{pmatrix}$$

is the general solution, and

$$\mathbf{y} = \begin{pmatrix} (-e^t\cos 2t + 5e^t\sin 2t - 3e^t)/2 \\ (e^t\sin 2t + 5e^t\cos 2t - e^t)/2 \end{pmatrix}$$

is the solution to the initial value problem.

(b) $$\mathbf{y}_g = c_1\begin{pmatrix} \sin t \\ \frac{1}{5}\cos t + \frac{2}{5}\sin t \end{pmatrix} + c_2\begin{pmatrix} \cos t \\ -\frac{1}{5}\sin t + \frac{2}{5}\cos t \end{pmatrix} + \begin{pmatrix} t\sin t + (\cos t)[\ln(\cos t)] \\ t\cos t - (\sin t)[\ln(\cos t)] \end{pmatrix}$$

is the general solution, and

$$\mathbf{y} = \begin{pmatrix} (\sin t)(11 + t) + (\cos t)[2 + \ln(\cos t)] \\ (\sin t)[4 - \ln(\cos t)] + (\cos t)(3 + t) \end{pmatrix}$$

is the solution to the initial value problem.

(c) $$\mathbf{y}_g = c_1\begin{pmatrix} e^{-t} \\ e^{-t} \\ 2e^{-t} \end{pmatrix} + c_2\begin{pmatrix} e^t \\ e^t \\ e^t \end{pmatrix} + c_3\begin{pmatrix} e^{2t} \\ 2e^{2} \\ e^{2t} \end{pmatrix} - \begin{pmatrix} 2e^{3t} \\ 3e^{3t} \\ 3e^{3t} \end{pmatrix}$$

is the general solution, and

$$\mathbf{y} = \begin{pmatrix} e^{-t} + e^{2t} - 2e^{3t} \\ e^{-t} + 2e^{2t} - 3e^{3t} \\ 2e^{-t} + e^{2t} - 3e^{3t} \end{pmatrix}$$

is the solution to the initial value problem.

8.7-4. (a) $\mathbf{y} = \begin{pmatrix} 0 \\ -e^t/2 \end{pmatrix} + \begin{pmatrix} 6 \sin t - \cos t \\ -5 \sin t + 7 \cos t \end{pmatrix}$

(b) $\mathbf{y} = \begin{pmatrix} \frac{3}{2} \\ -\frac{7}{2} \\ 7 \end{pmatrix} e^{-t} + \begin{pmatrix} \frac{1}{5} \\ \frac{2}{5} \\ -\frac{9}{5} \end{pmatrix} e^{2t}$

8.7-6. (a) $\mathbf{y} = \begin{pmatrix} \frac{11}{16} \\ -t/4 \end{pmatrix}$

Index